GEOGRAPHIC THOUGHT

Without social movements and wider struggles for progressive social change, the field of Geography would lack much of its contemporary relevance and vibrance. Moreover, these struggles and the geographical scholarship that engages with them have changed the philosophical underpinnings of the discipline and have inflected the quest for geographical knowledge with a sense not only of urgency but also hope. This reader, intended for advanced undergraduate and beginning graduate courses in Geographic Thought, is at once an analysis of Geography's theoretical and practical concerns and an encounter with grounded political struggles.

This reader offers a fresh approach to learning about Geographic Thought by showing, through concrete examples and detailed editorial essays, how the discipline has been forever altered by the rise of progressive social struggles. Structured to aid student understanding, the anthology presents substantive main and part introductory essays and features more than two dozen unabridged published works by leading scholars that emphatically articulate geographic thought to progressive social change. Each section is introduced with an explanation of how the following pieces fit into the broader context of geographic work amidst the socially progressive struggles that have altered social relations in various parts of the world over the last half-century or so. Doubly, it places this work in the context of the larger goals of social struggles to frame or reframe rights, justice, and ethics. *Geographic Thought* provides readers with insights into the encounters between scholarship and practice and aims to prompt debates over how social and geographical knowledges arise from the context of social struggles and how these knowledges might be redirected at those contexts in constructive, evaluative ways.

The reader is unique not only in knowing Geographic Thought through its progressive political attachments, instead of through a series of abstract "isms," but in gathering together salient works by geographers as well as scholars in cognate fields, such as Nancy Fraser, Chantal Mouffe, Iris Marion Young, and Jack Kloppenburg, whose own engagements have proved lasting and influential. For researchers and students interested in the connections between theoretically informed work and the possibilities for bettering people's everyday lives, this book provides an innovative and compelling argument for why Geographic Thought is valuable and necessary.

George Henderson is a human geographer who teaches and writes about the political economy of American capitalism. He is the author of the book *California and the Fictions of Capital* (Temple University Press paperback, 2003) and is Associate Professor of Geography at the University of Minnesota.

Marv Waterstone is Associate Professor of Geography at the University of Arizona. He was also the Director of the University of Arizona's Interdisciplinary Graduate Program in Comparative Cultural and Literary Studies. His current teaching includes History of Geographic Thought, Risk and Society, Radical Geography, Geography and Social Justice, Environmental and Resource Geography, and Governing Science and Technology.

Geographic Thought
A praxis perspective

Edited by

George Henderson

and

Marvin Waterstone

Routledge
Taylor & Francis Group
LONDON AND NEW YORK

First published 2009
by Routledge
2 Park Square, Milton Park, Abingdon, Oxon, OX14 4RN

Simultaneously published in the USA and Canada
by Routledge
270 Madison Avenue, New York, NY 10016

Routledge is an imprint of the Taylor & Francis Group, an informa business

Typeset in Amasis MT Lt and Akzidenz Grotesk by
Keystroke, 28 High Street, Tettenhall, Wolverhampton
Printed and bound in Great Britain by
MPG Books Ltd, Bodmin

British Library Cataloguing in Publication Data
A catalogue record for this book is available from the British Library

Library of Congress Cataloging in Publication Data
Geographic thought: a praxis perspective/[edited by] George Henderson and
Marvin Waterstone.
p. cm.
Includes bibliographical references and index.
1. Geography—Philosophy. I. Henderson, George L., 1958– II. Waterstone, Marvin.
G70.G4346 2008
910.01—dc22 2008004335

ISBN 10: 0–415–47169–9 (hbk)
ISBN 10: 0–415–47170–2 (pbk)
ISBN 10: 0–203–89307–7 (ebk)

ISBN 13: 978–0–415–47169–5 (hbk)
ISBN 13: 978–0–415–47170–1 (pbk)
ISBN 13: 978–0–203–89307–4 (ebk)

"The philosophers have only *interpreted* the world,
in various ways; the point, however, is to change it."

Karl Marx, *Theses on Feuerbach* (1845)

Contents

Acknowledgements

In putting together this book we have incurred numerous debts. We wish to thank, first, colleagues far and wide who read the early drafts of our proposal to put together a new book on geographic thought. Everyone seemed to feel that an approach reflecting the interweaving of modern Geography with the recent history of actually existing social–spatial struggles and their drive to understand and change the world was warranted. We benefited enormously from the support, friendly criticisms, and suggestions we received. Among this crew we count Stuart Aitken, Bruce Braun, Michael Brown, Susan Craddock, Jody Emel, Vinay Gidwani, Matt Hannah, J. P. Jones, Sallie Marston, Don Mitchell, Eric Sheppard, David M. Smith, Matt Sparke, Barney Warf, and Penny Waterstone. The anonymous reviewers of this book also helped us to sharpen our focus and our resolve that this project was worth undertaking. Marv Waterstone also thanks two separate cohorts of students in the history of geographic thought course at the University of Arizona, with whom he tried out several of the ideas that have culminated in this book. George Henderson apologizes to the students in his geographic thought courses for waiting too long.

Scanning and converting to Word files the articles and book chapters that we have reprinted here was a labor-intensive process. We owe especially large debts of gratitude to John McEwen, Ellen Frick, and Jodi Larson, at the University of Minnesota, and Linda Koski at the University of Arizona. Mark Lindberg, also at Minnesota, saved the day in preparing illustrations.

We must thank David McBride, our original editor at Routledge, for endorsing enthusiastically this project at the outset. His departure from Routledge is surely a loss.

Our partners and families have been especially supportive, forgiving our times away from home, listening patiently to our worries over the subjects and authors we did not include in the anthology, but also pushing and prodding at several junctures when we could not clearly describe just what we were up to.

Kyong Halverson at the University of Minnesota handled reprint permission requests like the pro she is. We are in her debt for now being able to write: We are grateful to the British Library and to the American Geographical Society Library at the University of Wisconsin-Milwaukee, for permission to reproduce several map images (see Chapter 8 by J. Brian Harley), and to the following publishers for permission to reprint articles and essays: Blackwell, Sage, Duke University Press, Elsevier Science and Technology Journals, the Rural Sociological Society, Pion Limited, London, University of Texas Press, Princeton University Press, and Cambridge University Press.

The chapters anthologized here appeared originally in the following venues:

Harvey, David 1972 "Revolutionary and Counter-Revolutionary Theory in Geography and the Problem of Ghetto Formation." *Antipode* 4(2): 1–13.

Blaut, James M. 1970 "Geographic Models of Imperialism." *Antipode* 2: 65–85.

Monk, Janice and Susan Hanson 1982 "On Not Excluding Half of the Human in Human Geography." *Professional Geographer* 34(1): 11–23.

Young, Iris Marion 1990 "Five Faces of Oppression," in *Justice and the Politics of Difference*. Princeton: Princeton University Press, pp. 39–65.

Fraser, Nancy 1999 "Social Justice in the Age of Identity Politics: Redistribution, Recognition, and Participation," in Larry Ray and Andrew Sayer, eds, *Culture and Economy after the Cultural Turn*. Thousand Oaks, CA: Sage, pp. 25–52.

Smith, David M. 2000 "Moral Progress in Human Geography: Transcending the Place of Good Fortune." *Progress in Human Geography* 24(1): 1–18.

Whatmore, Sarah 1997 "Dissecting the Autonomous Self: Hybrid Cartographies for a Relational Ethics," *Environment and Planning D: Society and Space* 15: 37–53.

Harley, J. Brian 1988. "Maps, Knowledge and Power." In D. Cosgrove and S. Daniels, eds, *The Iconography of Landscape*. Cambridge: Cambridge University Press, pp. 277–312. Reprinted with the permission of Cambridge University Press.

Nagar, Richa, Farah Ali *et al.* 2003 "Collaboration across Borders: Moving beyond Positionality," *Singapore Journal of Tropical Geography* 24(3): 356–372.

Heyman, Rich 2000 "Research, Pedagogy, and Instrumental Geography." *Antipode* 32(3), 292–307.

Merrifield, Andy 1995 "Situated Knowledge through Exploration: Reflections on Bunge's 'Geographic Expeditions'" *Antipode* 27(1): 49–70.

Blomley, Nicholas K. 1994 "Mobility, Empowerment and the Rights Revolution" *Political Geography* 13(5): 407–422, Copyright Elsevier.

Mohan, Giles and Jeremy Holland 2001 "Human Rights and Development in Africa: Moral Intrusion or Empowering Opportunity?" *Review of African Political Economy* 88: 177–196.

Gallaher, Carolyn and Oliver Froehling 2002 "New World Warriors: 'Nation' and 'State' in the Politics of Zapatista and US Patriot Movements" *Social & Cultural Geography* 3(1): 81–102.

Kloppenburg, Jr., Jack 1991 "Social Theory and the De/Reconstruction of Agricultural Science: Local Knowledge for an Alternative Agriculture" *Rural Sociology* 56(4): 519–548.

Pulido, Laura 1994 "Restructuring and the Contraction and Expansion of Environmental Rights in the United States" *Environment and Planning A: Society and Space* 26: 915–936.

Ishiyama, Noriko 2003 "Environmental Justice and American Indian Tribal Sovereignty: Case Study of a Land-Use Conflict in Skull Valley, Utah" *Antipode* 35(1): 119–139.

Glassman, James F. 2003 "Structural Power, Agency, and National Liberation: The Case of East Timor" *Transactions of the Institute of British Geographers* NS 28: 264–280.

Mouffe, Chantal 1995 "Post-Marxism: Democracy and Identity" *Environment and Planning A: Society and Space* 13: 259–265.

Sandoval, Chela 1991 "U.S. Third World Feminism: The Theory and Method of Oppositional Consciousness in the Postmodern World" *Genders* 10: 1–24.

Gibson-Graham, J. K. 2003 "An Ethics of the Local" *Rethinking Marxism* 15(1): 49–74.

Preface

If you are holding this book in your hands there is a strong possibility that you either teach or are taking a course in Geographic Thought (or Philosophy of Geography, or History of Geographic Thought, or some such designation). If this is not you but you have picked it up curious about what "geographic thought" might mean, well, this book is for you too. Virtually all attempts to characterize geographic thought and its histories are in our view a response to the question: "Is geographic thought good to think?" Reviews of both programmatic and substantive work by geographers over time have documented the changing contents and directions of the discipline as various scholars have contended to put forward their particular perspectives on what makes the discipline relevant and compelling; that is, their versions of "What makes geography good to think, for whom, and for what purposes?" This book is our answer to this crucial question. And the answer, in short, puts a premium on Geography's and geographers' relationship to social and political struggles. Just why we take this approach we hope will become clear in the pages to follow.

We begin here by offering a brief guide to the contents of the book. Next, we offer, for two reasons, our thoughts about what differentiates our approach from others that have preceded us. First, we want to make clear our own sense of a particular stand *vis-á-vis* an important role for scholarship in general and geographic scholarship in particular. Second, we want to take the opportunity at the outset to alert readers to a number of excellent works that pursue different paths in exploring the history of geographic thought and scholarship.

A BRIEF GUIDE TO THE CONTENTS OF THE BOOK

This volume is, on the one hand, an anthology of essays, all previously published (some are classics) and, on the other hand, a series of our own "editorial" essays that describe how we envision Geography. You will see that our sense of the discipline, especially our sense of what makes it "good to think," depends utterly on the fruits of those whose work is strongly connected to struggles for social change. We can put this very succinctly: It is not possible to describe what Geography has become in the last thirty years or so without acknowledging that it has come of age with movements for progressive political and social change. Apart from writing the editorial essays our efforts therefore have been directed at locating published exemplars that help to spell out why progressive social change is a good thing; why struggle is necessary; and why struggle is not only a material process but involves developing and refining concepts through which to understand social-political strategies and goals. (That action also begets thought is, then, an important supplement to the quote from Marx in the beginning pages of this book.) There are two notes about these readings. First, because it takes time for authors to develop the connections among these things, and because we think readers have the right to witness and be affected by how one aspect of an argument leads on to the next, we opted early to include whole, rather than abridged essays. This means fewer but perhaps more satisfying selections. Second, we make no claim to a comprehensive representation of the social struggles documented or participated in by geographers. What we did do is attempt to include works that clearly argue why and how social struggle, as such, is so prevalent and how and why the struggle to change the world is intimately linked to the struggle to understand the world in new ways. As for our own editorial essays stitching the book together, these are the result of our joint work—discussion, debate, writing, and revision—though one or the other of us at various times took the lead in writing drafts.

The book is divided into three main sections, two of these are broken into distinct parts. These are all preceded by introductory essays which we hope helpfully develop the particular themes that comprise the book and set into motion discussions that you can move forward in your own ways. In Section 1, we explicitly lay out the conditions for an argument that geographical thought is good to think *as part of a project of progressive social change, struggle, and activism*. We establish these conditions by examining a number of key pieces of scholarship that both constituted and elaborated a key turning point in the field in the late 1960s and early 1970s (variously termed the "interpretive" or "normative" turn). Section 1 (our editorial essay and the readings themselves) articulates our notion of geographic thought as always political. We include in this section three seminal pieces (and a discussion of a fourth) that represent early statements about essential elements of interpretive/normative geographic scholarship: the nature of knowledge and knowledge production, the politics of scholarship, the personnel involved in knowledge production, and the practices and methods of geographic work.[1]

The interpretative essay and readings in Section 2 take up the inter-related issues/themes concerning what is at stake in such a politicized conception of geographic thought. Here, working on the recognition that scholarly work is always (and inevitably) closely intertwined in political and social contexts, we turn to the kinds of questions that must be raised in order to help insure that scholarly work serves progressive rather than regressive purposes. The essay and readings address such matters as: How is progressive social change to be recognized and evaluated? How might scholarship and activism be helpful in enabling such change? What are the implications for the practice of geography?

In the essay and readings for Section 3 of the book, we are interested in moving the theoretical and conceptual orientations developed in Sections 1 and 2 into the realm of geographic practice in order to begin to understand the myriad potential links between knowledge and progressive politics. We utilize a three-part analytical framework to distinguish pieces of empirical scholarship and the social change struggles with which they intersect. The framework is based upon three different (though complexly inter-related) "worldviews" of both sources of oppression, as well as mechanisms for remedy and redress: those based upon rights, upon justice, and upon ethics. Readers anxious to know what we mean by these may jump ahead to the introduction to Section 3.

OUR DEPARTURE FROM CONVENTIONAL APPROACHES

There are numerous accounts of the rise of modern Geography and its changing fortunes over the last 150 or so years. We enumerate several of these here, in part to guide readers to these excellent sources, and in part to help differentiate explicitly our approach to the topic of geographic thought from those taken in these other efforts. Many of these works take what has been termed an "isms" approach (positivism, humanism, structuralism, post-structuralism, etc.), a research paradigm approach (models or theories in urban, social, cultural geography, biophysical, geographical information science [GIS], etc.), or the two approaches in combination. Examples of books that take these approaches are Peet's *Modern Geographic Thought* (Blackwell, 1998), Holt-Jensen's *Geography: History and Concepts* (Sage, 1999, 3rd edn), Johnston's *Geography and Geographers* (Arnold, 1997, 5th edn), Cloke *et al.*'s *Approaching Human Geography* (Guilford, 1991), Unwin's *The Place of Geography* (Longman, 1992), and Stoddart's *On Geography and Its History* (Blackwell, 1986).

An alternate route, however, is taken in Hubbard *et al.*'s *Thinking Geographically* (Continuum, 2002), Gregory's *Geographical Imaginations* (Blackwell, 1994), and Gillian Rose's *Feminism and Geography* (Cambridge, 1993). These works, just to give a few examples, inquire into how geography's discursive practices differentially constitute its objects of study: map, text, region, city, place, landscape, body, etc. We note that most existing readers in geographic thought take one or all of the above approaches—e.g. Agnew *et al.*'s *Human Geography: An Essential Anthology* (Blackwell, 1996) or Barnes and Gregory's *Reading Human Geography* (Oxford University Press, 1998).

In addition, geographic thought has been written, per Livingstone's *The Geographical Tradition* (Blackwell, 1992), as a critical, social-spatial history of ideas, focusing on the relations between science, scientific practice, and society and drawing upon the field of "science studies." More specialized books on the disciplinary history of geography are also available, especially those placing the professionalization of the field in the context of nationalisms, imperialism, and militarism (e.g. Godlewska and Smith's *Geography and Empire* [Blackwell, 1994]; see also Livingstone's *The Geographical Tradition* [Blackwell, 1992]).

These books are all excellent in devising ways of imagining geographic thought, and they have charted expertly the shifting intellectual terrain of the discipline: (1) from its early concerns with an organicist view of society growing out of Darwin's, Spenser's, and Lamarck's new formulations of biology, evolution, and natural selection (e.g. Ratzel, 1896); (2) through encounters with environmental determinism (e.g. Semple, 1911; Huntington, 1924; see also Peet, 1985 for an assessment of environmental determinism); (3) the early forms of the man/land (human/environment, nature/society) interaction studies (e.g. Marsh, 1864; Geddes 1898; Barrows, 1911); (4) the rise of cultural and regional geographies (e.g. Sauer, 1925; Hartshorne, 1939); (5) the so-called "quantitative revolution" and its aftermath (e.g. Schaefer, 1953; Burton, 1963; also see Gould, 1979 for an assessment of this critical transition period), including our main concerns here.

While these works provide very useful descriptions of the changing nature and content of the discipline, and while a number of them take up the more recent scholarship in Geography, none take the "interpretive turn" referred to above (which touches virtually all of them) into deep enough territory so as to: (1) explicitly and in a sustained way draw the connections amongst knowledge, social struggle, goals and strategies, and processes of moral justification; and (2) de-purify and adulterate the epistemologies through which we usually characterize the field (i.e. we want to make clear in what follows the constant interplay between scholarly knowledge and the political, social, and cultural contexts in which it is produced and put to use).

In asserting this, what we are basically asking is the following: *Instead of an imaginary that organizes geographic thought around different theoretical and conceptual approaches, what happens if we say it is organized around different sorts of political and moral commitments and is transected by movements for social change?* In asking this question we do not deny that there is a tradition of work in geography that theorizes positivism or humanism, or Marxisms or feminisms, and so on. Clearly there is such work, just as there is work that scrutinizes ontologies of space, landscape, scale, ecology, body, subject, city, region, nation, etc. But we do not seek to simply describe all that can be found in the discipline. We seek, like virtually anyone who attempts to craft "geographic thought," a useful imaginary. Geography's paradigms, concepts, methods, and/or history have all served as that imaginary. We want to alter the playing field so as to look at how processes of social struggle, including geographic research allied (implicitly and explicitly) to various struggles, constitute (geographic) knowledge as part and parcel of their politics. In depicting this kind of praxical imaginary, we would add that, in fact, knowledges constituted out of struggle are not bulwarks: feminisms, Marxisms, humanisms, environmentalisms, postcolonialisms, and so on, travel through these knowledges, and are reworked by them. The same is true for the ontological reworking of geographical "key concepts." It is our expectation that the approach to geographical thought that we take brushes up against, and recontextualizes, other approaches; it does not utterly forsake them.

Having said all this, there are two recent books with which ours might be compared, both having a family resemblance to our project. The first of these is David M. Smith's *Moral Geographies: Ethics in a World of Difference* (Edinburgh, 2000). In this work Smith does three important things. First, he spatializes and historicizes (i.e. materially situates) moral theory. He then devotes a chapter to each of several basic geographic concepts (e.g. territory, distance, environment), each signaling an always-already differentiated world, as opposed to the homogenous plain demanded of certain strains of abstracted moral reasoning, and develops the implications that these have for moral thought. Third, he works out these implications through a series of international case studies, sometimes of movements and sometimes of contentious policy. We conceive of our project as augmenting Smith's and also offering a different point of emphasis. As should be clear, this book concerns the issue of what moral turns are called upon by different sorts of collective politics and movements. In the works we have assembled that engage this issue (including a pertinent paper by Smith), we observe that geographical (and other) concepts are also summoned up. We therefore offer a different point of entry.

The second comparable work is Alison Blunt and Jane Wills' *Dissident Geographies: An Introduction to Radical Ideas and Practice* (Prentice Hall, 2000). While this book is organized around a set of radical epistemologies (e.g. anarchism, Marxism, feminism, queer theory, postcolonialism) it devotes considerable attention to the social and political movements and events that gave these epistemologies their warrant. It therefore goes much further in this direction than other "isms" books on geographic thought. We believe that our project complements Blunt and Wills' work by giving perhaps greater play to the idea that the politics of becoming cut across both different epistemologies and different social collectives and alliances. Also, given the fact that this book offers both primary readings and editorial essays we hope students will have greater opportunity to grapple with more of the existing literature and the subtleties that lie therein.

George Henderson, Minneapolis
Marv Waterstone, Tucson

NOTE

1 Readers will also find in Section 1 the first of several "text boxes" that point toward the work of additional scholars engaged with the issues under discussion.

SECTION 1

The politics of geographic thought

Introduction

Why is geographic thought always political?

A FEW WORDS ON THOUGHT ITSELF

Before we begin our substantive discussion of geographic thought, it is useful to spend some time on the notion of thought itself, on how we might distinguish scholarly (including geographic) thought from other modes of thinking, and finally, how we might distinguish geographic thought from other types of scholarship.

Is there something that characterizes scholarly thought, and distinguishes it from non-scholarly thought? It is clear that scholarly thought can be distinguished in terms of who produces it (quite circularly, scholars, on which more momentarily), for whom it is produced, and with what intent. The producers of scholarly thought are often conscious of themselves as engaged in producing "scholarship" or intellectual thought, and are usually aware of the likely audience for such products. In addition, such producers (scholars) typically have a purpose in mind when they engage in scholarly work, from such general notions as "advancing knowledge" to more concrete attempts to address specific problems. Usually, though by no means always, societies have authorized some members to produce scholarship, and have developed more or less formal mechanisms for determining who is so authorized. One predominant mechanism is the educational system, and the credentialing that typically accompanies this form of legitimation. Once designated as "scholars" these individuals are often accorded the time necessary for cogent reflection, and their output (at this point often evaluated by their peers) can be designated scholarship. We are also cognizant of, and want to note here, the exclusionary implications of these kinds of credentialing processes. As we examine below, the production of knowledge and its designation and acceptance as legitimate and useful often constitute important sites of struggle.

This kind of orientation to knowledge production then leads to thinking about scholarship (including geographic) in a way that emphasizes its embeddedness (always-already) in a Gordian knot of knowledge, practice, politics, and personnel. Considered in this way, it is clear that knowledge always has a social component and that it is always for something and arising in particular material contexts. One way of looking at the trajectory of scholarship in any field, then, is to trace the shifting struggles over who constitutes an authorized scholarly voice, what counts as scholarship, and what scholarship is for. In accounts that simply describe the leading figures in a field, the major "schools of thought," the evolution of sub-disciplinary specialties, or the changing constellation of big concepts, these contested aspects of the historiography of a field often remain hidden or under-developed. Here we pay special attention to geography as a formal discipline historically linked to global, national, and regional projects that have both inclusive and exclusive components and regressive and liberatory moments. Thus, we make the point that Geography, including the content of geographic thought, has been contingently related to the rise of a discipline that is in some sense understandable as a kind, or kinds, of "movement politics" (at scales from the body to the globe) that brings into being the (contested) content of its thought. We begin to elaborate the specifics of these matters in the following sections. As we begin to contemplate geographic thought more specifically, we are mindful of the social as irreducibly spatial and power-laden and of individuals as irreducibly social.

WHY IS GEOGRAPHIC THOUGHT ALWAYS POLITICAL?

In the most general and ineluctable sense, then, all scholarly thought is always political. This does not necessarily mean political in the narrow sense of partisan (although this can often be the case), but rather in the sense that what such thought is about, and who and what it is for are always the result of the interplay of power within disciplines, and the embeddedness of scholarly work in the material and discursive contexts in which it is produced. As such, scholarly/intellectual thought always either supports the existing *status quo* (whether this be in terms of internal disciplinary matters, and/or in terms of the various "outsides" to which any discipline is inevitably connected), or works to subvert the *status quo* for either progressive or regressive purposes.

For much of its modern history (i.e. from the late 19th century until the last third of the 20th), the discipline of geography has been irrefutably, though often tacitly, supportive of the *status quo*. Although there have been tumultuous struggles within the discipline over how best to accomplish this purpose, geographic practitioners have often looked to the powerful within societies for legitimation of the discipline. Securing that legitimacy (or failing to do so) has been crucial to certain measures of disciplinary (as well as individual, scholarly) success, and has influenced the varying relationship between Geography and society over that period. It should be clear, from comments made thus far, that scholarly work in support of the *status quo* is, by no means, apolitical. Indeed, intellectual activity that helps to maintain existing conditions and power relations is often a key support and source of credible authority for those who are benefited by such conditions. This situation obtained (with a few notable exceptions, e.g. Reclus, 1876–94; Kropotkin, 1885, 1899, 1902; Vidal de la Blanche, 1926) in geography, as it did in many other disciplines, until the late 1960s (and to a large extent is still the case at present). The major dimensions of intra-disciplinary struggle up until this period (as many of the sources cited above document quite well) concerned questions of what constituted proper objects of geographic inquiry, the primacy of description or explanation as geography's goal, or the best methods to accomplish either or both of these ends.

At that point (i.e. by the late 1960s), a number of geographers, responding both to conditions within the field and to material and intellectual circumstances in society more generally, grew quite restive with many facets of the discipline. These scholars were becoming more aware of (and more responsive to) a number of important social movements that were beginning to coalesce around key issues of the time, including: (1) rising opposition to the Vietnam War (and its characterization as part of ongoing imperialist and neo-colonialist projects against the global south by the global north); (2) the early stirrings of so-called "second wave" feminism and mounting resistance to the structures and strictures of patriarchy (and, by extension, other forms of traditionally constituted "normativity"); (3) an ongoing struggle for the expansion of civil rights to a variety of minorities who saw themselves excluded from the post-World War II prosperity that had lifted many other segments of the U.S. society; and (4) a newly energized environmental movement given its impetus by overt signs of an environment polluted and overburdened to the point of crisis.

In this first section, we include several readings that represent early articulations of what has come to be called the "normative" turn within the discipline. This phrase has taken on various meaning over time, but here we take it to signify several inter-related dimensions. First, and foremost, it means that (for those who take the critiques of the normative turn seriously), geographical scholarship must be concerned not only with description and explanation of what the world *is*, but must equally be concerned with questions of what the world *should be*. Second, it has meant, and continues to mean, coming to grips with such questions as what is scholarship, who is authorized to produce it, under what circumstances, and for what purposes? This normative turn consisted of both a negative critique of existing responses to such questions within geography (and in academia more generally), and a positive critique that offered alternative questions, methods, and purposes. Not surprisingly, these critiques engendered intense struggles and debates within the discipline, and have been quite influential in shaping its subsequent trajectory.

The essays in this first section demonstrate the complex and continuous inter-relationships among the various elements that make up scholarly/intellectual thought, and highlight the then-emerging contentions within geography over knowledge, politics, personnel, and practices and methods. These pieces represent formative statements (some would say early, incendiary salvoes!) in debates that continue to resonate strongly within the discipline to this day, and in each case the pieces have contributed to very productive, multi-directional conversations within the discipline and with cognate fields. The themes raised by these early works have matured and evolved over the past 35 or so years in scholarly terms, and (as we shall explore in later sections of the book) in their ability to inform progressive practices as well.

We begin this section with a 1972 article by David Harvey (Chapter 1), appropriately titled (for our, and other, purposes) "Revolutionary and counter-revolutionary theory in geography and the problem of ghetto formation." In this paper, Harvey (currently a Distinguished Professor at the Graduate Center of the City University of New York), who just three years earlier had published one of the landmark monographs in the positivist geographic cannon (Harvey, 1969), enumerates several themes that will become pivotal to changing notions of geographic thought and its internal and external relations of knowledge, and provides anchors for a vivid sense of what the discipline ought to be about. Harvey's own biographical trajectory in this short period is emblematic of broader changes within geography (and other fields as well), and is worth a slight detour before delving into the specifics of the article.

In a recent interview, Harvey discusses this transition as follows:

> Well, my politics at that time were closer to a Fabian progressivism, which is why I was very taken with the ideas of planning, efficiency and rationality . . . there was no real conflict between a rational scientific approach to geographical issues [which Harvey sought to elucidate in *Explanation in Geography*], and an efficient application of planning to political issues. But I was so absorbed in writing the book that I didn't notice how much was collapsing around me. I turned in my magnum opus to the publishers in May 1968, only to find myself acutely embarrassed by the change of political temperature at large . . . Just at that moment, I got a job in the US, arriving in Baltimore a year after much of the city had burnt down in the wake of the assassination of Martin Luther King. In the States, the anti-war movement and the civil rights movement were really fired up; and here was I, having written this neutral tome that seemed somehow or other just not to fit. I realized I had to rethink a lot of things I had taken for granted in the sixties.
>
> (NLR, 2000)

Some of that formative rethinking is reflected in the piece included here, which Harvey begins with the question "How and why would we bring about a revolution in geographic thought?" In answering this rhetorical query over the next 13 pages of the recently inaugurated (1969) radical geographic journal *Antipode*, Harvey takes up three key, intertwined issues. First he critiques a prevalent argument offered by Kuhn (1962) regarding the ways in which the nature of knowledge production goes through periodic reformulations (or revolutions, as Kuhn argued) within and across disciplines. Second he offers an assessment that places changes within geography over the previous decade (the period of the so-called "quantitative revolution") within this framework, and concludes that that "revolution" had now run its course, and was itself ripe for overthrow. And finally, Harvey presents both a critique of the ways in which positivism had become irrelevant within geography specifically, and in the academy more generally, as well as pointing a way forward that would allow such a positivist orientation to be both recuperated and made pertinent. In sum, then, this piece articulates the incipient concerns of the "normative" turn, formulates a cogent critique of the then-current state of geographic scholarship from this normative perspective, and describes (by means of both argument and an abbreviated case study) what Harvey is then groping toward as a more engaged, productive and progressive form of such scholarship.

One striking feature of this paper is that only a few years after publication of the *status quo Explanation in Geography*, Harvey is advocating Marxism as the analytical framework most promising for advancing geographical knowledge and, importantly, social change. As he later explained himself:

> In America, I would then [in 1973, when he published *Social Justice and the City*, his next major monograph after *Explanation*] have been termed a card-carrying liberal. So I set out along these [i.e. liberal] lines. Then I found out they weren't working. So I turned to Marxist formulations to see if they yielded better results. The shift from one approach to the other wasn't premeditated—I stumbled on it . . . I wasn't a Marxist at the time, and knew very little of Marx . . . The [Marx] reading group [composed mostly of graduate students and Harvey at Johns Hopkins University, begun in 1971] was a wonderful experience, but I was in no position to instruct anybody. As a group, we were the blind leading the blind. That made it all the more rewarding.
>
> (NLR, 2000: 80)

As Harvey notes in the "Revolutionary . . ." piece, his engagement with Marxian thought grew out of a conviction that it provided a useful overlap among approaches that he thought productive: positivism, materialism, and phenomenology. This position is congruent with his notions that paradigm shifts, when they occur, do not wholly displace what came before, but rather incorporate what is useful from previous orientations into formulations that are more relevant to current situations and problems. As Harvey explains in this piece, his turn to Marxism was undertaken as a corrective to what he (and many other younger geographers at the time) saw as a sterile positivism, divorced from material reality, as well as problems that might come from the main alternatives then being proposed (abandoning positivism altogether or moving in the direction of phenomenology). As Harvey argues, either of these latter approaches held the risk of a move away from materialism (a concrete connection to particular times, places, and contexts) toward an abstract idealism. It was just this lack of a materialist basis that critics found so problematic about the discipline's preoccupation with abstract quantification, model building and law-seeking.

In Harvey's view Marxism provided the corrective for one other aspect of the existing *status quo* geography, an issue fundamental to the "normative" turn. As he states, following Marx directly, "positivism simply seeks to understand the world, whereas Marxism seeks to change it" (see p. 18). In pursuing this orientation, Harvey argues strongly that not only is *status quo* (and, even more extremely, counter-revolutionary) theorizing unlikely to lead to progressive change, it actually contributes to oppressive conditions. It accomplishes this important legitimizing function by supplying support, if only tacitly, to existing circumstances. The kind of evidentiary work being carried out at the time by most mainstream geographers (and other social scientists), even when working on crucial social issues, in Harvey's view, provided a sense that "bleeding heart liberals" were contributing to solutions when, in fact, they were merely perpetuating the problems themselves. Since the underlying causes of these problems reside in the capitalist system itself, and since that system was never a subject of analysis, Harvey argues that mainstream scholars were constantly doomed to treating symptoms and missing the underlying issues entirely.

Finally, Harvey argues in the piece for a new role for geographers and other scholars, as intellectuals and academics. This consists in developing arguments of such persuasive strength that "all opposition to that system of thought" will be made to look ludicrous. And he includes the caution that academics, in such matters, are often "our own worst opponents." Here Harvey is clearly referring to the difficulty of challenging the taken-for-granted categories of thought and scholarly practice that often constrain the shift to new paradigms, particularly those that also challenge the political *status quo* within which academics do their work. This, as Harvey concludes, becomes especially difficult when the intersections between theory and practice are also a part of the changing mix. He lays down a gauntlet in advocating the need for "real" as opposed to "merely liberal" commitment to social change, and in the taunt that it is "indeed very comfortable to be a mere liberal." Many of these challenges continue to reverberate through much of the work included in the remainder of the book.

Jim Blaut (1927–2000) examines another dimension of the critique embodied in the "normative" turn, the politics of scholarship, in Chapter 2. As the critique matured, it incorporated an increasing historical sensitivity to the ways in which geographical knowledge and scholarship had been intertwined with systems of power, particularly those of imperialism and colonialism. This paper by Blaut is an early formulation of this analysis, and Blaut, like many of his contemporaries, saw the necessity for this kind of work as the U.S. engaged in yet another round of neo-imperialism in the Vietnam War.

As in the piece by Harvey, Blaut is interested in the paradigmatic nature of scholarship, and particularly with its taken for granted, naturalized elements. In the series of papers Blaut wrote at the time (1969, 1975, 1976), he is interrogating the largely unexamined, and necessarily congruent, relationship between what he terms "ethnoscience" (the system of beliefs, values, methods, and objects of inquiry characteristic of a specific culture) and the interests of dominant élites within that culture. His specific focus in this series of papers, including in Chapter 1, is the way in which Western ethnoscience over the previous 500 years (including geographic knowledge) has been effectively utilized to justify centuries of oppressive imperialism, colonialism, and neo-colonialism; and, at the same time, to illustrate how those connections to the interests of the powerful in the West have served to legitimate those scholarly disciplines (and individuals within them) and establish their value to society. Indeed, one of the interesting themes that Blaut elaborates here (and which becomes a touchstone for subsequent theory building in the "normative" vein) is this dialectical (mutually constituting) relationship between knowledge and power, and between each of these dimensions and the material conditions in which they are complexly produced and altered.

Before turning specifically to the insights provided by Chapter 2, it is worth taking a moment to examine some of Blaut's other work at the time, which will help set the context for this paper and its concerns. In a piece published the year before in *Antipode*, and provocatively titled "Jingo Geography," Blaut lays out the programmatic elements that will infuse his subsequent scholarship: the nature of imperialism, its connection with and support by/of scholarly activity (most especially geographic scholarship), the geographic dimensions of the emergence of capitalism, and the transformation of scholarship to aid in the reduction and elimination of oppressions built into imperialist and neo-imperialist projects.

He begins by defining imperialism as follows:

> I use the term "imperialism" in a sense of pure opprobrium. It designates the subjection and exploitation of non-whites through colonial domination or some other, more subtle and modern, device—but always with the aid of latent or manifest force . . . To clear the air further: my condemnation of imperialism in geography is directed at no individual; the science as a whole is to blame . . . The field after all, was born and raised in the homelands of imperialism . . . Thus, the modal academic geographer is white, Western, and probably an honest believer in the rightness of some form of imperialism (perhaps under a different name). If he [the predominance of the masculinist view of the world was so taken for granted by Blaut at the time that there was no need to include "male" in his description of the "modal" geographer or the language used to designate "him"] disagrees on certain subjects with his [*sic*] colleagues in other Western countries, he [*sic*] nevertheless shares with them a common set of values and beliefs concerning the non-white, non-Western world. He [*sic*] therefore purveys a science that has the imperialistic affliction at its very core.
>
> (Blaut, 1969: 10)

In the paper, Blaut then goes on to discuss "the symptomolgy of imperialism in our field—the forms, occurrences, and effects of this attitude—and its treatment" (1969: 10). While he is concerned with geography's long-standing connection to imperialism (for example he quotes Strabo to the effect that "it is plain that geography as a whole has a direct bearing on the activities of commanders," and comments on geography's utility in early moments of empire building), Blaut is most interested here in "the symptoms of imperialistic geography in the modern world" (1969: 11). The paper explores in some detail the manifestations of the imperialist bias in geographic education and in both "pure" and "applied" geographic research. The goal of the paper is to trace out the largely unconscious (or at least largely taken-for-granted) imperialist notions that permeated geographic practices to produce the kinds of mindsets in students, policy makers, and other consumers of geographic education and scholarship conducive to constructing the "First World" as the norm, and the "Third World" as homogeneous and uniformly inferior, and therefore justifiably susceptible to all forms of exploitation.

In one additional related paper, "Where was Capitalism Born?" published in 1976 in *Antipode*, Blaut takes up the question of the origins of European world dominance, in order to investigate whether current Euro-American hegemony can be understood in more contingent terms, which would then undermine the by-now naturalized hierarchies that place "the West" (Europe, and more recently the U.S.) on top, in the center, or in other advantageous positions relative to "the Rest." This concern with contingency (and understanding phenomena in historico-spatio-material terms) is congruent with other elements of the evolving "normative" turn, and particularly its growing reliance on Marxian thought as an explanatory framework (which is further explored in Blaut, 1975). This concern also prefigures what will come to be crucial questions in subsequent geographic inquiries into (under-)development and North/South relations in particular, and into many other forms of difference and normativity more generally.

In attempting to answer the question of the title, Blaut puts forth a hypothesis that capitalism was actually born

> [i]n Asia, Africa and Europe. Countless centers of incipient capitalism were springing up across the Old World during the two or three centuries prior to 1492. Fourteen-ninety-two is the key date. Before that date, capitalism was growing *evenly* among the old world continents. After that date, capitalism was growing *unevenly*: Europe was ascendant.
>
> (Blaut, 1976: 1, emphasis in original)

He then offers two propositions to further explore this basic hypothesis: (1) prior to 1492 there were multiple centers of incipient capitalism throughout the world, any of which might have become predominant; but (2) after 1492, Europe, based upon its exploitation of the resources of the "new world," gained a decisive advantage over other areas, and moved much more quickly toward a fully articulated capitalist economy. Blaut attributes this emerging European supremacy to an important, but thoroughly contingent factor: "The fact that European merchants reached the New World first is due solely to the factor of location: among mercantile-maritime centers in 1492, the Iberian centers were by far the closest to America" (1976: 1). Based upon what Marx termed "primitive accumulation," the European merchants were able to amass very significant wealth, which Blaut argues was useful for both undermining the feudal systems in Europe itself, and for taking control of the long-distance trading systems with Asia and Africa. Once begun, the process based upon the competitive advantage achieved through exploitation of new world resources contributed to an increasing European dominance of the evolving global capitalist system: "Thereafter the dialectic of development and underdevelopment intensified, and the world economic system fixed itself in place" (Blaut, 1976: 1).

Blaut's argument makes a taken-for-granted (and natural) superiority of Euro-American culture implausible, and connects directly to the arguments made in Chapter 2, which point to the complex intersections between knowledge (including scholarship) and politics. As Blaut argues in this paper, Western ethnoscience has been committed to a rationalization of imperialism and colonialism, but the larger point is contained in his statement regarding the axiomatic character of this relationship:

> a fundamental belief in the ethnoscientific system associated with a given society is not likely to fall into or remain in conflict with a fundamental value or norm that is held by the members of the society or by the policy-making elite if the society is highly stratified. In other words, crucial beliefs should conform to crucial precepts: the true should also be the good.
>
> (see p. 25)

If we reverse this final phrase (i.e. the good should also be the true), we get a clearer sense of the ways in which knowledge and belief systems are constantly policed and maintained to produce a concordance with dominant worldviews.

What Blaut is articulating here is an early formulation of what has come to be termed "situated knowledge," which we take up more extensively in our discussion of the next paper. Blaut's contention that Western ethnoscience has been the underpinning of imperialism can be enlarged to allow us to consider the important connections between scholarship and the interests of the powerful under other historical and geographical circumstances. But it should be clear that recognition of these connections is also a first, crucial step in understanding that different relationships between knowledge and power can be constructed. Without such recognition (i.e. the bias disguised "behind a façade of spurious objectivity"; see p. 27), a particular form of (ethno)science is accorded an explanatory superiority based on putative impartiality and a mystified relationship to the interests of the powerful. Blaut's paper is an early call to geographers to develop and nurture this kind of sensitivity to the interplay between their scholarly activity and the contexts within which it is conducted and interpreted.

A third dimension of the "normative" turn concerns the personnel of geographic scholarship, and is explored in the piece below by Janice Monk and Susan Hanson (Chapter 3) (currently professors of geography at the University of Arizona and Clark University, respectively). Though not the first paper to take up this important issue (see, e.g., Hayford, 1974; Tivers, 1978), it is one of the earliest programmatic statements to do so in a relatively comprehensive manner. Drawing upon the emerging feminist scholarship both within and outside of geography at the time, Monk and Hanson elucidate the implications of these developments both for geography as a discipline, and for progressive change more generally. It is impossible to overstate the importance of the insights developed through feminist scholarship for challenging the kinds of *status quo* scholarship critiqued by both Harvey and Blaut (and others involved in the normative turn).

Monk and Hanson are certainly concerned with the absence of women in the field of geography, both as practitioners *per se*, and as objects of study. Interestingly, in this regard, later scholars have often interpreted the paper as an example of a relatively simplistic "add women and stir" formulation (Monk, 2007). As should be clear, however, they are primarily interested in bringing the central perspectives of feminism to bear on the discipline in a transformative way.

Perhaps the single most important of these insights is the notion of situated knowledge (or, more formally, the idea of standpoint epistemology), which Monk and Hanson raise in several different ways. The basic notion, which now seems quite reasonable and straightforward, is that all knowledge is produced by actors who are themselves inescapably situated in particular historical and geographic circumstances, and that these circumstances have important (if often unrecognized) effects on both the means of knowledge production and on the kinds of knowledge produced. It is perhaps difficult to imagine that this idea was ever controversial. As we have already seen in the pieces by Harvey and Blaut, however, the power of "normal" science and of an unquestioned ethnoscience often resided in the presumed objectivity of the scholars and the impartial universality of the scholarship thus produced. To question such objectivity threatens the entire *status quo* scholarly enterprise, and requires a re-examination of what and whom scholarship is for; exactly the questions raised by Monk and Hanson (and Harvey and Blaut).

Situated knowledge undermines the unexamined and naturalized privilege inherent in the various hierarchies created by powerful interests, and thereby opens up possibilities for progressive change. To see how this works, let us examine two key sentences from Monk and Hanson's Chapter 3:

> The kind of knowledge that emerges from a discipline depends very much upon who produces that knowledge, what methods are used to procure knowledge, and what purposes knowledge is acquired for . . . The number of *women* involved in generating knowledge in a given discipline appears to be important in determining the degree to which *feminism* is absorbed in that discipline's research tradition.
>
> (see p. 35, emphasis added)

While Monk and Hanson are especially interested in undermining the ubiquitous sexism and patriarchy of early 1980s geography (and academia), this formulation immediately points to further openings and yet other viewpoints. Instead of, or in addition to, women in this sentence, it also now possible to include the concerns of the working class (owning class hegemony), people of color (racism), people outside of the "First World" (colonialism or neo-colonialism), people with differing sexual orientations (heterosexism), and people with differing physical and mental capabilities (arbitrary standards of "normalcy"), among many others. In other words, situated knowledge makes any particular notion of "normal" or "superior" extremely difficult to justify and maintain. It also unmasks the connections between knowledge and power, and the mechanisms that are utilized to obscure these relationships. This is a critical and necessary (though rarely sufficient) first step toward the elimination or reduction of various forms of oppression.

Monk and Hanson also point to the ways in which the pervasive sexism embodied in the dominant paradigms of the day track through the objects, methods and purposes of scholarship, which clearly echoes the domination of élite class interests in *status quo* and counter-revolutionary scholarship discussed by Harvey, and the justification of ethnocentrism described by Blaut. Destabilizing and dismantling these structures of privilege was an early concern of the normative turn, as was an explicit recognition of the need for scholarship to be connected to progressive social transformation. These are themes that we will see constantly revisited in the papers in subsequent parts of the book.

As noted earlier, Harvey concludes the piece below with the following comments: "the emergence of a true revolution in geographic thought is bound to be tempered by commitment to revolutionary practice. Certainly the general acceptance of revolutionary theory will depend upon the strengths and accomplishments of revolutionary practice" (see p. 21). Blaut's piece ends on a note of impending crisis, produced through the blinders inherent in Western ethnoscience's woeful misunderstanding of the world situation, and a similar, if somewhat less explicit, call for new geographic practices. Finally, Monk and Hanson issue a similar call in their paper, when they ask:

> Is the purpose of geographic research to accumulate facts and knowledge in order to improve our understanding of current events or to formulate policy within the context of the status quo, or is the purpose to go beyond asking why things are the way they are to consider the shapes of possible futures? Feminist scholars emphasize the need for research to define alternative structures in which the lot of women is improved.
>
> (see p. 42)

BOX 1: THE "DETROIT GEOGRAPHICAL EXPEDITION": PRACTICING WHAT WE PREACH

An early attempt to restructure this fourth element of geographical scholarship, its practices, is reviewed in a 1971 paper by Ronald Horvath (currently at the University of Sydney, Australia): "The 'Detroit Geographical Expedition' Experience." The Detroit Geographical Expedition and Institute (DGEI) was an important collaborative effort begun in 1969 by William Bunge and a number of colleagues at Michigan State University, the University of Michigan, and Wayne State University. At the time, Bunge, like Harvey, had been a leading scholar in the previous "quantitative revolution" in geography and, also like Harvey, had produced a well-respected monograph in this area (Bunge, 1966).

As Horvath's chronicle observes, the DGEI was begun in a sustained period of intellectual and political upheaval, and was intended to confront directly two areas of neglected scholarly engagement: (1) higher educational opportunities for poor and minority students; and (2) research relevant to the needs of poor and minority communities. In addition to reviewing the DGEI's accomplishments and failures in these areas, Horvath's paper points us to a number of key themes that we have already encountered regarding the transformation of geographical practice. First, Bunge remained committed to a "science" of geography. The production of an influential school decentralization plan relied heavily on traditional geographical skills and practices (data collection, map-making, the analysis of spatial interaction patterns). The key, however, for Bunge and his colleagues, was to employ rigorous methods of analysis, not simply for description, or even for understanding and explanation, but with an intent to define (normatively) and maximize (prescriptively) social justice.

Two additional themes are embodied in the case method of instruction (and related research as well). As Horvath describes, the success of such courses depended heavily on the familiarity of the (white) faculty with inner-city conditions. As Bunge observed, such familiarity and experience were not to be acquired easily or quickly, but necessitated sufficient immersion to allow faculty to take on "insider" sensibilities. Another way to put this is that expeditionary faculty had to have the ability to assume new (or altered) positionalities or standpoints; they had to have their knowledge newly (or differently) situated. A related theme concerns the necessary and desirable dialectical relationship between theory and practice. In interacting with those whose positions, and therefore knowledges, were differently situated, DGEI faculty were constantly forced to examine the same fundamental questions of scholarship that others engaged in the normative turn were facing: Who produces knowledge, for what purposes, for whom, what counts as appropriate evidence, and on what bases is scholarship to be evaluated?

A final theme is one that undergirds this entire discussion, but is brought out most explicitly, to this point, by Horvath's paper, though the others hint at it as well. This is the critical notion of contestation and struggle, both within the discipline over the questions just posed, and between the discipline and the contexts to which it is connected. It involves issues of power, status, stake, and risk. As Horvath's paper makes clear, there was tremendous resistance to the organization, implementation, and products of the DGEI, both within the various academic universities of Michigan's higher education system, and within the broader community. The DGEI challenged the *status quo* of university bureaucracy, of local government, and of fundamental elements of social injustice. These challenges are not taken lightly by those who benefit from the *status quo*, and, as Horvath further describes, several of the participants paid heavily for their involvement. Bunge and several of his colleagues made the "hard personal choices" described by Harvey, and were clearly unwilling or unable to tolerate the comfortable position of being merely liberal. We turn more specifically to this question of what is at stake in a politicized conception of geographical thought in the next sections of the book, and return specifically to the ideas embodied in the geographical expedition when we discuss Andy Merrifield's updating of the idea in Chapter 11.

The questions raised in the pieces in this section represent a defining moment in the trajectory of geographic thought. In an important sense, once asked, the questions, as well as their implications, can never be unasked. The kinds of sensibilities and commitments that these issues open up must now remain an important part of the discipline. In the next section, we turn to some of these implications, and move our conversation forward to consider more explicitly how scholarship and progressive social change might fruitfully interact.

1

Revolutionary and counter-revolutionary theory in geography and the problem of ghetto formation

David Harvey

from *Antipode*, 1972, 4(2): 1–13

How and why would we bring about a revolution in geographic thought? To gain some insight into this question it is worth examining how revolutions and counter-revolutions occur in all branches of scientific thought. Thomas Kuhn provides an interesting analysis of this phenomenon as it occurs in the natural sciences. He suggests that most scientific activity is what he calls normal science, which amounts to the investigation of all facets of a particular paradigm (a paradigm being thought of as a set of concepts, categories, relationships, and methods, which are generally accepted throughout the scientific community at a given point in time). During the process of normal science certain anomalies arise, observations or paradoxes which cannot be resolved within an existing paradigm. These anomalies increasingly become the focus of attention until science is plunged into a period of crisis in which speculative attempts are made to solve the problems posed by the anomalies. Eventually there arises out of these attempts a new set of concepts, categories, relationships, and methods, which successfully resolve the existing dilemmas as well as successfully incorporating the worthwhile aspects of the old paradigm. Thus a new paradigm is born, to be followed once more by the onset of normal scientific activity.[1] Kuhn's schema is open to criticism on a number of grounds. I shall discuss two problems very briefly. Firstly, there is no explanation as to how anomalies arise and how, once they have arisen, they generate crises. This criticism can be met by distinguishing between significant and insignificant anomalies. Thus it was known for many years that the orbit of Mercury did not fit into Newton's calculations yet this anomaly was insignificant because it had no relevance when it came to the use of the Newtonian system in an everyday context. If, on the other hand, certain anomalies had arisen in, say, bridge construction, then they obviously would have been highly significant. Thus the Newtonian paradigm remained satisfactory and unchallenged until something of practical importance and relevance could not be accomplished using the Newtonian system. Secondly, there is the question, never satisfactorily answered by Kuhn, concerning the way in which a new paradigm comes to be accepted. Kuhn admits that acceptance is not a matter of logic and he therefore suggests that it involves a leap of faith. A leap of faith based on what? Underlying Kuhn's analysis is a guiding force which is never explicitly examined. This guiding force amounts to a fundamental belief in the virtues of control and manipulation of the natural environment and the leap of faith, then, is based on the belief that the new system will allow an extension of manipulability and control over some aspect of nature. Which aspect of nature? Presumably once again it will be an aspect of nature which is important in terms of everyday activity and everyday life as it exists at a particular point in history.

The central criticism of Kuhn which these two cases point to is his abstraction of scientific knowledge from its materialistic basis. Kuhn provides an idealist interpretation of scientific advancement when it is clear that scientific thought is fundamentally geared to material activities. This materialistic basis for the advancement of scientific knowledge has been explored by J. D. Bernal.[2] Material activity involves the manipulation of nature in the interests of man and scientific understanding cannot be interpreted independent of that general thrust. But at this juncture we are forced to add a further perspective because "the interest of man" is subject to a variety of interpretations depending upon which group of men we are thinking of. Bernal thus points out that the sciences in the West have, until very recently, been the preserve of a middle-class group and even recently, with the rise of what is often called the "meritocracy", the scientist is invariably drawn into middle-class ways of life and thought during the course of his career. We must thus expect the natural sciences tacitly to reflect a drive for manipulation and control over those aspects of nature which are relevant to capitalist entrepreneurs. Far more important, however, is the harnessing of scientific activity, by a process of patronage and funded research, to the special interests of those who are in control of the means of production. The coalition of industry and government heavily directs scientific activity. Thus manipulation and control mean manipulation and control in the interests of a particular group in society rather than in the interests of society as a whole.[3] With these perspectives we are far better able to understand the general thrust of scientific advancement hidden within the recurrent scientific revolutions which Kuhn so perceptively described.

It has frequently been questioned whether or not Kuhn's analysis could be extended to the social sciences. Kuhn appears to take the view that the social sciences are "pre-scientific" in the sense that no one social science has really established that corpus of generally accepted concepts, categories, relationships, and methods which form a paradigm. This view of the social sciences as being pre-scientific is in fact quite general among philosophers of science.[4] But a quick survey of the history of thought in the social sciences shows that revolutions do indeed occur and that such occurrences are marked by many of the same features which Kuhn identified in the natural sciences. There is no question that Adam Smith provided a paradigmatic formulation for economic thought, which was subsequently built upon by Ricardo. In modern times Keynes succeeded in doing something essentially similar. Johnson, in a recent article, explores such revolutions in thought in economics and his analysis in many respects parallels that of Kuhn's, with, however, a couple of extra twists to it. At the heart of the Keynesian revolution, Johnson asserts, was a crisis generated by the failure of pre-Keynesian economics to deal with the most pressing and significant problem of the 1930s—namely, unemployment. Unemployment provided a significant anomaly. Thus Johnson suggests that:

> by far the most helpful circumstance for the rapid propagation of a new and revolutionary theory is the existence of an established orthodoxy which is clearly inconsistent with the most salient facts of reality, and yet is sufficiently confident of its intellectual power to attempt to explain those facts, and in its efforts to do so exposes its incompetence in a ludicrous fashion.[5]

Thus objective social realities of the time overtook the conventional wisdom and served to expose its failings:

> In this situation of general confusion and obvious irrelevance of orthodox economics to real problems, the way was open for a new theory that offered a convincing explanation of the nature of the problem and a set of policy prescriptions based on that explanation.

So far, the similarity to Kuhn is quite remarkable. But Johnson then adds certain new considerations, some of which really stem from the sociology of science itself. He thus suggests that to be accepted a theory needs to possess five main characteristics:

> First, it had to attack the central proposition of conservative orthodoxy . . . with a new but academically acceptable analysis that reversed the proposition . . . Second, the theory had to appear to be new, yet absorb as much as possible of the valid or at least not readily disputable components of existing orthodox theory. In this process, it helps greatly to give old concepts new and confusing names, and to emphasize as crucial analytical steps that have previously been taken as platitudinous . . . Third, the new theory had to have the appropriate degree of difficulty to understand . . . so that senior

> academic colleagues would find it neither easy nor worthwhile to study, so that they would waste their efforts on peripheral theoretical issues, and so offer themselves as easy marks for criticism and dismissal by their younger and hungrier colleagues. At the same time the new theory had to appear both difficult enough to challenge the intellectual interest of younger colleagues and students, but actually easy enough for them to master adequately with sufficient investment of intellectual endeavor . . . Fourth, the new theory had to offer to the more gifted and less opportunistic scholars a new methodology more appealing than those currently available . . . Finally, (it had to offer) an important empirical relationship . . . to measure.[6]

The history of geographic thought in the last ten years is exactly mirrored in this analysis. The central proposition of the old geography was the qualitative and the unique and this clearly could not resist the drive in the social sciences as a whole towards tools of social manipulation and control which required an understanding of the quantitative and the general. There can be no doubt either that during the transition process old concepts were given new and confusing names and that fairly platitudinous assumptions were subject to rigorous analytical investigation. Nor can it be denied that the so-called quantitative revolution allowed the opportunity to pillory the elder statesmen in the discipline particularly whenever they ventured into issues related to the newly emerging orthodoxy. Certainly, the quantitative movement provided a challenge of appropriate difficulty and opened up the prospect for new methodologies, many of which were to be quite rewarding in terms of the analytic insights they generated. Lastly, new things to measure were in abundance and in the distance decay function, the threshold and the range of a good, and the measurement of spatial pattern, we found three apparently crucial new empirical topics which we could spend an inordinate amount of time investigating. The quantitative movement can thus be interpreted partly in terms of a challenging new set of ideas to be answered, partly as a rather shabby struggle for power and status within a disciplinary framework, and partly as a response to outside pressures to come up with means for manipulation and control in what may broadly be defined as "the planning field." In case anyone misinterprets my remarks as pointing a finger at one particular group, let me say that all of us were involved in this process and that there was and is no way in which we could and can escape such involvement.

Johnson also introduces the term "counter-revolution" into his analysis. In this regard his thought is not very enlightening since he clearly has an axe to grind against the monetarists whom he designates as counter-revolutionaries even though a significant anomaly (the combination of inflation and unemployment) exists as a pressing challenge to the Keynesian orthodoxy. But there is something very important to this notion which requires analysis, for it seems intuitively plausible to think of the movement of ideas in the social sciences as a movement based on revolution and counter-revolution in contrast to the natural sciences to which such a notion does not appear to be so immediately applicable. We can analyze the phenomena of counter-revolution by using our insight into paradigm formation in the natural sciences. That paradigm formation is based on the extension of the ability to manipulate and control naturally occurring phenomena. Similarly, we can anticipate that the driving force behind paradigm formation in the social sciences is the manipulation and control of human activity and social phenomena in the interest of man. Immediately the question arises as to who is going to control whom, in whose interest is the controlling going to be, and if control is exercised in the interest of all, who is going to take it upon themselves to define that public interest? We are thus forced to confront directly in the social sciences what arises only indirectly in the natural sciences, namely, the social bases and implications of control and manipulation. We would be extraordinarily foolish to presuppose that these bases are equitably distributed throughout society. Our history up until the present time shows that they are usually highly concentrated within a few key groupings in society. These groups may be benevolent or exploitative with respect to other groups. This, however, is not the issue. The point is that social science formulates concepts, categories, relationships, and methods, which are not independent of the existing social relationships which exist in society. Thus the concepts used are themselves the product of the very phenomena they are designed to describe. A revolutionary theory upon which a new paradigm is based will only gain general acceptance if the nature of the social relationships embodied in the theory is actualized in the real world. A counter-revolutionary theory is one which is deliberately proposed to deal with a proposed revolutionary theory in such a manner that the threatened social changes which general acceptance of the revolutionary theory would generate are, either by cooptation or subversion, prevented from being realized.

This process of revolution and counter-revolution in social science can most explicitly be examined by studying the relationship between the political economy of Adam Smith and Ricardo on the one hand, and Karl Marx on the other. In this regard Engels, in the Preface to Volume II of *Capital*, provides some quite extraordinary insights. At issue was the charge that Marx had plagiarized the theory of surplus value. Marx, however, had clearly acknowledged that both Adam Smith and Ricardo had discussed and partially understood the nature of surplus value. Thus Engels sets out to explain what was new in Marx's utterances on surplus value and how it was that Marx's theory of surplus value "struck home like a thunderbolt out of a clear sky."[7] To explain this Engels resorted to an analogy with an incident in the history of chemistry which, quite coincidentally, turns out to be one of the inspirations for Kuhn's thesis regarding the structure of revolutions in natural science.[8] The incident concerns the relationship between Lavoisier and Priestley in the discovery of oxygen. Both ran similar experiments and produced similar results. The essential difference between them was, however, that Priestley insisted for the rest of his life on seeking to interpret his results in terms of the old phlogiston theory and he therefore called his discovery "dephlogisticated air." Lavoisier, however, recognized that his discovery could not be reconciled with the old phlogiston theory as it was and, as a consequence, was able to reconstruct the theoretical framework of chemistry on a completely new basis. Thus both Engels and Kuhn suggest that Lavoisier was the "real discoverer of oxygen *vis-a-vis* the others who had only produced it without knowing what they had produced."

Engels continues:

> Marx stands in the same relation to his predecessors in the theory of surplus value as Lavoisier stood to Priestley . . . The existence of that part of the value of products which we now call surplus-value had been ascertained long before Marx. It had also been stated with more or less precision what it consisted of . . . But one did not get any further . . . (all economists) remained prisoners of the economic categories as they had come down to them. Now Marx appeared upon the scene. And he took a view directly opposite to that of all his predecessors. What they had regarded as a *solution*, he considered but a *problem*. He saw that he had to deal neither with dephlogisticated air nor with fireair, but with oxygen—that here it was not simply a matter of stating an economic fact or of pointing out the conflict between this fact and eternal justice and morality, but of explaining a fact which was destined to revolutionize all economics, and which offered to him who knew how to use it the key to an understanding of all capitalist production. With this fact as his starting point he examined all the economic categories which he found at hand, just as Lavoisier proceeding from oxygen had examined the categories of phlogistic chemistry.[9]

The Marxist theory was clearly dangerous in that it appeared to provide the key to understanding capitalist production from the point of view of those not in control of the means of production and consequently the categories, concepts, relationships, and methods which had the potential to form a paradigm were an enormous threat to the power structure of the capitalist world. The subsequent emergence of the marginal theory of value did away with much of the basics of Smith's and Ricardo's analysis (in particular the labor theory of value) and also incidentally served to turn back the Marxist challenge in economics. The counter-revolutionary cooptation of Marxist theory in Russia after Lenin's death, and the similar counter-revolutionary cooptation of much of the Marxist language into Western sociology (so much so that some sociologists suggest that "we are all Marxists now") without conveying the essence of Marxist thinking, has effectively prevented the true flowering of Marxist thought and concomitantly the emergence of that humanistic society which Marx envisaged. Both the concepts and the projected social relationships embodied in the concepts were frustrated.

Revolution and counter-revolution in thought are therefore characteristic of the social sciences in a manner which is not apparently characteristic of natural science. Revolutions in thought cannot ultimately be divorced from revolutions in practice. This may point to the conclusion that social sciences are indeed in a pre-scientific state. This conclusion is ill founded, however, since the natural sciences have never been wrested for any length of time out of the control of a restricted interest group and it is this fact rather than anything inherent in the nature of natural science knowledge itself which accounts for the lack of counter-revolutions in the natural sciences. In other words the revolutions of thought that are accomplished pose no threat to the existing order since they are constructed with the requirements of that existing order broadly in mind. This is not to say that there are not some uncomfortable social problems to resolve *en route*, for scientific discovery is not predictable and it can therefore be the source of social tension. What this suggests, however, is that the natural sciences are in a pre-social state. Thus questions of social action and social control, which the techniques of natural science frequently help to resolve, are not incorporated into natural science itself. In fact there is a certain fetishism about keeping them out since incorporating them will supposedly "bias" research conducted at the behest of the existing social order. The consequent moral dilemmas for those scientists who take their social responsibilities seriously are real indeed. Contrary to popular opinion, therefore, it seems appropriate to conclude that the philosophy of social science is in general much superior to that of natural science and that the eventual fusion of the two fields of study will not come about through attempts to "scientize" social science, but will instead require the socialization of natural science.[10] This may mean the replacement of manipulation and control by the realization of human potential as the basic criterion for paradigm acceptance. In such an event all aspects of science will experience both revolutionary and counter-revolutionary phases of thought which will undoubtedly be associated with revolutions and counter-revolutions in social practice.

Let us return now to the initial question. How and why would we bring about a revolution in geographic thought? The quantitative revolution has run its course and diminishing marginal returns are apparently setting in as yet another piece of factorial ecology, yet another attempt to measure the distance decay effect, yet another attempt to identify the range of a good, serve to tell us less and less about anything of great relevance. In addition there are younger people now, ambitious as the quantifiers were in the early 1960s, a little hungry, somewhat starved of interesting things to do. So there are murmurs of discontent within the social structure of the discipline as the quantifiers establish a firm grip on the "production" of graduate students and on the curricula of various departments. This sociological condition within the discipline is not sufficient to justify a revolution in thought (nor should it be) but the condition is there. More important, there is a clear disparity between the sophisticated theoretical and methodological framework which we are using and our ability to say anything really meaningful about events as they unfold around us. There are too many anomalies between what we purport to explain and manipulate and what actually happens. There is an ecological problem, an urban problem, an international trade problem, and yet we seem incapable of saying anything of any depth or profundity about any of them. When we do say something it appears trite and rather ludicrous. In short, our paradigm is not coping well. It is ripe for overthrow. The objective social conditions demand that we say something sensible and coherent or else forever (through lack of credibility or, even worse, through the further deterioration of the objective social conditions) remain silent. It is the emerging objective social conditions and our patent inability to cope with them which essentially explain the necessity for a revolution in geographic thought.

How should we accomplish such a revolution? There are a number of paths we could take. We could, as some appear to suggest, abandon the positivist basis of the quantitative movement for an abstract idealism and hope that objective social conditions will improve of their own accord or that concepts forged through idealist modes of thought will eventually achieve enough content to facilitate the creative change of objective social conditions. It is, however, a characteristic of idealism that it is forever doomed to search fruitlessly for real content. We could also reject the positivist basis of the 1960s for a phenomenological basis. This appears more attractive since it at least serves to keep us in contact with the concept of man as a being in constant sensuous interaction with the social and natural realities which surround him. Yet phenomenological approaches can lead us into idealism or back into naïve positivist empiricism just as easily as they can into a socially aware form of materialism.—The so-called behavioral revolution in geography is pointed in all of these directions. The most fruitful strategy at this juncture is therefore to explore that area of understanding in which certain aspects of positivism, materialism, and phenomenology overlap to provide adequate interpretations of the social reality in which we find ourselves. This overlap is most clearly explored in Marxist thought. Marx, in the *Economic and Philosophic*

Manuscripts of 1844 and in the *German Ideology* gave his system of thought a powerful and appealing phenomenological basis.[11] There are also certain things which Marxism and positivism have in common. They both have a materialist base and both resort to an analytic method. The essential difference of course is that positivism simply seeks to understand the world, whereas Marxism seeks to change it. Put another way, positivism draws its categories and concepts from an existing reality with all of its defects while Marxist categories and concepts are formulated through the application of dialectical method to history as it is written here and now through events and actions. The positivist method involves, for example, the application of traditional bi-valued Aristotelian logic to test hypotheses (the null hypothesis of statistical inference is purely an Aristotelian device). Thus hypotheses are either true or false and once categorized ever remain so. The dialectic on the other hand proposes a process of understanding which allows the interpenetration of opposites, incorporates contradictions and paradoxes, and points to the processes of resolution. Insofar as it is at all relevant to talk of truth and falsity, truth lies in the dialectical process rather than in the statements derived from the process, which can be designated "true" only at a given point in time and which in any case are contradicted by other "true" statements. This method allows us to invert analyses if necessary, to regard solutions as problems, to regard questions as solutions.[12]

I shall briefly summarize an extended argument on urban land use theory to provide an example of how the strategy described above works.

Geographers drew much of their initial inspiration from the Chicago school of sociologists (particularly Park and Burgess), who noted that cities exhibited certain regularities in spatial structure. This spatial structure was held together by some culturally derived form of social solidarity which Park called "the moral order."[13] Engels, writing some 80 years before Park and Burgess, noted the phenomenon of concentric zoning, interpreted it in economic class terms, and identified the market mechanism operating under capitalist institutions as the generating force behind the urban structure. His description of Manchester is insightful and worth quoting:

> "Manchester contains, at its heart, a rather extended commercial district, perhaps half a mile long and about as broad, and consisting almost wholly of offices and warehouses. Nearly the whole district is abandoned by dwellers, and is lonely and deserted at night . . . The district is cut through by certain main thoroughfares upon which the vast traffic concentrates, and in which the ground level is lined with brilliant shops. In these streets the upper floors are occupied, here and there, and there is a good deal of life upon them until late at night. With the exception of this commercial district, all Manchester proper, all Salford and Hulme . . . are all unmixed working people's quarters, stretching like a girdle, averaging a mile and a half in breadth, around the commercial district. Outside, beyond this girdle, lives the upper and middle bourgeoisie, the middle bourgeoisie in regularly laid out streets in the vicinity of working quarters . . . the upper bourgeoisie in remoter villas with gardens . . . in free, wholesome country air, in fine, comfortable homes, passed every half or quarter hour by omnibuses going into the city. And the finest part of the arrangement is this, that the members of the money aristocracy can take the shortest road through the middle of all the labouring districts without ever seeing that they are in the midst of the grimy misery that lurks to the right and left. For the thoroughfares leading from the Exchange in all directions out of the city are lined, on both sides, with an almost unbroken series of shops, and are so kept in the hands of the middle and lower bourgeoisie . . . (that) they suffice to conceal from the eyes of the wealthy men and women of strong stomachs and weak nerves the misery and grime which form the complement of their wealth . . . I know very well that this hypocritical plan is more or less common to all great cities; I know, too, that the retail dealers are forced by the nature of their business to take possession of the great highways; I know that there are more good buildings than bad ones upon such streets everywhere, and that the value of land is greater near them than in remote districts; but at the same time, I have never seen so systematic a shutting out of the working class from the thoroughfares, so tender a concealment of everything which might affront the eye and the nerves of the bourgeoisie, as in Manchester. And yet, in other respects, Manchester is less built according to plan after official regulations, is more outgrowth of accident, than any other city; and when I consider in this connection the eager assurances of the middle class, that the working class is doing

> famously, I cannot help feeling that the liberal manufacturers, the Big Wigs of Manchester, are not so innocent after all, in the matter of this sensitive method of construction.[14]

The description provided by Engels can, without too much adaptation, be applied to the contemporary American city, which suggests that capitalist cities tend towards a similarity of structure because the basic forces modifying them are the same. Certain passages written by Engels, for example, compare with those typically contained in contemporary governmental reports on urban problems (such as the Kerner Commission Report).[15] It therefore seems a pity that we continue to look to Park and Burgess for inspiration (as do the Chicago geographers) instead of following up the approach adopted by Engels. In fact the tradition that most closely relates to that of Engels arises from von Thünen's analysis which has been applied by Alonso and Muth[16] to the urban land market. In these models urban land use is determined through a process of competitive bidding for the land. Different groups in the population have different resources with which to bid and a variety of city structures can emerge depending upon the preferences of the rich groups, who can always use their resources to dominate the preferences of poor groups. This is the natural outcome of models built on neo-classical marginalist principles—models which are generally regarded as Pareto optimal.

Deviations from the normative model can be taken as an indication of disequilibrium. It is generally conceded that there is considerable disequilibrium in the American city at the present time as employment has become suburbanized but poor populations have been excluded from suburban locations by a variety of devices (such as zoning). It is interesting to note that many of the policies proposed by liberal groups (planners, civil rights groups, etc.) amount to advocating a return to equilibrium of the sort identified in the Alonso–Muth formulation. This is supported by large corporations who are in some cases suffering labor shortages in suburban locations. All of these proposals indicate returning to an equilibrium in which the poor still live where they can least afford to live—in other words a return to the *status quo* of the sort described by Engels is being advocated. How can we identify more revolutionary solutions?

Muth sought to show that the normative model he devised had empirical relevance. He tested it and found it broadly correct as a model of residential land use in Chicago. Let us assume the theory is true, in the sense used by logical positivists. This truth can be used to help us identify what the problem is. What for Muth would be regarded as a successful test of a theory we regard as an indicator of what the problem is. The theory predicts that the poor groups will live where they can least afford to live. Therefore, the only valid policy is to eliminate the conditions which give rise to the truth of the theory. In other words we want the von Thünen model of the urban land market to become not true. The simplest approach to this is to eliminate the mechanism which gives rise to the truth of the theory. The mechanism in this case is competitive bidding for the use of the land. If we eliminate this mechanism we will presumably eliminate the result. Competitive bidding should therefore be replaced by a socially controlled urban land market and a socialized control of the housing sector. We would thus render the von Thünen theory irrelevant to our understanding of spatial structure of cities. This process has begun in Cuba and in Havana competitive bidding has been completely eliminated, as have rental payments on many dwellings.[17]

We ought not to accept this argument too readily, for it is often the case that the mechanism which is assumed for the purpose of the theory is not necessarily the same as the real mechanisms which generate results in accord with the theory. We should merely be alerted to the possibility that the market mechanism is at fault and look for further proof of the contention. This proof can be gained from an argument stemming from the general characteristics of capitalism and market behavior. A market system becomes possible under conditions of resource scarcity for only under these conditions can price-fixing commodity exchange markets arise. The extension of market exchange has allowed an immense increase in the production of wealth. We therefore find a paradox, namely that wealth is produced under a system which relies upon scarcity for its functioning. It follows that if scarcity is eliminated then the market economy which is the source of productive wealth under capitalism is liable to collapse. Yet capitalism is always increasing its productive capacity. To resolve this dilemma many institutions and mechanisms are formed to ensure that scarcity does not disappear. In fact many institutions are geared to the maintenance of scarcity (universities being a prime example, although this is always done in the name of "quality"). A general analysis of capitalism and market exchange economies would indicate that a major barrier to the elimination of scarcity in advanced productive societies like the U.S.A. lies in the complicated set of interlocking institutions (financial, judicial, political, educational, and so on) which support the market process.

If we look very carefully we can identify manifestations of this general condition in the urban housing market. Commercial operators in the housing market (landlords, banks and other financial institutions, developers, and so on) are not interested in housing *per se* but are interested in maximizing their returns (rents, interest, profit—or, as Marx called it, surplus value). Even if each operator behaves ethically, according to the usual norms of capitalist entrepreneurial behavior, the net output of the interactions among them all is to write off use-values in housing in one part of the city in order to reap exchange-values in another part of the city. In other words, scarcity is being created in one part of the city so that the market can function (at a certain level of profit) at the other end. This process can be detailed. If this process is general, and the evidence suggests that it is, then we must anticipate that the market process will naturally counteract any policies designed to eliminate scarcity in the housing market. Again, there are some disturbing similarities between the accounts provided by Engels and contemporary urban policy problems. Here is how Engels described the attempts at urban renewal in the nineteenth century:

> In reality the bourgeoisie has only one method of solving the housing question after its fashion—that is to say, of solving it in such a way that the solution continually reproduces itself anew. This method is called "Haussmann" . . . By "Haussmann", I mean the practice which has now become general of making breaches in the working class quarters of our big towns, and particularly in areas which are centrally situated, quite apart from whether this is done from considerations of public health and for beautifying the town, or owing to the demand for big centrally situated business premises, or owing to traffic requirements, such as the laying down of railways, streets (which sometimes appear to have the strategic aim of making barricade fighting more difficult) . . . No matter how different the reasons may be, the result is everywhere the same; the scandalous alleys disappear to the accompaniment of lavish self-praise from the bourgeoisie on account of the tremendous success, but they appear again immediately somewhere else and often in the immediate neighborhood! . . . The breeding places of disease, the infamous holes and cellars in which the capitalist mode of production confines our workers night after night, are not abolished; they are merely *shifted elsewhere*! The same economic necessity which produced them in the first place, produces them in the next place also. As long as the capitalist mode of production continues to exist, it is folly to hope for an isolated solution to the housing question or of any other social question affecting the fate of the workers. The solution lies in the abolition of the capitalist mode of production and the appropriation of all the means of life and labour by the working class itself.[18]

It is difficult to avoid concluding from the accumulated evidence that Engels was probably right. There is good reason to believe that the market mechanism is the culprit in a sordid drama. And yet it is curious that although all serious analysts concede the seriousness of certain of our contemporary urban problems, few call into question the forces which rule at the very heart of our economic system. We thus discuss everything except the basic characteristics of the capitalist market economy. We devise all manner of solutions except those which might challenge the continuance of that economy. Such discussions and solutions which so avoid the central issue serve only to make us look foolish, for they eventually lead us to discover, rather belatedly, what Engels was only too well aware of in 1872—that capitalist solutions provide no foundation for dealing with deteriorated social conditions which are structurally necessary for the perpetuation of capitalism. Such solutions are mere "dephlogisticated air." We can, if we will, discover oxygen and all that goes with it by subjecting the very basis of our capitalist society (with all its institutionalized scarcities) to a rigorous and critical examination. It is this task which a revolutionary theory must address itself to. What does this task entail?

First, let me say what it does not entail. It does not entail yet another empirical investigation of the social conditions in the ghettos. We have enough information already and it is a waste of energy and resources to spend our time on such work. In fact, mapping even more evidence of man's patent inhumanity to man is counter-revolutionary in the sense that it allows the bleeding-heart liberal to pretend he is contributing to a solution when he in fact is not. This kind of empiricism is irrelevant. There is already enough information in congressional reports, daily newspapers, books, articles, and so on to provide us with all the evidence we need. Our task does not lie here. Nor does it lie in what can only be termed moral masturbation of the sort which accompanies the masochistic assemblage of some huge dossier on the daily injustices to the populace of the ghetto, over which we beat our breasts, commiserate with each other, before retiring to our fireside comforts. This, too, is counter-revolutionary for it merely serves to expiate guilt without our ever being forced to face the fundamental issues, let alone do anything about them. Nor is it a solution to indulge in that emotional tourism which attracts us to live and work with the poor "for a while" in the hope that we can really help them improve their lot. This, too, is counter-revolutionary, for so what if we help a community win a playground in one summer of work to find the school deteriorates in the fall? These are the paths we should not take. They merely serve to divert us from the essential task at hand.

This immediate task is nothing more nor less than the self-conscious and aware construction of a new paradigm for social geographic thought through a deep and profound critique of our existing analytical constructs. This is what we are best equipped to do. We are academics, after all, working with the tools of the academic trade. Our task is therefore to mobilize our powers of thought to formulate concepts and categories, theories and arguments, which we can apply in the process of bringing about a humanizing social change. These concepts and categories cannot be formulated in abstraction. They must be forged realistically with respect to the events and actions as they unfold around us. Certainly, empirical evidence, the already assembled dossiers, and the experiences gained in the community can be made use of here. But all of those experiences and all of that information means nothing unless we synthesize it into powerful patterns of thought. But our thought cannot rest merely on existing reality. It has to embrace alternatives creatively. We cannot afford to plan for the future on the basis of positivist theory, for to do so would merely be to reinforce the *status quo*. Yet, as in the formation of any new paradigm, we must be prepared to incorporate and reassemble all that is useful and valuable within that corpus of theory. We can restructure the formulation of existing theory in the light of possible lines of future action. We can critique existing theories as "mere apologetics" for the dominant force in our society—the capitalist market system and all its concomitant institutions. In this manner we will be able to establish the circumstances under which location theory can be used to create better futures and the circumstances in which it reinforces modes of thought conducive to the maintenance of the *status quo*. The problem in many cases is not the marginalist method *per se* or optimizing techniques *per se*, but that these methods are being applied in the wrong context. Pareto optimality as it enters location theory is a counter-revolutionary concept and so is any formulation which calls for the maximization of any one of the partial manifestations of surplus value (such as rent or return on capital investment). Yet programming solutions are clearly extremely relevant devices for understanding how resources can best be mobilized for the production of surplus value.[19] Formulations based on the achievement of equality in distribution are also counter-revolutionary unless they are derived from an understanding of how production is organized to create surplus value.[20] By examining questions such as these we can at least begin to evaluate existing theory and in the process (who knows?) perhaps begin to derive the lineaments of new theory.

A revolution in scientific thought is accomplished by marshaling concepts and ideas, categories and relationships into such a superior system of thought, when judged against the realities which require explanation, that we succeed in making all opposition to that system of thought look ludicrous. Since we are, for the most part, our own worst opponents in this matter, many of us will find that a first initial step on this path will be to discomfort ourselves, to make ourselves look ludicrous to ourselves. This is not easy, particularly if we are possessed of intellectual pride. Further, the emergence of a true revolution in geographic thought is bound to be tempered by commitment to revolutionary practice. Certainly the general acceptance of revolutionary theory will depend upon the strengths and accomplishments of revolutionary practice. Here there will be many hard personal decisions to make. Decisions that require "real" as opposed to "mere liberal" commitment, for it is indeed very comfortable to be a mere liberal. But if conditions are as serious as many of us believe, then increasingly we will come to recognize that nothing much can be lost by that kind of commitment and that almost everything stands to be gained should we make it and succeed.

NOTES

1 Kuhn, T. S., 1962, *The Structure of Scientific Revolutions* (Chicago).
2 Bernal, J. D., 1971 edition, *Science in History* (MIT Press, Cambridge, Mass.).
3 *Ibid.*; Rose, H. and Rose, S., 1969, *Science and Society* (Penguin Books, Harmondsworth, Middx.).
4 Kuhn, *op. cit.*, p. 37; Nagel, E., 1961, *The Structure of Science* (Free Press, New York).
5 Johnson, H. G., "The Keynesian Revolution and the Monetarist Counter-revolution," *American Economic Review* 61, no. 2, pp. 1–14. I would like to thank Gene Mumy for drawing my attention to this reference.
6 *Ibid.*
7 Marx, K., 1967 edition, *Capital*, Volume 2 (International Publishers, New York), preface (by F. Engels); this whole incident is discussed in depth in Althusser, L. and Balibar, E., 1970, *Reading Capital* (NLB, London).
8 Kuhn, *op. cit.*, pp. 52–6.
9 Marx, *op. cit.*, pp. 11–18.
10 Marx clearly envisaged this kind of resolution of the conflict between the natural and social sciences. See Marx, K., 1964 edition, *The Economic and Philosophic Manuscripts of 1844* (International Publishers, New York), p. 164.
11 Marx, K., 1971 edition, *The German Ideology* (International Publishers, New York); *The Economic and Philosophic Manuscripts of 1844, op. cit.* Marx derived his phenomenological position from Hegel, G., 1967 edition, *The Phenomenology of Mind* (Harper Torchbooks, New York).
12 Marx also derived his dialectical method from Hegel. See *The Economic and Philosophic Manuscripts of 1844, op. cit.*, pp. 170–93.
13 Park, R. E., "The Urban Community as a Spatial Pattern and a Moral Order," in Burgess, E. W. 1926, *The Urban Community* (Chicago University Press, Chicago).
14 Engels, F., 1962 edition, *The Condition of the Working Class in England in 1844* (Allen and Unwin, London), pp. 46–7.
15 Kerner Commission, 1968, *Report of the National Advisory Commission on Civil Disorders* (Bantam Books, New York).
16 Alonso, W., 1964, *Location and Land Use* (Harvard U.P., Cambridge, Mass.); Muth, R., 1969, *Cities and Housing* (Chicago University Press, Chicago).
17 Valdes, N. P., 1971, "Health and Revolution in Cuba," *Science and Society* 35, pp. 311–35.
18 Engels, F., 1935 edition, *The Housing Question* (International Publishers, New York), p. 23.
19 The experience of Russian central planning in this provides some interesting lessons as Kantrovitch developed programming solutions to many of the allocation distribution problems which arose in the economy—see Ellman, M., 1971, *Soviet Planning Today* (Cambridge University Press, Cambridge). This suggests that some of the programming solutions to urban location problems may be more useful than not as the basis for a revolutionary urban land use theory.
20 Marx, K., *Capital*, Volume 3 (*op. cit.*), pp. 876–86. It is on this ground that I now reject my attempt in a previous paper to examine questions of distribution in a manner distinct from the problem of production. See Harvey, D., 1971, "Social Justice and Spatial Systems," in Peet, R. (ed.), *Geographical Perspectives on American Poverty*, Antipode Monographs in Social Geography, no. 1, 1972, pp. 87–106.

2

Geographic models of imperialism

James M. Blaut

from *Antipode*, 1970, 2: 65–85

I

Imperialism, as I speak of it here, is white exploitation of the non-white world, a plague that began some 500 years ago on the West African coast and spread across the globe. It has not been cured by emancipation, by decolonialization, or by economic development (which suffers from the same disease). It *has* been cured at times by revolution, for instance in China and Cuba. One such cure is now underway in Indochina. But a deadly pattern has emerged which we see in Indochina and elsewhere: no revolution may run its course without armed intervention by the white world, the West. This pattern is grounded in the logic and beliefs of imperialism. Here are two allegories:

"Those gooks *can't* win." If P, then not Q. This statement enjoys the status of axiomatic certainty. There is no possible logic of withdrawal that follows from such a self-verifying axiom. "The gooks can't win," so we escalate. Thus we come to fight the gooks and the chinks and the niggers as well. "But those gooks didn't win, did they? Now those chinks and gooks and niggers *together* can't win . . ." So the air-tight logic flows on, and so we enter World War III.

The second logical sequence begins benignly enough, "No sane man wants violence." Is it therefore insane to sanction the incessant violence that a Black South African endures? But this, of course, is not violence. It is merely a high mortality rate from disease, starvation, and suicide. We blame it on the Population Bomb or on Their Own Stupidity, never on our own Chase Manhattan debentures. But when their revolution begins—that is termed "violence," and violence is insane. So we send in the marines: "peacekeepers" who never even heard of *Apartheid*. When Black troops arrive from East and West Africa, we defend the territorial integrity of South Africa against these *invaders*, these perpetrators of "violence." Next to arrive are the gooks and the chinks . . . And so we enter World War III.

II

These allegories express a proposition that is the foundation stone of this paper. I can state the proposition in two ways, one of which will seem trite and the other perhaps mystical or foolish. First: all things can be rationalized. Second: all of Western science and historiography is so closely interwoven with Western imperialism that the former can only describe and justify the latter, not predict it or explain it or control it—not even when human survival is at stake, as may now be the case. The second form is easily confused with the "East is East and West is West" form of cultural relativism, an argument which has some predictive use in linguistics, but otherwise merely expresses the fact that cross-cultural communication is always difficult, always imperfect, but never truly impossible. I am trying to say something rather more specific. At this point I need a felicitous term.

The word *ethnoscience* has been used for the past few years to designate an interdisciplinary field on the common border of anthropology, linguistics, geography, and psychology. That field tries to analyze the cognitive systems—the beliefs about reality—which are characteristic of a given cultural-linguistic universe, and to theorize about such matters on a cross-cultural basis.[1] I will speak of "an ethnoscience" and mean thereby the total set of explicit and implicit terms, relations, and propositions which circulate among the members of a culture or group of cultures.

Some propositions will be axiomatically true by common consent. Some will carry different truth-functions for different individuals. Some will contradict some others. Concepts will vary in much the same way. This universe of discourse will be said to possess the following defining characteristics:

1 It includes propositions about unique events as well as general propositions. Thus it extends over all of history, all of science, and all of practical knowledge as understood by the members.
2 It is absolutely comprehensive. If a given phenomenon is known to the members of two such groups, it will be incorporated into both ethnosciences.
3 An ethnoscience does *not* include the judgements of value, preference, virtue, or taste which the members hold. This is of course a tricky point. In Western science and history, for instance, the notion of value-free statements is viewed as a fiction, sometimes useful and more often not. In epistemology it is difficult even to imagine a statement or thought, however mundane, that is value-free. All I require here is the possibility of crudely splitting off a value-statement from a corresponding knowledge-statement, and *only* to the extent that the latter can be stated as a distinct proposition, *relatively* clear of valuation. The statement, "I see the lovely chair" must then become "I see a chair" + "I judge the chair to be lovely." The former is explicitly within the ethnoscience. Though still not wholly value-free, it can be used as a relatively independent variable (i.e. the fly on Napoleon's horse's nose).

   This separation is needed to distinguish the universe called "an ethnoscience" from the larger universe embracing all thought and expressed by terms like "world view," *Weltanschauung*, cosmology, and so on. Only in this way can we get at the interactions between ethnoscientific variables and non-ethnoscientific variables. The crucial interaction is the effect of Western imperialism, as a set of interests and norms, upon the two most relevant portions of Western ethnoscience; first, historical "truisms," or conventional beliefs about what happened in history; and, second, social-science "paradigms," or clusters of accepted social-science theory.[2]
4 Any two ethnosciences can be mapped on one another, by way of comparing them. Each can be a different state-of-knowledge for the same culture. Each can be from a different culture. One can be from *a* specific culture and the other from a group of related cultures in which the first is included. The pair with which I will be concerned in this paper is, first, the whole of Western science and history and, second, a theoretical ethnoscience that I create by modifying the first in one respect: I withdraw the more glaring rationalizations for imperialism.

An ethnoscience has two additional properties which are axiomatic but testable. The first describes the relations among statements within the system. The second describes the relation between an ethnoscience and a corresponding value system.

We can think of Western science and history at a given time as containing a certain number of persistent theoretical paradigms and historical reconstructions. This population of scientific and historical beliefs can be assumed to have an overall structure, however loose it may be. I will speak of a relation of "compatibility" between pairs of beliefs, meaning simply that they can co-exist. A pair in widely separated disciplines can no doubt contradict one another and still co-exist, and there are rare cases of this sort within single disciplines, e.g. particles vs. waves.

The general rule would seem to be that accepted paradigms are likely to reinforce one another—by using common elements, for example—or at the very least be essentially unrelated. Compatibility on these terms is no problem. The same should hold true for pairs of historical beliefs and for mixed pairs, as in psychoanalytic history, for instance. This should also hold for paramount beliefs in public policy, e.g. the putative views of the electorate.

Obviously, the concept of compatibility is probabilistic in specific cases and becomes axiomatic only when we deal with beliefs in aggregate. The axiom is best stated in the same form: in a given ethnoscience, through a given epoch, it is unlikely that any basic, important beliefs, scientific, historical, or public-policy, will be sharply and embarrassingly incompatible with any other such belief without a resolution of the conflict taking place relatively quickly. This axiom is closely analogous to the theory of cognitive dissonance, i.e. incompatible beliefs tend to get in one another's way.

The axiom will let us deal with each ethnoscience as a system, and it lets us connect together various distant beliefs—distant in subject, time, and space.

The second axiom is more crucial to my argument: a fundamental belief in the ethnoscientific system associated with a given society is not likely to fall into or remain in conflict with a fundamental value or norm that is held by the members of the society or by the policy-making élite if the society is highly stratified. In other words, crucial beliefs should conform to crucial precepts: the true should also be the good. If there were no such conformality between ethnoscientific system and value system, we would have science proving that religion is false, history undercutting patriotism and the like—dissonances that a culture certainly cannot tolerate in high degree.

III

I think I can identify a single ethnoscience that is characteristic of the European nations (or élites) which have participated directly or indirectly in the imperializing process. This ethnoscience spans the entire European culture world through five centuries of its history. This level of generality would be too broad to be useful in most other contexts, but that is not the case here, for two reasons. First, the span is quite normal for studies in the history of scholarly ideas. Second, whatever the variations among the national (or national-élite) ethnosciences, all should have a basic similarity in matters pertaining to imperialism, to relationships between Europe, with its set of wants, and the rest of the world, where the wants are to be fulfilled.

This White, or Western, or European, ethnoscience is the intellectual underpinning of imperialism. It includes within it the varying paradigms of Western science and the propositions of history. Allowing for necessary variations, this is the common, general system of scientific and historical ideas in which we White, Western social scientists are working. Its growth has paralleled and supported the growth of imperialism, and it has become for us an almost irresistibly strong current of thought, pulling each new theory and interpretation in the same direction as the old: toward compatibility with the policies and goals of Europe and empire. There is nothing mysterious about this force, and some of us succeed in swimming against its pull (else there would be no *Antipode*!). But it has produced a general drift of bias in those parts of Western ethnoscience which are closely involved with imperialism. I will show in later paragraphs how this bias works its way through the chinks in scientific and historiographic method.

European ethnoscience, like every other, is perfectly comprehensive in scope. Hence it contains a set of historical beliefs and social-science generalizations about the non-White world as well as the White. I noted earlier that long-run consistency must be maintained between the ethnoscientific system and the value system. The governing system of values in European ethnoscience is that of the White world alone: the imperialists, not the imperialized. To fit this ethnoscience to an anti-imperialist value structure requires quite drastic changes, even if this structure is a limited set of normative propositions and not the entire value system of a specific non-European culture. Even more drastic changes are required to incorporate the findings of Third-World social scientists and historians (to the extent practicable).

The European model has to be examined very closely for biased and questionable historical statements dealing with the Third World, with imperialism, with European affairs relating to the Third World, and for social-science models and generalizations which are comparably biased or questionable. Each of these must be deleted. In some cases I will very deliberately introduce alternative statements with biases tending in a Third-World direction and build models to generate new hypotheses, however improbable these may seem. If I call the end product a sketch of Third-World ethnoscience, nothing impressive is implied. This is not the ethnoscience of an entire culture. Nor is it that of a synthesis of cultures. It is best described as an attempt to broaden Western ethnoscience by removing its more flagrant imperialistic biases and enriching it with Third-World data. Many Third-World social scientists and historians, perhaps most of them, are engaged in essentially the same enterprise.

IV

The argument turns now to the sphere of method, since a crucial point is the vulnerability of scientific method and historiography to bias. If the drift of bias in Western scholarship is in the direction of congruency with imperialism, the one most pervasive and persistent interest of Western culture as a whole, then methodology will not check that drift. Even the most careful, disciplined, expert, and perceptive scholarship will be unable to do so, although without such care and expertise matters would be worse.

The problem is most acute in history, but most easily diagnosed. Visualize the historian's job of pursuing information along a chain of documentary reports, each document adding its quantum of value bias, imperfect perception, incomplete description, subjective categorization, and so on. These are normal hazards of the historian's trade, and he negotiates them as best he can. Always he seeks to overcome these difficulties of concrete, artifactual data and, as it were, enter the subject's mind.

Consider now the problem faced by a European historian trying to gather data on, say, the history of a colonial possession of his own country. One set of sources derives from those individuals whose participation in the events under study would ordinarily persuade the historian to rely most heavily on them for primary data. But they write in non-Western language and script, convey the beliefs and values of a non-Western culture, and are likely to evince rather consistently negative bias against the occupying power, its agents, and their actions. By contrast, there is an abundance of easily available records written in the historian's own language by a group of his own countrymen whose ordinary bias is inflated by racial, cultural, class, and patriotic prejudice. The historian must thus choose between two kinds of account, each with an opposing bias. Not surprisingly, he is likely to accept the bias of his own countrymen, whose material he can deal with in terms of known means of judging evidence. He can, as it were, enter their minds. However carefully he may avoid contaminating his work with his own attitudes, the bias has entered it even so.

The shelves of colonial and non-Western history contain rather few works by Europeans who are familiar with the non-Western language and culture of their area, and fewer still by colonials themselves. Instead, we have a vast literature written by colonial administrators-turned-historians, with titles like "My Ten Years among the Dyaks." There is also an abundance of strongly biased writings by trained historians from the occupying nations, works very aptly described by Van Leur as history "written from the deck of the ship."[3] Hence we see the importance of a handful of studies written by non-Western historians, mostly after independence has been attained, studies which present a counterfoil to the European view, a different interpretation, and often a body of new and important data.[4] Until this literature has grown very much larger, we must assume that the basis for reasonably objective judgment of the colonial and pre-colonial past has not as yet been laid. Thus we must garnishee the bulk of existing data and review all the evidence behind the most crucial generalizations about the nature and evolution of imperialism.

Science has proven as vulnerable as history in the matter of resisting imperialistic bias—of breaking out of the paradigms which assert that, for any hypothesis P, P is true of the Third World if P is useful to imperialistic policies and false if not. My concern is with lacunae in scientific method, specifically in the methodology of pure and applied social-science research being carried out today in neo-colonial countries, independent areas which maintain one or another form of colonial economy. (Much more will be said about neo-colonialism later.) Nearly all of this research has the stated goal of contributing directly or indirectly to economic and social development, and those who conduct the research accept by and large the goal and honestly follow the canons of scientific method. Yet the results continue to fit the old paradigms.

We notice to begin with that the probable direction of bias can be predicted from the roles, values, and reward system that are typical of the investigators. Much of the research is carried out by White social scientists from North America or Europe with financial support from their own government or a corporate foundation. Most are inclined to accept the ideology of their own culture in at least skeletal form—would not receive the financial support if they did not, in most cases—and this ideology is compatible with the paradigms in question. These paradigms assert that any P is untrue, unworkable, or wrong if P leads to radical, social, and political change, or merely to political instability. Investigators who are nationals of the neo-colonial country itself usually are government employees, participants in a system that provides faster promotion and like rewards for findings which do not point to governmental errors and contradictions at any level, from the small development project to the national policy of neo-colonialism. University research is only slightly less constrained, and professors in any case have little opportunity to do any. It should be added that government and university social scientists are usually recruited from the class that benefits from neo-colonialism. If many of them oppose it nonetheless, they tend not to do so in the context of sponsored research. Hence we find the potential for bias toward the paradigms of imperialism at the start of research. The potential is enhanced by the fact that nearly all investigators ingested these paradigms at the time they were trained.

Scientific method is relied on to ward off systematic bias, but it cannot do so in a neo-colonial situation such as we are describing. It merely disguises the bias behind a façade of spurious objectivity. The façade is most impressive when formal models are used and when masses of quantitative data are processed. However, these approaches seem to confer no greater immunity than do others.[5] Most of the models are drawn from Western theory. Given that other models are likely to provide equally good fit in a typically complex system, the systematic choice of a Western model adds to the probability that the system will be wrongly subsumed under an inappropriate paradigm. The same systematic error recurs in the choice of assumptions.

A special problem arises when simulation models are developed specifically for mass data-processing. The choice of variables is conditioned by the availability of statistics. These, in turn, reflect the information needs of the prior colonial epoch (or present large-scale commerce); hence, the simulation becomes a caricature. One must see this problem in the context of theories that explain so little of the variance—when they are tested at all—that a bad model or bizarre assumption is almost never rejected for reasons that have anything to do with scientific method or results. The favored models are congruent with views, values, and interests which would not be abandoned in any case.

Empirical research fares no better than theoretical: it is hobbled by the same biases. Implicit Western models tend to govern the selection of problem, field-work area, sample design, data categories, and the like. Interview biases are monotonous in their congruence with the hypothesis and purpose of a study. Perhaps the most serious problem in empirical research is the tendency to read into a given situation some truism dredged up from European history—about which more will be said later.

Thus it appears that Western science, like Western history, has been methodologically incapable of controlling its own tendency to interpret the Third World in terms of the paradigms of Western ethno-science and the interests of imperialism. For this reason, one must adopt an attitude of systematic skepticism toward theories about the dynamics of the Third World. The predictions from such theories suggest strategies, e.g. for economic planning, but one finds very often that the predictions are merely restatements of the assumptions originally chosen for a model that has not really been tested. The assumptions themselves turn out to be epigrams of imperialism. So economic development can become a way of phrasing imperialist strategy, not an alternative to imperialism.

Few of us believe in the possibility of a perfectly objective science or history, so the foregoing argument should not be, in principle, unpalatable. If it gives some discomfort, this may be a symptom of the difficulty we have in swallowing the proposition that our lack of objectivity is not a random error, nor even a class or national bias, but a systematic tendency of Western thought, tied to the common Western imperialism. The tendency is rather slight at the level of individual research: an unconscious value-loading of adjectives; a not-absolutely-random sample; a project selected because research funds are available for this sort of thing and not that. The cumulative effect, like the Coriolus [*sic*] force and the solar wind, is no less powerful for being unnoticed. Hypotheses that clash with imperialism simply do not rise to the status of paradigms or truisms. Hence, over the decades and centuries, we maintain a body of belief that is truly the ethnoscience of the Western world; not, as it claims to be, the universal science and history of the world as a whole.

V

We can now compare the geographic models of imperialism which emerge from Western and Third-World systems of belief. I have said nothing thus far about the content of these ethnoscientific systems. Instead I gave an elaborate methodological fanfare, the aim of which was to raise some doubts in advance about the seeming self-evidence of the one set of beliefs and the seeming improbability of the other. I will discuss these beliefs only to the extent that each underlies or enters into its respective geographic model, but their basic form will emerge quite clearly as we proceed. I speak of the models as "geographic" because space and resources are perhaps their main dimensions. They span some 500 years of human history, but they also span the globe. This scope is routine in historical geography. Note that the Western model is non-Marxist. A Marxist model of the classical or European type—something of an intermediate case—will be examined briefly at a later point. The Western model will be given rather cursory discussion in any event, since it is a collection of all-too-familiar truisms. It will in fact be treated very shabbily, and used mainly as a foil for the Third-World model, toward which I admit a favorable bias.

It should be said at this point that my use of the term "model" in this discussion is intended to not only emphasize the fact that we are simplifying process to a bare structure for analytic purposes, but also to emphasize a property of models which is vital to this kind of discourse. Models are not reality. They can be as improbable and outlandish as one may desire, so long as the model world and real world remain separated. Some of the historical statements in the Third-World model are so thoroughly contradictory to the truisms of Western history that they may not even seem plausible. My task is to clothe them with just [such] supporting evidence as conveys their plausibility. Historians must carry the burden from there.

"European civilization arose and flowered, until in the end it covered the face of the earth."[6] These words by Marc Bloch sum up the Western model quite nicely. They convey the root belief in an ineffable European spirit, a *sui generis* cause of European evolution and expansion. A small number of such beliefs are the basic truisms of imperialism, generating those arguments which justified imperialism during its evolution and those which (I claim) disguise it today. The following truisms seem to be crucial:

1 Europe is a spatio-temporal individual, clearly demarcated and internally coherent—a "civilization." It has been such since the Middle Ages or before, although the boundaries have extended to Anglo-America and beyond. This conception gives the model a simple domocentric form,[7] with a distinctive geometry: an inner space, closed and undifferentiated (all portions have the property "European"); an outer space, also closed on the spherical surface; and a boundary between them which has finite width and important internal properties.
2 The rise of European civilization throughout this period has been generated mainly by inner processes. Non-European peoples and areas have had no crucial role in epochal events: the Reformation, the Enlightenment, the Industrial Revolution, and so on. Whenever events outside of (topological) Europe assume significance, as during the ages of exploration and mercantilism, Europeans themselves play the active role. Here we have the first property of the boundary: selective permeability. Major forces in cultural evolution cannot filter through it in a centripetal direction, although raw materials can do so; likewise Aztec gold.
3 All non-European cultures are more or less primitive, at the time of colonization, by comparison with Europeans in the abstract and by comparison with the particular Europeans who colonize a given area and pass judgement on its inhabitants. All such cultures are unprogressive. All are either standing still or declining at the time of colonization. (China is usually conceded to have barely reached the "European" civilization level of pre-Enlightenment times, but is the very model of decadence.) All such cultures are barbarous and heathen. In sum, the non-European world is less strong, less intelligent, and less virtuous than Europe. Hence there is a kind of osmotic differential in power, knowledge, and righteousness.
4 The outward expansion of Europe, like the rise of Europe itself, is, *sui generis*, a product of internal forces and motives. It is a "striving outward," an "urge to expand." There is self-generated evolution within the boundaries of Europe and there is osmotic pressure across the boundary. The result is unidirectional flow: a diffusion process, not an equilibrating system. (By no coincidence, classic diffusionism in European social science was imperialistic, ethnocentric, and often racist in tone. Does this perhaps hold true for some of diffusion theory today?) As a corollary, any given part of the non-European world gains its important attributes from the European impact. Therefore the non-European world as a whole—excepting the areas depopulated and settled by Europeans, thus becoming pseudopoda of Europe itself—displays the pattern created by a decay function; the farther one gets from Europe (in the sense of connectivity, not true distance), the less intense the attribute. This can be described in part as a series of age-area or wave-diffusion bands, and in part as a continuous cline. Thus, whenever non-European areas display qualities indicative of importance, progress, and the

like, they do so as a result of Europeanization and in proportion as they have received the European impact. Thus also, the process of imperialism becomes *a* matter of giving civilization while taking resources.

These historical truisms provide some of the major elements in a structure of ideas that underlay European imperialism throughout its course and underlies it still. A double standard of morality was accepted by which privacy, brigandage, privateering, slave-raiding, slave-trading, and slavery itself were permitted so long as the venue were extra-European—indeed, the Enlightenment in Europe rather coincided with the age of slavery beyond the boundary. Colonialism acquired the status of a natural and inevitable process, almost foreordained by the internal evolution of Europe and developing smoothly and continuously from the first voyages of Henry the Navigator to the final partition of Africa. The establishment of large-scale capitalist enterprise in colonies and former colonies was equally inevitable, a matter of finding better and higher uses for land and labor than the natives themselves were capable of achieving. In the twentieth century, discomforting events like Japanese militarism and the growth of Chinese communism were cognized as effects of European ideas. Were it my intention to elaborate fully the Western model, I would attempt to show how these and like processes are, and have been in the past, cognized in terms of the few basic persistent truisms (not all of which have been mentioned, of course) which serve as assumptions in the model. At all stages in modern history, Europeans have drawn these truisms from Western ethnoscience whenever the need arose to justify events past, current, or planned.

The same holds true today. The conception of non-European peoples as inferior in strength, intelligence, and virtue—that is, in national power, technological innovativeness, and justifiable aspirations—is still basic to the international policies of the West, although rarely stated in explicit terms and perhaps not even drawn out into the conscious decision-making process. Before pursuing this matter further, we had best present the alternative model, that of the Third World.

VI

Somewhat earlier, I characterized Third-World ethnoscience in a highly simplistic way, calling it Western ethnoscience with one constraint removed: its support of imperialism. But if that one constraint is removed, the shape of historiography and social science must change. Fanon makes this blunt assertion: "What the West has in truth not understood is that today . . . a new theory of man is coming into being."[8] Whether he is literally right or not, this is clearly the program for a Third-World ethnoscience.

The model, as I build it here, goes well beyond existing scholarship (as any model should). Where my own speculations are woven into the fabric, the design remains in harmony with the whole. This design has one basic motif: basic skepticism with regard to any truism or paradigm of Western ethnoscience which seems to reinforce imperialism: by derogating a part or property of the Third World; by asserting a dependency on the West; by claiming that some form of Western enterprise in the Third World is *a priori* logical, proper, or necessary; by rationalizing the use or potential use of Western force in a Third-World region; or simply by asserting that European history is *sui generis*—that cultural evolution is a European monopoly. Skepticism leads to criticism, and thus the fabric is unraveled and rewoven.

This model asserts a body of propositions, most of which are sample denials of one or another part of one or another Western truism. I will draw these propositions together into a schema of the historical geography of imperialism, somewhat arbitrarily divided into three space-time stages. Very little will be said about the geometry of this model since its most fundamental quality is (or can be expressed as) the absence of ethnocentrism, domocentrism, and therefore nodality.

Stage I can be called the Slave-Based Industrial Revolution, with bounds extending from Atlantic Europe to the West African coast; thence to the Atlantic coast of South America; thence northward along that coast to the West Indies; thence back to Europe. This period lasts very roughly from 1450 to 1750. To deal with it adequately, one would have to discuss events occurring in Asia at the same time, but I will content myself with one proposition: the impact of Europe on Asia throughout this period was very light; the model itself suggests why this was so.[9]

The period begins with privateering—chartered piracy—on the Guinea coast by Portuguese merchant and naval vessels. Small-scale slave-raiding occurs. Equally small-scale sugar planting begins on several Atlantic islands (e.g. *Sâo Thomé*), using captured slaves and producing for the European market. Iberian seamen continue to probe southward, less concerned with rounding Africa than with preying on her coast. As profits and experience increase, ships grow sturdier; eventually the orbit enlarges to the point where a landing is made in the New World, and Iberian interest shifts in part to New World privateering ("conquest"). Meanwhile, the Portuguese open a lucrative trade in the Indian Ocean, beating Arab competition mainly through high-seas piracy on the smaller Arab vessels. This trade interests us mainly because it leads to settlement on the Brazilian coast, whose warfare and European diseases, combined with slavery, quickly destroy Amerindian competitors for land. Planters move across the Atlantic, vastly increasing plantation acreage and stimulating the slave trade. Now Europe hits the jackpot and commences the true explosion of imperialism: massive expansion of commercial, slave-based agriculture in the New World.

The acquisition of disease-emptied land in limitless quantities was the *one* major advantage which Europeans managed to wrest over the civilizations of Africa and Asia. In this model Europe had no "urge to expand" not shared by these other civilizations; nor did Europe have any technological advantage, save ships that were slightly more sophisticated as a result of the prior epoch of African piracy; nor did Europe display any other distinguishing sign of cultural advancement or achievement-motivation in the fifteenth and early sixteenth century. She merely got to the New World first, and obtained its lands in exchange for genocide.

The Third-World model can therefore postulate that Europeans had *no* innate superiority, nor even the power to conquer Old-World civilizations. This power was only gained in the eighteenth century, 250 years after Columbus, as a result of the industrial revolution, which began not in Europe but in the West Indies, thereafter diffusing (outward? inward?) to Europe.[10] To make this last assertion plausible, we need a subsidiary model, a microgeographic system-model of the slave-based plantation, showing its homology to subsequent factory industry. Such a model would show that, in level of machine technology (mill and field), level of capital, scale of labor input and production, organizational complexity, multiplier-generating effects, and other attributes, the sugar plantation was equal to the level of the early English textile mill, which it long preceded. The overwhelming difference was slavery.

The Third-World model here invokes alternative propositions. The first is an anthropological truism: every culture has rules, or laws, which limit the degree to which any one participant in that culture can exploit any other, but no such rules need apply to outsiders. Beyond the bounds of Europe, a heathen alien could be murdered or enslaved at will. With fresh land and a brisk market, moral scruples were brushed aside. The second proposition is cultural-geographic: new technology is extremely costly in its earliest phase. Third is the basic Marxist proposition that power must be employed to extract surplus from labor—surely most feasible under the guns of a slave colony. Thus we have rounded out the argument for a slave-based, extra-European origin of the factory system—indeed, of capitalism itself if one accepts Marx's distinction between merchants and capitalists.[11] In our model, the factory system had to evolve under slavery to the point where labor's return was, literally, enough to keep the laborer alive; only then could the system be transferred from the colonies to Europe, and from the southern U.S. to the north; then reconstituted as a new kind of mill, with semi-free labor and child labor forming a transition.

This model of the origins of industry and industrial capitalism is reasonably strange in the context of Western ethnoscience. In the Third-World system, the model may be poor but it is far from strange. C. L. R. James, one of the greatest of Third-World historians, says simply: "There is no question today that the resources which initiated and established [the factory system] . . . resulted from the Atlantic slave-trade and the enslavement of Africans in the Americas."[12] The slave-trade itself, in this model, was mostly brigandage on the West African coast; involvement of African kingdoms came relatively late, responding to one of the fruits of the European industrial revolution: cheap guns.[13] One must add the input of profits from Spanish enterprise in the New World—plunder, mining, and a bit of agriculture—as well as the profits from mercantilism in Asia. The Asian element, however, is much overrated. Europeans controlled no significant territory prior to the mid-eighteenth century, and had nothing much to sell the Asians until slave-based industry had triggered off European industrial revolution.

Stage II in the evolution of imperialism, as portrayed in this model, is classical Colonialism, or large-scale territorial conquest, mainly in areas with sophisticated state organization. This period commences slowly in the early eighteenth century. By the end of the nineteenth, colonial control has spread at least nominally over nearly all of Africa and Asia, Japan has joined the colonizers, and China has become a giant colonial condominium. The stage is brought to an end almost everywhere, by a formal grant of independence and a change of color on world maps, in the two decades following World War II. In the model, it ends by a gradual re-occupation of territory and gradual crumbling of colonial political control over a period of perhaps eighty years.

The initiating conditions for Stage II are in essence also the terminating conditions for Stage I. These conditions can best be understood in terms of the geometry of the model. Initially, the Old World is a single uniform region dotted with mercantile cities, not a nodal region centered on Europe. We assume no initiating condition within Europe itself which might destroy the relative spatial equilibrium among Old-World cities, and we assume (with Polanyi) that merchants throughout the archaic world are subservient to or portions of the state apparatus—that dual, competing power poles are unstable and hence inadmissible.[14] We explain the acquisition of power by European merchants as a boundary process between the Old World and the New. Two vital ingredients of a capitalist society were thus drawn into Europe; first, the system of industrial capitalism which (in this model) evolved under slavery; and, second, the profits—from plunder, slavery, Asiatic trade, and new markets in North America and other pseudopoda of Europe—which were needed by the merchant (now capitalist) class to acquire political power and thus legitimize itself. By the end of Stage I, capitalism and the State were again a single focus of power, but West European states were adapting themselves to capitalism, not vice versa. In this way political and legal authority was obtained to create a semi-slave proletariat in Europe itself by widening the limits of allowable exploitation within the society, and to redefine the territorial conquest of non-European areas as an affair of state, not of private Chartered Company (although the transition from one to the other was slow).[15]

Thus we derive the three initiating conditions of Stage II, or Colonialism. First, the European country has already become a capitalist—not simply mercantilist—society, and colonies are therefore sought as extensions of the European marketing and raw-material sphere. Second, industrialization and mercantilism together have advanced to the point where both the technology and capital needed for large-scale conquest are at hand. And third, since the first condition is best encountered in a potential colony which possesses a strong pre-existing state organization and large population, formal and informal procedures will be followed to forge an alliance between the imperialists and the colonial power structure—an alliance with genuine advantages for the colonial participant since it offers him a share of the economic and political spoils. This, you will note, is a redefinition of the "divide and conquer" precept. In our model, the precept reads: imperialism should be fitted into the pre-existing forms of exploitation so that the same classes are exploited, only more vigorously. I shall have occasion to return to this proposition in a modern context.

At the close of World War II, the colonial system was dismantled with the speed of a traveling circus, and in most places independence was achieved without violent revolution and at the stated initiative of the imperial power, rather than on presentation of an ultimatum. The Western model uses these facts as evidence that imperialism is on its deathbed. The Third-World model interprets these facts quite differently: an area can be exploited more efficiently in the late twentieth century without overt political control. This is neocolonialism, Stage III of our model. To understand the homology between these two stages, and to understand the nature of neocolonialism—a crucial matter for the Third-World analysis of economic and social development programs—we must focus for a moment on certain structural properties of an ideal-typical colony as delineated by our model.

The colony as a whole can be described as a feudal state, a three-tiered society consisting of a gentry or landlord class above whom are the state aristocracy and below whom are the peasants, with an imperialist superstructure added as another tier above—and indeed trickling through all the economic interstices at the state and gentry level as well. The colonial bureaucracy is usually self-sustaining with tax revenues which are generally paid directly or indirectly by the peasants. The bureaucracy supports a large corps of not always vitally needed Europeans. It has the additional tacit function of providing a source of decently paid employment for the sons of lesser gentry. It is indeed a significant element in the spoils system; hence the bureaucracy may be an exploitative structure if its technical services do not redress the balance, as may or may not occur.

The main exploitation, however, comes from private business, the functions of which are kept scrupulously separate from those of government. A large, diversified colony, however densely populated, will have at least some European-owned plantations (if only tea gardens above the food cultivation zone), some European-owned mines, a host of European-owned import–export agencies which usually feed into locally managed distribution networks, and other such enterprises. Almost all manufactured goods are brought in from the colonizing country. Local manufacturing may be suppressed overtly, as happened to India's cotton textile production and export during the company era and later to sugar refining in the West Indies, or local industry may be unable to face competition from factory-produced imports.[16] In any case, a massive loss of income occurs as a result of this process, wiping out incipient manufacturing industries, depriving urban artisans of a livelihood (but benefitting merchants), and reducing peasant family income.[17]

To sum up these structural features: the colonial bureaucracy provides welfare and infrastructural services and maintains an exceedingly large police or military establishment, roughly at cost. The private European sector profits by exporting plantation and mine products and importing manufactured goods. The landowning gentry continue as always to collect rent or share from the peasants, while their kinsmen maintain an élite status in new roles, bureaucratic or business. The peasants find their burden growing slowly but steadily, generation by generation, and the class of landless laborers increases in proportion. Population growth may be a contributing cause to their problem—I will argue to the contrary below—but the primary cause is exploitation: heavier charges placed on farmers who always have limited land supplies and rarely have access to yield-raising inputs.

The colony thus created is dualistic, but not in the sense of an economy in which European enterprise is distinct from and cannot integrate with peasant economy, if such is even possible.[18] The dualism here is between government and economy. This is a fundamental feature of capitalist societies; it permits unrestricted economic activity within a broad area of legally permissible actions. The same dualism is constructed, for the same reason, in our colony. After a firm network of ties has been established between metropolitan and colonial enterprise, and the latter has acquired a corps of managers and shopkeepers from the local élite, then, in theory, independence need not interfere with business as usual. In pre-colonial society, this kind of dualism is rarely seen. Land ownership, for instance, may have as many political-territorial attributes as it does economic. But colonialism bribes the old élite into assuming a new role, part economic bourgeoisie, part bureaucratic functionary. If peasants and laborers have not been pushed to the point of full-scale revolution, then the new élite will almost always be found in the vanguard of the less extreme independence movement. But if the colonizers refuse to leave, many from this group will turn revolutionary overnight, as happened in Indonesia, Algeria, and elsewhere. One need not entirely dismiss the significance of imperialistic beliefs when noting that almost all the imperialist countries chose to ignore such beliefs and free their colonies without fuss. So the terminating conditions for Stage II and initiating conditions for Stage III are bound up in the colonial process, which creates an economic fief and secures it against all political dangers short of socialism.

"Neo-Colonialism" is the most widely used term for what our model identifies as the third stage of imperialism. Nkrumah characterizes an independent nation as neo-colonial if "its economic system and thus its political policy is directed from outside."[19] In the present model, a neo-colonial state is given five defining properties. First, its economy is connected to European capitalism in the colonial manner. Second, its internal political structure is effectively controlled by an élite group of the sort I described for Stage II colonies, a group of businessmen, civil servants, or both, committed by self-interest or ideology to maintaining the colonial economy. Third, it may have economic connections with more than one European power. Fourth, the state lies under a permanent threat of invasion or some other hostile act if European economic interests within its territory are not properly protected; hence, its sovereignty is contingent. And fifth, since exploitation in the neo-colonial mode is much like that of classical colonialism, neo-colonialism has much the same need to be backed up by military power if the citizens grow restless. Accordingly, the neo-colony is customarily given substantial military assistance to insure internal security, while its European partner stands ready to airlift troops into the country if they are needed.

Note that the first two of these properties, a colonial economy and a bureaucratic-mercantile élite, are the two most fundamental features of Stage II colonies, aside from European rule itself. Note also that the third of these properties suggests a plausible reason why imperialist powers might find their interests best served by granting independence to their colonies and converting them into neo-colonies. To begin with, if all the major powers de-colonize at about the same time—exactly as happened—then the Common Market principle takes effect: each gives up its trade protections in a small colonial market, gains access instead to a vastly larger one, and still retains the fat pickings of neo-colonialism in the original. The second imperialist excuse for de-colonization is a military one. Given the state of war technology in the 1950s as compared, say, with the 1900s, military airlifts and roving navies may have come to provide greater reserve power than colonial garrisons maintained (at great expense) throughout the empire. From a Third-World point of view there is every reason to believe that imperialism is still very much alive.

The proposition that imperialism still holds sway in the new guise of neo-colonialism leads to a pervasive skepticism about Western policy in the Third World today. It also raises doubts about the pronouncements of Western ethnoscience on matters of peace, equality, and economic development in the Third World. Take the following propositions as examples. First, given the propensity to define non-European movements in European terms, will it ever be possible to stage a revolution against any form of exploitation in the Third World without having that revolution defined and responded to as Marxist—as a subset of Western thought? Second, will the West be able to accept the possibility that Third-World nations can defeat it militarily—that conflict resolution in places like Southeast Asia must follow the same principles as elsewhere? Third, can the Third World accept the thesis that any major economic-development program is not merely a part of the process of imperialism? For instance, is there any real difference between economic aid programs and former colonial technical services? Do both serve as pattern-maintenance or welfare services to permit smooth functioning of private exploitation? Fourth, are population-control efforts really designed to assist the Third World or are they simply another dimension of imperialism? After all, it is as reasonable to argue that high peasant birth rates are a function of exploitation as it is to argue the current Western view, which assumes incredible ignorance on the part of peasant families and, to many non-Whites, carries overtones of racism. If time permitted, I would continue with many other skeptical propositions of the same sort. Perhaps the Third World is truly coming to the conviction that peace, justice, and development must emerge from a new ethnoscience: "a new theory of man."

VII

Systems of belief are by no means immune to change, but they are less likely to foretell external events than to explain them after the fact. This is notably true when a culture is losing control over such events. Reality, for that culture, is changing; the belief-system is signaling "no change"; the members of the culture believe and act on the signal; and sooner or later the gap between belief and verification becomes too great to be ignored. Unfortunately, that discovery may occur during the millisecond before a nuclear holocaust. Let me be more specific: Western ethnoscience defines the geography of the present-day world in a way that is so grossly unrealistic that we can only hope for a change in belief that occurs in time to save us, or a slow enough intrusion of reality so that beliefs may somehow respond in time. The Western model has persuaded the West that imperialism is under control, that economic development is just around the corner, and that peace is only a matter of right thinking. The Third-World model, on the other hand, describes a world in which imperialism is far from dead—instead, it has changed from colonialism into neo-colonialism, a cooperative enterprise of the European world as a whole—and that resistance to imperialism is mounting throughout the world. If the real world bears any resemblance to this model, then we are on the brink of disaster.

NOTES

1 Conklin, H., "Lexicographical Treatment of Folk Taxonomies," *International Journal of American Linguistics*, vol. 28, 1962.
2 Kuhn, T. S., *The Structure of Scientific Revolutions*, Chicago, 1962. Scriven, M., "Truisms as the Grounds for Historical Explanations," in *Theories of History*, P. L. Gardiner, ed., Glencoe, 1959.
3 Van Leur, J. C., *Indonesian Trade and Society: Essays in Asian Social and Economic History*, The Hague, 1955.
4 The following may be mentioned as examples written in or translated into English: A. Aguilar, *Pan-Americanism,*

From Monroe to the Present, New York, 1968; R. C. Dutt, *Economic History of India under Early British Rule*; H.-T. Fei, *China's Gentry*, Chicago, 1953; Gilberto Freyre, *The Masters and the Slaves*, New York, 1946; S. Gopal, *British Policy in India, 1858–1905*, Cambridge, 1965; R. Guerra y Sanchez, *Sugar and Society in the Caribbean*, New Haven, 1965; C. L. R. James, *The Black Jacobins: Toussaint L'Ouverture and the San Domingo Revolution*, New York, 1962, 2nd edition; *ibid.*, *A History of Pan-African Revolt*, Washington, 1969, 2nd edition; F. Ortiz, *Cuban Counterpoint: Tobacco and Sugar*, New York, 1947; K. M. Panikkar, *Asia and Western Dominance*, London, 1959; R. Thapar, *A History of India*, Hammondsworth, 1966; E. Williams, *Capitalism and Slavery*, New York, 1961 (1944).

5 See Myrdal, G., *Asian Drama*, New York, 1968, pp. 16–26.

6 Bloch, M., *Feudal Society*, London, 1962, p. xx.

7 The term "domocentric" is adapted from Shemyakin's usage.

8 Fanon, F., *Toward the African Revolution*, New York, 1967, p. 125.

9 Panikkar, *op. cit.*; van Leur, *op. cit.*

10 James, C. L. R., "The Atlantic Slave Trade and Slavery: Some Interpretations of Their Significance in the Development of the United States and the Western World," Williams, J. and Harris, C., eds., *Amistad* 1, New York, 1970, pp. 119–64, especially pp. 122–3; Williams, E., *op. cit.*; Ortiz, F., *op. cit.*

11 Marx, K., *Capital*, New York, 1967, vol. 3, p. 325.

12 James, *op. cit.*, p. 122. Contrast the views presented in, e.g., Ashton, T., *The Industrial Revolution, 1760–1830*, London, 1948; Deane, P., *The First Industrial Revolution*, Cambridge, 1969; Hartwell, R., ed., *The Causes of the Industrial Revolution in England*, London, 1967. E. Wrigley, a geographer, discusses "The Supply of Raw Materials in the Industrial Revolution" in Hartwell, *op. cit.*, pp. 97–120, without mentioning a single non-European area, and the southern United States only once—a feat of myopia.

13 Cf. Polanyi, K., *Dahomey and the Slave Trade*, Seattle, 1966, for a discussion of this economic relation.

14 Polanyi, K., ed., *Trade and Markets in the Early Empires*, Glencoe, 1957.

15 On the Chartered Companies and colonialism, see Coornaert, E., "European Economic Institutions and the New World: The Chartered Companies," Rich, E. and Wilson, C., eds., *The Economy of Expanding Europe in the 16th and 17th Centuries* (The Cambridge Economic History of Europe, vol. 4), Cambridge, 1967, pp. 223–75. On colonialism, see Memmi, A., *The Colonizer and the Colonized*, Boston, 1965; Puthucheary, J., *Ownership and Control in the Malayan Economy*, Singapore, 1960; and various works by G. Padmore.

16 Jathar, G., and Beri, S., *Indian Economics*, Madras, 1945; Dutt, *op. cit.*; Gadgil, D., *The Industrial Evolution of India*; Myrdal, G., *Economic Theory and Underdeveloped Regions*, London, 1957, p. 60.

17 Fei, *op. cit.*; Jathar and Beri, *op. cit.*

18 Boeke, J. , *Economics and Economic Policy of Dual Societies*, New York, 1953. A more realistic analysis of indigenous–alien dualism is given by Myint, H., in "An Interpretation of Economic Backwardness", *Oxford Economic Papers*, vol. 6, 1954.

19 Nkrumah, K., *Neo-Colonialism; The Last Stage of Imperialism*, New York, 1965, p. ix; Fanon, F., *The Wretched of the Earth*; Andre Gunder Frank, *Capitalism and Underdevelopment in Latin America*, New York, 1967.

3

On not excluding half of the human in human geography

Janice Monk and Susan Hanson

from *Professional Geographer*, 1982, 34(1): 11–23

Recent challenges to the acceptability of traditional gender roles for men and women have been called the most profound and powerful source of social change in this century [76], and feminism is the "ism" often held accountable for instigating this societal transformation. One expression of feminism is the conduct of academic research that recognizes and explores the reasons for and implications of the fact that women's lives are qualitatively different from men's lives. Yet the degree to which geography remains untouched by feminism is remarkable, and the dearth of attention to women's issues, explicit or implicit, plagues all branches of human geography.

Our purpose here is to identify some sexist biases in geographic research and to consider the implications of these for the discipline as a whole. We do not accuse geographers of having been actively or even consciously sexist in the conduct of their research, but we would argue that, through omission of any consideration of women, most geographic research has in effect been passively, often inadvertently, sexist. It is not our primary purpose to castigate certain researchers or their traditions, but rather to provoke lively debate and constructive criticism on the ways in which a feminist perspective might be incorporated into geography.

There appear to us to be two alternative paths to this goal of feminizing the discipline. One is to develop a strong feminist strand of research that would become one thread among many in the thick braid of geographic tradition. We support such research as necessary, but not sufficient. The second approach, which we favor, is to encourage a feminist perspective within all streams of human geography. In this way, issues concerning women (some of which are discussed later in this paper) would become incorporated in all geographic research endeavors. Only in this way, we believe, can geography realize the promise of the profound social change that would be wrought by eliminating sexism. In this paper we first briefly consider the reasons for the meager impact of feminism on the field to date, and review the nature of feminist scholarship in other social sciences and the humanities. We then examine the nature of sexist bias in geographic research and, through examples of this, demonstrate ways in which a nonsexist geography might evolve.

WHY THE NEGLECT OF WOMEN'S ISSUES?

Why has geography for the most part assiduously avoided research questions that embrace half the human race? We believe the answer lies very simply in the fact that knowledge is a social creation. The kind of knowledge that emerges from a discipline depends very much upon who produces that knowledge, what methods are used to procure knowledge, and what purposes knowledge is acquired for [78]. The number of women involved in generating knowledge in a given discipline appears to be important in determining the degree to which feminism is absorbed in that discipline's research tradition. Although the number of women researchers in geography is growing, women still constitute only 9.6 percent of the college and university faculty who are members of the Association of American Geographers. The characteristics of researchers influence the kinds of issues a discipline focuses upon. Geographers have, for instance, been more concerned with studying the spatial dimensions of social class than of social roles, such as gender roles. Yet for many individuals and groups, especially women, social roles are likely to have a greater impact than social class on spatial behavior.

Geography's devotion to strict logical positivism in recent years can also help to account for the lack of attention to women's issues. As King has pointed out, positivism has not been particularly concerned with social relevance or with social change [48]. It is a method that tends to preserve the *status quo*. The separation of facts from values and of subject from object are elements of positivism that would prevent positivist research from ever guiding, much less leading, social change [15, 48]. Researchers in the positivist tradition have tended to ask normative questions that have little to do with defining optimal social conditions (e.g. the traveling salesman problem). This is not to say that positivism is incapable of asking socially relevant normative questions, but only to point out that the *status quo* orientation of positivism has not fostered the sort of normative thinking that challenges existing social conditions.

Although strict logical positivism no longer has a life-threatening grip on the discipline, alternative paradigms have done little to incorporate a feminist perspective. Marxists have championed social change but, with a few exceptions [14, 37, 58], they have not explored the effects of capitalism on women. Phenomenologists have promised a more humanistic geography, a geography that would increase self-knowledge and would focus on the full range of human experience [15, 84], but even this research stream has produced few insights into the lives of women.

Finally, the purpose of much geographic research has been to provide a rational basis for informed decision making. Insofar as planners are committed to maintaining the *status quo* [29], and insofar as both researcher and decision maker were, especially in the past, likely to belong to the male power establishment, a focus on women, or even a recognition of women, was unlikely. In sum, most academic geographers have been men, and they have structured research problems according to their values, their concerns, and their goals, all of which reflect their experience. Women have not been creatures of power or status, and the research interests of those in power have reflected this fact.

FEMINIST CRITICISM IN OTHER DISCIPLINES

Although scholarship on women has, to date, made little impact on mainstream geography, much of relevance to our discipline can be learned from a decade of research and feminist criticism in other social sciences and in the humanities. Characterizing the development of this research, Stimpson notes an initial stage in which researchers responded to an urgently felt need to document women's sufferings, invisibility, and subordination, and to explore causes of women's secondary status. Later focus shifted to examining "the relationship of two interdependent, intersecting worlds . . . the male world of production, public activity, formal cultures, and power . . . [and] the world of the female—of reproduction, domestic activity, informal culture, and powerlessness . . ." [80, p. 187]. There have been demands for recognizing the diversity among women and for developing a sense of woman as an active force rather than a passive or marginal being. Most recently, the debate over the nature, permanence, and significance of sexual differences has revived [80].

Paralleling these changing emphases in work on women have been changes in feminist critiques of traditional disciplines. Early work was concerned mainly with correcting stereotypes and filling in omissions, but this has been followed by recognition of the need for basic transformations of the disciplines if women's experiences and actions are to be incorporated into enriched interpretations and analyses of human experience [32, 50, 65]. Inadequacies were identified not only in content, but in critical concepts and categorizations [18], in methodologies, and in the very purposes of scholarly research [86]. For example, among many new content themes identified for research were the relationship between language and power, the psychology of rape, and the history of sexuality and reproduction [53, 62]. In some fields, these new endeavors stimulated and enhanced important disciplinary trends, such as the shift in social history toward a focus on ordinary people rather than on the élite [54] or a shift in anthropology from emphasizing formal structures in society to developing and refining models of adaptive behaviors within social systems [79].

The need for revisions of concepts and categories has included broad issues such as the concept of genres and canons of masterpieces in literature or the appropriateness of using historical periods based on political or military activities for conceptualizing historical changes in women's lives [53, 65]. Feminist social scientists have questioned the prevailing definitions of concepts such as status, class, work, labor force, and power because, in current use, these concepts reflect male spheres of action [32, 65, 67, 72, 73]. How can work, for example, be defined and measured so that the concept incorporates nonmarket production and the maintenance activities involved in housework? Does social class, if derived from stratifications of male occupations, serve as an appropriate frame of reference in examining women's behavior and attitudes?

Critiques of disciplinary methodologies have focused on the implications of positivism and social scientists' applications of the scientific method. Some critics (for example, [47]) consider that revisions are needed in defining problems and hypotheses and in interpreting results, but argue that there is still a place for research that is objective/rational as opposed to subjective, involving naturalistic observation and qualitative patterning. Other scholars, examining the sociology of knowledge, have emphasized difficulties with the concept of objectivity, pointing out the crucial role subjectivity plays in the production and validation of knowledge. They discuss problems with the assumption that the object of knowing is completely separate from the knower, and they see knowledge as a dialogue that is "an unpredictable emergent rather than a controlled outcome" [86, p. 426]. These critics go beyond advocating a new orthodoxy in which subjectivity is valued. Instead of accepting explanations developed and validated by male experience as the complete and only truth, they propose recognizing all explanations as only partial and temporary truths, and they point to the importance of women researchers in creating a fuller vision of human possibilities [78, 86].

Other strands in the criticism have taken aim at the ahistorical nature of positivist work and at neglect of contextual variations in behavior [30, 62], both of which are shown to contribute to inadequate and stereotyped interpretations of women's lives. Although these various criticisms have much in common with positions advanced by advocates of hermeneutic, structuralist, and Marxist approaches, they are clearly different in their attention to the implications of patriarchal culture for scholarship.

Associated with the new methodological directions have also been reorientations in techniques of data collection, partly on philosophical grounds and partly because of gaps in recorded data on women. Thus we see more attention to naturalistic observation, oral histories, and analysis of documents produced by women such as diaries, mémoirs, and literary works.

Reflection on content and methodological issues has led ultimately to questioning the purposes of research. Distinctions are drawn between work on women, by women, and for women. It is suggested that research for women will be informed by visions of a transformed and equitable society [86]. With such a purpose, research oriented toward recording and modeling the *status quo* is seen as counterproductive. In the following section we examine some of the ways in which women have been excluded from consideration in geographic research. By pointing to omissions we implicitly suggest ways in which issues that affect women can be fruitfully incorporated in geographers' research designs.

SOME EXAMPLES OF SEXIST BIAS IN GEOGRAPHIC RESEARCH

Following Westkott [86], we consider sexist biases in the content, method, and purpose of geographic research. We do not imply that all human geography is sexist, but aim to demonstrate the pervasive nature of the problem by drawing illustrative examples from many areas of geographic endeavor. Neither the examples given nor the topic areas covered are intended as an exhaustive exposé of the problems we address. We have also not included extensive references to the feminist research emerging in geography, which we have reviewed elsewhere [87]. Our purpose here is merely to suggest the dimensions and sketch out the character of sexist bias in geographic research.

Content

Perhaps the most numerous examples of sexist bias in geographic research concern content. Problems relating to content include inadequate specification of the research problem, construction of gender-blind theory, the assumption that a population adheres to traditional gender roles, avoidance of research themes that directly address women's lives, and denial of the significance of gender or of women's activities.

Inadequate specification of research problems

Many geographic research questions apply to both men and women, but are analyzed in terms of male experiences only. We see this in two recent historical studies involving immigration of families from Europe to North America. Ostergren neglected to identify the farm woman's role in his analysis of economic activities in Sweden and Minnesota [61]. Cumber restricted his treatment of working-class institutions in Fall River to lodges, unions, working-men's societies, taverns, and sporting organizations. Study of the women's lives might have supported or weakened these authors' conclusions. As it stands, generalizations about communities were drawn from data on men only.

The omission of women's experience from Muller's text on suburbanization [60] is more surprising than are similar omissions from the historical studies, because women might be assumed to spend more of their lives in suburbia than do men. Yet his section on the social organization of contemporary suburbia and its human consequences fails to address women's lives directly. He identifies post-World War II migrants to the suburbs as "earnest young war veterans, possessing strong familistic values, who desired to educate themselves, work hard and achieve the good life" [60, p. 54]. He writes, "any major salary increase or promotion was immediately signified by a move to a better neighborhood, with the move governed by aggressive, achievement-oriented behavior" [60, p. 35]. Are women only passive followers to the suburbs? There is research suggesting that women are ambivalent about suburban life, and that husbands and wives evaluate residential choices differently [59, 68, 69].

Inadequate specification can involve male as well as female exclusion when neither type of misspecification seems warranted. Studies of shopping behavior, for example, have assumed a female consumer and have analyzed data collected for samples of women only (e.g. [23]). A problem that seems to be related to the researcher's perception of shoppers as female is the assumption, implicit in models of consumer store choice (e.g. [17]), that all shopping trips originate at home, rather than, say, being chained to the journey to work. Hence such models employ a home-to-store distance variable rather than some other, possibly more important, variable such as workplace-to-store.

Gender-blind theory

A concern stemming from inadequate problem specification is the emergence of gender-blind theory. Such theory may be dangerously impoverished if gender is an important explanatory variable and is omitted. Geographers interested in theories of development have drawn extensively on work outside the discipline [9, 10, 11, 21, 25, 36]. Nevertheless, these writers have not cited the significant quantity of literature on women and development that followed the publication of Boserup's *Women's Role in Economic Development* [7]. Thus geographers address the political economy of the international division of labor, but ignore the theoretical implications of the sexual division of labor. Study of the literature on women would extend the range of development issues worth considering. For example, is development enhanced if women have access to wage incomes or only if they are increasingly involved in decision making with regard to income allocation? Should theories focus on production or give more attention than previously to family maintenance activities?

Geographic theories aimed at problems in industrialized countries also suffer when they are gender blind. Attempts to build theories of urban travel demand have largely overlooked the importance of gender roles in determining travel patterns [81], but recent work suggests the seriousness of this omission [34]. Theories of the residential location-decision process have likewise failed to take gender roles into account, yet Madden has recently shown the necessity of incorporating such elements in any successful theory of residential choice [58]. Similarly Howe and O'Connor demonstrate the importance of gender to any insightful theory of intraurban industrial location [41].

Gender-blind theory is also emerging in research on issues of social well-being [19, 49, 74, 75] and equity [8]. Although sexual discrimination receives passing mention, few of the welfare indicators refer specifically to women, nor are data disaggregated by gender. Yet, as Lee and Schultz demonstrate, there are marked differences in the spatial patterns of relative versus absolute well-being of males and females in the United States [51]. On a topic related to social well-being, Bourne's discussion of equity issues in housing focuses upon race and class as important factors, but does not mention discrimination on the basis of sex [8]. The result of the general omission of gender in welfare and equity research is that race, class, and the political economy dominate explanations, while the contributions of gender and the patriarchal organization of society to the creation of disadvantage remain invisible. So long as gender remains a variable that is essential to understanding geographic processes and spatial form and to outlining alternative futures, explanations that omit gender are in many cases destined to be ineffective. Clearly, theoretical work along diverse lines of inquiry could benefit from becoming gender sighted rather than remaining gender blind.

The assumption of traditional gender roles

Explicit geographic writing on women, though rare, is likely to assume traditional gender (social) or sexual (biological) roles. Sauer's hypothesis about women's role in the origins of sedentary settlement and social life relies on his concept of the "nature of women," the "maternal bond," and associated assumed restrictions on spatial mobility [70]. The assumption that women universally (and perhaps historically) are primarily engaged in home and child care may reflect stereotypes of Western culture in the recent past, but can lead to inaccurate generalizations. Hoy, for example, referring to "the diverse cultures of most poor nations" stated that "women may work with men in the fields during times of peak labor requirements, but their major role is in the home where they may engage in some craft industry such as weaving for household use and for sale and barter" [42, p. 84]. Urban women have options as "domestics, secretaries, and more recently in industry" [42, p. 84]. He thus ignored women's central roles in agriculture in much of Africa and in many Asian countries, their provision of fuel and water, and their extensive roles in marketing and petty trading [7]. Pfeifer also assigned marginal roles to women peasants in central Europe, whom he described as a reserve *labor force* (our emphasis) that performs an estimated 50 percent of the work [63]!

Traditional urban land use theory, assuming as it does that each household has only one wage earner and therefore need be concerned with only one journey to work, seems also to be founded upon traditional gender roles (e.g. [1]). As we have pointed out elsewhere [87], models and theories that simply assume that all households are "traditional" nuclear families are not particularly useful for understanding changing urban spatial structure as a function of fundamental demographic or social changes. An additional example of gender stereotyping is the practice originating with Shevky and Bell [71], and continued in factorial ecologies [38], of identifying women's participation in the paid labor force as part of an index of urbanization or familism. Work outside the paid labor force is not recognized, and within the labor force is not broken down by type of occupation as it is for the male head of household on whom the social status index is therefore based. The implications appear to be that nonurban women do not work and that knowing simply that a woman works outside the home is more important than knowing how she is employed. Neither seems conceptually sound.

Review of such examples highlights the need for rethinking the concepts of work and labor force if research is to treat women accurately. Normally such concepts are used to refer to the formal sector of the economy traditionally connected with male activity. Yet women also work in the informal sector (for example in marketing food and crafts or as baby sitters or domestic servants), in home production for the market (food processing, sewing), in subsistence production (keeping domestic animals, raising gardens), and in unpaid service work (housework, child care, community volunteer work). Among partial solutions proposed for incorporating women's work are a Japanese indicator "net national welfare," which includes the contributions of housework (at female wage rates) [19], and estimates of work in terms of time or energy expended. Certainly more attention to this problem is warranted.

Avoidance of research themes that directly address women's lives

Women are generally invisible in geographic research, reflecting the concentration on male activity and on public spaces and landscapes. Work in recent issues of the *Journal of Cultural Geography* (1980, 1981), for example, deals with farm silos, farmsteads, housing exteriors, gasoline stations, a commercial strip, and country music (identified as a male WASP form). The massive *Man's Role in Changing the Face of the Earth* [82] is aptly named. Women make only cameo appearances in three papers in the entire volume [26, 63, 70]. A sampling of research on regional cultural landscapes and historical landscape perception, such as studies of the Mormon landscape and the Great Plains, discloses a preoccupation almost entirely with public spaces and men's perceptions [5, 6, 27, 44r 45]. Hudson's Great Plains country town has streets, businesses and businessmen, railroad depots, and men marketing livestock and making the trip to the elevator [43]. We see little of the churches, schools, homes, and other social settings where women passed their lives.

Not surprisingly, the only mention of women's lives in the Great Plains studies reviewed is by a woman historian. She described not only the hardships that space brought to men, but the loneliness and isolation of women separated from kin and friends, the oppression of emptiness, and women's terror of injury, disease, and childbirth remote from doctors. She also compared barriers to social interaction for ranch and farm wives [35]. Such insights suggest how research on women, the family, and social spheres would enrich our understanding of place. Beginning research on domestic interiors and symbolic uses of space similarly indicates how the horizons of cultural geography might be extended by attention to places closer to women's lives [33, 39, 54, 66].

In the urban realm, geographic research could profit from assessing the effects of the availability of such facilities as shopping areas, day care, medical services, recreation, and transportation on female labor-force participation and on labor in the home. Take, for example, the provision of child care, a topic practically untouched by geographic researchers yet one of great consequence in the lives of women. Compare the trickle of research on this issue with the virtual torrent of material produced in the past few years on the provision of mental health care, an area that touches the lives of fewer people. Pursuing research themes that directly address the lives of women will do more than merely flesh out a bony research agenda: such research should also provide needed insights on the diversity of women's experiences and needs.

Dismissing the significance of gender or women's activities

Preconceived notions of significance lead some authors to dismiss women's activities or to overlook gender as a variable, despite evidence to the contrary. Gosal and Krishnan, for example, discussing the magnitude of internal migration in India, pointed out that females account for two-thirds of migrants [31]. Because they interpret this as marriage migration, they used male migration as the "true index" of economic mobility [31, p. 198], thereby dismissing the economic implications of marriage-related movement. Later, they noted that women make up 75 percent of rural-to-rural migrants but wrote "a more realistic picture will be obtainable if only males are taken into account" [31, p. 199].

Another interesting example comes from incomplete interpretations of the findings of Bederman and Adams that Atlanta's unemployed are mainly black female heads of families [2]. Both Smith [75] and Muller [60] reported this aspect of the study, but in drawing conclusions from it focused on racial [75] or "racial and other" [60] discrimination. Both missed the double bind of gender and race.

A corollary of discounting the significance of women's activities may be a tendency to notice women primarily when they enter the male sphere or disrupt the traditional society. Hoy's few index references to women cover female participation in the (paid) labor force and related population and social policies in the USSR, Eastern Europe, and China, and the presumed association between women's liberation and urban ills in Japan [42].

Method

Sexist bias can afflict geographic research in the methods used as well as in content. A number of specific methodological concerns enter into empirical research design and execution regardless of the general approach (e.g. positivist or humanist) of the researcher. Here we address a few of these concerns and the ways in which they are susceptible to sexist bias.

Variable selection

We have identified several inappropriate or inadequate practices in the selection and interpretation of variables in studies in which women are or should be included. One problem is the use of data on husbands to describe wives. For example, two of eight variables included by Lee in a study of housewives' perceptions of neighborhoods in Cambridge, England, were "location of husband's work" and "husband's occupation" [51]. A third variable, "car ownership," may also have been inappropriate, because Lee did not report if women drove. Such use of husband's occupation as a surrogate for social class is problematic. Its appropriateness and the identification of alternatives are a concern of feminist sociologists as well as geographers insofar as geographers use measures of social class in their own research.

The assumption that data on males adequately describes the entire population is also suspect. For example, Soja [77] measured "minimal adult literacy" in Kenya and Lycan [55] measured education of "persons" in the US and Canada by using only data on men. Yet we know there are gender differences in educational access and attainment, and that this varies spatially [87].

The diversity among women and the range of women's needs often go unrecognized in variable selection. Male occupational categories are invariably differentiated, but women are recorded only by "female labor force participation" (e.g. [60]) or "female acitivity rate" (e.g. [49]). Social welfare studies would better reflect women's condition if indicators were included on such topics as women's legal situation, rape rates, or the provision of services such as day care.

Lack of awareness of women is also evident in variable interpretation and factor naming. For example, Knox chose "old age" as the salient feature to name a factor that had high loadings on female divorce rate, illegitimate birth rate, high proportions of persons over sixty, low proportions in younger age groups, small households and shared dwellings [49]. Without denying the significance of the elderly, the factor could be identified more comprehensively as "female-headed households." Such gender-blind naming of factors has theoretical and policy implications.

Respondent selection

There is a need to rethink the unit of observation in survey research [83]. Frequently data are collected on one individual yet reported as representative of the household; in particular, researchers like to rely upon responses from the "head of household" [12, 46]. This practice presents several problems. First, it assumes one person represents the household, which is questionable. Second, aggregation by head of household may mask important gender differences, given that there are substantial and increasing numbers of female-headed households throughout much of the world [16]. Third, cultural custom may lead to an assumption of male headship, even when the male does not have principal responsibilities for household support [16]. Collection of data on individuals (or appropriately varying combinations of individuals) would help to avoid this male bias in data. Problems also arise when authors indicate that the sampling unit was the head of household but do not indicate whether or not other household members were surveyed [22], or when the sex composition of the sample is not given despite the clear theoretical importance of considering gender differences in that research context (e.g. [40]). Clear, complete reporting of research methodology and disaggregating samples by gender would alleviate these problems.

Interviewing practices

Research results can be colored by interviewing practices such as having other members of a household present when one member is being interviewed. Interpretation of survey responses may raise problems, particularly on topics relating to women's role in family support or decision making. Either subjects or interviewers may discount or underestimate the importance of women's involvement. Elmendorf noted that rural Mexican women described themselves as "helping" the family, rather than working for its support, despite substantial activity in planting, harvesting, animal care, and food processing [24]. Bedford, studying population mobility, commented that New Hebridean women offered passive reasons for moves, described as largely directed by parents or husbands [3]. This may be, but we might question whether his interpretation reflected the cultural expectations of a foreign male researcher or of the women themselves.

Inadequate secondary data sources

Convenience or the nature of secondary data sources can contribute to the omission of women from research. Migration studies by Poulson *et al.* and Wareing demonstrate this problem [64, 85]. They drew, respectively, on electoral registrations (women could not be traced because of name changes) and male apprenticeship registrations. The US Census definition of household head prior to the 1980 census [13, pp. 100–1] makes difficult the use of census data for investigating certain research questions related to women.

Purpose

One purpose of geographic research has been to provide a basis for informed policy and decision making. Yet policy-oriented research that ignores women cannot help to form or guide policy that will improve women's conditions. In fact, there are numerous examples of the results of policies that have overlooked or have minimized the needs of women. One is the urban transportation system that is organized to expedite the journey to work for the full-time worker but not travel for other purposes.

Is the purpose of geographic research to accumulate facts and knowledge in order to improve our understanding of current events or to formulate policy within the context of the status quo, or is the purpose to go beyond asking why things are the way they are to consider the shapes of possible futures? Feminist scholars emphasize the need for research to define alternative structures in which the lot of women is improved [28, 86].

A geography that avoids or dismisses women and their activities, that is gender blind, or that assumes traditional gender roles can never contribute to the equitable society feminists envision [28, 86]. For such purposes we need a cultural and historical geography that would permit women to develop the sense of self-worth and identity that flows from awareness of heritage and relationship to place and a social and economic geography that goes beyond describing the *status quo*. Blaikie recognized this implication of his studies of family planning in India [4]. Policies developed from his diffusion research may improve dissemination of contraceptive information to socially and spatially isolated women, but more radical social change in that context requires research addressing the conditions leading to women's isolation.

TOWARD A MORE FULLY HUMAN GEOGRAPHY

A more sensitive handling of women's issues is essential to developing a nonsexist, if not a feminist, human geography. Moreover, we believe that eliminating sex biases would create a more policy-relevant geography. As long as gender roles significantly define the lives of women and men, it will be fruitful to include gender as a potentially important variable in many research contexts. Through examples of sexist bias in the content, method, and purpose of geographic research, we have attempted to indicate some of the ways in which women's issues can be included in research designs. Many of the problems we have identified are problems that are easily solved (e.g. the need to disaggregate samples by gender), but others, such as the need for nonsexist measures of social class, are more challenging. Although we encourage an awareness of gender differences and of women's issues throughout the discipline now (so that the geography of women does not become "ghettoized"), we would like to see gender blurred and then erased as a line of defining inequality.

REFERENCES

1 Alonso, W. *Location and Land Use*. Cambridge, Mass.: Harvard University Press, 1963.

2 Bederman, S., and J. Adams. "Job Accessibility and Under-employment." *Annals of the Association of American Geographers*, 64 (1974), 378–86.

3 Bedford, R. "A Transition in Circular Mobility: Population Movement in the New Hebrides. 1800–1970." In *The Pacific in Transition*, pp. 187–227. Edited by H. C. Brookfield. New York: St. Martin's Press, 1973.

4 Blaikie, P. "The Theory of the Spatial Diffusion of Innovations: A Spacious Cul-de-Sac." *Progress in Human Geography*, 2 (1978), 268–95.

5 Blouet, B. W., and M. P. Lawson, eds. *Images of the Great Plains: The Role of Human Nature in Settlement*. Lincoln: University of Nebraska Press, 1975.

6 Blouet, B. W., and F. C. Luebke, eds. *The Great Plains: Environment and Culture*. Lincoln: University of Nebraska Press, 1979.

7 Boserup, E. *Women's Role in Economic Development*. London: George Allen and Unwin, 1970.

8 Bourne, L. "Housing Supply and Housing Market Behavior in Residential Development." In *Social Areas in Cities: Spatial Processes and Form*, pp. 111–58. Edited by D. T. Herbert and R. J. Johnston. New York: Wiley, 1978.

9 Brookfield, H. *Interdependent Development*. Pittsburgh: University of Pittsburgh Press, 1975.

10 —— "Third World Development." *Progress in Human Geography*, 2 (1978), 121–32.

11 Browett, J. "Development, the Diffusionist Paradigm and Geography." *Progress in Human Geography*, 4 (1980), 57–79.

12 Brunn, S. D., and R. N. Thomas. "The Migration System of Tegucigalpa, Honduras." In *Population Dynamics of Latin America: A Review and Bibliography*, pp. 63–82. Edited by R. N. Thomas. East Lansing, Mich.: Conference of Latin Americanist Geographers, 1973.

13 Bureau of the Census. *Census User's Guide, Part I.* Washington: U.S. Department of Commerce, 1976.

14 Burnett, P. "Social Change, the Status of Women and Models of City Form and Development." *Antipode*, 5 (1973), 57–62.

15 Buttimer, A. "Grasping the Dynamism of Lifeworld." *Annals of the Association of American Geographers*, 66 (1976), 277–92.

16 Buvinic, M., and N. Youssef. *Women-Headed Households: The Ignored Factor in Development.* Washington: Agency for International Development, Office of Women in Development, 1978.

17 Cadwallader, M. "A Behavorial Model of Consumer Spatial Decision Making." *Economic Geography*, 51 (1975), 339–49.

18 Carroll, B. A. "Introduction." In *Liberating Women's History*, pp. ix–xiv. Edited by B. A. Carroll. Urbana: University of Illinois Press, 1976.

19 Coates, B. E., R. J. Johnston, and P. L. Knox. *Geography and Inequality*. Oxford: Oxford University Press, 1977.

20 Cumber, J. T. "Transatlantic Working Class Institutions." *Journal of Historical Geography*, 6 (1980), 275–90.

21 DeSouza, A., and P. W. Porter. *The Underdevelopment and Modernization of the Third World.* Commission on College Geography Resource Paper No. 28. Washington: Association of American Geographers, 1974.

22 Downes, J. D., and R. Wroot. "1971 Repeat Survey of Travel in the Reading Area." Crowthorne, England. *Transport and Road Research Laboratory Supplementary Report 43* L/C, 1974.

23 Downs, R. M. "The Cognitive Structure of an Urban Shopping Center." *Environment and Behavior*, 2 (1970), 13–39.

24 Elmendorf, M. "The Dilemma of Peasant Women: A View from a Village in Yucatan." In *Women and World Development*, pp. 88–94. Edited by I. Tinker and M. B. Bramsen. Washington: Overseas Development Council, 1976.

25 Ettema, W. A. "Geographers and Development." *Tidjschrift voor Economische en Sociale Geografie*, 70 (1979), 66–73.

26 Evans, E. E. "The Ecology of Peasant Life in Western Europe." In *Man's Role in Changing the Face of the Earth*, pp. 217–39. Edited by W. L. Thomas. Chicago: University of Chicago Press, 1956.

27 Francaviglia, R. "The Mormon Landscape: Definition of an Image in the American West." *Proceedings, Association of American Geographers*, 2 (1970), 59–61.

28 Gamarnikow, E. "Introduction to Special Issue." *International Journal of Urban and Regional Research*, 2 (1978), 390–403.

29 Goodman, R. *After the Planners*. New York: Simon & Schuster, 1971.

30 Gordon, L., P. Hunt, E. Pleck, R. Goldberg, and M. Scott. "Historical Phallacies: Sexism in American Historical Writing." In *Liberating Women's History*, pp. 55–74. Edited by B. A. Carroll. Urbana: University of Illinois Press, 1976.

31 Gosal, G. S., and G. Krishnan. "Patterns of Internal Migration in India." In *People on the Move*, pp. 193–206. Edited by L. A. Kosinski and R. M. Prothero. London: Methuen, 1975.

32 Gould, M. "The New Sociology." *Signs*, 5 (1980), 459–67.

33 Greenbaum, J. "Kitchen Culture/Kitchen Dialectic." *Heresies*, 11, 3 (1981), 59–61.

34 Hanson, S., and P. Hanson. "The Impact of Married Women's Employment on Household Travel Patterns: A Swedish Example." *Transportation*, 10 (1981), 165–83.

35 Hargreaves, M. W. M. "Space: Its Institutional Impact in the Development of the Great Plains." In *The Great Plains: Environment and Culture*, pp. 205–23. Edited by B. W. Blouet and F. C. Luebke. Lincoln: University of Nebraska Press, 1979.

36 Harriss, B., and J. Harriss. "Development Studies." *Progress in Human Geography*, 4 (1980), 577–88.

37 Hayford, A. M. "The Geography of Women: An Historical Introduction." *Antipode*, 5 (1973), 26–33.

38 Herbert, D. T. *Urban Geography: A Social Perspective*. New York: Praeger, 1973.

39 Hess, J. E. "Domestic Interiors in Northern New Mexico." *Heresies*, 11, 3 (1981), 30–3.

40 Horton, F., and D. Reynolds. "Effects of Urban Spatial Structure on Individual Behavior." *Economic Geography*, 47 (1971), 36–48.

41 Howe, A., and K. O'Connor. "Travel to Work and Labor Force Participation of Men and Women in an Australian Metropolitan Area." *The Professional Geographer*, 34 (1982), xx.

42 Hoy, D. R., ed. *Geography and Development: A World Regional Approach*. New York: Macmillan, 1978.

43 Hudson, J. C. "The Plains Country Town." In *The Great Plains: Environment and Culture*, pp. 99–118. Edited by B. W. Blouet and F. C. Luebke. Lincoln: University of Nebraska Press, 1979.

44 Jackson, R. H. "The Use of Adobe in the Mormon Cultural Region." *Journal of Cultural Geography*, 1 (1980), 82–95.

45 Jackson, R. H., and R. Layton. "The Mormon Village: An Analysis of a Settlement Type." *The Professional Geographer*, 23 (1976), 136–41.

46 Johnston, R. J. "Mental Maps of the City: Suburban Preference Patterns." *Environment and Planning*, 3 (1971), 63–72.

47 Kelly, A. "Feminism and Research." *Women's Studies International Quarterly*, 1 (1978), 225–32.

48 King, L. J. "Alternatives to a Positive Economic Geography." *Annals of the Association of American Geographers*, 66 (1976), 293–308.

49 Knox, P. L. "Levels of Living in England and Wales in 1961." *Transactions, Institute of British Geographers*, 62 (1974), 1–24.

50 Kolodny, A. "Dancing through the Minefield: Some Observations on the Theory, Practice and Politics of a Feminist Literary Criticism." *Feminist Studies*, 6 (1980), 1–25.

51 Lee, D., and R. Schultz. "Regional Patterns of Female Status in the United States." *The Professional Geographer*, 34 (1982), xx.

52 Lee, T. R. "Urban Neighborhoods as a Socio-Spatial Scheme." *Human Relations*, 21 (1968), 241–68.

53 Lerner, G. "Placing Women in History: A 1975 Perspective." In *Liberating Women's History*, pp. 357–67. Edited by B. A. Carroll. Urbana: University of Illinois Press, 1976.

54 Lewis, J. "Women Lost and Found: The Impact of Feminism on History." In *Men's Studies Modified*, pp. 55–72. Edited by D. Spender. Oxford: Pergamon Press, 1981.

55 Loyd, B. "Woman's Place: Man's Place." *Landscape*, 20 (1975), 10–13.

56 Lycan, D. R. "Interregional Migration in the United States and Canada." In *People on the Move*, pp. 207–21. Edited by L. A. Kosinski and R. M. Prothero. London: Methuen, 1975.

57 MacKenzie, S. "Women's Place—Women's Space: A Perspective on the Geographical Study of Women." *Area*, 12 (1980), 47–49.

58 Madden, J. "Urban Land Use and the Growth in Two-Earner Households." *American Economic Review*, 70 (1980), 191–97.

59 Michelson, W. *Environmental Choice, Human Behavior and Residential Satisfaction*. New York: Oxford University Press, 1977.

60 Muller, P. O. *Contemporary Suburban America*. Englewood Cliffs, N.J.: Prentice-Hall, 1981.

61 Ostergren, R. "A Community Transplanted: The Formative Experience of a Swedish Immigrant Community in the Upper Middle West." *Journal of Historical Geography*, 5 (1979), 189–212.

62 Parlee, M. B. "Psychology and Women." *Signs*, 5 (1979), 121–33.

63 Pfeifer, G. "The Quality of Peasant Living in Central Europe." In *Man's Role in Changing the Face of the Earth*, pp. 240–77. Edited by W. L. Thomas. Chicago: University of Chicago Press, 1956.

64 Poulson, J. F., D. J. Rowland, and R. J. Johnston. "Patterns of Maori Migration in New Zealand." In *People on the Move*, pp. 309–24. Edited by L. A. Kosinski and R. M. Prothero. London: Methuen, 1975.

65 Reuben, E. "In Defiance of the Evidence: Notes on Feminist Scholarship." *Women's Studies International Quarterly*, 1 (1978), 215–18.

66 Rock, C., S. Torre, and G. Wright. "The Appropriation of the House: Changes in House Design and Concepts of Domesticity." In *New Space for Women*, pp. 83–100. Edited by G. R. Wekerle, R. Peterson, and D. Morley. Boulder, Colo.: Westview Press, 1980.

67 Rogers, S. C. "Women's Place: A Critical Review of Anthropological Theory." *Comparative Studies in Society and History*, 20 (1978), 123–62.

68 Rothblatt, D. N., D. J. Garr, and J. Sprague. *The Suburban Environment and Women*. New York: Praeger, 1979.

69 Saegert, S., and G. Winkel. "The Home: A Critical Problem for Changing Sex Roles." In *New Space for Women*, pp. 41–64. Edited by G. R. Wekerle, R. Peterson, and D. Morley Boulder, Colo.: Westview Press, 1980.

70 Sauer, C. O. "The Agency of Man on Earth." In *Man's Role in Changing the Face of the Earth*, pp. 49–69. Edited by W. L. Thomas. Chicago: University of Chicago Press, 1956.

71 Shevky, E., and W. Bell. *Social Area Analysis: Theory, Illustrative Application, and Computational*

Procedures. Stanford, Calif.: Stanford University Press, 1955.

72 Smith, D. "Women's Perspective as a Radical Critique of Sociology." *Sociological Inquiry*, 44 (1974), 7–13.

73 —— "Some Implications of a Sociology for Women." In *Women in a Man-Made World*, pp. 15–29. Edited by N. Glazer-Malbin and H. Youngelson Waehrer. Chicago: Rand McNally, 1977.

74 Smith, D. M. *The Geography of Social Well-Being in the United States.* New York: McGraw-Hill, 1973.

75 —— *Where the Grass Is Greener: Geographic Perspectives on Inequality.* London: Croom Helm, 1979.

76 Smith, R. "The Movement of Women into the Labor Force." In *The Subtle Revolution: Women at Work*, pp. 1–29. Edited by R. Smith. Washington: The Urban Institute, 1979.

77 Soja, E. *The Geography of Modernization in Kenya.* Syracuse, N.Y.: Syracuse University Press, 1968.

78 Spender, D. "Introduction." In *Men's Studies Modified*, pp. 1–9. Edited by D. Spender, Oxford: Pergamon Press, 1981.

79 Stack, C., M. D. Caulfield, V. Estes, S. Landes, K. Larson, P. Johnson, J. Rake, and J. Shirek. "Anthropology." *Signs*, 1 (1975), 147–59.

80 Stimpson, C. R., J. N. Burstyn, D. C. Stanton, E. Dixler, and L. N. Dwight. "Editorial." *Signs*, 6 (1980), 187–8.

81 Stopher, P., and A. Meyburg. *Behavioral Travel-Demand Models.* Lexington, Mass.: Lexington Books, 1976.

82 Thomas, W. L., ed. *Man's Role in Changing the Face of the Earth.* Chicago: University of Chicago Press, 1956.

83 Tivers, J. "How the Other Half Lives: The Geographical Study of Women." *Area*, 10 (1978), 302–6.

84 Tuan, Y.-F. "Humanistic Geography." *Annals of the Association of American Geographers*, 66 (1976), 266–76.

85 Wareing, J. "Changes in the Geographical Distribution of the Recruitment of Apprentices to the London Companies 1486–1750." *Journal of Historical Geography*, 6 (1980), 241–9.

86 Westkott, M. "Feminist Criticism of the Social Sciences." *Harvard Educational Review*, 49 (1979), 422–30.

87 Zelinsky, W., J. Monk, and S. Hanson. "Women and Geography: A Review and Prospectus." *Progress in Human Geography* (in press).

SECTION 2

Staking claims

Introduction

Moral knowledge/geographical knowledge – what does it mean to claim moral ground, or how is oppression to be recognized?

Following the recognition of scholarship (including geographic scholarship) as always political in the ways elaborated in Section 1, this section explores further the necessary, though rarely sufficient, steps that must be taken to help ensure that scholarship works for progressive rather than regressive purposes (i.e. in Harvey's terminology in Chapter 1, that scholarship is revolutionary rather than *status quo* or counter-revolutionary). Here, we take up the following discussion. What we commonly understand to be critical scholarship (i.e. "revolutionary" social theoretic scholarly work) is always underlaid by at least two assumptions, whether stated or implied: individuals and groups are subjected to oppressions of various kinds; and critical scholarship can play an important role in documenting, describing, explaining, and/or possibly ameliorating such oppressions. In other words, critical scholars engage in the kinds of work they do because they believe that such scholarship may be a vitally important component of enabling "progressive" social change. This orientation immediately raises several important questions that the remainder of the book will take up:

1 How is oppression to be recognized and evaluated? Who is oppressed, in what ways, and with what effects? What are the implications for practicing geography? These essentially diagnostic questions inform Section 2 of the book.
2 In a broader sense, what do progressive social and political struggles aim to do and why? How might progressive change be defined and assessed, and how might more emancipatory practices be advanced? What are some instances of progressive struggles and how have geographers understood them? How can we understand the wide landscape of social change and struggle as such? These questions are for Section 3 of the book, which we have parsed into three interrelated goals or appeals that seem to be apparent (implicitly or explicity) in social struggles: rights-based appeals, social justice-based appeals, and appeals based on ethical/moral conceptions of the "good."

Our introduction to Section 3 will explain this framework. For now we want to signal that people's struggles have led to real accomplishments, that these ought to be noticed, and indebtedness to them recognized. We also want to signal that appeals based on "rights" or "justice" or "ethics" embody the criteria to determine progress, stasis, or retrenchment. For example, within a rights-based approach it may be possible to document an enlarged or diminished set of entities to whom rights are accorded, or to determine whether the suite of accorded rights are expanding or contracting over time or across space. Similarly, within a justice-based approach it may be possible to observe changes in the conceptualizations of justice itself as well as the extent of its application.

By employing the mix of these approaches that are evident in recent critical scholarship it is possible to understand (in a multi-dimensional way) that "progressive" social change has occurred in a number of arenas of struggle. By this we mean that some battles no longer need to be fought constantly, or that the terms of the contest have changed, and some matters are (temporarily, at least) settled. This does not mean that progress has been achieved evenly, and certainly does not mean that struggles in these arenas are over.

PART 1

Characterizing oppressions and recognizing injustice

Introduction

The term "oppression" has many meanings, both denotative and connotative. In the main sense that we want to convey here, the term is intimately bound up with notions of injustice and inequity, and we seek to understand the various ways in which oppression works in the world to make life "heaven on earth" for some and "a living hell" for others. In order to attain this understanding we need to give the term "oppression" some precision. In common parlance, we take the term to describe a condition in which some individuals or groups are constrained in some way by other individuals or groups, most often without the willing consent of the former. But this formulation is insufficient, by itself, to make clear the important connections between oppression and injustice. It is also too vague to allow us to analyze how particular oppressions/injustices arise, or to suggest appropriate and effective remedies or redress. The first two readings in this section are aimed at improving our understanding of these intersecting ideas of oppression and injustice.

In Chapter 4, political and feminist social theorist and philosopher Iris Marion Young (1949–2006) undertakes three major tasks: (1) to elaborate a broad conception of justice; (2) to tie that conception to the conditions that either enable or constrain the attainment of justice; and (3) to develop an understanding of subject formation and politics in the context of moral knowledge. In other words, she is asking what kinds of injustice should we be aware of and how do these intersect? And how shall subjects (a "we") be formed for struggle? It is important, at the outset, to understand that Young's view on these issues is structural and systemic. She is clear that what we come to think of as (in)justice is built into the fabrics of the societies in which we live, and that these conceptions are the result of ongoing struggles. This is clear throughout the piece as she broadens out the notion of justice beyond distribution of life's goods and bads (a frequently used definition), through her dissection of the various forms that oppression can take, and in her assessment of how social groups are formed and take on meaning.

Young's discussion of social groups and the relationship between such collectivities and individual identity is central to her understanding of justice and to the kinds of situations that promote or constrain it. For Young, groups (which she differentiates from aggregates and associations, the former as mere assemblages of characteristics, the latter as primarily voluntary conjunctures), into which individuals are largely assigned by the mechanisms of society and culture, are in constant mutual interplay with the production of individual subject identities. Through this formulation she argues against the methodological individualism that sees individuals as existing prior to, and autonomously from, the social and cultural contexts in which they develop. Groups, as Young argues, constitute individuals, and her insistence on this point is to establish firmly that individuals are the product of social processes. This allows her, then, to advance her arguments that justice and/or injustice are not, by and large, the products of individual actions, but rather derive out of the complex of systemic, structural social interactions.

One other dimension of Young's discussion of groups merits some additional comment. Drawing on Martin Heidegger (1962), Young describes group affiliation as having the character of "thrownness," or an element of being assigned to the group by others. This is congruent with Young's formulation of identity being a relational matter, i.e. "to how others identify us, and they do so in terms of groups which are always already associated with specific attributes, stereotypes and norms" (see p. 59). It is this assignment or "thrownness" that distinguishes groups from associations, affiliations that one chooses on one's own volition. And it is on this basis that Young begins to establish the case that matters of social (as opposed to merely individual) justice or injustice are structured into the encounters among contending groups in society. As we will see shortly, there are important connections between Young's formulation of group assignment and the arbitrariness of good fortune that David Smith discusses in Chapter 6.

The central element in Young's piece is her careful explication of the precise nature of oppression. She conceives of oppression, first of all as "*structural* phenomena that immobilize or diminish a group" (see p. 57). By structural, Young means that these systems of oppression are not necessarily, or even primarily, the result of individual acts to repress the actions of others, but rather are built into the everyday practices of society. One implication of this orientation is that assessing oppression forces us to look not to the "good" or "bad" motives or intentions of individuals, but instead to the ways in which powerful norms and hierarchies of both privilege and injustice are built into our everyday practices. A further implication is that remedies for injustice must also be aimed at changing these structural elements rather just the behavior of individuals. None of this rules out the possibility of individual acts of oppression, but these are not Young's primary interest, at least as individual acts. Rather these individual acts are to be understood within structures of injustice that allow them to be rationalized or at least explained. Finally, this conceptualization points out Young's need to begin with an operational understanding of groups.

The need to better understand the nature of oppression stems from Young's expanded notion of justice, and her commitment to furthering the emancipatory goals of social movement groups. In Young's work, the concept of justice moves beyond the equitable (though not necessarily equal, as we will discuss later in this introduction) distribution of life's necessities, comforts, luxuries, and burdens, to include enabling people to participate fully in the conditions, situations, and decision processes that give rise to particular distributions in the first place. As Young makes clear, fair and equitable distribution of goods and bads is inevitably a key component of justice, but for some groups to be always and only (i.e. systemically) on the receiving end (rather than participants in the construction of the distribution itself) of these distributional processes (even if equitable) is itself an injustice.

The bulk of the paper is taken up with Young's explication of her "five faces" of oppression: exploitation, marginalization, powerlessness, cultural imperialism, and violence. Her descriptions, differentiations, and explorations of overlap among the five faces are quite clear and, we think, illuminating. The disaggregation of the notion of oppression into these five facets is useful both as a clarifying diagnostic to indicate how particular kinds of injustice arise out of specific forms of social organization (especially under capitalism in its current form), as well as to point to means of remedy and redress. At this point, therefore, we simply want to highlight a few key insights that emerge from Young's schema.

As Young elaborates, the first three faces of oppression (exploitation, marginalization, and powerlessness) emerge from the social division of labor and the unequal power relations embedded in that division. Groups burdened with these forms of oppression clearly face obstacles in their material lives as well as in their ability to control and deploy their own creative and other capabilities. The two other faces of oppression (cultural imperialism and violence), Young argues, operate in a somewhat different way. Young uses the notion of cultural imperialism to describe the systematic and structural ways in which a dominant group constructs a social hierarchy of difference, with their own experiences and cultural products at the top (and superior), and those of all other groups as subordinate. The worldview of the dominant group is taken as the norm, and all other viewpoints as not only different, but inferior. In Young's argument cultural imperialism is the principle mechanism through which a dominant group's perspectives become taken for granted and naturalized as not only the way things are (descriptively), but the way things should be (normatively and prescriptively). Of course, these are never settled matters, but are sites for intense struggle, as Chapter 5 by Nancy Fraser will make clear momentarily.

Finally, Young is concerned with systematic (as opposed presumably to "random") violence as a form of direct oppression. While her discussion is edifying, the potential connections between violence and the other faces of oppression need a bit of elaboration. To provide this, we believe it is useful to draw on Antonio Gramsci's (1971) notions of hegemony and coercion. By and large, the four other faces of oppression work very much in accordance with Gramsci's formulation of hegemony as governance largely (though not completely or evenly) with the consent of the governed. Under hegemonic conditions, the interests of subordinated groups are made to seem, through various apparatuses (e.g. the media and the educational system to name just two), congruent with those of dominant groups or élites. Those in dominant positions are seen to hold them legitimately since they are presumably acting in the interest of all. As long as the hegemony holds, governance produces little resistance or opposition. It is when hegemony begins to break down, when the legitimacy and credibility of those in dominant positions begin to be questioned, that other means of social control become necessary to maintain the *status quo*. One of these other means is systemic violence.

One crucial intersection between the other faces of oppression and violence is the social context produced by exploitation, marginalization, powerlessness, and cultural imperialism in which some groups are significantly devalued and deligitimized relative to others. These processes not only mark out differences from the dominant "norm," but hierarchize such differences in a social pecking order. This establishes a set of cultural and societal patterns that make violence against members of such groups both "possible and acceptable" according to Young (see p. 68).

It is also important to note, as Young makes clear, that this careful dissection of oppression into its variety of etiologies and effects has quite profound implications for political coalition and movements for justice. To the extent that oppressions can be shown to be variously produced, but systemic nonetheless, it might be possible to reduce internecine claims that some oppressions are more fundamental (or authentic or worthy) than others, and that differing bases for calls for justice can be used to join struggles together. Two examples will help to clarify the utility of this analysis. While seemingly forming quite separate political factions, coalitions might be formed between the elderly, the poor, and the differently abled on the basis of their shared marginalization. Similarly, the struggles of women, gay men and lesbians, and people of color might be united through the recognition of their common subjugation under varying manifestations of cultural imperialism.

The second paper in this section (Chapter 5), by Nancy Fraser (currently Professor of Philosophy and Politics at the New School for Social Research in New York), takes up similar questions to those of Young, and for quite similar reasons. Fraser, like Young, is vitally concerned with matters of social justice, and seeks to understand justice in ways that go beyond the typical and traditional focus on (re)distribution. Fraser constructs her analysis along two important axes of claims for justice, neither of which, she argues, is reducible to the other: (1) *redistribution*, understood as redress for existing maldistributions of goods and resources, but also presumably of life's bads and burdens as well; and (2) *recognition*, understood as redress for cultural domination and impositions of dominant culture as the norm.

As Fraser describes elsewhere (1997), her interest in these intersecting dimensions of justice grew out of empirical observations that the rise of post-socialist political culture and of identity politics seemed to put these two bases for claims for justice into conflict or competition. The paper here is an attempt to reconcile these appeals for justice and to demonstrate their fundamental compatibility within the realms of both analysis and politics.

In many respects, the papers by Fraser and Young are quite similar. Fraser combines Young's first three faces of oppression (exploitation, marginalization, and powerlessness) into the axis of redistribution, and equates Young's notion of cultural imperialism (and, by extension, Young's category of violence) with the axis of recognition. Her argument is that both maldistribution and misrecognition are distinct categories of injustice, that they arise through different mechanisms, and that they require different forms of remedy and redress. Although there are some similarities between the papers, the inclusion of Fraser's piece here, with its explicit focus on the issue of recognition, allows us to examine several critical components of justice in more detail.

A useful route into this examination is a brief discussion of a debate between Fraser and Young over the course of the 1990s (Fraser, 1989, 1995, 1997; Young, 1997). Though there are many interesting elements to this debate, here we single out one main theme for its salience to our subsequent concerns in the rest of the book. This is the question of the relationships among oppression, liberation, and justice. Young and Fraser are in substantial agreement on these matters when thinking about justice as fair distribution of material goods and bads (i.e. those elements of both schemas that relate directly to the political economy and the division of labor: Young's first three faces of oppression and Fraser's axis of distribution). Where they diverge is over the matter of recognition (Fraser) and cultural imperialism and violence (Young). The nub of the argument is that Young sees recognition (or the redress for cultural imperialism and violence) as a means (one among several) to the end of a just distribution, while Fraser seems to see recognition primarily as an end itself. Two critical questions arise from this element of the debate. First, what is the metric to be used to assess justice? Put another way, how do we know when oppression (in Young's terms) or misrecognition (in Fraser's terms) has been eliminated or reduced? Young's answer is when distributions are more equitable and remaining inequalities can be explained not as the result of invidious comparisons among stereotyped groups, but rather due largely to the arbitrariness of life's lottery. (It is exactly in this sense that not all inequality is injustice, as we shall discuss further in the introduction to Part Two.)

Fraser's analysis provides no clear answer to this question. It is somewhat difficult to see how remedies of misrecognition could be assessed meaningfully except as they result in more equitable distributions. Indeed, elsewhere Fraser herself recognizes that such is the case. In a more recent piece than the one included here, Fraser wonders why so many contemporary conflicts take the form of claims to recognition. Her conclusion:

> To pose this question is also to note the relative decline in claims for egalitarian redistribution. Once the hegemonic grammar of political contestation, the language of distribution is less salient today. The movements that not long ago boldly demanded an equitable share of resources and wealth have not, to be sure, wholly disappeared. But thanks to the sustained neoliberal rhetorical assault on egalitarianism, to the absence of any credible model of 'feasible socialism' and to widespread doubts about the viability of state-Keynesian social democracy in the face of globalization, their role has been greatly reduced . . . In this context, questions of recognition are serving less to supplement, complicate and enrich redistributive struggles than to marginalize, eclipse and displace them.
>
> (Fraser, 2000: 107–108)

In other words, failing to achieve more parity in distributional terms, misrecognition remedies take the form of symbolic compromises. As Fraser goes on to note in this vein, "insofar as the politics of recognition displaces the politics of redistribution, it may actually promote economic inequality" (2000: 108).

All of this then leads to the second question, and perhaps this helps to resolve the dilemma: What is to be included in the notion of distribution? It is clear that Young's conception of justice goes beyond fair distribution of material goods, and includes some control over the decision processes that govern distributions. Fraser's position here is similar, and is made explicit with her concept of "parity of participation." In a similar vein, James O'Connor (1998: 338) makes a distinction between productive and distributive justice. For O'Connor productive justice operates in just the spheres of decision making, capacity enablement, communication, and participation that concern Young and Fraser. This includes real (as opposed to merely token) involvement in the processes that help determine life chances for oneself and others. Are there ways, in this light, to think about distribution as including more than material goods and bads? Productive justice (or Young's expanded notion of justice, or Fraser's parity of participation) is about control over one's own decisions and choices. But to what end? The fair distribution of all of life's goods and bads, including material as well as such non-material goods as respect, security from harm, and the elimination (or at least reduction) of hierarchies of difference.

This formulation is responsive to Young's critique of Fraser, and helps to resolve the dilemma that Fraser presents. By thinking about recognition as a necessary, though often insufficient, step toward fair distribution, and by thinking about distribution in this expanded way, it is possible to reconcile these two axes of justice, and accord them their due status in both theoretical and political spheres. These issues will be illuminated further in this section.

4
Five faces of oppression

Iris Marion Young

from I. M. Young *Justice and the Politics of Difference*. Princeton: Princeton University Press, 1990, pp. 39–65

> Someone who does not see a pane of glass does not know that he does not see it. Someone who, being placed differently, does see it, does not know the other does not see it.
>
> When our will finds expression outside ourselves in actions performed by others, we do not waste our time and our power of attention in examining whether they have consented to this. This is true for all of us. Our attention, given entirely to the success of the undertaking, is not claimed by them as long as they are docile. . . .
>
> Rape is a terrible caricature of love from which consent is absent. After rape, oppression is the second horror of human existence. It is a terrible caricature of obedience.
>
> —Simone Weil

I have proposed an enabling conception of justice. Justice should refer not only to distribution, but also to the institutional conditions necessary for the development and exercise of individual capacities and collective communication and cooperation. Under this conception of justice, injustice refers primarily to two forms of disabling constraints, oppression and domination. While these constraints include distributive patterns, they also involve matters which cannot easily be assimilated to the logic of distribution: decision-making procedures, division of labor, and culture.

Many people in the United States would not choose the term "oppression" to name injustice in our society. For contemporary emancipatory social movements, on the other hand—socialists, radical feminists, American Indian activists, Black activists, gay and lesbian activists—oppression is a central category of political discourse. Entering the political discourse in which oppression is a central category involves adopting a general mode of analyzing and evaluating social structures and practices which is incommensurate with the language of liberal individualism that dominates political discourse in the United States.

A major political project for those of us who identify with at least one of these movements must thus be to persuade people that the discourse of oppression makes sense of much of our social experience. We are ill prepared for this task, however, because we have no clear account of the meaning of oppression. While we find the term used often in the diverse philosophical and theoretical literature spawned by radical social movements in the United States, we find little direct discussion of the meaning of the concept as used by these movements.

In this chapter I offer some explication of the concept of oppression as I understand its use by new social movements in the United States since the 1960s. My starting point is reflection on the conditions of the groups said by these movements to be oppressed: among others, women, Blacks, Chicanos, Puerto Ricans and other Spanish-speaking Americans, American Indians, Jews, lesbians, gay men, Arabs, Asians, old people, working-class people, and the physically and mentally disabled. I aim to systematize the meaning of the concept of oppression as used by these diverse political movements, and to provide normative argument to clarify the wrongs the term names.

Obviously the above-named groups are not oppressed to the same extent or in the same ways. In the most general sense, all oppressed people suffer some inhibition of their ability to develop and exercise their capacities and express their needs, thoughts, and feelings. In that abstract sense all oppressed people face a common condition. Beyond that, in any more specific sense, it is not possible to define a single set of criteria that describe the condition of oppression of the above groups. Consequently, attempts by theorists and activists to discover a common description or the essential causes of the oppression of all these groups have frequently led to fruitless disputes about whose oppression is more fundamental or more grave. The contexts in which members of these groups use the term oppression to describe the injustices of their situation suggest that oppression names in fact a family of concepts and conditions, which I divide into five categories: exploitation, marginalization, powerlessness, cultural imperialism, and violence.

In this chapter I explicate each of these forms of oppression. Each may entail or cause distributive injustices, but all involve issues of justice beyond distribution. In accordance with ordinary political usage, I suggest that oppression is a condition of groups. Thus before explicating the meaning of oppression, we must examine the concept of a social group.

OPPRESSION AS A STRUCTURAL CONCEPT

One reason that many people would not use the term oppression to describe injustice in our society is that they do not understand the term in the same way as do new social movements. In its traditional usage, oppression means the exercise of tyranny by a ruling group. Thus many Americans would agree with radicals in applying the term oppression to the situation of Black South Africans under apartheid. Oppression also traditionally carries a strong connotation of conquest and colonial domination. The Hebrews were oppressed in Egypt, and many uses of the term oppression in the West invoke this paradigm.

Dominant political discourse may use the term oppression to describe societies other than our own, usually Communist or purportedly Communist societies. Within this anti-Communist rhetoric both tyrannical and colonialist implications of the term appear. For the anti-Communist, Communism denotes precisely the exercise of brutal tyranny over a whole people by a few rulers, and the will to conquer the world, bringing hitherto independent peoples under that tyranny. In dominant political discourse it is not legitimate to use the term oppression to describe our society, because oppression is the evil perpetrated by the Others.

New left social movements of the 1960s and 1970s, however, shifted the meaning of the concept of oppression. In its new usage, oppression designates the disadvantage and injustice some people suffer not because a tyrannical power coerces them, but because of the everyday practices of a well-intentioned liberal society. In this new left usage, the tyranny of a ruling group over another, as in South Africa, must certainly be called oppressive. But oppression also refers to systemic constraints on groups that are not necessarily the result of the intentions of a tyrant. Oppression in this sense is structural, rather than the result of a few people's choices or policies. Its causes are embedded in unquestioned norms, habits, and symbols, in the assumptions underlying institutional rules and the collective consequences of following those rules. It names, as Marilyn Frye puts it, "an enclosing structure of forces and barriers which tends to the immobilization and reduction of a group or category of people" (Frye, 1983, p. 11). In this extended structural sense oppression refers to the vast and deep injustices some groups suffer as a consequence of often unconscious assumptions and reactions of well-meaning people in ordinary interactions, media and cultural stereotypes, and structural features of bureaucratic hierarchies and market mechanisms—in short, the normal processes of everyday life. We cannot eliminate this structural oppression by getting rid of the rulers or making some new laws, because oppressions are systematically reproduced in major economic, political, and cultural institutions.

The systemic character of oppression implies that an oppressed group need not have a correlate oppressing group. While structural oppression involves relations among groups, these relations do not always fit the paradigm of conscious and intentional oppression of one group by another. Foucault (1977) suggests that to understand the meaning and operation of power in modern society we must look beyond the model of power as "sovereignty," a dyadic relation of ruler and subject, and instead analyze the exercise of power as the effect of often liberal and "humane" practices of education, bureaucratic administration, production and distribution of consumer goods, medicine, and so on. The conscious actions of many individuals daily contribute to maintaining and reproducing oppression, but those people are usually simply doing their jobs or living their lives, and do not understand themselves as agents of oppression.

I do not mean to suggest that within a system of oppression individual persons do not intentionally harm others in oppressed groups. The raped woman, the beaten Black youth, the locked-out worker, the gay man harrassed on the street, are victims of intentional actions by identifiable agents. I also do not mean to deny that specific groups are beneficiaries of the oppression of other groups, and thus have an interest in their continued oppression. Indeed, for every oppressed group there is a group that is *privileged* in relation to that group.

The concept of oppression has been current among radicals since the 1960s partly in reaction to Marxist attempts to reduce the injustices of racism and sexism, for example, to the effects of class domination or bourgeois ideology. Racism, sexism, ageism, homophobia, some social movements asserted, are distinct forms of oppression with their own dynamics apart from the dynamics of class, even though they may interact with class oppression. From often heated discussions among socialists, feminists, and antiracism activists in the last ten years a consensus is emerging that many different groups must be said to be oppressed in our society, and that no single form of oppression can be assigned causal or moral primacy (see Gottlieb, 1987). The same discussion has also led to the recognition that group differences cut across individual lives in a multiplicity of ways that can entail privilege and oppression for the same person in different respects. Only a plural explication of the concept of oppression can adequately capture these insights.

Accordingly, I offer below an explication of five faces of oppression as a useful set of categories and distinctions which I believe is comprehensive, in the sense that it covers all the groups said by new left social movements to be oppressed and all the ways they are oppressed. I derive the five faces of oppression from reflection on the condition of these groups. Because different factors, or combinations of factors, constitute the oppression of different groups, making their oppression irreducible, I believe it is not possible to give one essential definition of oppression. The five categories articulated in this chapter, however, are adequate to describe the oppression of any group, as well as its similarities with and differences from the oppression of other groups. But first we must ask what a group is.

THE CONCEPT OF A SOCIAL GROUP

Oppression refers to structural phenomena that immobilize or diminish a group. But what is a group? Our ordinary discourse differentiates people according to social groups such as women and men, age groups, racial and ethnic groups, religious groups, and so on. Social groups of this sort are not simply collections of people, for they are more fundamentally intertwined with the identities of the people described as belonging to them. They are a specific kind of collectivity, with specific consequences for how people understand one another and themselves. Yet neither social theory nor philosophy has a clear and developed concept of the social group (see Turner *et al.*, 1987).

A social group is a collective of persons differentiated from at least one other group by cultural forms, practices, or way of life. Members of a group have a specific affinity with one another because of their similar experience or way of life, which prompts them to associate with one another more than with those not identified with the group, or in a different way. Groups are an expression of social relations; a group exists only in relation to at least one other group. Group identification arises, that is, in the encounter and interaction between social collectivities that experience some differences in their way of life and forms of association, even if they also regard themselves as belonging to the same society.

As long as they associated solely among themselves, for example, an American Indian group thought of themselves only as "the people." The encounter with other American Indians created an awareness of difference; the others were named as a group, and the first group came to see themselves as a group. But social groups do not arise only from an encounter between different societies. Social processes also differentiate groups within a single society. The sexual division of labor, for example, has created social groups of women and men in all known societies. Members of each gender have a certain affinity with others in their group because of what they do or experience, and differentiate themselves from the other gender, even when members of each gender consider that they have much in common with members of the other, and consider that they belong to the same society.

Political philosophy typically has no place for a specific concept of the social group. When philosophers and political theorists discuss groups, they tend to conceive them either on the model of aggregates or on the model of associations, both of which are methodologically individualist concepts. To arrive at a specific concept of the social group it is thus useful to contrast social groups with both aggregates and associations.

An aggregate is any classification of persons according to some attribute. Persons can be aggregated according to any number of attributes—eye color, the make of car they drive, the street they live on. Some people interpret the groups that have emotional and social salience in our society as aggregates, as arbitrary classifications of persons according to such attributes as skin color, genitals, or age. George Sher, for example, treats social groups as aggregates, and uses the arbitrariness of aggregate classification as a reason not to give special attention to groups. "There are really as many groups as there are combinations of people and if we are going to ascribe claims to equal treatment to racial, sexual, and other groups with high visibility, it will be mere favoritism not to ascribe similar claims to these other groups as well" (Sher, 1987, p. 256).

But "highly visible" social groups such as Blacks or women are different from aggregates, or mere "combinations of people" (see French, 1975; Friedman and May, 1985; May, 1987, chap. 1). A social group is defined not primarily by a set of shared attributes, but by a sense of identity. What defines Black Americans as a social group is not primarily their skin color; some persons whose skin color is fairly light, for example, identify themselves as Black. Though sometimes objective attributes are a necessary condition for classifying oneself or others as belonging to a certain social group, it is identification with a certain social status, the common history that social status produces, and self-identification that define the group as a group.

Social groups are not entities that exist apart from individuals, but neither are they merely arbitrary classifications of individuals according to attributes which are external to or accidental to their identities. Admitting the reality of social groups does not commit one to reifying collectivities, as some might argue. Group meanings partially constitute people's identities in terms of the cultural forms, social situation, and history that group members know as theirs, because these meanings have been either forced upon them or forged by them or both (cf. Fiss, 1976). Groups are real not as substances, but as forms of social relations (cf. May, 1987, pp. 22–23).

Moral theorists and political philosophers tend to elide social groups more often with associations than with aggregates (e.g. French, 1975; May, 1987, chap. 1). By an association I mean a formally organized institution, such as a club, corporation, political party, church, college, or union. Unlike the aggregate model of groups, the association model recognizes that groups are defined by specific practices and forms of association. Nevertheless it shares a problem with the aggregate model. The aggregate model conceives the individual as prior to the collective, because it reduces the social group to a mere set of attributes attached to individuals. The association model also implicitly conceives the individual as ontologically prior to the collective, as making up, or constituting, groups.

A contract model of social relations is appropriate for conceiving associations, but not groups. Individuals constitute associations, they come together as already formed persons and set them up, establishing rules, positions, and offices. The relationship of persons to associations is usually voluntary, and even when it is not, the person has nevertheless usually entered the association. The person is prior to the association also in that the person's identity and sense of self are usually regarded as prior to and relatively independent of association membership.

Groups, on the other hand, constitute individuals. A person's particular sense of history, affinity, and separateness, even the person's mode of reasoning, evaluating, and expressing feeling, are constituted partly by her or his group affinities. This does not mean that persons have no individual styles, or are unable to transcend or reject a group identity. Nor does it preclude persons from having many aspects that are independent of these group identities.

The social ontology underlying many contemporary theories of justice, [I point out in my previous] chapter,[1] is methodologically individualist or atomist. It presumes that the individual is ontologically prior to the social. This individualist social ontology usually goes together with a normative conception of the self as independent. The authentic self is autonomous, unified, free, and self-made, standing apart from history and affiliations, choosing its life plan entirely for itself.

One of the main contributions of poststructuralist philosophy has been to expose as illusory this metaphysic of a unified self-making subjectivity, which posits the subject as an autonomous origin or an underlying substance to which attributes of gender, nationality, family role, intellectual disposition, and so on might attach. Conceiving the subject in this fashion implies conceiving consciousness as outside of and prior to language and the context of social interaction, which the subject enters. Several currents of recent philosophy challenge this deeply held Cartesian assumption. Lacanian psychoanalysis, for example, and the social and philosophical theory influenced by it, conceives the self as an achievement of linguistic positioning that is always contextualized in concrete relations with other persons, with their mixed identities (Coward and Ellis, 1977). The self is a product of social processes, not their origin.

From a rather different perspective, Habermas indicates that a theory of communicative action also must challenge the "philosophy of consciousness" which locates intentional egos as the ontological origins of social relations. A theory of communicative action conceives individual identity not as an origin but as a product of linguistic and practical interaction (Habermas, 1987, pp. 3–40). As Stephen Epstein describes it, identity is "a socialized sense of individuality, an internal organization of self-perception concerning one's relationship to social categories, that also incorporates views of the self perceived to be held by others. Identity is constituted relationally, through involvement with—and incorporation of—significant others and integration into communities" (Epstein, 1987, p. 29). Group categorization and norms are major constituents of individual identity (see Turner *et al.*, 1987).

A person joins an association, and even if membership in it fundamentally affects one's life, one does not take that membership to define one's very identity, in the way, for example, being Navaho might. Group affinity, on the other hand, has the character of what Martin Heidegger (1962) calls "thrownness": *one finds oneself* as a member of a group, which one experiences as always already having been. For our identities are defined in relation to how others identify us, and they do so in terms of groups which are always already associated with specific attributes, stereotypes, and norms.

From the thrownness of group affinity it does not follow that one cannot leave groups and enter new ones. Many women become lesbian after first identifying as heterosexual. Anyone who lives long enough becomes old. These cases exemplify thrownness precisely because such changes in group affinity are experienced as transformations in one's identity. Nor does it follow from the thrownness of group affinity that one cannot define the meaning of group identity for oneself; those who identify with a group can redefine the meaning and norms of group identity. Indeed, in [my Chapter 6 I show] how oppressed groups have sought to confront their oppression by engaging in just such redefinition. The present point is only that one first finds a group identity as given, and then takes it up in a certain way. While groups may come into being, they are never founded.

Groups, I have said, exist only in relation to other groups. A group may be identified by outsiders without those so identified having any specific consciousness of themselves as a group. Sometimes a group comes to exist only because one group excludes and labels a category of persons, and those labeled come to understand themselves as group members only slowly, on the basis of their shared oppression. In Vichy France, for example, Jews who had been so assimilated that they had no specifically Jewish identity were marked as Jews by others and given a specific social status by them. These people "discovered" themselves as Jews, and then formed a group identity and affinity with one another (see Sartre, 1948). A person's group identities may be for the most part only a background or horizon to his or her life, becoming salient only in specific interactive contexts.

Assuming an aggregate model of groups, some people think that social groups are invidious fictions, essentializing arbitrary attributes. From this point of view problems of prejudice, stereotyping, discrimination, and exclusion exist because some people mistakenly believe that group identification makes a difference to the capacities, temperament, or virtues of group members. This individualist conception of persons and their relation to one another tends to identify oppression with group identification. Oppression, on this view, is something that happens to people when they are classified in groups. Because others identify them as a group, they are excluded and despised. Eliminating oppression thus requires eliminating groups. People should be treated as individuals, not as members of groups, and allowed to form their lives freely without stereotypes or group norms.

[My] book takes issue with that position. While I agree that individuals should be free to pursue life plans in their own way, it is foolish to deny the reality of groups. Despite the modern myth of a decline of parochial attachments and ascribed identities, in modern society group differentiation remains endemic.

As both markets and social administration increase the web of social interdependency on a world scale, and as more people encounter one another as strangers in cities and states, people retain and renew ethnic, locale, age, sex, and occupational group identifications, and form new ones in the processes of encounter (cf. Ross, 1980, p. 19; Rothschild, 1981, p. 130). Even when they belong to oppressed groups, people's group identifications are often important to them, and they often feel a special affinity for others in their group. I believe that group differentiation is both an inevitable and a desirable aspect of modern social processes. Social justice, [I argue in my] later chapters, requires not the melting away of differences, but institutions that promote reproduction of and respect for group differences without oppression.

Though some groups have come to be formed out of oppression, and relations of privilege and oppression structure the interactions between many groups, group differentiation is not in itself oppressive. Not all groups are oppressed. In the United States Roman Catholics are a specific social group, with distinct practices and affinities with one another, but they are no longer an oppressed group. Whether a group is oppressed depends on whether it is subject to one or more of the five conditions I shall discuss below.

The view that groups are fictions does carry an important antideterminist or antiessentialist intuition. Oppression has often been perpetrated by a conceptualization of group difference in terms of unalterable essential natures that determine what group members deserve or are capable of, and that exclude groups so entirely from one another that they have no similarities or overlapping attributes. To assert that it is possible to have social group difference without oppression, it is necessary to conceptualize groups in a much more relational and fluid fashion.

Although social processes of affinity and differentiation produce groups, they do not give groups a substantive essence. There is no common nature that members of a group share. As aspects of a process, moreover, groups are fluid; they come into being and may fade away. Homosexual practices have existed in many societies and historical periods, for example. Gay men or lesbians have been identified as specific groups and so identified themselves, however, only in the twentieth century (see Ferguson, 1989, chap. 9; Altman, 1982).

Arising from social relations and processes, finally, group differences usually cut across one another. Especially in a large, complex, and highly differentiated society, social groups are not themselves homogeneous, but mirror in their own differentiations many of the other groups in the wider society. In American society today, for example, Blacks are not a simple, unified group with a common life. Like other racial and ethnic groups, they are differentiated by age, gender, class, sexuality, region, and nationality, any of which in a given context may become a salient group identity.

This view of group differentiation as multiple, crosscutting, fluid, and shifting implies another critique of the model of the autonomous, unified self. In complex, highly differentiated societies like our own, all persons have multiple group identifications. The culture, perspective, and relations of privilege and oppression of these various groups, moreover, may not cohere. Thus individual persons, as constituted partly by their group affinities and relations, cannot be unified, themselves are heterogeneous and not necessarily coherent.

THE FACES OF OPPRESSION

Exploitation

The central function of Marx's theory of exploitation is to explain how class structure can exist in the absence of legally and normatively sanctioned class distinctions. In precapitalist societies domination is overt and accomplished through directly political means. In both slave society and feudal society the right to appropriate the product of the labor of others partly defines class privilege, and these societies legitimate class distinctions with ideologies of natural superiority and inferiority.

Capitalist society, on the other hand, removes traditional juridically enforced class distinctions and promotes a belief in the legal freedom of persons. Workers freely contract with employers and receive a wage; no formal mechanisms of law or custom force them to work for that employer or any employer. Thus the mystery of capitalism arises: when everyone is formally free, how can there be class domination? Why do class distinctions persist between the wealthy, who own the means of production, and the mass of people, who work for them? The theory of exploitation answers this question.

Profit, the basis of capitalist power and wealth, is a mystery if we assume that in the market goods exchange at their values. The labor theory of value dispels this mystery. Every commodity's value is a function of the labor time necessary for its production. Labor power is the one commodity which in the process of being consumed produces new value. Profit comes from the difference between the value of the labor performed and the value of the capacity to labor which the capitalist purchases. Profit is possible only because the owner of capital appropriates any realized surplus value.

In recent years Marxist scholars have engaged in considerable controversy about the viability of the labor theory of value this account of exploitation relies on (see Wolff, 1984, chap. 4). John Roemer (1982), for example, develops a theory of exploitation which claims to preserve the theoretical and practical purposes of Marx's theory, but without assuming a distinction between values and prices and without being restricted to a concept of abstract, homogeneous labor. My purpose here is not to engage in technical economic disputes, but to indicate the place of a concept of exploitation in a conception of oppression.

Marx's theory of exploitation lacks an explicitly normative meaning, even though the judgment that workers are exploited clearly has normative as well as descriptive power in that theory (Buchanan, 1982, chap. 3). C. B. Macpherson (1973, chap. 3) reconstructs this theory of exploitation in a more explicitly normative form. The injustice of capitalist society consists in the fact that some people exercise their capacities under the control, according to the purposes, and for the benefit of other people. Through private ownership of the means of production, and through markets that allocate labor and the ability to buy goods, capitalism systematically transfers the powers of some persons to others, thereby augmenting the power of the latter. In this process of the transfer of powers, according to Macpherson, the capitalist class acquires and maintains an ability to extract benefits from workers. Not only are powers transferred from workers to capitalists, but also the powers of workers diminish by more than the amount of transfer, because workers suffer material deprivation and a loss of control, and hence are deprived of important elements of self-respect. Justice, then, requires eliminating the institutional forms that enable and enforce this process of transference and replacing them with institutional forms that enable all to develop and use their capacities in a way that does not inhibit, but rather can enhance, similar development and use in others.

The central insight expressed in the concept of exploitation, then, is that this oppression occurs through a steady process of the transfer of the results of the labor of one social group to benefit another. The injustice of class division does not consist only in the distributive fact that some people have great wealth while most people have little (cf. Buchanan, 1982, pp. 44–49; Holmstrom, 1977). Exploitation enacts a structural relation between social groups. Social rules about what work is, who does what for whom, how work is compensated, and the social process by which the results of work are appropriated operate to enact relations of power and inequality. These relations are produced and reproduced through a systematic process in which the energies of the have-nots are continuously expended to maintain and augment the power, status, and wealth of the haves.

Many writers have cogently argued that the Marxist concept of exploitation is too narrow to encompass all forms of domination and oppression (Giddens, 1981, p. 242; Brittan and Maynard, 1984, p. 93; Murphy, 1985; Bowles and Gintis, 1986, pp. 20–24). In particular, the Marxist concept of class leaves important phenomena of sexual and racial oppression unexplained. Does this mean that sexual and racial oppression are non-exploitative, and that we should reserve wholly distinct categories for these oppressions? Or can the concept of exploitation be broadened to include other ways in which the labor and energy expenditure of one group benefits another, and reproduces a relation of domination between them?

Feminists have had little difficulty showing that women's oppression consists partly in a systematic and unreciprocated transfer of powers from women to men. Women's oppression consists not merely in an inequality of status, power, and wealth resulting from men's excluding them from privileged activities. The freedom, power, status, and self-realization of men is possible precisely because women work for them. Gender exploitation has two aspects, transfer of the fruits of material labor to men and transfer of nurturing and sexual energies to men.

Christine Delphy (1984), for example, describes marriage as a class relation in which women's labor benefits men without comparable remuneration. She makes it clear that the exploitation consists not in the sort of work that women do in the home, for this might include various kinds of tasks, but in the fact that they perform tasks for someone on whom they are dependent. Thus, for example, in most systems of agricultural production in the world, men take to market the goods women have produced, and more often than not men receive the status and often the entire income from this labor.

With the concept of sex-affective production, Ann Ferguson (1984; 1989, chap. 4) identifies another form of the transference of women's energies to men. Women provide men and children with emotional care and provide men with sexual satisfaction, and as a group receive relatively little of either from men (cf. Brittan and Maynard, 1984, pp. 142–48). The gender socialization of women makes us tend to be more attentive to interactive dynamics than men, and makes women good at providing empathy and support for people's feelings and at smoothing over interactive tensions. Both men and women look to women as nurturers of their personal lives, and women frequently complain that when they look to men for emotional support they do not receive it (Easton, 1978). The norms of heterosexuality, moreover, are oriented around male pleasure, and consequently many women receive little satisfaction from their sexual interaction with men (Gottlieb, 1987).

Most feminist theories of gender exploitation have concentrated on the institutional structure of the patriarchal family. Recently, however, feminists have begun to explore relations of gender exploitation enacted in the contemporary workplace and through the state. Carol Brown argues that as men have removed themselves from responsibility for children, many women have become dependent on the state for subsistence as they continue to bear nearly total responsibility for childrearing (Brown, 1981; cf. Boris and Bardaglio, 1983; A. Ferguson, 1984). This creates a new system of the exploitation of women's domestic labor mediated by state institutions, which she calls public patriarchy.

In twentieth-century capitalist economies the workplaces that women have been entering in increasing numbers serve as another important site of gender exploitation. David Alexander (1987) argues that typically feminine jobs involve gender-based tasks requiring sexual labor, nurturing, caring for others' bodies, or smoothing over workplace tensions. In these ways women's energies are expended in jobs that enhance the status of, please, or comfort others, usually men; and these gender-based labors of waitresses, clerical workers, nurses, and other caretakers often go unnoticed and undercompensated.

To summarize, women are exploited in the Marxist sense to the degree that they are wage workers. Some have argued that women's domestic labor also represents a form of capitalist class exploitation insofar as it is labor covered by the wages a family receives. As a group, however, women undergo specific forms of gender exploitation in which their energies and power are expended, often unnoticed and unacknowledged, usually to benefit men by releasing them for more important and creative work, enhancing their status or the environment around them, or providing them with sexual or emotional service.

Race is a structure of oppression at least as basic as class or gender. Are there, then, racially specific forms of exploitation? There is no doubt that racialized groups in the United States, especially Blacks and Latinos, are oppressed through capitalist superexploitation resulting from a segmented labor market that tends to reserve skilled, high-paying, unionized jobs for whites. There is wide disagreement about whether such superexploitation benefits whites as a group or only benefits the capitalist class (see Reich, 1981), and I do not intend to enter into that dispute here.

However one answers the question about capitalist superexploitation of racialized groups, is it possible to conceptualize a form of exploitation that is racially specific on analogy with the gender-specific forms just discussed? I suggest that the category *of menial* labor might supply a means for such conceptualization. In its derivation "menial" designates the labor of servants. Wherever there is racism, there is the assumption, more or less enforced, that members of the oppressed racial groups are or ought to be servants of those, or some of those, in the privileged group. In most white racist societies this means that many white people have dark- or yellow-skinned domestic servants, and in the United States today there remains significant racial structuring of private household service. But in the United States today much service labor has gone public: anyone who goes to a good hotel or a good restaurant can have servants. Servants often attend the daily—and nightly—activities of business executives, government officials, and other high-status professionals. In our society there remains strong cultural pressure to fill servant jobs—bellhop, porter, chambermaid, busboy, and so on—with Black and Latino workers. These jobs entail a transfer of energies whereby the servers enhance the status of the served.

Menial labor usually refers not only to service, however, but also to any servile, unskilled, low-paying work lacking in autonomy, in which a person is subject to taking orders from many people. Menial work tends to be auxiliary work, instrumental to the work of others, where those others receive primary recognition for doing the job. Laborers on a construction site, for example, are at the beck and call of welders, electricians, carpenters, and other skilled workers, who receive recognition for the job done. In the United States explicit racial discrimination once reserved menial work for Blacks, Chicanos, American Indians, and Chinese, and menial work still tends to be linked to Black and Latino workers (Symanski, 1985). I offer this category of menial labor as a form of racially specific exploitation, as a provisional category in need of exploration.

The injustice of exploitation is most frequently understood on a distributive model. For example, though he does not offer an explicit definition of the concept, by "exploitation" Bruce Ackerman seems to mean a seriously unequal distribution of wealth, income, and other resources that is group based and structurally persistent (Ackerman, 1980, chap. 8). John Roemer's definition of exploitation is narrower and more rigorous: "An agent is exploited when the amount of labor embodied in *any* bundle of goods he could receive, in a feasible distribution of society's net product, is less than the labor he expended" (Roemer, 1982, p. 122). This definition too turns the conceptual focus from institutional relations and processes to distributive outcomes.

Jeffrey Reiman argues that such a distributive understanding of exploitation reduces the injustice of class processes to a function of the inequality of the productive assets classes own. This misses, according to Reiman, the relationship of force between capitalists and workers, the fact that the unequal exchange in question occurs within coercive structures that give workers few options (Reiman, 1987; cf. Buchanan, 1982, pp. 44–49; Holmstrom, 1977). The injustice of exploitation consists in social processes that bring about a transfer of energies from one group to another to produce unequal distributions, and in the way in which social institutions enable a few to accumulate while they constrain many more. The injustices of exploitation cannot be eliminated by redistribution of goods, for as long as institutionalized practices and structural relations remain unaltered, the process of transfer will re-create an unequal distribution of benefits. Bringing about justice where there is exploitation requires reorganization of institutions and practices of decision-making, alteration of the division of labor, and similar measures of institutional, structural, and cultural change.

Marginalization

Increasingly in the United States racial oppression occurs in the form marginalization rather than exploitation. Marginals are people the system of labor cannot or will not use. Not only in Third World capitalist countries, but also in most Western capitalist societies, there is a growing underclass of people permanently confined to lives of social marginality, most of whom are racially marked—Blacks or Indians in Latin America, and Blacks, East Indians, Eastern Europeans, or North Africans in Europe.

Marginalization is by no means the fate only of racially marked groups, however. In the United States a shamefully large proportion of the population is marginal: old people, and increasingly people who are not very old but get laid off from their jobs and cannot find new work; young people, especially Black or Latino, who cannot find first or second jobs; many single mothers and their children; other people involuntarily unemployed; many mentally and physically disabled people; American Indians, especially those on reservations.

Marginalization is perhaps the most dangerous form of oppression. A whole category of people is expelled from useful participation in social life and thus potentially subjected to severe material deprivation and even extermination. The material deprivation marginalization often causes is certainly unjust, especially in a society where others have plenty. Contemporary advanced capitalist societies have in principle acknowledged the injustice of material deprivation caused by marginalization, and have taken some steps to address it by providing welfare payments and services. The continuance of this welfare state is by no means assured, and in most welfare state societies, especially the United States, welfare redistributions do not eliminate large-scale suffering and deprivation.

Material deprivation, which can be addressed by redistributive social policies, is not, however, the extent of the harm caused by marginalization. Two categories of injustice beyond distribution are associated with marginality in advanced capitalist societies. First, the provision of welfare itself produces new injustice by depriving those dependent on it of rights and freedoms that others have. Second, even when material deprivation is somewhat mitigated by the welfare state, marginalization is unjust because it blocks the opportunity to exercise capacities in socially defined and recognized ways. I shall explicate each of these in turn.

Liberalism has traditionally asserted the right of all rational autonomous agents to equal citizenship. Early bourgeois liberalism explicitly excluded from citizenship all those whose reason was questionable or not fully developed, and all those not independent (Pateman, 1988, chap. 3; cf. Bowles and Gintis, 1986, chap. 2). Thus poor people, women, the mad and the feebleminded, and children were explicitly excluded from citizenship, and many of these were housed in institutions modeled on the modern prison: poorhouses, insane asylums, schools.

Today the exclusion of dependent persons from equal citizenship rights is only barely hidden beneath the surface. Because they depend on bureaucratic institutions for support or services, the old, the poor, and the mentally or physically disabled are subject to patronizing, punitive, demeaning, and arbitrary treatment by the policies and people associated with welfare bureaucracies. Being a dependent in our society implies being legitimately subject to the often arbitrary and invasive authority of social service providers and other public and private administrators, who enforce rules with which the marginal must comply, and otherwise exercise power over the conditions of their lives. In meeting needs of the marginalized, often with the aid of social scientific disciplines, welfare agencies also construct the needs themselves. Medical and social service professionals know what is good for those they serve, and the marginals and dependents themselves do not have the right to claim to know what is good for them (Fraser, 1987a; K. Ferguson, 1984, chap. 4). Dependency in our society thus implies, as it has in all liberal societies, a sufficient warrant to suspend basic rights to privacy, respect, and individual choice.

Although dependency produces conditions of injustice in our society, dependency in itself need not be oppressive. One cannot imagine a society in which some people would not need to be dependent on others at least some of the time: children, sick people, women recovering from childbirth, old people who have become frail, depressed or otherwise emotionally needy persons, have the moral right to depend on others for subsistence and support.

An important contribution of feminist moral theory has been to question the deeply held assumption that moral agency and full citizenship require that a person be autonomous and independent. Feminists have exposed this assumption as inappropriately individualistic and derived from a specifically male experience of social relations, which values competition and solitary achievement (see Gilligan, 1982; Friedman and May, 1985). Female experience of social relations, arising both from women's typical domestic care responsibilities and from the kinds of paid work that many women do, tends to recognize dependence as a basic human condition (cf. Hartsock, 1983, chap. 10). Whereas on the autonomy model a just society would as much as possible give people the opportunity to be independent, the feminist model envisions justice as according respect and participation in decisionmaking to those who are dependent as well as to those who are independent (Held, 1987). Dependency should not be a reason to be deprived of choice and respect, and much of the oppression many marginals experience would be lessened if a less individualistic model of rights prevailed.

Marginalization does not cease to be oppressive when one has shelter and food. Many old people, for example, have sufficient means to live comfortably but remain oppressed in their marginal status. Even if marginals were provided a comfortable material life within institutions that respected their freedom and dignity, injustices of marginality would remain in the form of uselessness, boredom, and lack of self-respect. Most of our society's productive and recognized activities take place in contexts of organized social cooperation, and social structures and processes that close persons out of participation in such social cooperation are unjust. Thus while marginalization definitely entails serious issues of distributive justice, it also involves the deprivation of cultural, practical, and institutionalized conditions for exercising capacities in a context of recognition and interaction.

The fact of marginalization raises basic structural issues of justice, in particular concerning the appropriateness of a connection between participation in productive activities of social cooperation, on the one hand, and access to the means of consumption, on the other. As marginalization is increasing, with no sign of abatement, some social policy analysts have introduced the idea of a "social wage" as a guaranteed socially provided income not tied to the wage system. Restructuring of productive activity to address a right of participation, however, implies organizing some socially productive activity outside of the wage system (see Offe, 1985, pp. 95–100), through public works or self-employed collectives.

Powerlessness

As I have indicated, the Marxist idea of class is important because it helps reveal the structure of exploitation: that some people have their power and wealth because they profit from the labor of others. For this reason I reject the claim some make that a traditional class exploitation model fails to capture the structure of contemporary society. It remains the case that the labor of most people in the society augments the power of relatively few. Despite their differences from nonprofessional workers, most professional workers are still not members of the capitalist class. Professional labor either involves exploitative transfers to capitalists or supplies important conditions for such transfers. Professional workers are in an ambiguous class position, it is true, because, as [I argue in my] Chapter 7, they also benefit from the exploitation of nonprofessional workers.

While it is false to claim that a division between capitalist and working classes no longer describes our society, it is also false to say that class relations have remained unaltered since the nineteenth century. An adequate conception of oppression cannot ignore the experience of social division reflected in the colloquial distinction between the "middle class" and the "working class," a division structured by the social division of labor between professionals and nonprofessionals. Professionals are privileged in relation to nonprofessionals, by virtue of their position in the division of labor and the status it carries. Nonprofessionals suffer a form of oppression in addition to exploitation, which I call powerlessness.

In the United States, as in other advanced capitalist countries, most workplaces are not organized democratically, direct participation in public policy decisions is rare, and policy implementation is for the most part hierarchical, imposing rules on bureaucrats and citizens. Thus most people in these societies do not regularly participate in making decisions that affect the conditions of their lives and actions, and in this sense most people lack significant power. At the same time, [as I argue in my] Chapter 1, domination in modern society is enacted through the widely dispersed powers of many agents mediating the decisions of others. To that extent many people have some power in relation to others, even though they lack the power to decide policies or results. The powerless are those who lack authority or power even in this mediated sense, those over whom power is exercised without their exercising it; the powerless are situated so that they must take orders and rarely have the right to give them. Powerlessness also designates a position in the division of labor and the concomitant social position that allows persons little opportunity to develop and exercise skills. The powerless have little or no work autonomy, exercise little creativity or judgment in their work, have no technical expertise or authority, express themselves awkwardly, especially in public or bureaucratic settings, and do not command respect. Powerlessness names the oppressive situations Sennett and Cobb (1972) describe in their famous study of working-class men.

This powerless status is perhaps best described negatively: the powerless lack the authority, status, and sense of self that professionals tend to have. The status privilege of professionals has three aspects, the lack of which produces oppression for nonprofessionals.

First, acquiring and practicing a profession has an expansive, progressive character. Being professional usually requires a college education and the acquisition of a specialized knowledge that entails working with symbols and concepts. Professionals experience progress first in acquiring the expertise, and then in the course of professional advancement and rise in status. The life of the nonprofessional by comparison is powerless in the sense that it lacks this orientation toward the progressive development of capacities and avenues for recognition.

Second, while many professionals have supervisors and cannot directly influence many decisions or the actions of many people, most nevertheless have considerable day-to-day work autonomy. Professionals usually have some authority over others, moreover—either over workers they supervise, or over auxiliaries, or over clients. Nonprofessionals, on the other hand, lack autonomy, and in both their working and their consumer–client lives often stand under the authority of professionals.

Though based on a division of labor between "mental" and "manual" work, the distinction between "middle class" and "working class" designates a division not only in working life, but also in nearly all aspects of social life. Professionals and nonprofessionals belong to different cultures in the United States. The two groups tend to live in segregated neighborhoods or even different towns, a process itself mediated by planners, zoning officials, and real estate people. The groups tend to have different tastes in food, décor, clothes, music, and vacations, and often different health and educational needs. Members of each group socialize for the most part with others in the same status group. While there is some inter-group mobility between generations, for the most part the children of professionals become professionals and the children of nonprofessionals do not.

Thus, third, the privileges of the professional extend beyond the workplace to a whole way of life. I call this way of life "respectability." To treat people with respect is to be prepared to listen to what they have to say or to do what they request because they have some authority, expertise, or influence. The norms of respectability in our society are associated specifically with professional culture. Professional dress, speech, tastes, demeanor, all connote respectability. Generally professionals expect and receive respect from others. In restaurants, banks, hotels, real estate offices, and many other such public places, as well as in the media, professionals typically receive more respectful treatment than nonprofessionals. For this reason nonprofessionals seeking a loan or a job, or to buy a house or a car, will often try to look "professional" and "respectable" in those settings.

The privilege of this professional respectability appears starkly in the dynamics of racism and sexism. In daily interchange women and men of color must prove their respectability. At first they are often not treated by strangers with respectful distance or deference. Once people discover that this woman or that Puerto Rican man is a college teacher or a business executive, however, they often behave more respectfully toward her or him. Working-class white men, on the other hand, are often treated with respect until their working-class status is revealed. In Chapter 5 [I explore] in more detail the cultural underpinnings of the ideal of respectability and its oppressive implications.

I have discussed several injustices associated with powerlessness: inhibition in the development of one's capacities, lack of decisionmaking power in one's working life, and exposure to disrespectful treatment because of the status one occupies. These injustices have distributional consequences, but are more fundamentally matters of the division of labor. The oppression of powerlessness brings into question the division of labor basic to all industrial societies: the social division between those who plan and those who execute. [I examine] this division in more detail in [my] Chapter 7.

Cultural imperialism

Exploitation, marginalization, and powerlessness all refer to relations of power and oppression that occur by virtue of the social division of labor—who works for whom, who does not work, and how the content of work defines one institutional position relative to others. These three categories refer to structural and institutional relations that delimit people's material lives, including but not restricted to the resources they have access to and the concrete opportunities they have or do not have to develop and exercise their capacities. These kinds of oppression are a matter of concrete power in relation to others—of who benefits from whom, and who is dispensable.

Recent theorists of movements of group liberation, notably feminist and Black liberation theorists, have also given prominence to a rather different form of oppression, which following Lugones and Spelman (1983) I shall call cultural imperialism. To experience cultural imperialism means to experience how the dominant meanings of a society render the particular perspective of one's own group invisible at the same time as they stereotype one's group and mark it out as the Other.

Cultural imperialism involves the universalization of a dominant group's experience and culture, and its establishment as the norm. Some groups have exclusive or primary access to what Nancy Fraser (1987b) calls the means of interpretation and communication in a society. As a consequence, the dominant cultural products of the society, that is, those most widely disseminated, express the experience, values, goals, and achievements of these groups. Often without noticing they do so, the dominant groups project their own experience as representative of humanity as such. Cultural products also express the dominant group's perspective on and interpretation of events and elements in the society, including other groups in the society, insofar as they attain cultural status at all.

An encounter with other groups, however, can challenge the dominant group's claim to universality. The dominant group reinforces its position by bringing the other groups under the measure of its dominant norms. Consequently, the difference of women from men, American Indians or Africans from Europeans, Jews from Christians, homosexuals from heterosexuals, workers from professionals, becomes reconstructed largely as deviance and inferiority. Since only the dominant group's cultural expressions receive wide dissemination, their cultural expressions become the normal, or the universal, and thereby the unremarkable. Given the normality of its own cultural expressions and identity, the dominant group constructs the differences which some groups exhibit as lack and negation. These groups become marked as Other.

The culturally dominated undergo a paradoxical oppression, in that they are both marked out by stereotypes and at the same time rendered invisible. As remarkable, deviant beings, the culturally imperialized are stamped with an essence. The stereotypes confine them to a nature which is often attached in some way to their bodies, and which thus cannot easily be denied. These stereotypes so permeate the society that they are not noticed as contestable. Just as everyone knows that the earth goes around the sun, so everyone knows that gay people are promiscuous, that Indians are alcoholics, and that women are good with children. White males, on the other hand, insofar as they escape group marking, can be individuals.

Those living under cultural imperialism find themselves defined from the outside, positioned, placed, by a network of dominant meanings they experience as arising from elsewhere, from those with whom they do not identify and who do not identify with them. Consequently, the dominant culture's stereotyped and inferiorized images of the group must be internalized by group members at least to the extent that they are forced to react to behavior of others influenced by those images. This creates for the culturally oppressed the experience that W. E. B. Du Bois called "double consciousness"—"this sense of always looking at one's self through the eyes of others, of measuring one's soul by the tape of a world that looks on in amused contempt and pity" (Du Bois, 1969 [1903], p. 45). Double consciousness arises when the oppressed subject refuses to coincide with these devalued, objectified, stereotyped visions of herself or himself. While the subject desires recognition as human, capable of activity, full of hope and possibility, she receives from the dominant culture only the judgment that she is different, marked, or inferior.

The group defined by the dominant culture as deviant, as a stereotyped Other, *is* culturally different from the dominant group, because the status of Otherness creates specific experiences not shared by the dominant group, and because culturally oppressed groups also are often socially segregated and occupy specific positions in the social division of labor. Members of such groups express their specific group experiences and interpretations of the world to one another, developing and perpetuating their own culture. Double consciousness, then, occurs because one finds one's being defined by two cultures: a dominant and a subordinate culture. Because they can affirm and recognize one another as sharing similar experiences and perspectives on social life, people in culturally imperialized groups can often maintain a sense of positive subjectivity.

Cultural imperialism involves the paradox of experiencing oneself as invisible at the same time that one is marked out as different. The invisibility comes about when dominant groups fail to recognize the perspective embodied in their cultural expressions as a perspective. These dominant cultural expressions often simply have little place for the experience of other groups, at most only mentioning or referring to them in stereotyped or marginalized ways. This, then, is the injustice of cultural imperialism: that the oppressed group's own experience and interpretation of social life find little expression that touches the dominant culture, while that same culture imposes on the oppressed group its experience and interpretation of social life.

In several of [my] following chapters [I explore] more fully the consequences of cultural imperialism for the theory and practice of social justice. Chapter 4 expands on the claim that cultural imperialism is enacted partly through the ability of a dominant group to assert its perspective and experience as universal or neutral. In the sphere of the polity, [I argue], claim to universality operates politically to exclude those understood as different. In Chapter 5 [I trace] the operations of cultural imperialism in nineteenth-century scientific classifications of some bodies as deviant or degenerate. [I explore] how the devaluation of the bodies of some groups still conditions everyday interactions among groups, despite our relative success at expelling such bodily evaluation from discursive consciousness. In Chapter 6, finally, [I discuss] recent struggles by the culturally oppressed to take over definition of themselves and assert a positive sense of group difference. There [I argue] that justice requires us to make a political space for such difference.

Violence

Finally, many groups suffer the oppression of systematic violence. Members of some groups live with the knowledge that they must fear random, unprovoked attacks on their persons or property, which have no motive but to damage, humiliate, or destroy the person. In American society women, Blacks, Asians, Arabs, gay men, and lesbians live under such threats of violence, and in at least some regions Jews, Puerto Ricans, Chicanos, and other Spanish-speaking Americans must fear such violence as well. Physical violence against these groups is shockingly frequent. Rape Crisis Center networks estimate that more than one-third of all American women experience an attempted or successful sexual assault in their lifetimes. Manning Marable (1984, pp. 238–41) catalogues a large number of incidents of racist violence and terror against Blacks in the United States between 1980 and 1982. He cites dozens of incidents of the severe beating, killing, or rape of Blacks by police officers on duty, in which the police involved were acquitted of any wrongdoing. In 1981, moreover, there were at least five hundred documented cases of random white teenage violence against Blacks. Violence against gay men and lesbians is not only common, but has been increasing in the last five years. While the frequency of physical attack on members of these and other racially or sexually marked groups is very disturbing, I also include in this category less severe incidents of harrassment, intimidation, or ridicule simply for the purpose of degrading, humiliating, or stigmatizing group members.

Given the frequency of such violence in our society, why are theories of justice usually silent about it? I think the reason is that theorists do not typically take such incidents of violence and harrassment as matters of social injustice. No moral theorist would deny that such acts are very wrong. But unless all immoralities are injustices, they might wonder, why should such acts be interpreted as symptoms of social injustice? Acts of violence or petty harrassment are committed by particular individuals, often extremists, deviants, or the mentally unsound. How then can they be said to involve the sorts of institutional issues I have said are properly the subject of justice?

What makes violence a face of oppression is less the particular acts themselves, though these are often utterly horrible, than the social context surrounding them, which makes them possible and even acceptable. What makes violence a phenomenon of social injustice, and not merely an individual moral wrong, is its systemic character, its existence as a social practice.

Violence is systemic because it is directed at members of a group simply because they are members of that group. Any woman, for example, has a reason to fear rape. Regardless of what a Black man has done to escape the oppressions of marginality or powerlessness, he lives knowing he is subject to attack or harrassment. The oppression of violence consists not only in direct victimization, but in the daily knowledge shared by all members of oppressed groups that they are *liable* to violation, solely on account of their group identity. Just living under such a threat of attack on oneself or family or friends deprives the oppressed of freedom and dignity, and needlessly expends their energy.

Violence is a social practice. It is a social given that everyone knows happens and will happen again. It is always at the horizon of social imagination, even for those who do not perpetrate it. According to the prevailing social logic, some circumstances make such violence more "called for" than others. The idea of rape will occur to many men who pick up a hitchhiking woman; the idea of hounding or teasing a gay man on their dorm floor will occur to many straight male college students. Often several persons inflict the violence together, especially in all-male groupings. Sometimes violators set out looking for people to beat up, rape, or taunt. This rule-bound, social, and often premeditated character makes violence against groups a social practice.

Group violence approaches legitimacy, moreover, in the sense that it is tolerated. Often third parties find it unsurprising because it happens frequently and lies as a constant possibility at the horizon of the social imagination. Even when they are caught, those who perpetrate acts of group-directed violence or harrassment often receive light or no punishment. To that extent society renders their acts acceptable.

An important aspect of random, systemic violence is its irrationality. Xenophobic violence differs from the violence of states or ruling-class repression. Repressive violence has a rational, albeit evil, motive: rulers use it as a coercive tool to maintain their power. Many accounts of racist, sexist, or homophobic violence attempt to explain its motivation as a desire to maintain group privilege or domination. I do not doubt that fear of violence often functions to keep oppressed groups subordinate, but I do not think xenophobic violence is rationally motivated in the way that, for example, violence against strikers is.

On the contrary, the violation of rape, beating, killing, and harrassment of women, people of color, gays, and other marked groups is motivated by fear or hatred of those groups. Sometimes the motive may be a simple will to power, to victimize those marked as vulnerable by the very social fact that they are subject to violence. If so, this motive is secondary in the sense that it depends on a social practice of group violence. Violence-causing fear or hatred of the other at least partly involves insecurities on the part of the violators; its irrationality suggests that unconscious processes are at work. In [my] Chapter 5 [I discuss] the logic that makes some groups frightening or hateful by defining them as ugly and loathsome bodies. I offer a psychoanalytic account of the fear and hatred of some groups as bound up with fears of identity loss. I think such unconscious fears account at least partly for the oppression I have here called violence. It may also partly account for cultural imperialism.

Cultural imperialism, moreover, itself intersects with violence. The culturally imperialized may reject the dominant meanings and attempt to assert their own subjectivity, or the fact of their cultural difference may put the lie to the dominant culture's implicit claim to universality. The dissonance generated by such a challenge to the hegemonic cultural meanings can also be a source of irrational violence.

Violence is a form of injustice that a distributive understanding of justice seems ill equipped to capture. This may be why contemporary discussions of justice rarely mention it. I have argued that group-directed violence is institutionalized and systemic. To the degree that institutions and social practices encourage, tolerate, or enable the perpetration of violence against members of specific groups, those institutions and practices are unjust and should be reformed. Such reform may require the redistribution of resources or positions, but in large part can come only through a change in cultural images, stereotypes, and the mundane reproduction of relations of dominance and aversion in the gestures of everyday life. [I discuss] strategies for such change in [my] Chapter 5.

APPLYING THE CRITERIA

Social theories that construct oppression as a unified phenomenon usually either leave out groups that even the theorists think are oppressed, or leave out important ways in which groups are oppressed. Black liberation theorists and feminist theorists have argued persuasively, for example, that Marxism's reduction of all oppressions to class oppression leaves out much about the specific oppression of Blacks and women. By pluralizing the category of oppression in the way explained in this chapter, social theory can avoid the exclusive and oversimplifying effects of such reductionism.

I have avoided pluralizing the category in the way some others have done, by constructing an account of separate systems of oppression for each oppressed group: racism, sexism, classism, heterosexism, ageism, so on. There is a double problem with considering each group's oppression a unified and distinct structure or system. On the one hand, this way of conceiving oppression fails to accommodate the similarities and overlaps in the oppressions of different groups. On the other hand, it falsely represents the situation of all group members as the same.

I have arrived at the five faces of oppression—exploitation, marginalization, powerlessness, cultural imperialism, and violence—as the best way to avoid such exclusions and reductions. They function as criteria for determining whether individuals and groups are oppressed, rather than as a full theory of oppression. I believe that these criteria are objective. They provide a means of refuting some people's belief that their group is oppressed when it is not, as well as a means of persuading others that a group is oppressed when they doubt it. Each criterion can be operationalized; each can be applied through the assessment of observable behavior, status relationships, distributions, texts and other cultural artifacts. I have no illusions that such assessments can be value-neutral. But these criteria can nevertheless serve as means of evaluating claims that a group is oppressed, or adjudicating disputes about whether or how a group is oppressed.

The presence of any of these five conditions is sufficient for calling a group oppressed. But different group oppressions exhibit different combinations of these forms, as do different individuals in the groups. Nearly all, if not all, groups said by contemporary social movements to be oppressed suffer cultural imperialism. The other oppressions they experience vary. Working-class people are exploited and powerless, for example, but if employed and white do not experience marginalization and violence. Gay men, on the other hand, are not *qua* gay exploited or powerless, but they experience severe cultural imperialism and violence. Similarly, Jews and Arabs as groups are victims of cultural imperialism and violence, though many members of these groups also suffer exploitation or powerlessness. Old people are oppressed by marginalization and cultural imperialism, and this is also true of physically and mentally disabled people. As a group women are subject to gender-based exploitation, powerlessness, cultural imperialism, and violence. Racism in the United States condemns many Blacks and Latinos to marginalization, and puts many more at risk, even though many members of these groups escape that condition; members of these groups often suffer all five forms of oppression.

Applying these five criteria to the situation of groups makes it possible to compare oppressions without reducing them to a common essence or claiming that one is more fundamental than another. One can compare the ways in which a particular form of oppression appears in different groups. For example, while the operations of cultural imperialism are often experienced in similar fashion by different groups, there are also important differences. One can compare the combinations of oppressions groups experience, or the intensity of those oppressions. Thus with these criteria one can plausibly claim that one group is more oppressed than another without reducing all oppressions to a single scale.

Why are particular groups oppressed in the way they are? Are there any causal connections among the five forms of oppression? Causal or explanatory questions such as these are beyond the scope of this discussion. While I think general social theory has a place, causal explanation must always be particular and historical. Thus an explanatory account of why a particular group is oppressed in the ways that it is must trace the history and current structure of particular social relations. Such concrete historical and structural explanations will often show causal connections among the different forms of oppression experienced by a group. The cultural imperialism in which white men make stereotypical assumptions about and refuse to recognize the values of Blacks or women, for example, contributes to the marginalizaion and powerlessness many Blacks and women suffer. But cultural imperialism does not always have these effects.

[My succeeding chapters explore] the categories explicated here in different ways. Chapters 4, 5, and 6 explore the effects of cultural imperialism. Those chapters constitute an extended argument that modern political theory and practice wrongly universalize dominant group perspectives, and that attention to and affirmation of social group differences in the polity are the best corrective to such cultural imperialism. Chapters 7 and 8 also make use of the category of cultural imperialism, but focus more attention on social relations of exploitation and powerlessness.

NOTE

1 References to chapters here are to other chapters in the book by Young from which this extract is taken.

REFERENCES

Ackerman, Bruce. 1980. *Social Justice and the Liberal State.* New Haven: Yale University Press.

Alexander, David. 1987. "Gendered Job Traits and Women's Occupations." Ph.D. dissertation, Economics, University of Massachusetts.

Altman, Dennis. 1982. *The Homosexualization of America: The Americanization of Homosexuals.* Boston: Beacon.

Boris, Ellen and Peter Bardaglio. 1983. "The Transformation of Patriarchy: The Historic Role of the State." In Irene Diamond, ed., *Families, Politics and Public Policy.* New York: Longman.

Bowles, Samuel and Herbert Gintis. 1986. *Democracy and Capitalism.* New York: Basic.

Brittan, Arthur and Mary Maynard. 1984. *Sexism, Racism and Oppression.* Oxford: Blackwell.

Brown, Carol. 1981. "Mothers, Fathers and Children: From Private to Public Patriarchy." In Lydia Sargent, ed., *Women and Revolution.* Boston: South End.

Buchanan, Allen. 1982. *Marxism and Justice.* Totowa, N.J.: Rowman and Allanheld.

Coward, Rosalind and John Ellis. 1977. *Language and Materialism.* London: Routledge and Kegan Paul.

Delphy, Christine. 1984. *Close to Home: A Materialist Analysis of Women's Oppression.* Amherst: University of Massachusetts Press.

Du Bois, W. E. B. 1969 [1903]. *The Souls of Black Folk.* New York: New American Library.

Easton, Barbara. 1978. "Feminism and the Contemporary Family." *Socialist Review* 39 (May–June): 11–36.

Epstein, Steven. 1987. "Gay Politics, Ethnic Identity: The Limits of Social Constructionism." *Socialist Review* 17 (May–August): 9–54.

Ferguson, Ann. 1984. "On Conceiving Motherhood and Sexuality: A Feminist Materialist Approach." In Joyce Trebilcot, ed., *Mothering: Essays in Feminist Theory.* Totowa, N.J.: Rowman and Allanheld.

——. 1989. *Blood at the Root.* London: Pandora.

Ferguson, Kathy. 1984. *The Feminist Case against Bureaucracy.* Philadelphia: Temple University Press.

Fiss, Owen. 1976. "Groups and the Equal Protection Clause." *Philosophy and Public Affairs* 5 (Winter): 107–76.

Foucault, Michel. 1977. *Discipline and Punish.* New York: Pantheon.

Fraser, Nancy. 1987a. "Women, Welfare, and the Politics of Need Interpretation." *Hypatia: A Journal of Feminist Philosophy* 2 (Winter): 103–22.

——. 1987b. "Social Movements vs. Disciplinary Bureaucracies: The Discourse of Social Needs." CHS Occasional Paper No. 8. Center for Humanistic Studies, University of Minnesota.

French, Peter. 1975. "Types of Collectivities and Blame." *The Personalist* 56 (Spring): 160–69.

Friedman, Marilyn and Larry May. 1985. "Harming Women as a Group." *Social Theory and Practice* 11 (Summer): 297.

Frye, Marilyn. 1983. "Oppression." In *The Politics of Reality.* Trumansburg, N.Y.: Crossing.

Giddens, Anthony. 1981. *A Contemporary Critique of Historical Materialism.* Berkeley and Los Angeles: University of California Press.

Gilligan, Carol. 1982. *In a Different Voice.* Cambridge: Harvard University Press.

Gottlieb, Roger. 1987. *History and Subjectivity.* Philadelphia: Temple University Press.

Habermas, Jürgen. 1987. *The Theory of Communicative Competence.* Vol. 2: *Lifeworld and System.* Boston: Beacon.

Hartsock, Nancy. 1983. *Money, Sex and Power.* New York: Longman.

Heidegger, Martin. 1962. *Being and Time.* New York: Harper and Row.

Held, Virginia. 1987. "A Non-Contractual Society." In Marsha Hanen and Kai Nielsen, eds, *Science, Morality and Feminist Theory.* Calgary: University of Calgary Press.

Holmstrom, Nancy. 1977. "Exploitation." *Canadian Journal of Philosophy* 7 (June): 353–69.

Lugones, Maria C. and Elizabeth V. Spelman. 1983. "Have We Got a Theory for You! Feminist Theory, Cultural Imperialism and the Demand for 'the Woman's Voice.'" *Women's Studies International Forum* 6: 573–81.

Macpherson, C. B. 1973. *Democratic Theory: Essays in Retrieval.* Oxford: Oxford University Press.

Marable, Manning. 1984. *Race, Reform and Rebellion: The Second Construction in Black America, 1945–82.* Jackson: University Press of Mississippi.

May, Larry. 1987. *The Morality of Groups: Collective Responsibility, Group-Based Harm, and Corporate Rights.* Notre Dame: Notre Dame University Press.

Murphy, Raymond. 1985. "Exploitation or Exclusion?" *Sociology* 19 (May): 225–43.

Offe, Claus. 1985. *Disorganized Capitalism.* Cambridge: MIT Press.

Pateman, Carole. 1988. *The Sexual Contract.* Stanford: Stanford University Press.

Reich, Michael. 1981. *Racial Inequality.* Princeton: Princeton University Press.

Reiman, Jeffrey. 1987. "Exploitation, Force and the Moral Assessment of Capitalism: Thoughts on Roemer and Cohen." *Philosophy and Public Affairs* 16 (Winter): 3–41.

Roemer, John. 1982. *A General Theory of Exploitation and Class.* Cambridge: Harvard University Press.

Ross, Jeffrey. 1980. Introduction to Jeffrey Ross and Ann Baker Cottrell, eds, *The Mobilization of Collective Identity.* Lanham, MD: University Press of America.

Rothschild, Joseph. 1981. *Ethnopolitics.* New York: Columbia University Press.

Sartre, Jean-Paul. 1948. *Anti-Semite and Jew.* New York: Schocken.

Sennett, Richard and Jonathan Cobb. 1972. *The Hidden Injuries of Class.* New York: Vantage.

Sher, George. 1987. "Groups and the Constitution." In Gertrude Ezorsky, ed., *Moral Rights in the Workplace.* Albany: State University of New York Press.

Symanski, Al. 1985. "The Structure of Race." *Review of Radical Political Economy* 17 (Winter): 106–20.

Turner, John C., Michael A. Hogg, Penelope V. Oakes, Stephen D. Rucher, and Margaret S. Wethrell. 1987. *Rediscovering the Social Group: A Self-Categorization Theory.* Oxford: Blackwell.

Wolff, Robert Paul. 1984. *Understanding Marx.* Princeton: Princeton University Press.

5

Social justice in the age of identity politics

Redistribution, recognition, and participation*

Nancy Fraser

from Larry Ray and Andrew Sayer, eds, *Culture and Economy after the Cultural Turn.* Thousand Oaks, CA: Sage, 1999, pp. 25–52.

In today's world, claims for social justice seem increasingly to divide into two types. First, and most familiar, are redistributive claims, which seek a more just distribution of resources and goods. Examples include claims for redistribution from the North to the South, from the rich to the poor, and (not so long ago) from the owners to the workers. To be sure, the recent resurgence of free-market thinking has put proponents of egalitarian redistribution on the defensive. Nevertheless, egalitarian redistributive claims have supplied the paradigm case for most theorizing about social justice for the past 150 years.

Today, however, we increasingly encounter a second type of social-justice claim in the "politics of recognition." Here the goal, in its most plausible form, is a difference-friendly world, where assimilation to majority or dominant cultural norms is no longer the price of equal respect. Examples include claims for the recognition of the distinctive perspectives of ethnic, "racial," and sexual minorities, as well as of gender difference. This type of claim has recently attracted the interest of political philosophers, moreover, some of whom are seeking to develop a new paradigm of justice that puts recognition at its center.

In general, then, we are confronted with a new constellation. The discourse of social justice, once centered on distribution, is now increasingly divided between claims for redistribution, on the one hand, and claims for recognition, on the other. Increasingly, too, recognition claims tend to predominate. The demise of communism, the surge of free-market ideology, the rise of "identity polities" in both its fundamentalist and progressive forms—all these developments have conspired to de-center, if not to extinguish, claims for egalitarian redistribution.

In this new constellation, the two kinds of justice claims are often dissociated from one another—both practically and intellectually. Within social movements such as feminism, for example, activist tendencies that look to redistribution as the remedy for male domination are increasingly dissociated from tendencies that look instead to recognition of gender difference. And the same is true of their counterparts in the US academy, where feminist social theorizing and feminist cultural theorizing maintain an uneasy arm's-length co-existence. The feminist case exemplifies a more general tendency in the United States (and elsewhere) to decouple the cultural politics of difference from the social politics of equality.

In some cases, moreover, the dissociation has become a polarization. Some proponents of redistribution reject the politics of recognition outright, casting claims for the recognition of difference as "false consciousness," a hindrance to the pursuit of social justice. Conversely, some proponents of recognition approve the relative eclipse of the politics of redistribution, construing the latter as an outmoded materialism, simultaneously blind to and complicit with many injustices. In such cases, we are effectively presented with what is constructed as an either/or choice: redistribution or recognition? Class politics or identity politics? Multiculturalism or social democracy?

These, I maintain, are false antitheses. It is my general thesis that justice today requires *both* redistribution *and* recognition. Neither alone is sufficient. As soon as one embraces this thesis, however, the question of how to combine them becomes paramount. I contend that the emancipatory aspects of the two paradigms need to be integrated in a single, comprehensive framework. Theoretically, the task is to devise a two-dimensional conception of justice that can accommodate both defensible claims for social equality and defensible claims for the recognition of difference. Practically, the task is to devise a programmatic political orientation that integrates the best of the politics of redistribution with the best of the politics of recognition.

My argument proceeds in four steps. In the first section below, I outline the key points of contrast between the two political paradigms, as they are presently understood. Then, in the second section, I problematize their current dissociation from one another by introducing a case of injustice that cannot be redressed by either one of them alone, but that requires their integration. Finally, I consider some normative philosophical questions (in the third section) and some social-theoretical questions (fourth section) that arise when we contemplate integrating redistribution and recognition in a single comprehensive framework.

REDISTRIBUTION OR RECOGNITION? ANATOMY OF A FALSE ANTITHESIS

I begin with some denotative definitions. The paradigm of redistribution, as I shall understand it, encompasses not only class-centered orientations, such as New Deal liberalism, social-democracy, and socialism, but also those forms of feminism and anti-racism that look to socio-economic transformation or reform as the remedy for gender and racial-ethnic injustice. Thus, it is broader than class politics in the conventional sense. The paradigm of recognition, in contrast, encompasses not only movements aiming to revalue unjustly devalued identities—for example, cultural feminism, black cultural nationalism, and gay identity politics—but also deconstructive tendencies, such as queer politics, critical "race" politics, and deconstructive feminism, which reject the "essentialism" of traditional identity politics. Thus, it is broader than identity politics in the conventional sense.

With these definitions, I mean to contest one widespread misunderstanding of these matters. It is often assumed that the politics of redistribution means class politics, while the politics of recognition means "identity politics," which in turn means the politics of sexuality, gender, and "race." This view is erroneous and misleading. For one thing, it treats recognition-oriented currents within the feminist, anti-heterosexist, and anti-racist movements as the whole story, rendering invisible alternative currents dedicated to righting gender-specific, "race"-specific, and sex-specific forms of economic injustice that traditional class movements ignored. For another, it forecloses the recognition dimensions of class struggles. Finally, it reduces what is actually a plurality of different kinds of recognition claims (including universalist claims and deconstructive claims) to a single type, namely, claims for the affirmation of difference.

For all these reasons, the definitions I have proposed here are far preferable. They take account of the complexity of contemporary politics by treating redistribution and recognition as *dimensions of justice that can cut across all social movements.*

Understood in this way, the paradigm of redistribution and the paradigm of recognition can be contrasted in four key respects. First, the two paradigms assume different conceptions of injustice. The redistribution paradigm focuses on injustices it defines as socio-economic and presumes to be rooted in the political economy. Examples include exploitation, economic marginalization, and deprivation. The recognition paradigm, in contrast, targets injustices it understands as cultural, which it presumes to be rooted in social patterns of representation, interpretation, and communication. Examples include cultural domination, non-recognition, and disrespect.

Second, the two paradigms propose different sorts of remedies for injustice. In the redistribution paradigm, the remedy for injustice is political-economic restructuring. This might involve redistributing income, reorganizing the division of labor, or transforming other basic economic structures. (Although these various remedies differ importantly from one another, I mean to refer to the whole group of them by the generic term "redistribution.") In the paradigm of recognition, in contrast, the remedy for injustice is cultural or symbolic change. This could involve upwardly revaluing disrespected identities, positively valorizing cultural diversity, or the wholesale transformation of societal patterns of representation, interpretation, and communication in ways that would change everyone's social identity. (Although these remedies, too, differ importantly from one another, I refer once again to the whole group of them by the generic term "recognition.")

Third, the two paradigms assume different conceptions of the collectivities who suffer injustice. In the redistribution paradigm, the collective subjects of injustice are classes or class-like collectivities, which are defined economically by a distinctive relation to the market or the means of production. The classic case in the Marxian variant is the exploited working class, whose members must sell their labor power in order to receive the means of subsistence. But the conception can cover other cases as well. Also included are racialized groups of immigrants or ethnic minorities that can be economically defined, whether as a pool of low-paid menial laborers or as an "underclass" largely excluded from regular waged work, deemed "superfluous" and unworthy of exploitation. When the notion of the economy is broadened to encompass unwaged labor, moreover, women become visible as a collective subject of economic injustice, as the gender burdened with the lion's share of unwaged carework and consequently disadvantaged in employment and disempowered in relations with men. Also included, finally, are the complexly defined groupings that result when we theorize the political economy in terms of the intersection of class, "race," and gender.

In the recognition paradigm, in contrast, the victims of injustice are more like Weberian status groups than Marxian classes. Defined not by the relations of production, but rather by the relations of recognition, they are distinguished by the lesser esteem, honor, and prestige they enjoy relative to other groups in society. The classic case in the Weberian paradigm is the low-status ethnic group, whom dominant patterns of cultural value mark as different and less worthy. But the conception can cover other cases as well. In the current constellation, it has been extended to gays and lesbians, who suffer pervasive effects of institutionalized stigma; to racialized groups, who are marked as different and lesser; and to women, who are trivialized, sexually objectified, and disrespected in myriad ways. It is also being extended, finally, to encompass the complexly defined groupings that result when we theorize the relations of recognition in terms of "race," gender, and sexuality simultaneously as intersecting cultural codes.

It follows, and this is the fourth point, that the two approaches assume different understandings of group differences. The redistribution paradigm treats such differences as unjust differentials that should be abolished. The recognition paradigm, in contrast, treats differences either as cultural variations that should be celebrated or as discursively constructed hierarchical oppositions that should be deconstructed.

Increasingly, as I noted at the outset, redistribution and recognition are posed as mutually exclusive alternatives. Some proponents of the former, such as Richard Rorty (1998) and Todd Gitlin (1995), insist that identity politics is a counterproductive diversion from the real economic issues, one that balkanizes groups and rejects universalist moral norms. They claim, in effect, that "it's the economy, stupid." Conversely, some proponents of the politics of recognition, such as Charles Taylor (1994), insist that a difference-blind politics of redistribution can reinforce injustice by falsely universalizing dominant group norms, requiring subordinate groups to assimilate to them, and misrecognizing the latters' distinctiveness. They claim, in effect, that "it's the culture, stupid."

This, however, is a false antithesis.

EXPLOITED CLASSES, DESPISED SEXUALITIES, AND BIVALENT CATEGORIES: A CRITIQUE OF JUSTICE TRUNCATED

To see why, imagine a conceptual spectrum of different kinds of social differentiations. At one extreme are differentiations that fit the paradigm of redistribution. At the other extreme are differentiations that fit the paradigm of recognition. In between are cases that prove difficult because they fit both paradigms of justice simultaneously.[1]

Consider, first, the redistribution end of the spectrum. At this end let us posit an ideal-typical social differentiation rooted in the economic structure, as opposed to the status order, of society. By definition, any structural injustices attaching to this differentiation will be traceable ultimately to the political economy. The root of the injustice, as well as its core, will be socio-economic maldistribution, while any attendant cultural injustices will derive ultimately from that economic root. At bottom, therefore, the remedy required to redress the injustice will be redistribution, as opposed to recognition.

An example that appears to approximate this ideal type is class differentiation, as understood in orthodox, economistic Marxism. In this conception, class is an artifact of an unjust political economy, which creates, and exploits, a proletariat. The core injustice is exploitation, an especially deep form of maldistribution in which the proletariat's own energies are turned against it, usurped to sustain a social system that disproportionately burdens it and benefits others. To be sure, its members also suffer serious cultural injustices, the "hidden (and not so hidden) injuries of class" (Sennett and Cobb, 1973). But far from being rooted directly in an autonomously unjust status order, these derive from the political economy, as ideologies of class inferiority proliferate to justify exploitation. The remedy for the injustice, consequently, is redistribution, not recognition. The last thing the proletariat needs is recognition of its difference. On the contrary, the only way to remedy the injustice is to restructure the political economy in such a way as to put the proletariat out of business as a distinctive group.

Now consider the other end of the conceptual spectrum. At this end let us posit an ideal-typical social differentiation that fits the paradigm of recognition. A differentiation of this type is rooted in the status order, as opposed to the economic structure, of society. Thus, any structural injustices implicated here will be traceable ultimately to the reigning patterns of cultural value. The root of the injustice, as well as its core, will be cultural misrecognition, while any attendant economic injustices will derive ultimately from that root. The remedy required to redress the injustice will be recognition, as opposed to redistribution.

An example that appears to approximate this ideal type is sexual differentiation, understood through the prism of the Weberian conception of status. In this conception, the social differentiation between heterosexuals and homosexuals is not grounded in the political economy, as homosexuals are distributed throughout the entire class structure of capitalist society, occupy no distinctive position in the division of labor, and do not constitute an exploited class. The differentiation is rooted, rather, in the status order of society, as cultural patterns of meaning and value constitute heterosexuality as natural and normative, while simultaneously constituting homosexuality as perverse and despised. When such heteronormative meanings are pervasively institutionalized, for example in law, state policy, social practices, and interaction, gays and lesbians become a *despised sexuality*. As a result, they suffer sexually specific forms of *status subordination*, including shaming and assault, exclusion from the rights and privileges of marriage and parenthood, curbs on their rights of expression and association, and denial of full legal rights and equal protections. These harms are injustices of misrecognition. To be sure, gays and lesbians also suffer serious economic injustices: they can be summarily dismissed from civilian employment and military service, are denied a broad range of family-based social-welfare benefits, and face major tax and inheritance liabilities. But far from being rooted directly in the economic structure of society, these injustices derive instead from the status order, as the institutionalization of heterosexist norms produces a category of despised persons who incur economic disadvantages as a by-product. The remedy for the injustice, accordingly, is recognition, not redistribution. Overcoming homophobia and heterosexism requires changing the sexual status order, dismantling the cultural value patterns (as well as their legal and practical expressions) that deny equal respect to gays and lesbians. Change these relations of recognition, and the maldistribution will disappear.

Matters are thus fairly straightforward at the two extremes of our conceptual spectrum. When we deal with groups that approach the ideal type of the exploited working class, we face distributive injustices requiring redistributive remedies. What is needed is a politics of redistribution. When we deal with groups that approach the ideal type of the despised sexuality, in contrast, we face injustices of misrecognition. What is needed *here* is a politics of recognition.

Matters become murkier, however, once we move away from these extremes. When we posit a type of social differentiation located in the middle of the conceptual spectrum, we encounter a hybrid form that combines features of the exploited class with features of the despised sexuality. I call such differentiations "bivalent." Rooted at once in the economic structure and the status order of society, they may entrench injustices that are traceable to both political economy and culture simultaneously. Bivalently oppressed groups, accordingly, suffer both maldistribution and misrecognition *in forms where neither of these injustices is an indirect effect of the other, but where both are primary and co-original*. In their case, neither the politics of redistribution alone nor the politics of recognition alone will suffice. Bivalently oppressed groups need both.

Gender, I contend, is a bivalent social differentiation. Neither simply a class, nor simply a status group, it is a hybrid category with roots in both culture and political economy. From the economic perspective, gender structures the fundamental division between paid "productive" labor and unpaid "reproductive" and domestic labor, as well as the divisions within paid labor between higher-paid, male-dominated, manufacturing, and professional occupations and lower-paid, female-dominated, "pink collar," and domestic service occupations. The result is an economic structure that generates gender-specific modes of exploitation, economic marginalization, and deprivation. Here, gender appears as a class-like differentiation. And gender injustice appears as a species of maldistribution that cries out for redistributive redress.

From the perspective of the status order, however, gender encompasses elements that are more like sexuality than class and that bring it squarely within the problematic of recognition. Gender codes pervasive patterns of cultural interpretation and evaluation, which are central to the status order as a whole. As a result, not just women, but all low-status groups, risk being feminized and thereby demeaned. Thus, a major feature of gender injustice is androcentrism: a pattern of culture value that privileges traits associated with masculinity, while pervasively devaluing things coded as "feminine"—paradigmatically, but not only, women. Institutionalized in law, state policies, social practices, and interaction, this value pattern saddles women with gender-specific forms of *status subordination*, including sexual assault and domestic violence; trivializing, objectifying, and demeaning stereotypical depictions in the media; harassment and disparagement in everyday life; and denial of full legal rights and equal protections. These harms are injustices of recognition. They cannot be remedied by redistribution alone but require additional independent remedies of recognition.

Gender, in sum, is a "bivalent" social differentiation. It encompasses a class-like aspect that brings it within the ambit of redistribution, while also including a status aspect that brings it simultaneously within the ambit of recognition. Redressing gender injustice, therefore, requires changing both the economic structure and the status order of society.

The bivalent character of gender wreaks havoc on the idea of an either/or choice between the paradigm of redistribution and the paradigm of recognition. That construction assumes that the collective subjects of injustice are either classes or status groups, but not both; that the injustice they suffer is either maldistribution or misrecognition, but not both; that the group differences at issue are either unjust differentials or unjustly devalued cultural variations, but not both; that the remedy for injustice is either redistribution or recognition, but not both.

Gender, we can now see, explodes this whole series of false antitheses. Here we have a category that is a compound of both status and class, that implicates injustices of both maldistribution and misrecognition, whose distinctiveness is compounded of both economic differentials and culturally constructed distinctions. Gender injustice can only be remedied, therefore, by an approach that encompasses both a politics of redistribution and a politics of recognition.

Gender, moreover, is not unusual in this regard. "Race", too, is a bivalent social differentiation, a compound of status and class. Rooted simultaneously in the economic structure and the status order of capitalist society, racism's injustices include both maldistribution and misrecognition. Yet neither dimension of racism is wholly an indirect effect of the other. Thus, overcoming racism requires both redistribution and recognition. Neither alone will suffice.

Class, too, is probably best understood as bivalent for practical purposes. To be sure, the ultimate cause of class injustice is the economic structure of capitalist society.[2] But the resulting harms include misrecognition as well as maldistribution (Thompson, 1963). And cultural harms that originated as byproducts of economic structure may have since developed a life of their own. Left unattended, moreover, class misrecognition may impede the capacity to mobilize against maldistribution. Thus, a politics of class recognition may be needed to get a politics of redistribution off the ground.[3]

Sexuality, too, is for practical purposes bivalent. To be sure, the ultimate cause of heterosexist injustice is the heteronormative value pattern that is institutionalized in the status order of contemporary society.[4] But the resulting harms include maldistribution as well as misrecognition. And economic harms that originate as byproducts of the status order have an undeniable weight of their own. Left unattended, moreover, they may impede the capacity to mobilize against misrecognition. Thus, a politics of sexual redistribution may be needed to get a politics of recognition off the ground.

For practical purposes, then, virtually all real-world axes of oppression are bivalent. Virtually all implicate both maldistribution and misrecognition in forms where each of those injustices has some independent weight, whatever its ultimate roots. To be sure, not all axes of oppression are bivalent in the same way, nor to the same degree. Some axes of oppression, such as class, tilt more heavily toward the distribution end of the spectrum; others, such as sexuality, incline more to the recognition end; while still others, such as gender and "race," cluster closer to the center. Nevertheless, in virtually every case, the harms at issue comprise both maldistribution and misrecognition in forms where neither of those injustices can be redressed entirely indirectly but where each requires some practical attention. As a practical matter, therefore, overcoming injustice in virtually every case requires both redistribution and recognition.

The need for this sort of two-pronged approach becomes more pressing, moreover, as soon as we cease considering such axes of injustice singly and begin instead to consider them together as mutually intersecting. After all, gender, "race," sexuality, and class are not neatly cordoned off from one another. Rather, all these axes of injustice intersect one another in ways that affect everyone's interests and identities. Thus, anyone who is both gay and working class will need both redistribution and recognition. Seen this way, moreover, virtually every individual who suffers injustice needs to integrate those two kinds of claims. And so, furthermore, will anyone who cares about social justice, regardless of their own personal social location.

In general, then, one should roundly reject the construction of redistribution and recognition as mutually exclusive alternatives. The goal should be, rather, to develop an integrated approach that can encompass, and harmonize, both dimensions of social justice.

NORMATIVE-PHILOSOPHICAL ISSUES: FOR A TWO-DIMENSIONAL THEORY OF JUSTICE

Integrating redistribution and recognition in a single comprehensive paradigm is no simple matter, however. To contemplate such a project is to be plunged immediately into deep and difficult problems spanning several major fields of inquiry. In moral philosophy, for example, the task is to devise an overarching conception of justice that can accommodate both defensible claims for social equality and defensible claims for the recognition of difference. In social theory, by contrast, the task is to devise an account of our contemporary social formation that can accommodate not only the differentiation of class from status, economy from culture, but also their mutual imbrication. In political theory, meanwhile, the task is to envision a set of institutional arrangements and associated policy reforms that can remedy both maldistribution and misrecognition, while minimizing the mutual interferences likely to arise when the two sorts of redress are sought simultaneously. In practical politics, finally, the task is to foster democratic engagement across current divides in order to build a broad-based programmatic orientation that integrates the best of the politics of redistribution with the best of the politics of recognition.

This, of course, is far too much to take on here. In the present section, I limit myself to some of the moral-theoretical dimensions of this project. (In the next, I turn to some issues in social theory.) I shall consider three normative philosophical questions that arise when we contemplate integrating redistribution and recognition in a single comprehensive account of social justice: First, is recognition really a matter of justice, or is it a matter of self-realization? Second, do distributive justice and recognition constitute two distinct, *sui generis,* normative paradigms, or can either of them be subsumed within the other? And third, does justice require the recognition of what is distinctive about individuals or groups, or is recognition of our common humanity sufficient? (I defer to a later occasion discussion of a fourth crucial question: How can we distinguish justified from unjustified claims for recognition?)

On the first question, two major theorists, Charles Taylor and Axel Honneth, understand recognition as a matter of self-realization. Unlike them, however, I propose to treat it as an issue of justice. Thus, one should not answer the question "What's wrong with misrecognition?" by reference to a thick theory of the good, as Taylor (1994) does. Nor should one follow Honneth (1995) and appeal to a "formal conception of ethical life" premised on an account of the "intersubjective conditions" for an "undistorted practical relation-to-self." One should say, rather, that it is unjust that some individuals and groups are denied the status of full partners in social interaction simply as a consequence of institutionalized patterns of cultural value in whose construction they have not equally participated and which disparage their distinctive characteristics or the distinctive characteristics assigned to them.

This account offers several advantages. First, it permits one to justify claims for recognition as morally binding under modern conditions of value pluralism.[5] Under these conditions, there is no single conception of self-realization or the good that is universally shared, nor any that can be established as authoritative. Thus, any attempt to justify claims for recognition that appeals to an account of self-realization or the good must necessarily be sectarian. No approach of this sort can establish such claims as normatively binding on those who do not share the theorist's conception of ethical value.

Unlike such approaches, I propose an account that is deontological and non-sectarian. Embracing the modern view that it is up to individuals and groups to define for themselves what counts as a good life and to devise for themselves an approach to pursuing it, within limits that ensure a like liberty for others, it appeals to a conception of justice that can be accepted by people with divergent conceptions of the good. What makes misrecognition morally wrong, in my view, is that it denies some individuals and groups the possibility of participating on a par with others in social interaction. The norm of *participatory parity* invoked here is non-sectarian in the required sense. It can justify claims for recognition as normatively binding on all who agree to abide by fair terms of interaction under conditions of value pluralism.

Treating recognition as a matter of justice has a second advantage as well. It conceives misrecognition as *status subordination* whose locus is social relations, not individual psychology. To be misrecognized, on this view, is not simply to be thought ill of, looked down on, or devalued in others' conscious attitudes or mental beliefs. It is rather to be denied the status of a full partner in social interaction and prevented from participating as a peer in social life as a consequence of *institutionalized* patterns of cultural value that constitute one as comparatively unworthy of respect or esteem. When such patterns of disrespect and disesteem are institutionalized, they impede parity of participation, just as surely as do distributive inequities.

Eschewing psychologization, then, the justice approach escapes difficulties that plague rival approaches. When misrecognition is identified with internal distortions in the structure of self-consciousness of the oppressed, it is but a short step to blaming the victim, as one seems to add insult to injury. Conversely, when misrecognition is equated with prejudice in the minds of the oppressors, overcoming it seems to require policing their beliefs, an approach that is authoritarian. On the justice view, in contrast, misrecognition is a matter of externally manifest and publicly verifiable impediments to some people's standing as full members of society. And such arrangements are morally indefensible whether or not they distort the subjectivity of the oppressed.

Finally, the justice account of recognition avoids the view that everyone has an equal right to social esteem. That view is patently untenable, of course, because it renders meaningless the notion of esteem. Yet it seems to follow from at least one prominent account of recognition in terms of the self-realization.[6] The account of recognition proposed here, in contrast, entails no such *reductio ad absurdum*. What it *does* entail is that everyone has an equal right to pursue social esteem under fair conditions of equal opportunity. And such conditions do not obtain when, for example, institutionalized patterns of interpretation pervasively downgrade femininity, "non-whiteness," homosexuality, and everything culturally associated with them. When that is the case, women and/or people of color and/or gays and lesbians face obstacles in the quest for esteem that are not encountered by others. And everyone, including straight white men, faces further obstacles if they opt to pursue projects and cultivate traits that are culturally coded as feminine, homosexual, or "non-white."

For all these reasons, recognition is better viewed as a matter of justice than as a matter of self-realization. But what follows for the theory of justice?

Does it follow, turning now to the second question, that distribution and recognition constitute two distinct, *sui generis* conceptions of justice? Or can either of them be reduced to the other? The question of reduction must be considered from two different sides. From one side, the issue is whether standard theories of distributive justice can adequately subsume problems of recognition. In my view, the answer is no. To be sure, many distributive theorists appreciate the importance of status over and above the allocation of resources and seek to accommodate it in their accounts.[7] But the results are not wholly satisfactory. Most such theorists assume a reductive economistic-cum-legalistic view of status, supposing that a just distribution of resources and rights is sufficient to preclude misrecognition. In fact, however, as we saw, not all misrecognition is a byproduct of maldistribution, nor of maldistribution plus legal discrimination. Witness the case of the African-American Wall Street banker who cannot get a taxi to pick him up. To handle such cases, a theory of justice must reach beyond the distribution of rights and goods to examine patterns of cultural value. It must consider whether institutionalized patterns of interpretation and valuation impede parity of participation in social life.[8]

What, then, of the other side of the question? Can existing theories of recognition adequately subsume problems of distribution? Here, too, I contend the answer is no. To be sure, some theorists of recognition appreciate the importance of economic equality and seek to accommodate it in their accounts.[9] But once again the results are not wholly satisfactory. Such theorists tend to assume a reductive culturalist view of distribution. Supposing that economic inequalities are rooted in a cultural order that privileges some kinds of labor over others, they assume that changing that cultural order is sufficient to preclude maldistribution (Honneth, 1995). In fact, however, as we saw, and as I shall argue more extensively later, not all maldistribution is a byproduct of misrecognition. Witness the case of the skilled white male industrial worker who becomes unemployed due to a factory closing as a result of a speculative corporate merger. In that case, the injustice of maldistribution has little to do with misrecognition. It is rather a consequence of imperatives intrinsic to an order of specialized economic relations whose *raison d'être* is the accumulation of profits. To handle such cases, a theory of justice must reach beyond cultural value patterns to examine the structure of capitalism. It must consider whether economic mechanisms that are relatively decoupled from cultural value patterns and that operate in a relatively impersonal way can impede parity of participation in social life.

In general, then, neither distribution theorists nor recognition theorists have so far succeeded in adequately subsuming the concerns of the other.[10] Thus, instead of endorsing either one of their paradigms to the exclusion of the other, I propose to develop what I shall call a two-dimensional conception of justice. Such a conception treats distribution and recognition as distinct perspectives on, and dimensions of, justice. Without reducing either one of them to the other, it encompasses both dimensions within a broader, overarching framework.

The normative core of my conception, which I have mentioned several times, is the notion of *parity of participation*.[11] According to this norm, justice requires social arrangements that permit all (adult) members of society to interact with one another as peers. For participatory parity to be possible, I claim, it is necessary but not sufficient to establish standard forms of formal legal equality. Over and above that requirement, at least two additional conditions must be satisfied.[12] First, the distribution of material resources must be such as to ensure participants' independence and "voice." This I call the "objective" precondition of participatory parity. It precludes forms and levels of material inequality and economic dependence that impede parity of participation. Precluded, therefore, are social arrangements that institutionalize deprivation, exploitation, and gross disparities in wealth, income, and leisure time, thereby denying some people the means and opportunities to interact with others as peers.[13]

In contrast, the second additional condition for participatory parity I call "intersubjective." It requires that institutionalized cultural patterns of interpretation and evaluation express equal respect for all participants and ensure equal opportunity for achieving social esteem. This condition precludes cultural patterns that systematically depreciate some categories of people and the qualities associated with them. Precluded, therefore, are institutionalized value schemata that deny some people the status of full partners in interaction—whether by burdening them with excessive ascribed "difference" from others or by failing to acknowledge their distinctiveness.

Both the objective precondition and the intersubjective precondition are necessary for participatory parity. Neither alone is sufficient. The objective condition brings into focus concerns traditionally associated with the theory of distributive justice, especially concerns pertaining to the economic structure of society and to economically defined class differentials. The intersubjective precondition brings into focus concerns recently highlighted in the philosophy of recognition, especially concerns pertaining to the status order of society and to culturally defined hierarchies of status. Thus, a two-dimensional conception of justice oriented to the norm of participatory parity encompasses both redistribution and recognition, without reducing either one to the other.

This brings us to the third question: Does justice require the recognition of what is distinctive about individuals or groups, over and above the recognition of our common humanity? Here it is important to note that participatory parity is a universalist norm in two senses. First, it encompasses all (adult) partners to interaction. And second, it presupposes the equal moral worth of human beings. But moral universalism in these senses still leaves open the question whether recognition of individual or group distinctiveness could be required by justice as one element among others of the intersubjective condition for participatory parity.

This question cannot be answered, I contend, by an *a priori* account of the kinds of recognition that everyone always needs. It needs rather to be approached in the spirit of pragmatism as informed by the insights of a critical social theory. From this perspective, recognition is a remedy for injustice, not a generic human need. Thus, the form(s) of recognition justice requires in any given case depend(s) on the form(s) of misrecognition to be redressed. In cases where misrecognition involves denying the common humanity of some participants, the remedy is universalist recognition. Where, in contrast, misrecognition involves denying some participants' distinctiveness, the remedy could be recognition of difference.[14] In every case, the remedy should be tailored to the harm.

This pragmatist approach overcomes the liabilities of two other views that are mirror opposites and hence equally decontextualized. First, it avoids the view, espoused by some distributive theorists, that justice requires limiting public recognition to those capacities all humans share. That approach dogmatically forecloses recognition of what distinguishes people from one another, without considering whether the latter might be needed in some cases to overcome obstacles to participatory parity. Second, the pragmatist approach avoids the opposite view, also decontextualized, that everyone always needs their distinctiveness recognized (Taylor, 1994; Honneth, 1995). Favored by recognition theorists, this anthropological view cannot explain why it is that not all, but only some, social differences generate claims for recognition, nor why only some of those that do, but not others, are morally justified. More specifically, it cannot explain why dominant groups, such as men and heterosexuals, usually shun recognition of their (gender and sexual) distinctiveness, claiming not specificity but universality. By contrast, the approach proposed here sees claims for the recognition of difference pragmatically and contextually—as remedial responses to specific harms. Putting questions of justice at the center, it appreciates that the recognition needs of subordinate groups differ from those of dominant groups; and that only those claims that promote participatory parity are morally justified.

For the pragmatist, accordingly, everything depends on precisely what currently misrecognized people need in order to be able to participate as peers in social life. And there is no reason to assume that all of them need the same thing in every context. In some cases, they may need to be unburdened of excessive ascribed or constructed distinctiveness. In other cases, they may need to have hitherto underacknowledged distinctiveness taken into account. In still other cases, they may need to shift the focus onto dominant or advantaged groups, outing the latter's distinctiveness, which has been falsely parading as universality. Alternatively, they may need to deconstruct the very terms in which attributed differences are currently elaborated. Finally, they may need all of the above, or several of the above, in combination with one another and in combination with redistribution. Which people need which kind(s) of recognition in which contexts depends on the nature of the obstacles they face with regard to participatory parity. That, however, cannot be determined by abstract philosophical argument. It can only be determined with the aid of a critical social theory, a theory that is normatively oriented, empirically informed, and guided by the practical intent of overcoming injustice.

SOCIAL-THEORETICAL ISSUES: AN ARGUMENT FOR "PERSPECTIVAL DUALISM"

This brings us to the social-theoretical issues that arise when we try to encompass redistribution and recognition in a single framework. Here, the principal task is to theorize the relations between class and status, and between maldistribution and misrecognition, in contemporary society. An adequate approach must allow for the full complexity of these relations. It must account *both for the differentiation of class and status and for the causal interactions between them*. It must accommodate, as well, *both the mutual irreducibility of maldistribution and misrecognition and their practical entwinement with one another.* Such an account must, moreover, be historical. Sensitive to shifts in social structure and political culture, it must identify the distinctive dynamics and conflict tendencies of the present conjuncture. Attentive both to national specificities and to transnational forces and frames, it must explain why today's grammar of social conflict takes the form that it does: why, that is, struggles for recognition have recently become so salient; why egalitarian redistribution struggles, hitherto central to social life, have lately receded to the margins; and why, finally, the two kinds of claims for social justice have become decoupled and antagonistically counterposed.[15]

First, however, some conceptual clarifications. The terms class and status, as I use them here, denote socially entrenched orders of domination. To say that a society has a class structure, accordingly, is to say that it institutionalizes mechanisms of distribution that systematically deny some of its members the means and opportunities they need in order to participate on a par with others in social life. To say, likewise, that a society has a status hierarchy is to say that it institutionalizes patterns of cultural value that pervasively deny some of its members the recognition they need in order to be full, participating partners in interaction. The existence of either a class structure or a status hierarchy constitutes an obstacle to parity of participation and thus an injustice.

In what follows, then, I assume an internal conceptual relation between class and status, on the one hand, and domination and injustice, on the other. I do not, however, present a full theory of class or status. Deferring that task to another occasion, I assume only that both orders of domination emerged historically with developments in social organization, as did the conceptual distinction between them and the possibility of their mutual divergence. I assume, too, that a society's class structure becomes distinguishable from its status order only when its mechanisms of economic distribution become differentiated from social arenas in which institutionalized patterns of cultural value regulate interaction in a relatively direct and unmediated way. Thus, only with the emergence of a specialized order of economic relations can the question arise, whether the society's class structure diverges from its status hierarchy or whether, alternatively, they coincide. Only then, likewise, can the question become politically salient whether the status hierarchy and/or the class structure are unjust.

What follows from this approach for our understanding of the categories economy and culture? Both of these terms, as I use them here, denote social processes and social relations.[16] Both, moreover, must be grasped historically. As I just noted, specifically economic processes and relations became differentiated from unmediatedly value-regulated processes and relations only with historical shifts in the structure of societies. Only with the rise of capitalism did highly autonomous economic institutions emerge, making possible the modern ideas of "the economic" and "the cultural," as well as the distinction between them.[17] To be sure, these ideas can be applied retrospectively to precapitalist societies—provided one situates one's usage historically and explicitly notes the anachronism. But this only serves to underline the key point: Far from being ontological or anthropological, economy and culture are *historically emergent categories of social theory*. What counts as economic and as cultural depends on the type of society in question. So, as well, does the relation between the economic and the cultural.

An analogous point holds for maldistribution and misrecognition. It is not the case that the former denotes a species of material harm and the latter one of immaterial injury. On the contrary, status injuries can be just as material as distributive injustices—witness gay-bashing, gang rape, and genocide.[18] Far from being ontological, this distinction, too, is historical. Distribution and recognition correspond to historically specific social-structural differentiations, paradigmatically those associated with modern capitalism. *Historically emergent normative categories*, they became distinguishable dimensions of justice only with the differentiation of class from status and of the economic from the cultural. Only, in other words, with the relative uncoupling of specialized economic mechanisms of distribution from broader patterns of cultural value did the distinction between maldistribution and misrecognition become thinkable. And only then could the question of the relation between them arise. To be sure, these categories too can be applied retrospectively, provided one is sufficiently self-aware. But the point, once again, is to historicize. The relations between maldistribution and misrecognition vary according to the social formation under consideration. It remains an empirical question in any given case whether and to what extent they coincide.

In every case, the level of differentiation is crucial. In some societies, conceivable or actual, economy and culture are not institutionally differentiated. Consider, for example, an ideal-typical pre-state society of the sort described in the classical anthropological literature, while bracketing the question of ethnographic accuracy.[19] In such a society, the master idiom of social relations is kinship. Kinship organizes not only marriage and sexual relations, but also the labor process and the distribution of goods; relations of authority, reciprocity, and obligation; and symbolic hierarchies of status and prestige. Of course, it could well be the case that such a society has never existed in pure form. Still, we can imagine a world in which neither distinctively economic institutions nor distinctively cultural institutions exist. A single order of social relations secures (what we would call) both the economic integration and the cultural integration of the society. Class structure and status order are accordingly fused. Because kinship constitutes the overarching principle of distribution, kinship status dictates class position. In the absence of any quasi-autonomous economic institutions, status injuries translate immediately into (what we would consider to be) distributive injustices. Misrecognition directly entails maldistribution.

This ideal-type of a fully kin-governed society represents an extreme case of non-differentiation, one in which cultural patterns of value dictate the order of economic domination. It is usefully contrasted with the opposite extreme of a fully marketized society, in which economic structure dictates cultural value. In such a society, the master determining instance is the market. Markets organize not only the labor process and the distribution of goods, but also marriage and sexual relations; political relations of authority, reciprocity, and obligation; and symbolic hierarchies of status and prestige. Granted, such a society has never existed, and it is doubtful that one ever could.[20] For heuristic purposes, however, we can imagine a world in which a single order of social relations secures not only the economic integration but also the cultural integration of society. Here, too, as in the fully kin-governed society, class structure and status order are effectively fused. But the determinations run in the opposite direction. Because the market constitutes the sole and all-pervasive mechanism of valuation, market position dictates social status. In the absence of any quasi-autonomous cultural value patterns, distributive injustices translate immediately into status injuries. Maldistribution directly entails misrecognition.

As mirror-opposites of each other, these two imagined societies share a common feature: the absence of any meaningful differentiation of the economy from the larger culture.[21] In both of them, accordingly (what we would call), class and status map perfectly onto each other. So, as well, do (what we would call) maldistribution and misrecognition, which convert fully and without remainder into one another. As a result, one can understand both these societies reasonably well by attending exclusively to a single dimension of social life. For the fully kin-governed society, one can read off the economic dimension of domination directly from the cultural; one can infer class directly from status and maldistribution directly from misrecognition. For the fully marketized society, conversely, one can read off the cultural dimension of domination directly from the economic; one can infer status directly from class, and misrecognition directly from maldistribution. For understanding the forms of domination proper to the fully kin-governed society, therefore, culturalism is a perfectly appropriate social theory.[22] If, in contrast, one is seeking to understand the fully marketized society, one could hardly improve on economism.[23]

When we turn to other types of societies, however, such simple and elegant approaches no longer suffice. They are patently inappropriate for the actually existing capitalist society that we currently inhabit and seek to understand. In this society, a specialized set of economic institutions has been differentiated from the larger social field. The paradigm institutions are markets, which operate by instrumentalizing the cultural value patterns that regulate some other orders of social relations in a fairly direct and unmediated way. Filtering meanings and values through an individual-interest-maximizing grid, markets decontextualize and rework cultural patterns. As the latter are pressed into the service of an individualizing logic, they are disembedded, instrumentalized, and resignified. The result is a specialized zone in which cultural values, though neither simply suspended nor wholly dissolved, do not regulate social interaction in a direct and unmediated way. Rather, they impact it indirectly, through the mediation of the "cash nexus."

Markets have always existed, of course, but their scope, autonomy, and influence attained a qualitatively new level with the development of modern capitalism. In capitalist society, these value-instrumentalizing institutions directly organize a significant portion of the labor process (the waged portion), the distribution of most products and goods (commodities), and the investment of most social surplus (profit). They do not, however, *directly* organize marriage, sexuality, and the family; relations of political authority and legal obligation; and symbolic hierarchies of status and prestige. Rather, each of these social orders retains distinctive institutional forms and normative orientations; each also remains connected to, and informed by, the general culture; some of them, finally, are regulated by institutionalized patterns of cultural value in a relatively direct and unmediated way.

Thus, in capitalist society, relations between economy and culture are complex. Neither devoid of culture, nor directly subordinated to it, capitalist markets stand in a highly mediated relation to institutionalized patterns of cultural value. They work through the latter, while also working over them, sometimes helping to transform them in the process. Thoroughly permeated by significations and norms, yet possessed of a logic of their own, capitalist economic institutions are neither wholly constrained by, nor fully in control of, value patterns.

To be sure, capitalist market processes heavily influence non-market relations. But their influence is indirect. In principle and, to a lesser degree, in practice, non-marketized arenas have some autonomy *vis-à-vis* the market, as well as *vis-à-vis* one another. It remains an empirical question exactly how far in each case market influence actually penetrates—and a normative question how far it should. The reverse is, by contrast, fairly clear: in capitalist societies, market processes generally have considerable autonomy *vis-à-vis* politics, although the precise extent varies according to the régime. In its Western European heyday, Keynesian social democracy sought with some success to use "politics to tame markets" within state borders. In the current climate of post-Keynesian, neoliberal, globalizing capitalism, the market's scope, autonomy, and influence are sharply increasing.

The key point here is that capitalist society is structurally differentiated. The institutionalization of specialized economic relations permits the partial uncoupling of economic distribution from structures of prestige. As markets instrumentalize value patterns that remain constitutive for non-marketized relations, a gap arises between status and class. The class structure ceases perfectly to mirror the status order, even as each of them influences the other. Because the market does not constitute the sole and all-pervasive mechanism of valuation, market position does not dictate social status. Partially cultural value patterns prevent distributive injustices from converting fully and without remainder into status injuries. Maldistribution does not directly entail misrecognition, although it may well contribute to the latter. Conversely, because no single status principle such as kinship constitutes the sole and all-pervasive principle of distribution, status does not dictate class position. Relatively autonomous economic institutions prevent status injuries from converting fully without remainder into distributive injustices. Misrecognition does not directly entail maldistribution, although it, too, may contribute to the latter.

In capitalist society, accordingly, class and status do not perfectly mirror each other, their interaction and mutual influence notwithstanding. Nor, likewise, do maldistribution and misrecognition convert fully and without remainder into one another, despite interaction and even entwinement. As a result, one cannot understand this society by attending exclusively to a single dimension of social life. One cannot read off the economic dimension of domination directly from the cultural, nor the cultural directly from the economic. Likewise, one cannot infer class directly from status, nor status directly from class. Finally, one cannot deduce maldistribution directly from misrecognition, nor misrecognition directly from maldistribution. It follows that neither culturalism nor economism suffices for understanding capitalist society. Instead, one needs an approach that can accommodate differentiation, divergence, and interaction at every level.

What sort of social theory can handle this task? What approach can theorize both the differentiation of status from class and the causal interactions between them? What kind of theory can accommodate the complex relations between maldistribution and misrecognition in contemporary society, grasping at once their conceptual irreducibility, empirical divergence, and practical entwinement? And what approach can do all this *without reinforcing the current dissociation of the politics of recognition from the politics of redistribution*? If neither economism nor culturalism is up to the task, what alternative approaches are possible?

Two possibilities present themselves, both of them species of dualism.[24] The first approach I call "substantive dualism." It treats redistribution and recognition as two different "spheres of justice," pertaining to two different societal domains. The former pertains to the economic domain of society, the relations of production. The latter pertains to the cultural domain, the relations of recognition. When we consider economic matters, such as the structure of labor markets, we should assume the standpoint of distributive justice, attending to the impact of economic structures and institutions on the relative economic position of social actors. When, in contrast, we consider cultural matters, such as the representation of female sexuality on MTV, we should assume the standpoint of recognition, attending to the impact of institutionalized patterns of interpretation and value on the status and relative standing of social actors.

Substantive dualism may be preferable to economism and culturalism, but it is nevertheless inadequate —both conceptually and politically. Conceptually, it erects a dichotomy that opposes economy to culture and treats them as two separate spheres. It thereby mistakes the differentiations of capitalist society for institutional divisions that are impermeable and sharply bounded. In fact, these differentiations mark orders of social relations that can overlap one another institutionally and are more or less permeable in different régimes. As just noted, the economy is not a culture-free zone, but a culture-instrumentalizing and -resignifying one. Thus, what presents itself as "the economy" is always already permeated with cultural interpretations and norms—witness the distinctions between "working" and "caregiving," "men's jobs" and "women's jobs," which are so fundamental to historical capitalism. In these cases, gender meanings and norms have been appropriated from the larger culture and bent to capitalist purposes, with major consequences for both distribution and recognition. Likewise, what presents itself as "the cultural sphere" is deeply permeated by "the bottom line"—witness global mass entertainment, the art market, and transnational advertising, all fundamental to contemporary culture. Once again, the consequences are significant for both distribution and recognition. *Contra* substantive dualism, then, nominally economic matters usually affect not only the economic position but also the status and identities of social actors. Likewise, nominally cultural matters affect not only status but also economic position. In neither case, therefore, are we dealing with separate spheres.[25]

Practically, moreover, substantive dualism fails to challenge the current dissociation of cultural politics from social politics. On the contrary, it reinforces that dissociation. Casting the economy and the culture as impermeable, sharply bounded separate spheres, it assigns the politics of redistribution to the former and the politics of recognition to the latter. The result is effectively to constitute two separate political tasks requiring two separate political struggles. Decoupling cultural injustices from economic injustices, cultural struggles from social struggles, it reproduces the very dissociation we are seeking to overcome. Substantive dualism is not a solution to, but a symptom of, our problem. It reflects, but does not critically interrogate, the institutional differentiations of modern capitalism.

A genuinely critical perspective, in contrast, cannot take the appearance of separate spheres at face value. Rather, it must probe beneath appearances to reveal the hidden connections between distribution and recognition. It must make visible, and *criticizable*, both the cultural subtexts of nominally economic processes and the economic subtexts of nominally cultural practices. Treating *every* practice as simultaneously economic and cultural, albeit not necessarily in equal proportions, it must assess each of them from two different perspectives. It must assume both the standpoint of distribution and the standpoint of recognition, without reducing either one of these perspectives to the other.

Such an approach I call "perspectival dualism." Here redistribution and recognition do not correspond to two substantive societal domains, economy and culture. Rather, they constitute two analytical perspectives that can be assumed with respect to any domain. These perspectives can be deployed critically, moreover, against the ideological grain. One can use the recognition perspective to identify the cultural dimensions of what are usually viewed as redistributive economic policies. By focusing on the production and circulation of interpretations and norms in welfare programs, for example, one can assess the effects of institutionalized maldistribution on the identities and social status of single mothers.[26] Conversely, one can use the redistribution perspective to bring into focus the economic dimensions of what are usually viewed as issues of recognition. By focusing on the high "transaction costs" of living in the closet, for example, one can assess the effects of heterosexist misrecognition on the economic position of gays and lesbians.[27] With perspectival dualism, then, one can assess the justice of any social practice, regardless of where it is institutionally located, from either or both of two analytically distinct normative vantage points, asking: Does the practice in question work to ensure both the objective and intersubjective conditions of participatory parity? Or does it, rather, undermine them?

The advantages of this approach should be clear. Unlike economism and culturalism, perspectival dualism permits us to consider both distribution and recognition, without reducing either one of them to the other. Unlike substantive dualism, moreover, it does not reinforce their dissociation. Because it avoids dichotomizing economy and culture, it allows us to grasp their imbrication and the crossover effects of each. And because, finally, it avoids reducing classes to statuses or vice versa, it permits us to examine the causal interactions between those two orders of domination. Understood perspectivally, then, the distinction between redistribution and recognition does not simply reproduce the ideological dissociations of our time. Rather, it provides an indispensable conceptual tool for interrogating, working through, and eventually overcoming those dissociations.

Perspectival dualism offers another advantage as well. Of all the approaches considered here, it alone allows us to conceptualize some practical difficulties that can arise in the course of political struggles for redistribution and recognition. Conceiving the economic and the cultural as differentiated but interpenetrating social orders, perspectival dualism appreciates that neither claims for redistribution nor claims for recognition can be contained within a separate sphere. On the contrary, they impinge on one another in ways that may give rise to unintended effects.

Consider, first, that redistribution impinges on recognition. Virtually any claim for redistribution will have some recognition effects, whether intended or unintended. Proposals to redistribute income through social welfare, for example, have an irreducible expressive dimension,[28] they convey interpretations of the meaning and value of different activities, for example "childrearing" versus "wage-earning," while also constituting and ranking different subject positions, for example "welfare mothers" versus "tax payers" (Fraser, 1993). Thus, redistributive claims invariably affect the status and social identities of social actors. These effects must be thematized and scrutinized, lest one end up fueling misrecognition in the course of remedying maldistribution.

The classic example, once again, is "welfare." Means-tested benefits aimed specifically at the poor are the most directly redistributive form of social welfare. Yet such benefits tend to stigmatize recipients, casting them as deviants and scroungers and invidiously distinguishing them from "wage-earners" and "tax-payers" who "pay their own way." Welfare programs of this type "target" the poor—not only for material aid but also for public hostility. The end result is often to add the insult of misrecognition to the injury of deprivation. Redistributive policies have misrecognition effects when background patterns of cultural value skew the meaning of economic reforms, when, for example, a pervasive cultural devaluation of female caregiving inflects Aid to Families with Dependent Children as "getting something for nothing."[29] In this context, welfare reform cannot succeed unless it is joined with struggles for cultural change aimed at revaluing caregiving and the feminine associations that code it.[30] In short, no redistribution without recognition.

Consider, next, the converse dynamic, whereby recognition impinges on distribution. Virtually any claim for recognition will have some distributive effects, whether intended or unintended. Proposals to redress androcentric evaluative patterns, for example, have economic implications, which work sometimes to the detriment of the intended beneficiaries. For example, campaigns to suppress prostitution and pornography for the sake of enhancing women's status may have negative effects on the economic position of sex workers, while no-fault divorce reforms, which appeared to dovetail with feminist efforts to enhance women's status, may have had at least short-term negative effects on the economic position of some divorced women, although their extent has apparently been exaggerated and is currently in dispute (Weitzman, 1985). Thus, recognition claims can affect economic position, above and beyond their effects on status. These effects, too, must be scrutinized, lest one end up fueling maldistribution in the course of trying to remedy misrecognition. Recognition claims, moreover, are liable to the charge of being "merely symbolic."[31] When pursued in contexts marked by gross disparities in economic position, reforms aimed at recognizing distinctiveness tend to devolve into empty gestures; like the sort of recognition that would put women on a pedestal, they mock, rather than redress, serious harms. In such contexts, recognition reforms cannot succeed unless they are joined with struggles for redistribution. In short, no recognition without redistribution.

The need, in all cases, is to think integratively, as in the example of comparable worth. Here a claim to redistribute income between men and women is expressly integrated with a claim to change gender-coded patterns of cultural value. The underlying premise is that gender injustices of distribution and recognition are so complexly intertwined that neither can be redressed entirely independently of the other. Thus, efforts to reduce the gender wage gap cannot fully succeed if, remaining wholly "economic," they fail to challenge the gender meanings that code low-paying service occupations as "women's work," largely devoid of intelligence and skill. Likewise, efforts to revalue female-coded traits such as interpersonal sensitivity and nurturance cannot succeed if, remaining wholly "cultural," they fail to challenge the structural economic conditions that connect those traits with dependency and powerlessness. Only an approach that redresses the cultural devaluation of the "feminine" precisely *within* the economy (and elsewhere) can deliver serious redistribution and genuine recognition.

CONCLUSION

Let me conclude by recapitulating my overall argument. I have argued that to pose an either/or choice between the politics of redistribution and the politics of recognition is to posit a false antithesis. On the contrary, justice today requires both. Thus, I have argued for a comprehensive framework that encompasses both redistribution and recognition so as to challenge injustice on both fronts.

I then examined two sets of issues that arise once we contemplate devising such a framework. On the plane of moral theory, I argued for a single, two-dimensional conception of justice that encompasses both redistribution and recognition, without reducing either one of them to the other. And I proposed the notion of *parity of participation* as its normative core. On the plane of social theory, I argued for a perspectival dualism of redistribution and recognition. This approach alone, I contended, can accommodate both the differentiation of class from status in capitalist society and also their causal interaction. And it alone can alert us to potential practical tensions between claims for redistribution and claims for recognition.

Perspectival dualism in social theory complements participatory parity in moral theory. Taken together, these two notions constitute a portion of the conceptual resources one needs to begin answering what I take to be the key political question of our day: How can we develop a coherent programmatic perspective that integrates redistribution and recognition? How can we develop a framework that integrates what remains cogent and unsurpassable in the socialist vision with what is defensible and compelling in the apparently "postsocialist" vision of multiculturalism?

If we fail to ask this question, if we cling instead to false antitheses and misleading either/or dichotomies, we will miss the chance to envision social arrangements that can redress both economic and cultural injustices. Only by looking to integrative approaches that unite redistribution and recognition can we meet the requirements of justice for all.

NOTES

* Portions of this chapter are adapted and excerpted from my Tanner Lecture on Human Values, delivered at Stanford University, 30 April to 2 May, 1996. The text of the Lecture appears in *The Tanner Lectures on Human Values*, volume 9, ed. Grethe B. Peterson (The University of Utah Press, 1998: 1–67). I am grateful to the Tanner Foundation for Human Values for permission to adapt and reprint this material. I thank Elizabeth Anderson and Axel Honneth for their thoughtful responses to the Tanner Lecture, and Rainer Forst, Theodore Koditschek, Eli Zaretsky, and especially Erik Olin Wright for helpful comments on earlier drafts.

1 The following discussion revises a subsection of my essay "From redistribution to recognition?" (Fraser, 1995: 68–93), reprinted in Fraser (1997a).

2 It is true that pre-existing status distinctions, for example between lords and commoners, shaped the emergence of the capitalist system. Nevertheless, it was only the creation of a differentiated economic order with a relatively autonomous life of its own that gave rise to the distinction between capitalists and workers.

3 I am grateful to Erik Olin Wright (personal communication, 1997) for several of the formulations in this paragraph.

4 In capitalist society, the regulation of sexuality is relatively decoupled from the economic structure, which comprises an order of economic relations that is differentiated from kinship and oriented to the expansion of surplus value. In the current "post-Fordist" phase of capitalism, moreover, sexuality increasingly finds its locus in the relatively new, late-modern sphere of "personal life," where intimate relations that can no longer be identified with the family are lived as disconnected from the imperatives of production and reproduction. Today, accordingly, the heteronormative regulation of sexuality is increasingly

removed from, and not necessarily functional for, the capitalist economic order. As a result, the economic harms of heterosexism do not derive in any straightforward way from the economic structure. They are rooted, rather, in the heterosexist status order, which is increasingly out of phase with the economy. For a fuller argument, see Fraser (1997c). For the counterargument, see Butler (1997).

5 I am grateful to Rainer Forst for help in formulating this point.

6 On Axel Honneth's account, social esteem is among the "intersubjective conditions for undistorted identity formation," which morality is supposed to protect. It follows that everyone is morally entitled to social esteem. See Honneth (1995).

7 John Rawls, for example, at times conceives "primary goods" such as income and jobs as "social bases of self-respect," while also speaking of self-respect itself as an especially important primary good whose distribution is a matter of justice. Ronald Dworkin, likewise, defends the idea of "equality of resources" as the distributive expression of the "equal moral worth of persons." Amartya Sen, finally, considers both a "sense of self" and the capacity "to appear in public without shame" as relevant to the 'capability to function," hence as falling within the scope of an account of justice that enjoins the equal distribution of basic capabilities. See Rawls (1971: §67 and §82; 1993: 82, 181 and 318ff), Dworkin (1981), and Sen (1985).

8 The outstanding exception of a theorist who has sought to encompass issues of culture within a distributive framework is Will Kymlicka. Kymlicka proposes to treat access to an "intact cultural structure" as a primary good to be fairly distributed. This approach was tailored for multinational polities, such as the Canadian, as opposed to polyethnic polities, such as the United States. It becomes problematic, however, in cases where mobilized claimants for recognition do not divide neatly (or even not so neatly) into groups with distinct and relatively bounded cultures. It also has difficulty dealing with cases in which claims for recognition do not take the form of demands for (some level of) sovereignty but aim rather at parity of participation within a polity that is crosscut by multiple, intersecting lines of difference and inequality. For the argument that an intact cultural structure is a primary good, see Kymlicka (1989). For the distinction between multinational and polyethnic politics, see Kymlicka (1996).

9 See especially Honneth (1995).

10 To be sure, this could conceivably change. Nothing I have said rules out *a priori* that someone could successfully extend the distributive paradigm to encompass issues of culture. Nor that someone could successfully extend the recognition paradigm to encompass the structure of capitalism, although that seems more unlikely to me. In either case, it will be necessary to meet several essential requirements simultaneously: first, one must avoid hypostatizing culture and cultural differences; second, one must respect the need for non-sectarian, deontological moral justification under modern conditions of value pluralism; third, one must allow for the differentiated character of capitalist society, in which status and class can diverge; fourth, one must avoid overly Unitarian or Durkheimian views of cultural integration that posit a single pattern of cultural values that is shared by all and that pervades all institutions and social practices. Each of these issues is discussed in my contribution to Fraser and Honneth (2000).

11 Since I coined this phrase in 1995, the term "parity" has come to play a central role in feminist politics in France. There, it signifies the demand that women occupy a full 50 percent of seats in parliament and other representative bodies. "Parity" in France, accordingly, means strict numerical gender equality in political representation. For me, in contrast, "parity" means the condition of being a *peer*, of being on a *par* with others, of standing on an equal footing. I leave the question open exactly to what degree or level of equality is necessary to ensure such parity. In my formulation, moreover, the moral requirement is that members of society be ensured the *possibility* of parity, if and when they choose to participate in a given activity or interaction. There is no requirement that everyone actually participate in any such activity.

12 I say "*at least* two additional conditions must be satisfied" in order to allow for the possibility of more than two. I have in mind specifically a possible third class of obstacles to participatory parity that could be called "political," as opposed to economic or cultural. Such obstacles would include decision-making procedures that systematically marginalize some people even in the absence of maldistribution and misrecognition: for example, single-district winner-take-all electoral rules that deny voice to quasi-permanent minorities. (For an insightful account of this example, see Guinier (1994).) The possibility of a third class of "political" obstacles to participatory parity adds a further Weberian twist to my use of the class/status distinction. Weber's own distinction was tripartite not bipartite: "class, status, and

party." I do not develop it here, however. Here I confine myself to maldistribution and misrecognition, while leaving the analysis of "political" obstacles to participatory parity character for another occasion.

13 It is an open question how much economic inequality is consistent with parity of participation. Some such inequality is inevitable and unobjectionable. But there is a threshold at which resource disparities become so gross as to impede participatory parity. Where exactly that threshold lies is a matter for further investigation.

14 I say the remedy *could* be recognition of difference, not that it must be. Elsewhere I discuss alternative remedies for the sort of misrecognition that involves denying distinctiveness. See my contribution to Fraser and Honneth (2000).

15 In this brief essay, I lack the space to consider these questions of contemporary historical sociology. See, however, my contributions in Fraser and Honneth (2000).

16 As I use it, the distinction between economy and culture is social-theoretical, not ontological or metaphysical. Thus, I do not treat the economic as an extra-discursive realm of brute materiality any more than I treat the cultural as an immaterial realm of disembodied ideality. For a reading of my work that mistakes economy and culture for ontological categories, see Butler (1997). For a critique of this misinterpretation, see Fraser (1997c).

17 This is not to deny the prior existence of other, premodern understandings of "economy," such as Aristotle's.

18 To be sure, misrecognition harms are rooted in cultural patterns of interpretation and evaluation. But this does not mean, *contra* Judith Butler (1997), that they are "merely cultural." On the contrary, the norms, significations, and constructions of personhood that impede women, racialized peoples, and/or gays and lesbians from parity of participation in social life are materially instantiated—in institutions and social practices, in social action and embodied ethereal realm, they are material in their existence and effects. For a rejoinder to Butler, see Fraser (1997c).

19 For example, Marcel Mauss, *The Gift*, and Claude Levi-Strauss, *The Elementary Structures of Kinship*.

20 For an argument against the possibility of a fully marketized society, see Polanyi (1957).

21 It is conceivable that our hypothetical fully marketized society could contain formal institutional differentiations, including, for example, a legal system, a political system, and a family structure. But these differentiations would not be meaningful. *Ex hypothesi*, institutions and arenas that were extra-market *de jure* would be *de facto* market-governed.

22 By culturalism, I mean a monistic social theory that holds that political economy is reducible to culture and that class is reducible to status. As I read him, Axel Honneth subscribes to such a theory. See Honneth (1995).

23 By economism, I mean a monistic social theory that holds that culture is reducible to political economy and that status is reducible to class. Karl Marx is often (mis)read as subscribing to such a theory.

24 In what follows, I leave aside a third possibility, which I call "deconstructive anti-dualism." Rejecting the economy/culture distinction as "dichotomizing," this approach seeks to deconstruct it altogether. The claim is that culture and economy are so deeply interconnected that it doesn't make sense to distinguish them. A related claim is that contemporary capitalist society is so monolithically systematic that a struggle against one aspect of it necessarily threatens the whole; hence, it is illegitimate, unnecessary, and counterproductive to distinguish maldistribution from misrecognition. In my view, deconstructive anti-dualism is deeply misguided. For one thing, simply to stipulate that all injustices, and all claims to remedy them, are simultaneously economic and cultural evacuates the actually existing divergence of status from class. For another, treating capitalism as a monolithic system of perfectly interlocking oppressions evacuates its actual complexity and differentiation. For two rather different versions of deconstructive anti-dualism, see Young (1997) and Butler (1997). For detailed rebuttals, see Fraser (1997b, 1997c).

25 For more detailed criticism of an influential example of substantive dualism, see "What's critical about critical theory? The case of Habermas and gender," in Fraser (1989).

26 See "Women, welfare, and the politics of need interpretation" and "Struggle over needs," both in Fraser (1989); also, Fraser and Gordon (1994), reprinted in Fraser (1997a).

27 Jeffrey Escoffier has discussed these issues insightfully in "The political economy of the closet: toward an economic history of gay and lesbian life before Stonewall", in Escoffier (1998: 65–78).

28 This formulation was suggested to me by Elizabeth Anderson in her comments on my Tanner Lecture, presented at Stanford University, 30 April to 2 May, 1996.

29 Aid to Families with Dependent Children (AFDC) is the major means-tested welfare programme in the United States. Claimed overwhelmingly by solo-mother families living below the poverty line, AFDC became a lightning

rod for racist and sexist anti-welfare sentiments in the 1990s. In 1997, it was "reformed" in such a way as to eliminate the federal entitlement that had guaranteed (some, inadequate) income support to the poor.

30 This formulation, too, was suggested to me by Elizabeth Anderson's comments on my Tanner Lecture, presented at Stanford University, 30 April to 2 May, 1996.

31 I am grateful to Steven Lukes for insisting on this point in conversation.

REFERENCES

Butler, J. (1997) "Merely cultural," *Social Text*, 52/53: 265–77.

Dworkin, R. (1981) "What is equality? Part 2: Equality of resources", *Philosophy and Public Affairs*, 10 (4): 283–345.

Escoffier, J. (1998) *American Homo: Community and Perversity*, Berkeley: University of California Press.

Fraser, N. (1989) *Unruly Practices: Power, Discourse and Gender in Contemporary Social Theory*, Minneapolis: University of Minnesota Press.

—— (1993) "Clintonism, welfare and the antisocial wage: the emergence of a neoliberal political imaginary," *Rethinking Marxism*, 6 (1): 9–23.

—— (1995) "From redistribution to recognition? Dilemmas of justice in a 'postsocialist' age," *New Left Review*, 212: 68–93.

—— (1997a) *Justice Interruptus: Critical Reflections on the "Postsocialist" Condition*, London: Routledge.

—— (1997b) "A rejoinder to Iris Young," *New Left Review*, 223: 126–9.

—— (1997c) "Heterosexism, misrecognition and capitalism: a response to Judith Butler," *Social Text*, 52/53: 278–89.

Fraser, N. and Gordon, L. (1994) "A genealogy of 'dependency': tracing a keyword of the US welfare state," *Signs*, 19 (2): 309–36.

Fraser, N. and Honneth, A. (2000) *Redistribution or Recognition? A Political-Philosophical Exchange*, London: Verso.

Gitlin, T. (1995) *The Twilight of Common Dreams: Why America Is Wracked by Culture Wars*, New York: Metropolitan Books.

Guinier, L. (1994) *The Tyranny of the Majority*, New York: The Free Press.

Honneth, A. (1995) *The Struggle for Recognition: The Moral Grammar of Social Conflicts*, trans J. Anderson, Cambridge: Polity.

Kymlicka, W. (1989) *Liberalism, Community and Culture*, Oxford: Oxford University Press.

—— (1996) "Three forms of group-differentiated citizenship in Canada," in S. Benhabib (ed.), *Democracy and Difference*, Princeton, NJ: Princeton University Press.

Polanyi, K. (1957) *The Great Transformation*, Boston: Beacon.

Rawls, J. (1971) *A Theory of Justice*, Cambridge, MA: Harvard University Press.

—— (1993) *Political Liberalism*, New York: Columbia University Press.

Rorty, R. (1998) *Achieving Our Country: Leftist Thought in Twentieth-Century America*, Cambridge, MA: Harvard University Press.

Sen, A. (1985) *Commodities and Capabilities*, Amsterdam: North-Holland.

Sennett, R. and Cobb, J. (1973) *The Hidden Injuries of Class*, New York: Knopf.

Taylor, C. (1994) "The politics of recognition," in A. Gutmann (ed.), *Multiculturalism: Examining the Politics of Recognition*, Princeton, NJ: Princeton University Press.

Thompson, E. P. (1963) *The Making of the English Working Class*, New York: Random House.

Weitzman, L. (1985) *The Divorce Resolution: The Unexpected Social Consequences for Women and Children in America*, New York: The Free Press.

Young, I. M. (1997) "Unruly categories: a critique of Nancy Fraser's dual systems theory," *New Left Review*, 222: 147–60.

PART 2

Making justice spatial

Introduction

As the pieces by Young (Chapter 4) and Fraser (Chapter 5) make clear, human beings are not isolated monads but like all entities (a freighted word, we understand) consist of relations with other entities (human, non-human, or both), and questions of injustice, oppression, and ethics ramify throughout those relations and the networks that support them. Young's piece particularly offers a critical discussion of the notion of the individual rights-bearing citizen, and posits the relational nature of identity. To take this argument further, we also propose that moral knowledge is inextricable from embodied and geographically specific experience. To know something is wrong, especially in an ongoing structural sense, can be a matter of accruing experience and insight, and also a matter of communal, discursive intelligence, not necessarily abstracted reason. The remaining pieces in this section investigate the ways in which this is so.

David Smith (an Emeritus Professor of Geography at Queen Mary College of the University of London) elaborates upon this theme in his paper (Chapter 6). A key notion for Smith in this piece, as in much of his other work on geography, social justice, and ethics (see, e.g., 1973, 1977, 1994), is the notion of moral arbitrariness, here termed good fortune, and its implications for the allocation of life's goods and bads. Smith adds a much-needed dimension to the importance of place and its multiple connections to morality, ethics, and justice, and also links back to an important element in Young's paper (Chapter 4). In highlighting the arbitrariness of place (as both status and geographic location) of birth, Smith recalls Young's argument about the assigned nature of group affiliation (and Heidegger's notion of "thrownness"). Here Smith establishes for readers the idea that inequalities are primarily morally arbitrary. This means that by and large the production, distribution, and consumption of life's benefits and burdens, including personal physical and mental attributes, are arbitrary with respect to the moral worth of the people involved. We can neither take credit nor be blamed for the circumstances (places and positions) into which we are born, uneven as these circumstances may be. As Smith then argues, because people are neither morally worthy nor unworthy of life's initial endowments, it is not equality that must be defended as both idea and practice, it is, rather, inequality.

Several fundamental issues are raised by this formulation by Smith of justice as equalization. First, and one which Smith deals with early on in the piece, is the pragmatic difference between equality and equalization; the former is clearly seen as the goal, impractical as it might be, the second as a process and a measure of progress (i.e. moving toward equality rather than away from it is "possible and morally justifiable" for Smith).

BOX 1: THE MANY MEANINGS OF LANDSCAPE: CONNECTING PLACE AND JUSTICE

In his important 1996 article, "Recovering the Substantive Nature of Landscape," Kenneth Olwig (presently a professor in the Department of Landscape Planning at the Swedish University of Agricultural Sciences), provides a useful starting point for this discussion with a detailed history and rehabilitation of the long-standing geographic concept of landscape. Through both an etymological excavation, and a philosophical recovery project, Olwig wants to re-establish a set of linkages between place and identity, including the links (non-deterministic to be sure) between the particularities embedded in the struggles over such geographic phenomena as landscapes, communities, and territories and the specific knowledges built through those struggles around such concepts as morality, ethics, justice, nature, and environmental equity.

Olwig's paper, which is also instructive on the contested nature of geographic thought itself as it tracks the waxing and waning importance of the landscape concept in the discipline, points to the important nexus between people and the places they inhabit. Through Olwig's analysis of landscape (in its various interpretations), we come to see how places and people are mutually constituted. Where one is in the world plays a crucial role in how and what one is in the world. The kinds of relationships and networks discussed by Fraser (new social movements) and Young (social groups) actually exist in particular times and places, and these material dimensions of their existence, as Olwig argues, matter a great deal to their make-up and to the acculturating effects they exert on their inhabitants. Knowing one's place means more than a knowledge of the physical landscape, and clearly implies knowing one's position in social, cultural, political, and power landscapes as well. Such knowledges are built in particular times and places through the performance of everyday practices in interaction with others, and such practices vary over time and among and between places. As Olwig's paper makes clear, it is through such grounded practices that we develop significant aspects of our notions of justice, ethics, and morality, for good or ill.

A second issue is more complicated: What is to be equalized? Smith discusses several possibilities: opportunity, primary goods, resources, capabilities, welfare outcomes? However this issue is decided, the choices Smith offers here reflect an understanding of justice as (equitable, or as equitable as possible) distribution. (He develops more elaborated notions of justice in his later works; see e.g. Smith, 2000; Lee and Smith, 2004.) This then leads Smith to a discussion of what it is that people require for life, and his search for commonalities across humankind. This formulation also seems to entail a hierarchy of needs (with basic subsistence needs to be met first) that flattens out both interpersonal, or intergroup, differences as well as the kinds of structured oppressions defined by Young. Smith's call for equalization in distribution (of whatever finally constitutes measurable dimensions of justice) leaves aside the calls by Young (to overcome such structured oppressions as cultural imperialism) or Fraser (for recognition). Unless, as suggested earlier, such matters as "parity of participation" and control over the decision processes that structure material distributions are included in the notion of distribution itself, Smith's formulation leaves these matters unattended.

In the next part of his paper, Smith is concerned with the critical question of what makes "good geography," and in light of his conceptions of "the place of good fortune" and justice as equalization, it is not surprising that Smith defines research that helps to illuminate the former and to further the process of the latter as "good" geography. What kinds of work does this entail? One illustrative example would point us to a fleshing out of the kind of model provided by Olwig. As Smith argues, "the justice and morality people actually practise, and the theories that ethicists devise, are embedded within specific sets of social and physical relationships manifest in geographical space, reflecting the particularity of place as well as time" (see pp. 102–3). This suggests, following Olwig's general assessment of the morally important dimension of landscape (understood in its broad sense), that comparative assessments that illuminate the variety of conceptions of justice, ethics and morality, as practiced, might open new imaginaries and approaches to such issues. Smith also suggests that much empirical work remains to be done "on spatial inequality and injustice in particular contexts." Without second-guessing precisely what Smith is suggesting in these two examples, it is important to keep Harvey's caution in mind to keep such work from slipping into either *status quo* or counter-revolutionary modes (an indeterminate cultural relativism in the first case, or a mere documenting of human injustice to humans in the second).

Smith's final sets of considerations address one additional dimension of moral progress in geography: professional ethics and the matter of "good" geographers. Here we simply emphasize Smith's point that a concern with progressive scholarship obligates scholars (including geographers) to think carefully about the products and processes of their work. Smith's concern with the ethical treatment of research "subjects" is just one of these considerations. We take these matters up in detail in the next section.

The final paper in this section by Sarah Whatmore (Chapter 7) (currently Professor of Environment and Public Policy at Oxford University) opens up key questions about identity, the ethical subject and the spatial relations of ethics, morality, and justice. Whatmore develops three key arguments in the paper: (1) in a discussion that draws upon recent work in feminist scholarship, and that resonates with major elements of the other papers in this section, Whatmore reframes and deepens our understanding of people as always constructed in relation to others, rather than as autonomous; (2) drawing upon work in environmental ethics, she makes an argument for extending moral considerability beyond the human; and (3) she considers the spatial implications for justice considered in these relational ways. The thread that ties the three arguments together is the possibility of expanding the notion of ethical community, and for extending the purview of justice.

Whatmore's first argument seeks to problematize the notion of the autonomous, independent self. The adherence to individualized notions of self and other, inherited from early Enlightenment thinkers (e.g. Locke), erects boundaries that prohibit the extension of moral considerability. Such formulations, Whatmore argues, contribute directly to a geography of proximity and homogeneity (i.e. a bounded "us" at a variety of scales: individual, neighborhood, nation) where care and justice obtain (though often quite unevenly), and a sharply demarcated, heterogeneous outside, where such considerations apply much less consistently, if at all. If the boundaries between self and other can be seen as always artificial, blurry, and dialectical, then one's obligations and sense of caring and justice might more readily be extended to an always-already co-present "we."

The second argument, drawing directly on evolving scholarship in environmental ethics, makes an analagous case for extending the boundary of moral considerability to the other entities (both human and non-human) with which we share the planet. By drawing out parallels based on the interconnectedness of human beings in relation to networks of other living and non-living "things," Whatmore's argument attempts to avoid the dilemmas that environmental ethicists have faced when trying to extend human-centered moral considerability to non-humans and the environment. These difficulties arise precisely from the same foundational concepts of autonomy and separateness that are typically encountered in extending care (morality, ethics, justice) to human "others." Whatmore argues that these problems are eased considerably when we are no longer able to postulate our identities (our subjecthood or self) in isolation from the networks in which we are always embedded. Just as we are always relationally defined by interaction with other people, we are materially (corporeally) immersed in the complicated relationships that comprise the biosphere. Understanding the always-already nature of these relationships both constrains us from thinking ourselves separately and enables the extension of moral considerability to other organisms and inanimate "nature."

Whatmore's final set of comments seeks to make all of this explicitly spatial. She is interested, here, in thinking through the possibilities of extending what feminist scholars (e.g. Gilligan, 1982) have called an "ethic of care." If notions of autonomy and independence allow for moral considerability largely on the basis of proximity (whether materially or affectively), Whatmore argues that notions of the self dialectically and continuously in relation to "others" should help us rethink our concepts of "nearness" and "distance." What she is articulating in these passages, we would suggest, is an alternative topology of justice. In this topology, we need an alternative metric. Linear distance can no longer be the measure of near or far "others" who always co-habit the networks we share. Perhaps such metrics as density and intensity of connections, necessity and frequency of reciprocity, or stability or ephemerality of affinities, could serve as the new measures that would allow us to map these alternative imaginaries of geographic justice and morality.

The readings in this section help to open up some of the critical theoretical and conceptual issues that are entailed in a commitment to progressive (or, in Harvey's terms, revolutionary) geographic scholarship. The pieces also point to (explicitly in the case of Smith and Whatmore) a variety of pressing issues concerning the practice of progressive geography. We turn to these issues in the next section.

6

Moral progress in human geography

Transcending the place of good fortune

David M. Smith

from *Progress in Human Geography*, 2000, 24(1):1–18

INTRODUCTION

> One part of mankind appears to have become captive of its own achievements in technology, economic growth, and the creation of an affluent materialistic society in which interpersonal relationships and some of the more simple intangible pleasures of life are becoming increasingly lost. Another part is still captive of the ills of an earlier age – poverty, ignorance, disease, economic exploitation, racial discrimination, and so on. We all have a personal interest in the process of liberation, for we are ourselves among the captives. As geographers we have a special role – a truly creative and revolutionary one – that of helping to reveal the spatial malfunctionings and injustices, and contributing to the design of a spatial order of society in which people can be really free to fulfil themselves in a secure social setting where the rights of all are respected. This, surely, would be 'progress in geography'.
>
> (Smith, 1973: 121).

A quarter of a century ago I made these observations on progress in geography, in concluding an outline of what was to become the welfare approach (Smith, 1977). Living in South Africa, I was arguing for engagement with contemporary moral problems, like apartheid, which attracted little attention among the majority of the profession. Apartheid was such an obvious target as to invite moral certainty, while the intellectual environment of the times encouraged faith in reason as a source of human betterment. Both these positions rest uneasily with the prevailing sentiments of this supposedly postmodern age, with its suspicion of truth claims and conceptions of progress associated with modernity. Yet, as I read these words again, I find nothing with which to quarrel. Some states of affairs are bad, and should be struggled against and changed. Such was apartheid in South Africa. Such is ethnic cleansing in former Yugoslavia. Such is mass starvation in central Africa. These are as close to moral truths as can be imagined, and those who deny them are wrong.

What I am prepared to concede now is that the notion of progress has become deeply problematic. I was writing at a time when faith in managerial rationality harnessed to the advance of technology supported a conception of human progress as almost linear inevitability, interrupted only by such occasional blips as localized urban insurrections and distant wars. This was potently evoked by the stages of economic growth theory of Walt Rostow (1960), with the take-off from traditional society to the age of high mass consumption depicted with the reliability of jet propulsion. Our understanding of progress is now more circumspect, tarnished as it is by the experience of abiding poverty, an environmental crisis, the demise of socialism, the instability of capitalism and repeated reminders of rampant human cruelty.

Nevertheless, it is a mistake to dismiss any notion of progress as Enlightenment error. While the affluent endure postmodern ambiguity and uncertainty in comfort, for those at the coal-face of human misery what constitutes progress is still likely to be self-evident. Indeed, in such contexts the very term 'progressive', as both adjective and noun, implies not only a moral stand but also a political commitment, as was the case with opposition to apartheid in South Africa, for example. To be (a) progressive means taking the side of the oppressed, the poor, the worst-off.

This article explores the question of what comprises moral progress in human geography. It makes explicit the understanding that progress in this or any other field of human endeavour is a normative issue. Progress in human geography will be examined in three senses: (1) geography as the world of human creation and experience; (2) geography as the intellectual project of attempting to comprehend and change this world; and (3) geography as professional practices and institutions.

Before proceeding, to situate very briefly what follows within my personal biography may make some of it less of a surprise. I have spent three decades trying to understand issues of human welfare and social justice in a geographical context. This has involved the ongoing interplay of theory and practice: working from the abstractions of social theory to field research on apartheid, for example, and back again to theory. It has now brought me to a new disciplinary interface: of geography with ethics (Smith, 2000), as I seek philosophical grounding for the continuing engagement with injustice. Hence my point of departure and underlying theme: an argument excavated from liberalism but with radical implications, referred to in an earlier publication as the place of good fortune (Smith, 1997a: 26). This is an argument for equality, and I can think of nothing more progressive.

THE PLACE OF GOOD FORTUNE

This expression incorporates three meanings of 'place': the role or part played by good fortune in people's lives, position in some social structure and place in its geographical sense. Each has an important bearing on human well-being. The crucial fact is that chance or luck are important elements in life. The crucial question to be explored is its moral significance.

That interest in this issue can be traced back to the ancient Greeks is explained by Williams (1985: 5): 'Impressed by the power of fortune to wreck what looked like the best-shaped lives, some of them, Socrates one of the first, sought a rational design of life which would reduce the power of fortune and would be to the greatest possible extent luck-free.' In those hazardous times, it was recognized that achievement of the good life might not be entirely a matter of individual volition. Williams (1985: 195) points out that most personal advantages and admired characteristics are distributed in ways which cannot be regarded as just, and that some people are simply luckier than others; morality is a value that transcends luck, and which has played a part in mobilizing power and social opportunity to compensate for misfortune.

The role of luck re-emerged in recent times in arguments about desert, central to the liberal egalitarian perspective on social justice initiated by John Rawls. He began with the conventional system of 'natural liberties' in which careers are open to the talented, with all persons having equal opportunities in the formal sense of the same legal rights. However, there is no attempt to promote equality in background social conditions; far from it:

> [T]he initial distribution of assets for any period of time is strongly influenced by natural and social contingencies. The existing distribution of income and wealth, say, is the cumulative effect of prior distributions of natural assets – that is, natural talents and abilities – as these have been developed or left unrealized, and their use favored or disfavored over time by social circumstances and such chance contingencies as accident and good fortune.
>
> (Rawls, 1971: 72)

The obvious injustice of such a system is that it permits access to positions of advantage and distributive shares to be influenced by factors so arbitrary from a moral point of view. He therefore invokes the principle of 'fair equality of opportunity', under which persons with the same talent and ability and the same willingness to use them should have the same prospects regardless of their initial place in the social system. However, this conception also appears defective:

> [E]ven if it works to perfection in eliminating the influence of social contingencies, it still permits the distribution of wealth and income to be determined by the natural distribution of abilities and talents . . . distributive shares are decided by the outcome of the natural lottery; and this outcome is arbitrary from a moral perspective. There is no more reason to permit the distribution of income and wealth to be settled by the distribution of natural assets than by historical and social fortune.
>
> (Rawls, 1971: 73–74)

Erasing the distinction between what may broadly be regarded as environmental effects and natural attributes achieves 'democratic equality', which strongly suggests equality of outcomes. Rawls has, in effect, made all sources of differential occupational achievement morally arbitrary. There is no case at the most basic level of justification for anything except equality in the distribution of Rawls's primary goods of liberty and opportunity, income and wealth, and the bases of self-respect. As Sandel (1982: 92–93) concludes: 'No one can be said to deserve anything (in the strong, pre-institutional sense), because no one can be said to possess anything (in the strong, constitutive sense).'

This argument from arbitrariness features prominently in subsequent work on social justice. For example, Miller (1992: 228, 240–41) points to the social determination of the effect of a difference in raw talent, to inheritance transmitting unequal competitive resources along family lines, and to inequalities guaranteed by the organization of education and production. He concludes: 'disadvantages in resources for social advancement are associated with generally inferior economic situations. It is as if the gamblers with the least funds were also dealt the fewest cards' (Miller, 1992: 255). He also argues that the unchosen risks of market competition stand in need of justification (Miller, 1992: 274), a point which has important implications for the morality (or otherwise) of capitalism.

Place in its geographical sense is readily added to the argument from arbitrariness. This is illustrated by Baker:

> So much of what people achieve is a matter of being in the right place at the right time, of having good luck in family, teachers, friends, and circumstances, that no one is in a strong position to take much credit for the way their lives turn out. There is no such thing as a literally self-made man [*sic*]. And so any judgement of desert will have to look closely at where responsibility really lies.
>
> (Baker, 1987: 60; see also Barry, 1989: 226)

Jones (1994: 167) points out that 'the distribution of resources across the world is entirely fortuitous and that it is morally unacceptable that people's lot in life should be determined by this accidental feature'. Barry (1989: 239) postulates Crusoe and Friday on two different islands, working equally hard and skilfully but with differences in production due to one island being fertile and the other barren, asserting that 'if anything can be called morally arbitrary – not reflecting any credit or discredit on the people concerned – it is this difference in the bounty of nature'.

The distribution of resources includes those created by humankind, like the local infrastructure, as well as those of the natural environment. It does not take a geographer to recognize the inequity of unequal access to facilities such as good schools (e.g. Barry, 1989: 220, 221) and of fiscal disparities between local governments (e.g. LeGrand, 1991: 108, 128). This is all part of the undeserved inheritance. As Miller explains:

> No one earns the right to be born to a family living in a spacious house in Armonk, New York, rather than on a straw mat in the slums of Calcutta. Yet the enormous differences at these starts include enormous differences in life prospects, given the same innate capacities and the same willingness to try.
>
> (Miller, 1992: 298)

The chance of birth in a particular place on the highly uneven surface of resources carries no greater moral credit than being born to a rich or poor family, male or female, black or white. And such initial advantage as arises from the place of good fortune is readily transferred to future generations, similarly devoid of moral justification.

As for the possibility of the disadvantaged seeking better opportunities elsewhere, for most people the capacity to change their place, from a poorly endowed to a richly resourced location (or state), may be as limited as it is to change their gender or skin pigmentation. Free movement is still 'the civil right we are not ready for' (Nett, 1971). Yet, in so far as rights of access to unevenly distributed resources are constrained by the boundaries of nation-states, as accidents of history, then this source of inequality might be considered morally irrelevant (Jones, 1994: 160). Indeed, the restrictive citizenship of Western liberal democracies has been described as the modern equivalent of feudal birthright privileges, and similarly hard to justify (Carens, 1987: 252).

The argument from arbitrariness, as outlined here, has attracted vigorous opposition (see, in particular, Anderson, 1999). Roemer (1996: 173) posits: 'Although we may agree that family background, natural talents, and inherited wealth are all morally arbitrary, perhaps there is such a thing as freely chosen effort.' There is a reluctance on the part of critics to concede to natural, social or chance circumstances everything about the individual, including responsibility for chosen life plans. If this worries some liberals comfortable with the notion of individual autonomy, it is anathema to communitarians with thicker conceptions of human identity. As Walzer (1983: 261) explains, if the effort they expend, like all their other capacities, is only the arbitrary gift of nature or nurture, 'while its purpose is to leave us with persons of equal entitlement, it is hard to see that it leaves us with persons at all'.

Those unwilling to assign everything about persons to morally arbitrary fortune face the challenge of how to draw the line among attributes. Dworkin (1981a, 1981b) argued that justice requires compensating individuals only for adverse aspects of their condition or situation for which they are not responsible, which excludes inclination to effort; outcomes should therefore be 'ambition-sensitive' but not 'endowment-sensitive', which leads him to equalize resources and not welfare. Further attempts to resolve the limits of individual responsibility include proposals for equal access to advantage (Cohen, 1989), equal opportunities for welfare (Arneson, 1989) and equalizing human capabilities (Sen, 1992). However, Roemer (1996, 1998) reveals conceptual and technical difficulties in sustaining particular cuts between circumstances for which persons cannot be held responsible and those for which they can.

Another line of critique is that initiated by Nozick (1974). He argued that persons have the moral right to use such natural endowments as intelligence and skill to their advantage, providing that this does no harm to others: the thesis of 'self-ownership'. Similarly, persons are entitled to hold and benefit from natural resources, provided that they acquired them justly by initial acquisition or by transfer (i.e. gift, inheritance or purchase). His criterion for the justice of initial acquisition is that no other persons are thereby made worse off (Nozick, 1974: 178), a modification of the proviso of John Locke that an individual is entitled to appropriate natural resources providing that there is as much and as good left in common for others.

Many objections have been raised to Nozick's entitlement theory (Smith, 1994a: 69–71; Roemer, 1996: 208–10). These include the difficulty of demonstrating that no one is worse off as a result of particular private ownerships of natural resources, and of tracing acquisition back through a series of transfers which, if unjust (e.g. involving deception, robbery or coercive acquisition) should be rectified. All this leaves Barry (1989: 218) to remark: 'From Locke to Nozick there is a long and disreputable tradition of using a fairy story about the way in which acquisition might have occurred as the basis for a defense of the status quo.'

An important issue arising from Nozick's similar treatment of natural endowments and acquired holdings is whether they may be different in some sense relevant to the place of good fortune. Reiman (1990: 173–75) proposes that the ownership of the external world is different from ownership of one's body, for the former can deprive others whereas the latter cannot. Another difference is that people cannot change their entire body, but may be able to change their place on the earth's surface. However, O'Neill (1991: 290) notes that the libertarian devotion to freedom does not extend to dismantling immigration laws: 'their stress on property rights entails an attrition of public space that eats into the freedom of movement and rights of abode of the unpropertied.' Any such system of exclusive ownership, which involves the differential power that some individuals have to compel others to work for them, is 'effectively a system of forced labor' (Reiman, 1990: 177). This can lead to an argument for collective ownership of the external world of natural resources (e.g. Cohen, 1986a, 1986b), whereas it is harder to envisage collective ownership of individual natural endowments.

Nevertheless, there is a strong supposition that groups of people are entitled to monopolize the resources of the territory which they occupy. This is encouraged by the modern concepts of national citizenship and sovereignty: The nation provides its members with an inalienable collective property: the land in which they have the right to live their lives' (Poole, 1991: 96). This is true in a formal, legal sense, but begs the questions of the morality of national boundaries and their closure to outsiders. Furthermore, the case of the territorially defined group is different from that of the individual, in an important respect explained by Sandel:

> [F]or the community as a whole to deserve the natural assets in its province and the benefits that flow from them, it is necessary to assume that society has some pre-institutional status that individuals lack, for only in this way could the community be said to possess its assets in the strong, constitutive sense of possession necessary to a desert base.
>
> (Sandel, 1982: 101)

And without this, to rephrase Sandel (1982: 92–93) in the individual context, no group anywhere deserves anything. A community, or nation, might claim a right to land in which they have mixed their labour, and even their blood, and to the advantages to be derived therefrom. A similar argument might be applied to the physical infrastructure, built to give future generations as well as present people a better life. But this would still leave unanswered the possible injustice of initial acquisition, and the moral arbitrariness of the good fortune of inheriting favourable conditions for sustaining a good life.

TOWARDS TERRITORIAL SOCIAL JUSTICE

The argument now proceeds to the first sense of moral progress in geography: that of the world of human creation and experience. It follows from recognition of the place of good fortune that there is a strong case for equality, by persons and territorially defined population aggregates, or at least for narrowing the gaps which have arisen from morally dubious if not arbitrary factors. The familiar and crucial practical question is: equality of what? Should it be opportunities (after liberal convention), primary goods (after Rawls), resources (after Dworkin), capabilities (after Sen) or welfare outcomes? And whichever is chosen, how is it to be defined and measured? These questions are complicated by the fact that the individual freedom to choose life plans so revered by liberals means that everyone might require a unique bundle of goods. Added to this is the postmodern respect for difference, which similarly works against some common conception of the good and of what is required to attain it.

Two arguments may be advanced to facilitate an approach to practice. One is to talk in terms not of equality but of equalization (Smith, 1994a: chap. 5). This strategy recognizes that achieving equality is virtually impossible, by any criteria, but that moves in this direction are both possible and morally justifiable. The process of equalization might be constrained by Rawls's 'difference principle', which requires that social and economic inequalities are to be arranged so that they are 'to the greatest benefit of the least advantaged' (Rawls, 1971: 302). Even if the place of good fortune is taken to undermine the moral credit for most if not all individual achievement, this principle is a defensible concession to the possibility that some inequalities can work to the advantage of everyone and especially the worst-off.

The practical pursuit of social justice as equalization requires the second argument, relating to the objects of distribution. This is that what people actually require for life is much the same, whoever and wherever they are, because they are themselves naturally much the same. Any suggestion these days that there may be such a thing as human nature attracts suspicion of essentialism. 'Any definition of human nature is dangerous because it threatens to devalue or exclude some acceptable individual desires, cultural characteristics, or ways of life', according to Young (1990: 36). However, there are increasing indications of dissatisfaction with this position, and its risk of relativism, for deciding what may be acceptable requires standards capable of transcending the here and now of specific individual, group or local practices.

Eagleton (1996) exemplifies the critique. He approves of postmodernism in challenging various kinds of oppression, but is critical of a form of reductionism which undervalues what persons have in common as natural, material creatures and overestimates the significance of cultural difference. 'Differences cannot fully flourish while men and women languish under forms of exploitation; and to combat those forms effectively implicates ideas of humanity which are necessarily universal' (Eagleton, 1996: 121–22). Similar positions are argued by others. For example, to the assertion by Rorty (1989) that the common traits of human beings are not substantial enough to constitute a useful notion, Geras responds:

> [T]hey are susceptible to pain and humiliation [Rorty's minimal concession], have the capacity for language and (in a large sense) poetry, have a sexual instinct, a sense of identity, integral beliefs – and then some other things too, like needs for nourishment and sleep, a capacity for laughter and for play, powers of reasoning and invention that are, by comparison with other terrestrial species, truly formidable.
>
> (Geras, 1995: 66)

These are not only natural facts, but also of moral significance.

While the sympathies of much contemporary human geography seem postmodernist, there are those who dissent. For example, Tuan (1986) recognizes that the meaning of the good life varies greatly among cultures, but claims that we do share some things. For Sack:

> [T]he encouragement of different and diverse viewpoints should not obscure the fact that human beings have much in common. We live in a concrete material environment and we share basic biological, social, intellectual, and perhaps even spiritual capacities; we also share the capacity to reason. Losing sight of this basic reality comes from too great an emphasis on difference and diversity.
>
> (Sack, 1997: 4)

He is unhappy about moves which 'deny the existence of anything essential and foundational that can lead to shared positions'. Harvey (1996: 360) emphasizes the importance of human similarity rather than difference, in alliance formation between seemingly disparate groups 'within an ethics of political solidarity built across different places'. Thus, diverse voices challenge the contemporary preoccupation with difference, and seek a universal perspective without abandoning the insights gained from recognition of the particularity of persons and places.

Having established the foundation of human similarity, the next step is to consider human needs. The notion of need implies some authority external to the individual, as opposed to a subjective personal want or desire. Particular needs are sometimes referred to as basic, to stress their urgency and thereby give them special moral force. However, attempts to define universal needs reveal differences. For example, Kekes identifies what he describes as context-independent requirements for human welfare, set by universal, historically constant and culturally invariant needs created by human nature, as follows:

> Many of these needs are physiological: for food, shelter, rest, and so forth; other needs are psychological: for companionship, hope, the absence of horror and terror in one's life, and the like; yet other needs are social: for some order and predictability in one's society, for security, for some respect, and so on.
>
> (Kekes, 1994: 49)

Compare this with the more restrictive view of O'Neill:

> It is not controversial that human beings need adequate food, shelter and clothing appropriate to their climate, clean water and sanitation, and some parental and health care. When these basic needs are not met they become ill and often die prematurely. It is controversial whether human beings need companionship, education, politics and culture, or food for the spirit – for at least some long and not evidently stunted lives have been lived without these goods.
>
> (O'Neill, 1991: 279)

Doyal and Gough (1991: 37) are closer to Kekes than O'Neill in asserting that our mammalian constitution shapes needs for such things as the food and warmth required to survive and maintain health, and that our cognitive attitudes and experience of childhood shape needs for supportive and close relationships. Their hostility to relativism is expressed in the notion that all people share one obvious need: to avoid serious harm. This goes beyond failure to survive in a physical sense, to include impaired participation in the prevailing social milieu. From this follow two basic needs (in their terms): for the physical health to continue living and functioning effectively, and for the personal autonomy or ability to make informed choices about what to do and how to do it in a given societal context. The actual need satisfiers, in the form of goods and services, may be culturally specific, as opposed to the universality of the basic needs themselves. This is similar to the approach adopted by Sen (1992) to poverty, which is absolute or universal in the sense of impairing people's capability to function, but relative with respect to the commodities required to alleviate it. O'Neill (1996: 191) now seems to accept that there is more to human life than mere physical survival or even longevity, a position endorsed by others who claim to derive sets of needs from human nature or the requirements for human flourishing (e.g. Brown, 1986: 159; Griffin, 1986: 86–87; Nussbaum, 1992: 222).

The human needs perspective, in theory and in development policy and practice (e.g. Friedmann, 1992; Corbridge, 1993), further strengthens the argument from essentialism. For example:

> We cannot jettison essentialism because we need to know among other things which needs are essential to humanity and which are not. Needs which are essential to our survival and well-being, such as being fed, keeping warm, enjoying the company of others and a degree of physical integrity, can then become political criteria: any social order which

> denies such needs can be challenged on the grounds that it is denying our humanity, which is usually a stronger argument against it than the case that it is flouting our contingent cultural conventions.
>
> (Eagleton, 1996: 104)

Even Young recognizes the significance of this perspective:

> [J]ustice in modern industrial societies requires a societal commitment to meeting the basic needs of all persons . . . If persons suffer material deprivation of basic needs for food, shelter, health care, and so on, then they cannot pursue lives of satisfying work, social participation, and expression.
>
> (Young, 1990: 91)

Arguments about the extent of 'and so on' will continue as long there are divergent views on what a truly human life might be. The more detailed the specification of human needs, the more difficult it is to sustain a universal position.

All this suggests a restricted set of criteria required universally to sustain a distinctively human form of life, and accepting that how they are interpreted and satisfied will be to some extent culturally relative. But it would be surprising if what was required differed very much, at the relevant level of living endured by the world's poor. 'Relief workers in Africa don't have to probe deep philosophical questions to discover that certain things are needed: those needs are immediate and obvious' (Baker, 1987: 15). For example, the major policy statement which guided development in South Africa in the immediate post-apartheid years identified lack of income, jobs, land, housing, water, electricity, telecommunications, transport, a clean environment, nutrition, health care and social welfare as basic unmet needs (ANC, 1994: 7). As Nelson Mandela (1994: 293) discovered, travelling beyond South Africa, 'poor people everywhere are more alike than they are different'. The argument concerning the equalization of the same or a closely similar package of the means of basic need satisfaction derives from the observation of human sameness or close similarity.

Of course, the moral argument for distribution according to need has a long history, going back at least to Karl Marx. Its penetration of mainstream economics, long impervious to distributional issues, is illustrated by LeGrand (1991: 88), in an echo of the argument from arbitrariness: 'distribution according to need can be viewed as compensating people for elements critical to survival that are beyond their control.' There are other arguments; for example, Fried (1983) rejects the egalitarianism of Rawls and Dworkin and the proposition that differences in talent are morally arbitrary, but invokes a duty to share and care based on the Kantian notion of the equal moral worth of individuals. The ethic of care, of which some feminists have made much in recent years (see, for example, Tronto, 1993: 162; Hekman, 1995; Clement, 1996; Bowden, 1997), has strongly egalitarian implications when interpreted as spatially extensive beneficence (Smith, 1998b).

An emphasis on basic need satisfaction has some radical implications for liberal egalitarianism. Rawls adopted liberal convention in prioritizing liberty over social and economic equality, but there is nothing sacred about this. In a reformulation from a Marxian perspective, Peffer proposes the following first priority:

> Everyone's basic security and subsistence rights are to be met: that is, everyone's physical integrity is to be respected and everyone is to be guaranteed a minimum level of material well-being including basic needs, i.e., those needs that must be met in order to remain a normal functioning human being.
>
> (Peffer, 1990: 14)

This takes precedence over Rawls's maximum system of equal basic liberties, as well as equal opportunity and an equal right to participate in social decision-making. Peffer's (1990:14) version of the difference principle is: 'Social and economic inequalities are to be justified if and only if they benefit the least advantaged . . . but are not to exceed levels that will seriously undermine the equal worth of liberty or the good of self-respect.' The priority given to economic and social security over liberty allows such hallowed tenets of liberalism as private property and freedom from imposed conceptions of the good to yield to the basic needs of the worst-off. The question of what kind of liberty some people in some places actually enjoy, if their major preoccupation in life is to survive rather than to flourish, might add weight to the prioritization of satisfaction of material needs at some expense to individual liberty.

Given limits to global resources, satisfying everyone's basic needs here and now, never mind provision for future generations, greatly limits the scope for inequality (Sterba, 1986: 15; 1998: 63). The wider the spatial reach of (re)distribution, as well as the more generous the conception of need, the more severely egalitarian its consequences. And the more egalitarian the outcomes, the greater the limitations on individual or group indulgences based on conceptions of the good which require disproportionate shares of sources of need satisfaction. Social justice as equalization clearly has implications for the good life (Smith, 1997a).

EQUALIZATION IN CONTEXT: GOOD GEOGRAPHY

It follows from the argument for (territorial) social justice as equalization that geographical research which helps to clarify and promote this process would qualify as moral progress. What kind of research might this be? The brief suggestions which follow, and the highly selective references, are not intended to be definitive, or exclusive of work not mentioned.

The return of social justice to the geographical agenda is the obvious starting point. Concern with social justice was an important part of the early radical geography movement, but little refinement was subsequently added to the outline of the 'just distribution justly arrived at' proposed by Harvey (1973: chap. 3). In the 1980s the emergence of a new social and cultural geography drawing attention to the disadvantage of various population groups resonated with the postmodern preoccupation with difference, so that when Harvey (1996) returned to social justice at book length it was to explore 'the just production of just geographical differences'. But by this time a massive new literature on social justice had accumulated outside geography. To the utilitarianism challenged by Rawls's contractarianism had been added libertarianism, Marxism, communitarianism and feminism (Kymlicka, 1990; see also Smith, 1994a). Harvey and others recognize that this plurality of theories has somehow to be transcended, to find a discourse of universality and generality uniting social and environmental justice. The key is to be found in a resurrection of the kind of egalitarianism sketched out in the previous section of this article, with its practical application grounded in a realistic recognition of the environmental context of resource constraints.

Harvey's extension of the discourse of social justice into the natural environment is also followed by Low and Gleeson (1998). The growing discovery of common ground spanning the old divide between human and physical geography, reflected in concerns with environmental ethics as well as justice (e.g. Light and Smith, 1997), is one of the most progressive moves in recent years.

The return of social justice is part of a broader 'moral turn' in human geography (Smith, 1997b), in social theory (Sayer and Storper, 1997) and in some other fields (Smith, 1999). Fertile common ground has been identified (Proctor, 1998; Smith, 1998a), along with some specifically geographical issues such as the spatial scope of care (Silk, 1998; Smith, 1998a). The centrality given to the moral dimension of human life by Tuan (1986, 1989, 1993) has been augmented by the moral perspective of Homo geographicus as elaborated by Sack (1997). And the accumulating research focused on moral geographies, landscapes and locations promises a distinctive contribution to descriptive ethics (e.g. Driver, 1988; Matless, 1994; Ogborn and Philo, 1994; Ploszajska, 1994; Holloway, 1998; Hubbard, 1998).

A further important dimension of geographical engagement with moral and political philosophy is in development ethics. Friedmann (1992) has set out a morally informed framework for development. Corbridge (1993, 1998) recognizes moral implications of the interdependence forged by globalization, and argues for a minimal universalism very much in keeping with the direction of this article:

> [P]oor people in poorer regions of the world are often lacking entitlements to and choices about 'development' for reasons that are in a very real sense random and accidental. To the extent that these Other people could have been 'Us' (the affluent), and to the extent that their lives are inextricably linked to our own, there are good reasons for attending to their needs and rights as fellow human beings.
>
> (Corbridge, 1998: 37)

The tension between the particularism encouraged in these postmodern times and the universalism of our Enlightenment heritage highlights the distinctive contribution which geography might make to the fields of social justice and ethics. Nussbaum (1998: 765) has commented: 'philosophy cannot do its job well unless it is informed by fact and experience: that is why the philosopher, while neither a field-worker nor a politician, should try to get close to the reality she describes.' In this, the geographer can help. For the justice and morality people actually practice, and the theories that ethicists devise, are embedded within specific sets of social and physical relationships manifest in geographical space, reflecting the particularity of place as well as time. It is this sensitivity to context that the geographer can provide. While stressing the imperative of getting closer to reality, Nussbaum (1998: 788) also points to the importance of theory: 'We won't learn much from what we see if we do not bring to our fieldwork such theories of justice and human good as we have managed to work out until then.' A weakness in current geographical work on moral issues is that it tends not to be closely linked to ethical theory.

Research at this new disciplinary interface will have to weave between theory and observation. This is how to collapse unhelpful dichotomies or dualisms – between absolutism and relativism, sameness and difference, universalism and particularism – in the creative process of scholarship. Walzer (1994: ix) has referred to the historical and cultural 'thickening' of those grand but 'thin' moral ideals like justice, suggesting that 'there are the makings of a thin and universalist morality inside every thick and particularist morality'. Thus, when we try to understand what social justice might mean in the specific geographical and historical circumstances of post-apartheid South Africa or postsocialist eastern Europe, for example (Smith, 1994a, 1994c, 1995b), we are working with both the particular and the universal. Attention to the particular involves careful empirical research, guided by the theory at our disposal. There is still enormous scope for work on spatial inequality or injustice in particular contexts (e.g. Laws, 1994; Black, 1996; Merrifield and Swyngedouw, 1997).

As to the impact of the place of good fortune, the British press has repeatedly featured such headlines as 'Lottery of life and death' above stories about the National Health Service explaining that how you are treated can depend on where you live (e.g. *Guardian*, October 1998, 6 January 1999). In this and other spheres the notion of a 'postcode lottery' has almost become conventional wisdom, within a society where the legitimacy of the lucky draw is celebrated twice a week on TV. So, when the governor of the Bank of England suggested that job losses in the North East are a price worth paying to curb inflation in the South, and a Labour minister prioritized wealth creation over redistribution, the scene was set for debating moral issues of fundamental importance. People in some places are losing their jobs through no fault of their own, victims of global market forces beyond their control, yet they are expected to bear the costs in terms of declining living standards and devastated communities: should not those who gain more fully compensate those who lose? And, what are the implications of the growing divorce of personal prospects from responsibility? Yet the British press soon settled down to the more titillating topic of outing MPs, aping its USA counterpart's preoccupation with President Clinton's prick. Meanwhile, the subject that generated the most electronic correspondence on the Critical Geography Forum in 1998 was gardening.

As the millennium dawns, closely followed by the year 2001, it seems appropriate to conclude these comments on what might be construed as progress in geographical research by a reminder of the next UK Research Assessment Exercise (RAE). Having spent a decade intermittently critiquing the process (e.g. Smith, 1986, 1988, 1995a), all I do here is link to the central theme of the place of good fortune. As in any process of production, research output is to some extent dependent on the local resource base, which is likely to reflect an inheritance over which present researchers had no control and for which they should not be held responsible. So, even in the unlikely event that there is a reliable way of rating university departments on the basis of the quality of their research, the result will to some (unknown) extent depend on the unequal endowment of consumables, laboratories, libraries, support staff, travel funds and so on. The conditions for Rawls's fair equality of opportunity are not met (Smith, 1996: 412). And it is not just that departments supposedly performing well may be rewarded on the basis of the morally arbitrary good fortune of working in a historically well-supported institution. It is also that the efficiency case for differential research funding is undermined by uneven starts.

Some things cannot be reduced to the calculus of money and markets. However progress in geography may be promoted, the RAE is not the way. Those who believe it is are wrong. The rest of us might take some comfort from an updating of an old saying: 'Those who can, do; those who can't, teach; those who can't even teach, appraise, assess or assure.'

PROFESSIONAL ETHICS: GOOD GEOGRAPHERS

The last of the three senses of moral progress in geography shifts the focus to the good geographer: to professional ethics. Interest in this subject has grown considerably in recent years (e.g. Brunn, 1989; Kirby, 1991; Rose, 1997; Hay, 1998), raising a wide range of issues that can be barely touched upon here. Some of them have been prompted by the changing societal context within which academic workers are required to operate, including increased pressure of performance assessment along with the growing commodification of knowledge. For example, the importance of having one's personal or institutional name on a publication raises questions of intellectual property rights (Curry, 1991). Innovations in the collection, display and dissemination of information, such as GIS and the Internet, pose ethical issues (Crampton, 1995); the more expensive the techniques, the more unequal access to them will be.

The construction (or production) of geographical knowledge is now part of our problematic. Ethical aspects of the treatment and representation of research subjects, first raised in a sustained way by Mitchell and Draper (1982), are of particular contemporary concern. Interviewing is becoming a more morally reflective practice (Winchester, 1996). Among the issues discussed by contributors to an edited collection (Proctor and Smith, 1999) are the importance of a communicative ethics in participatory research (Herman and Mattingly, 1999), the legitimacy of persons writing about a group (e.g. the disabled) of which they are not members (Kitchen, 1999), the conduct of cross-cultural research involving encounters with alternative views of the world (Rundstrom and Deur, 1999), and the relationship between research student and supervisor with different personal agendas (Gormley and Bondi, 1999). Such writers share a recognition that research ethics are relational and contextual, requiring reciprocity between researchers and researched which has to be negotiated in practice. However, Rundstrom and Deur (1999) stress that, although they emphasize contextuality, they are not willing to argue against ethical universals, recognizing that all people deserve respect, privacy, equitable treatment, and freedom from intrusion and oppression.

There are moves to impose formal codes of professional ethics, on the part of institutions such as the Association of American Geographers and the Royal Geographical Society. This leads to the question of what kind of institution(s) there are, and what kind of changes might constitute moral progress in this sphere of professional geography. It is tempting to identify institutional impediments as the four 'Ps': of privilege, patriarchy, patronage and parochialism. By privilege is meant association with a personal embodiment of undeserved good fortune (a monarchy). By patriarchy is meant the history of male domination reflected in 'fellowship'. By patronage is meant the subversive practice of private business sponsorship of a learned society. Parochialism refers to the national identity of a 'British' geography.

What remains to be said is confined to parochialism, as it relates most closely to the theme of the place of good fortune. Some thoughts on responsibility to distant colleagues were stimulated by a debate on the ethics of working in the Third World a few years ago (Sidaway, 1992; Madge, 1993; Potter, 1993; see also Paul, 1993), and are worth brief reiteration (following Smith, 1994b: 363–66). Our position of privilege as British geographers, in places well endowed with resources, carries no moral credit; it is merely a matter of good fortune. It is impossible to justify such gross inequalities with respect to the means of scholarship which exist across the world. A moral responsibility to less fortunate colleagues elsewhere surely follows. Scholars may not be the most deprived among poor populations, but it is the needs of those working in the academy that we are in a special position to understand and to assist. We can respond in various ways: involving them in our research, assisting their projects, helping them to publish, organizing seminars and so on. These things are being done, to some extent, by British geographers for others elsewhere. The question is whether we do enough. The attitude of some funding agencies is hardly encouraging, with research agendas increasingly focused on some conception of British national interest. The RAE has already become a special source of the 'self-interest and parochialism' to which Potter (1993: 291) has referred, promoting an exaggerated ethic of care for our own. If departmental pecuniary self-interest is now our predominant motivation, this may discourage us from doing things which are unlikely to appeal to those responsible for research rating, like facilitating the work of overseas colleagues and publish in foreign journals.

All this raises the question of what the 'international' standing supposedly associated with high research rating actually means. Being international in an academic context surely requires more than publishing work found interesting to a predominately Anglo-American, English-speaking audience. It also involves engagement with less fortunate others in a supportive way (as learners as well as teachers), enriching their geographical profession, narrowing the gaps between 'them' and 'us'. What is international, like what is good research, is normative and contestable, yet, as the UK Higher Education Funding Council (HEFC) seeks to define it, so this will further influence what we do, with (or for) whom, internationally.

CONCLUSION: TOWARDS A PROGRESSIVE HUMAN GEOGRAPHY

> [T]here is such a thing as moral progress . . . in the direction of greater human solidarity . . . the ability to see more and more traditional differences (of tribe, religion, race, custom, and the like) as unimportant when compared with similarities with respect to pain and humiliation – the ability to think of people wildly different from ourselves as included in the range of 'us'.
>
> (Rorty, 1989: 192)

I have offered views on what might be moral progress in (human) geography. Their foundational motivation is to transcend the place of good fortune. This involves the creation of a more equal world, in which people are less exposed to pain, humiliation and other ills arising from circumstances beyond their control and responsibility. Particular kinds of geographical research may contribute to this project: this would be good geography. And certain kinds of professional practice and institutions might encourage us to be good geographers in a moral sense.

I am conscious that my argument is far from complete. I have provided neither description nor analysis of the gross and growing inequalities which count as injustice. I have offered no blueprint for a new society, and no political project for its implementation. And I have set aside the question of moral motivation, of why we should care, reserving this for another publication (Smith, 2000). It is because I believe that understanding why inequality is wrong is a necessary condition for social change that I chose to prioritize this theme.

My central point is nevertheless one of moral responsibility: to other persons in places less fortunate than ours. We owe distant persons, including professional colleagues, far more than we give them. We may reject the notion of universal responsibility to the whole of humankind, as both a moral and practical proposition, but we should at least consider the possibility, in our personal scholarly practice, of contributing to the wider good of the potential 'world community' of professional geography. This could be one of the new international communities of mutuality invoked by Thompson (1992: 191), overlapping national boundaries and including individuals from wealthy and poor regions, with distribution of relevant goods and services according to principles which the members collectively endorse. We could try to transcend our narrowly self-interested parochialism, to (re)create the 'invisible college' of far-flung peers which Offer (1997: 463) identifies as one of his economies of regard.

When I originally addressed these issues (Smith, 1994b: 366), I expressed the fear that British geography may already be a lost cause, with competition for the money following research rating pitting department against department in a grotesque model of the business world. However, I did express some hope, and conclude with this:

> We could, as individuals and even (at some cost) as departments, reject this distortion of academic life, reaching out to others elsewhere with whom no corrosive competitive relationship exists. This could be a way of reforming, with distant others, the relationships which used to bind at least some of us in mutual collaborative endeavors in which questions of personal or departmental credit seemed inconsequential. It may provide a way of beginning to build a broader, collaborative structure, towards a universal professional-geographical ethic of care, to challenge and hopefully subvert those forces of darkness turning the practice of geography into an even more extreme expression of hierarchical domination and uneven development.
>
> (Smith, 1994b: 366)

This, surely, would be progress in geography.

ACKNOWLEDGEMENTS

This is a light revision of a lecture delivered at the Annual Conference of the Royal Geographical Society (with the Institute of British Geographers) at Leicester, 6 January 1999. A few paragraphs follow other published work, cited herein. Its preparation was assisted by a Leverhulme Fellowship. I have benefited from reactions to exposure of some of these ideas in presentations at the University of Iowa in November 1998 (thanks to Rex Honey), and to the research readings group in the Geography Department at Queen Mary and Westfield College in December 1998 (convened by Miles Ogborn). I delivered the lecture for Margaret.

REFERENCES

ANC. 1994: *The reconstruction and development program.* Johannesburg: African National Congress.

Anderson, E. S. 1999: 'What is the point of equality?' *Ethics* 109, 289–337.

Arneson, R. 1989: 'Equality and equal opportunity for welfare'. *Philosophical Studies* 56, 77–93.

Baker, J. 1987: *Arguing for equality.* London: Verso.

Barry, B. 1989: *Theories of justice.* London: Harvester-Wheatsheaf.

Black, R. 1996: 'Immigration and social justice: towards a progressive European immigration policy?' *Transactions, Institute of British Geographers* 21, 64–75.

Bowden, P. 1997: *Caring: gender-sensitive ethics.* London: Routledge.

Brown, A. 1986: *Modern political philosophy: theories of the just society.* Harmondsworth: Penguin.

Brunn, S. 1989: *Ethics in word and deed.* Professional Geographer 79, iii–iv.

Carens, J. H. 1987: 'Aliens and citizens: the case for open borders'. *The Review of Politics* 49, 251–73.

Clement, G. 1996: *Care, autonomy, and justice: feminism and the ethic of care.* London: Westview Press.

Cohen, G. A. 1986a: 'Self-ownership, world ownership, and equality'. In Lucash, E, ed., *Justice and equality here and now*, Ithaca, NY: Cornell University Press, 108–35.

——. 1986b: 'Self-ownership, world ownership, and equality'. Part II. *Social Philosophy and Policy* 2, 77–96.

——. 1989: 'On the currency of egalitarian justice'. *Ethics* 99, 906–44.

Corbridge, S. 1993: 'Marxism, modernities, and moralities: development praxis and the claims of distant strangers'. *Environment and Planning D: Society and Space* 11, 449–72.

——. 1998: 'Development ethics: distance, difference, plausibility'. *Ethics, Place and Environment* 1, 35–53.

Curry, M. R. 1991: 'On the possibility of ethics in geography. Writing, citing and the construction of intellectual property'. *Progress in Human Geography* 15, 125–47.

Crampton, J. 1995: 'The ethics of CIS'. *Cartography and Cartographic Information Systems* 22, 84–89.

Doyal, L. and Gough, I. 1991: *A theory of human need.* London: Macmillan.

Driver, F. 1988: 'Moral geographies: social science and the urban environment in mid-nineteenth century England'. *Transactions, Institute of British Geographers* 13, 275–87.

Dworkin, R. 1981a: 'What is equality? Part 1. Equality of welfare'. *Philosophy and Public Affairs* 10, 185–246.

——. 1981b: 'What is equality? Part 2. Equality of resources'. *Philosophy and Public Affairs* 10, 283–345.

Eagleton, T. 1996: *The illusions of postmodernism.* Oxford, Blackwell Publishers.

Fried, C. 1983: 'Distributive justice'. *Social Philosophy and Policy* 1, 45–59.

Friedmann, J. 1992: *Empowerment: the politics of alternative development.* Oxford: Blackwell.

Geras, N. 1995: *Solidarity in the conversation of humankind: the ungroundable liberalism of Richard Rorty.* London: Verso.

Gormley, N. and Bondi, L. 1999: 'Ethical issues in practical contexts'. In Proctor, J. D. and Smith, D. M., eds, *Geography and ethics: journeys in a moral terrain*, London: Routledge, 251–62.

Griffin, J. 1986: *Well-being: its meaning, measurement and moral importance.* Oxford: Clarendon Press.

Harvey, D. 1973: *Social justice and the city.* London: Edward Arnold.

——. 1996: *Justice, nature and the geography of difference.* Oxford: Blackwell.

Hay, I. 1998: 'Making moral imaginations. Research ethics, pedagogy, and professional human geography'. *Ethics, Place and Environment* 1(1), 55–75.

Hekman, S. 1995: *Moral voices, moral selves: Carol Gilligan and feminist moral theory.* Oxford: Polity Press.

Herman, T. and Mattingly, D. J. 1999: 'Community, justice and the ethics of research: negotiating reciprocal research relations'. In Proctor, J. D. and Smith, D. M., eds, *Geography and ethics: journeys in a moral terrain*, London: Routledge, 209–22.

Holloway, S. H. 1998: 'Local childcare cultures: moral geographies of mothering and the social organisation of pre-school education'. *Gender, Place and Culture* 5, 29–53.

Hubbard, P. 1998: 'Sexuality, immorality and the city: red-light districts and the marginalisation of female street prostitutes'. *Gender, Place and Culture* 5, 55–72.

Jones, P. 1994: *Rights*. London: Macmillan.

Kekes, J. 1994: 'Pluralism and the value of life'. In Paul, E. F., Miller, F. J. and Paul, J., eds, *Cultural pluralism and moral knowledge*, Cambridge: Cambridge University Press, 44–60.

Kirby, A. 1991: 'On ethics and power in higher education'. *Journal of Geography in Higher Education* 15, 75–77.

Kitchen, R. 1999: 'Morals and ethics in geographical studies of disability'. In Proctor, J. D. and Smith, D. M., eds, *Geography and ethics: journeys in a moral terrain*, London: Routledge, 223-36.

Kymlicka, W. 1990: *Contemporary political philosophy: an introduction*. Oxford: Clarendon Press.

Laws, G., ed. 1994: Special issue: social (in)justice in the city: theory and practice two decades later. *Urban Geography*, 15(7).

LeGrand, J. 1991: *Equity and choice: an essay in economics and applied philosophy*. London: HarperCollins.

Light, A. and Smith, J. M. 1997: *Space, place, and environmental ethics. Philosophy and Geography* London: Rowman & Littlefield.

Low, N. and Gleeson, B. 1998: *Justice, society and nature: an exploration of political ecology*. London: Routledge.

Madge, C. 1993: 'Boundary disputes: comments on Sidaway (1992)'. *Area* 25, 294–99.

Mandela, N. 1994: *Long walk to freedom*. London: Little, Brown & Co.

Matless, D. 1994: 'Moral geography in Broadland'. *Ecumene* 1, 127–56.

Merrifield, A. and Swyngedouw, E. 1997: *The urbanization of injustice*. Albany, NY: State University of New York Press.

Miller, R. W. 1992: *Moral differences: truth, justice and conscience in a world of conflict*. Princeton, NJ: Princeton University Press.

Mitchell, B. and Draper, D. 1982: *Relevance and ethics in geography*. London: Longman.

Nett, R. 1971: 'The civil right we are not ready for: the right of free movement of people on the face of the earth'. *Ethics* 81, 212–27.

Nozick, R. 1974: *Anarchy, state, and Utopia*. New York: Basic Books.

Nussbaum, M. C. 1992: 'Human functioning and social justice: in defense of Aristotelian essentialism'. *Political Theory* 20, 202–46.

——. 1998: 'Public philosophy and international feminism'. *Ethics* 108, 762–96.

Offer, A. 1997: 'Between the gift and the market: the economy of regard'. *Economic History Review* 50, 450–76.

Ogborn, M. and Philo, C. 1994: 'Soldiers, sailors and moral locations in nineteenth-century Portsmouth'. *Area* 26, 221–31.

O'Neill, O. 1991: 'Transnational justice'. In Held, D., ed., *Political theory today*, Cambridge: Polity Press, 276–304.

——. 1996: *Toward justice and virtue: a constructive account of practical reasoning*. Cambridge: Cambridge University Press.

Paul, B. K. 1993: 'A case for greater interaction between the geographers of developed and developing countries'. *Professional Geographer* 45, 461–65.

Peffer, R. G. 1990: *Marxism, morality, and social justice*. Princeton, NJ: Princeton University Press.

Ploszajska, T. 1994: 'Moral landscapes and manipulated spaces: gender, class and space in Victorian reformatory schools'. *Journal of Historical Geography* 20, 413–29.

Poole, R. 1991: *Morality and modernity*. London: Routledge.

Potter, R. 1993: 'Little England and little geography: reflections on third world teaching and research'. *Area* 25, 291–94.

Proctor, J. D. 1998: 'Ethics in geography: giving moral form to the geographical imagination'. *Area* 30, 8–18.

Proctor, J. D. and Smith, D. M., eds, 1999: *Geography and ethics: journeys in a moral terrain*. London: Routledge.

Rawls, J. 1971: *A theory of justice*. Cambridge, MA: Harvard University Press.

Reiman, J. 1990: *Justice and modern moral philosophy*. New Haven, CT, and London: Yale University Press.

Roemer, J. E. 1996: *Theories of distributive justice*. Cambridge, MA: Harvard University Press.

——. 1998: *Equality of opportunity*. Cambridge, MA: Harvard University Press.

Rorty, R. 1989: *Contingency, irony, and solidarity*. Cambridge: Cambridge University Press.

Rose, G. 1997: 'Situating knowledge: positionality, reflexivities and other tactics'. *Progress in Human Geography* 21, 305–20.

Rostow, W. W. 1960: *The stages of economic growth: a non-communist manifesto*. Cambridge: Cambridge University Press.

Rundstrom, R. and Deur, D. 1999: 'Reciprocal appropriation: toward an ethics of cross-cultural research'. In Proctor, J. D. and Smith, D. M., eds, *Geography and ethics: journeys in a moral terrain*. London: Routledge, 237–50.

Sack, R. D. 1997: *Homo geographicus: a framework for action, awareness, and moral concern*. Baltimore, MD, and London: Johns Hopkins University Press.

Sandel, M. 1982: *Liberalism and the limits of justice*. Cambridge: Cambridge University Press.

Sayer, A. and Storper, M. 1997: 'Ethics unbound: for a normative turn in social theory'. *Environment and Planning D: Society and Space* 15, 1–17.

Sen, A. 1992: *Inequality reexamined*. Oxford: Clarendon Press.

Sidaway, J. D. 1992: 'In other worlds: on the politics of research by "first world" geographers in the "third world"'. *Area* 24, 403–8.

Silk, J. 1998: 'Caring at a distance'. *Ethics, Place and Environment* 1, 165–82.

Smith, D. M. 1973: 'An introduction to welfare geography'. Department of Geography and Environmental Studies, Occasional Paper 11. Johannesburg: University of the Witwatersrand.

——. 1977: *Human geography: a welfare approach*. London: Edward Arnold.

——. 1986: 'UGC research ratings: pass or fail?' *Area* 18, 247–49.

——. 1988: 'On academic performance'. *Area* 20, 3–13.

——. 1994a: *Geography and social justice*. Oxford: Blackwell.

——. 1994b: 'On professional responsibility to distant others'. *Area* 26, 359–67.

——. 1994c: 'Social justice and the post-socialist city'. *Urban Geography* 15, 612–27.

——. 1995a: 'Against differential research funding'. *Area* 27, 79–83.

——. 1995b: 'Geography, social justice and the new South Africa'. *South African Geographical Journal* 77, 1–5.

——. 1996: 'To the mop closet! Response to Paul Curren'. *Area* 28, 410–13.

——. 1997a: 'Back to the good life: towards an enlarged conception of social justice'. *Environment and Planning D: Society and Space* 25,19–35.

——. 1997b: 'Geography and ethics: a moral turn?' *Progress in Human Geography* 21, 583–90.

——. 1998a: 'Geography and moral philosophy: some common ground'. *Ethics, Place and Environment* 1, 7–34.

——. 1998b: 'How far should we care? On the spatial scope of beneficence'. *Progress in Human Geography* 22, 15–38.

——. 1999: 'Geography and ethics: how far should we go?' *Progress in Human Geography* 23, 119–25.

——. 2000: *Moral geographies: ethics in a world of difference*. Edinburgh: Edinburgh University Press (in press).

Sterba, J. P. 1986: 'Recent work on alternative conceptions of justice'. *American Philosophical Quarterly* 23, 1–22.

——. 1998: *Justice for here and now*. Cambridge: Cambridge University Press.

Thompson, J. 1992: *Justice and world order: a philosophical inquiry*. London: Routledge.

Tronto, J. 1993: *Moral boundaries: a political argument for an ethic of care*. London: Routledge.

Tuan, Y.-F. 1986: *The good life*. Minneapolis, MN: University of Minnesota Press.

——. 1989: *Morality and imagination: paradoxes of progress*. Madison, WI: University of Wisconsin Press.

——. 1993: *Passing strange and wonderful: aesthetics, nature and culture*. Washington, DC: Island Press.

Walzer, M. 1983: *Spheres of justice: a defense of pluralism and equality*. Oxford: Blackwell.

——. 1994: *Thick and thin: moral argument at home and abroad*. Notre Dame, IN, and London: University of Notre Dame Press.

Williams, B. 1985: *Ethics and the limits of philosophy*. London: Fontana Press/Collins.

Winchester, H. P. M. 1996: 'Ethical issues in interviewing as a research method in human geography'. *Australian Geographer* 2, 117–31.

Young, I. M. 1990: *Justice and the politics of difference*. Princeton, NJ: Princeton University Press.

7

Dissecting the autonomous self

Hybrid cartographies for a relational ethics

Sarah Whatmore

from *Environment and Planning D: Society and Space*, 1997, 15: 37–53

THE PLACE OF ETHICS

The modernist ideals of universal democracy and justice realized through legislative régimes centered on individual rights have been the subject of sustained feminist and environmentalist critiques, reinvigorating political and philosophical interest in the question of ethics. Feminist writing has focused on deconstructing the discourse of rights, highlighting the gendered (and racialized) character of the autonomous self configured as rights-bearing citizen of a sovereign state (Cornell, 1985). By contrast, environmentalist work has centered on extending the political and discursive economy of rights to nonhuman beings; challenging established concepts of personhood and subject status (Callicott, 1979). These efforts share parallel concerns to establish relational, as opposed to individual, understandings of ethical agency and to recognize the significance of embodied, as against abstract, capacities in shaping ethical competence and considerability. Such concerns highlight the power of the geographical imaginaries of traditional ethical discourses and the difficulties of disrupting the entrenched cartographies of the nation, the neighborhood, and the individual in fashioning new possibilities for ethical community.

In this paper I explore what are, I think, creative tensions between feminist and environmentalist efforts to empower those eclipsed in orthodox ethical discourse, particularly at the embattled frontiers of the so-called "natural law" and "social contract" traditions. I trace some of the ways in which the conceptual and institutional parameters of notions of self (citizen), central to feminist concerns, intersect with those associated with notions of subject (person) at the heart of environmentalist concerns. In both cases, although for different reasons, I argue that dilemmas encountered by these attempts to construct alternative ethical orderings are intimately bound up with their adherence to what Latour has called the "purification" of nature and society as "distinct ontological zones" (1993: 10). This leads me to suggest a number of consequences for instituting a relational understanding of political and moral agency which centers on a recognition of the social embodiment and environmental embeddedness of the (re)configuration of "individuals" and "communities." In so doing, I aim to highlight the importance of *corporeality* and *hybridity* as concepts for rethinking the place of ethics.

Ethical discourse has conventionally been framed in terms of an opposition between natural law and social contract traditions, centered on competing accounts of the primacy of "human nature" as against civic order as the foundational claim to ethical competence and considerability (Poole, 1991). Commonly misunderstood as some kind of unchanging normative code inscribed in the heavens or the genes, natural law theories evoke the capacity for reason as the definitive basis of a distinctively human ethical standing. Early modern reinterpretations of a classical legacy, notably in the work of Locke, shifted accounts of this distinctively human capacity from the evocation of a "common good"—the cluster of obligations generated by the patterns of interdependence in human social life—to that of an "individual good"—the result of voluntary transactions between independent agents.[1] The most important implication of this shift was to elevate the "moral significance of the separateness of persons" (Buckle, 1991: 168).

The emergence of the individual as axiomatic of modern society is inscribed in legal, political, and religious institutions and discourses. Since Kant, this founding figure of the autonomous self has been most strongly associated with the social contract tradition of ethics (Kymlicka, 1991). However, it is worth emphasizing that it is less the significance accorded to this figure that marks out the social contract tradition than the resolution it reaches for the social regulation of such individuals. Natural law resolutions rely on some underlying uniformities (of reasonableness) that can sustain the idea of universal (natural) human goods and values. Social contract resolutions rest on particular social institutions of contract (market) and rights (law) as the basis for establishing universal (impartial) "laws of reason" as the precondition of ethical agency.

Contemporary elaborations of these debates can be seen in the philosophical and legal dilemmas of squaring claims to *human* rights with claims to *civil* rights. The one represents a species claim to the possession of reasoning faculties as the basis for the universal ethical considerability of individuals by virtue of their constitution as human beings: the other, a political claim to the possession of reasoning faculties as the basis for the ethical considerability of individuals by virtue of their constitution as civic persons (McHugh, 1992). Historical changes in the legal encoding of such claims underline the unstable and disputed social meaning of both "human" and "person" as ethical subjects, for example in the treatment of women and non-European peoples; instabilities which persist, also marking the unborn, children, and those deemed mentally "unfit". Despite these dilemmas, the figure of the Cartesian individual as an atomistic, presocial vessel of abstract reason and will continues to dominate contemporary ethical accounts.[2] Contingent moral commitments and norms associated with a particular individual's "life" context evaporate in the white heat of "enlightened self-interest." Ethical agency becomes cast in terms of the impartial and universal enactment of instrumental reason, institutionalized as a contractual polity of like individuals.[3] Such accounts of ethical agency rely upon spatially and temporally fixed conceptions of individual and collective social being—the sovereignty of self and state—etched in the cartographies of the citizen and the nation. Ironically, as Poole (1991) suggests, insofar as the modern world revolves around the autonomous self, it has also destroyed the conditions of its autonomy, reducing community to an infinitely expanded network of market interactions.

The commoditization of social (and environmental) relations has disrupted this configuration of political and ethical community on two fronts. First, it has done so by eroding the territorialized authority of the nation-state to govern increasingly global networks and mobilities of people and goods. Ethical communities bounded by national borders have become unsustainable because "the nation state is no longer able to resolve the contradictions between citizenship and humanity through claims to absolute authority" (Walker, 1991: 256). Second, the expansion of market relations has also undermined the personalized jurisdiction of the individual citizen over a coherent domain of the self (Giddens, 1991). As Haraway has observed,

> the proper state for a western person is to have ownership of the self, to have and hold a core identity, as if it were a possession . . . Not to have property of the self is not to be a subject and so not to have agency.
>
> (Haraway, 1991: 135)

However, this private domain of the rights-bearing citizen has long been exposed as masculine in conception. This has translated at different time-places into the dispossession of women, poor, and black people of political and ethical agency in their own right, through their "contractual" guises as wives, servants, and slaves (Pateman, 1989).[4] Moreover, this extended domain of the patriarchal self underpinning effective citizenship, the domain of the family and household, has itself become increasingly friable (Gobetti, 1992). In short, the disruption of this configuration of political and ethic community is centered on the instability of its spatial encoding as distinct realms of public and private (civic and domestic) competence, and the reordering of these competences by the invasive institutions of market and governance.

Recent work in the field of political philosophy is dominated by two divergent responses to the limitations of the liberal conception of political and ethical community sketched above.[5] One echoes a long-standing communitarian tradition which predicates the capacity to participate as ethically and politically competent subjects on the material satisfaction of basic human needs. As Porter has put it:

> A concern for persons in their own right is not possible where the primacy of rights relies on an atomist conception of the self-sufficient individual. This notion maintains that human capacities need no particular social context in which to develop and hence is not attached to other normative principles concerning what is good for humans or conducive to their development.
>
> (Porter, 1991: 127)

The more sophisticated communitarian accounts elaborated by political philosophers such as Sandel and Macintyre appeal to an intersubjective conception of the self as the basis of ethical agency. This conception centers on qualifying the absolute distinction between self and other associated with the figure of the sovereign individual "by allowing that, in certain moral circumstances, the relevant description of the self may embrace more than a single empirically-individuated human being" (Sandel, 1982: 79–80). This set of responses has become politically influential, with so-called "new communitarianism" coloring the rhetoric of conventional political opponents of free market liberalism, such as Blair's "New Labour" Party in Britain and Clinton's Democratic administration in the USA. In its concern with the material preconditions of a full human life, this perspective reengages with natural law arguments that ethical considerability precedes formal rights, requiring answers to the question "rights for what?" At the same time it readmits, in a limited way, nonhuman figures to the landscape of ethical community, as necessary material "resources" to service basic human needs. The environmental implications of this "new communitarian" perspective are rehearsed in US Vice President Al Gore's populist manifesto *Earth in Balance*, in which he argues that

> We have tilted so far toward individual rights and so far away from any sense of obligation that it is now difficult to muster an adequate defense of any rights vested in the community at large or in the nation—much less rights properly vested in all humankind.
>
> (Gore, 1992: 278)

A second response to contemporary dilemmas in the conception and practice of ethical community is that associated with a broader critique of the foundational coordinates of Modern society identified with "postmodernism" (Squires, 1993). Such critiques center on a radical deconstruction of the twin sovereignties of self and state. Here "the individual" is transformed into a site of heterogeneous and multiple identities which become performative resources in the creative enactment of new and "liberating" subject positions. Amongst the more sustained explorations of this postmodernist interpretation of political and ethical agency is Laclau and Mouffe's project of "radical democracy" characterized as "a polyphony of voices, each of which constructs its own irreducible discursive identity" (1985: 191). Far from breaking with the primacy of the individual as a foundational social unit, this approach inverts the Cartesian subject, replacing abstract reason with abstract desire as definitive of (human) social agency. It shifts the ground of ethical and political community from conventional practices of contract between universally equivalent agents to communicative practices of dialogue between radically different agents.[6] The biographing individual evoked in this postmodern vision liberates the possibilities of ethical community from the involuntary associations of birth or proximity, but it does so by obscuring the conditionality of dialogic engagement in terms of the mundane business of *living*.

The tensions between contractarian and natural law theories of ethical competence and considerability mark ongoing dilemmas over the relationship between social rationality and human mortality. The reified figure of the autonomous individual represents a cipher of abstract reason which inscribes the binaries of mind–body, self–other, subject–object onto the very possibility of ethical agency in Modern society. Recent critiques from communitarian and postmodernist positions open up new possibilities but are less radical departures than they sometimes appear. Communitarian approaches reassert the *situatedness of the individual* and point to the intersubjective constitution of ethical agency. However, they tend to do so by invoking highly conservative configurations of community, such as the family, the neighborhood, and the nation, without examining the power relations they enact. Moreover, this "situatedness" is predominantly defined in terms of social (human) relations. Where they are addressed at all, environmental (nonhuman) relations are treated as passive contextual extensions of human well-being. By contrast, a postmodernist insistence on the radical instability of the individual, divested of material fabric or context, tends to evoke highly *disembodied*, as well as disembedded, social agents (Levin, 1985; O'Neill, 1985; Pile and Thrift, 1995). In a world populated by such amorphous figures, constituted from cognitive and linguistic possibilities unshackled by the corporeal baggage of living, "the question of what human *be*-ing is" (Porter, 1991: 16) becomes unspeakable.

Emerging at the confluence of these various encounters with the intellectual and practical dilemmas of ethical agency is a recognition of formal justice as a derivative of some substantive moral propositions and ethical claims. Increasingly, this has been accompanied by a creative reengagement with ideas of human nature *not* in terms of any substance or essence of humanity, but in terms of the predicament of finitude, the inherent decay and mortality of all living beings. As Cornell has put it, only "by coming to terms with finitude can we gain the humility necessary to overcome the hubris of individualism" (1985: 338). Bauman's exploration of the ethical implications of mortality (1992), Giddens notion of "life polities" (1991), and Beck's account of "risk society" (1992) all exemplify the renewed interest in corporeal being for understanding ethical competence and considerability. Exploring issues such as the legal determination of the status and rights of the "unborn" fetus and the medical certification of the condition of death, these writers suggest that the more reflexively we "make ourselves" as persons the more significant bodily awareness becomes, heightening the sense of shared mortality as a mode of political association and ethical recognition. As a recent issue of this journal [i.e. *Environment and Planning D: Society and Space*] has illustrated, such efforts are echoed in popular concerns and everyday struggles which mobilize connectivities between environmental degradation, animal rights, human health, and scientific expertise (Wolch and Emel, 1995). These concerns are perhaps most graphically illustrated in the current political, economic, and animal carnage associated with bovine spongiform encephalopathy (BSE), so-called "mad cow disease", and its human form, Creutzfeldt–Jakob disease (CJD), in Britain.

But these themes have been taken up most persistently and powerfully by those most excluded from the humanist and masculinist presumptions of an abstracted world of equivalent moral agents, most notably in feminist and environmentalist critiques. These critiques center on concerns with the embodiment of difference and rationality and with the ethical significance of nonhuman life-forms and processes, respectively. In the next section I draw out what I see as key insights and tensions in these alternative discourses for the elaboration of a more relational understanding of ethical competence, before moving on to consider some of their spatial implications for the reconfiguration of ethical community.

FEMINIST ETHICS: THE EMBODIMENT OF CARE?

> When identities become pure, exclusive, innocent, the potential for diverse and democratic collectivities is threatened. We are all others of invention, otherness should not be reified but used as one fertile resource of feminist solidarity.
>
> (Caraway, 1992: 1)

The celebration of difference in postmodern theories has been highly influential, but also hotly contested, in feminist political thinking over recent years. A number of writers (for example Ebert, 1991; Hennessy, 1993) distinguish between two very different clusters of feminist engagements with this issue. The first, identified as *ludic postmodernism*, seeks to disrupt naturalized conceptions of identity as a model for political practice and locates the politics of difference in the discursive play of imagined possibilities in a theater of volatile subject positions (exemplified by the work of Mouffe, Young, and Flax). The second, identified as *resistance postmodernism*, locates the politics of difference not as the effect of rhetorical or textual strategies, but as the effect of social struggles which ground the meanings contested in such strategies in the materialities of everyday living (exemplified by the work of Benhabib, Cornell, and Grosz). Although the distinction between these feminist accounts of a politics of difference is overdrawn and even somewhat caricatured, it points up an important area of dispute about how difference (that is, the relation between "self and "other") and its political (and ethical) import are to be understood (Braidotti, 1992). Echoing tensions in Nietzsche's writing, Diprose outlines the parameters of this dispute in terms of whether we are more likely to "find our-selves" by looking inwards in an autonomous project of creative self-fabrication, or by looking outwards to our effects and relations with others which configure our place in the world (1994: 87).

The first of these approaches employs individualist theories of difference, or what Kruks has called "an epistemology of provenance" (1995: 4), to fashion self-exploration as a political process in its own right while relying on an unspoken normative claim to the ethical equivalence of all "subject positions" in this privatized polity. Collective claims to political agency and ethical considerability tend to be looked upon askance, as intrinsically "antidifference" (for example, see Young, 1989). This leaves feminism as a political project precariously positioned by what Anderson calls the "double gesture" of simultaneously asserting the theoretical universalism of decentered subjectivity whilst resorting to the practical lie of strategic essentialism to secure a space for women to identify common cause at all. Ironically, as she points out,

> the idea of subject-positions . . . precludes the possibility of an intersubjective perspective that would define the human subject not as purely autonomous and self-present, nor as a mere place on intersecting grids, but as constituted through its ongoing relations to others.
>
> (Anderson, 1992: 78)

It is the second of the feminist encounters with postmodern theories which is the more suggestive to me as a means of negotiating the impasse of individualism in reconstructions of ethical community. It centers on a notion of *difference in relation*, as intersubjectively constituted in the context of always/already existing configurations of self and community. In place of abstract or cognitive criteria, these always/already existing configurations of self and community are "defined by contingent and particular social attachments whose moral force consists partly in the fact that living by them is inseparable from understanding ourselves as the particular persons we are" (Friedman, 1989: 278). This approach to ethical and political community shares poststructuralist suspicions of the liberal ambition of value homogeneity but remains committed to a practice of participatory communalism enacted through particular economic, political, scientific, and civic orderings which condition individual capacities and arenas for action. As a feminist enterprise, it represents an attempt to understand the discursive construction of "woman" across multiple modalities of difference by adopting a problematic that can trace the connections between discursive practices and the exploitative social orderings of meaning, being, and struggle which permit and encode them (hooks, 1990).

The ethical dimensions of this approach are best captured in Benhabib's distinction between *generalized* and *concrete* others (1987). The generalized other stands for a universal principle of equal considerability in the right to be heard, to participate, to *make* a difference. The concrete other stands for more immediately realized ethical principles—of care, friendship, intimacy, solidarity, and empathy—which involve practical, though often asymmetrical, enactments of responsibility. However, Benhabib's elaboration of this intersubjective conception of ethical agency reproduces the Habermasian error of according a privileged status to the abstract qualities of rationality and language in the theory of "communicative action." More recently, Kruks has articulated an important step towards a more situated and practical approach to understanding ethical intersubjectivity which draws on Sartre's notion of dialectical reason. This approach

> begins from the situation of an embodied and practically engaged self; . . . from what human beings *do* in the world . . . so as to rediscover the totality of [her/his] practical bonds with others.
>
> (Kruks, 1995: 11–12)

Although this conception of a *materially* situated self has wider significance for the reconfiguration of ethical community, which I shall return to later in my consideration of environmental ethics, here I want to pursue two persistent themes in feminist ethical thinking with which it resonates most suggestively. These are the interconnected issues of corporeality (by which I mean both the finitude and embodiment of living being) and the praxis of care.

Feminist concerns with the material situatedness of social identity and of the particularity of sexed being have impelled a sustained consideration of the politics of embodiment and, more broadly, of what I have called the *corporeality* of living being. These concerns have centered on the specificities of women's experiences as (potential) childbearers, the objectification of women's bodies, and the cultural politics of the pejorative signification of "woman" as animal, natural, carnal. This is difficult terrain for feminists, with the specter of essentialism menacing any consideration of corporeal being in relation to gender and sexual identity. But there is a growing realization that "to separate the feminine from female morphology is misguided theoretically and politically even in strategic contexts" (Gross, 1986: 136). The concept of difference in relation requires a "theory of the flesh" (Moraga and Anzaldua, 1981: 23) to elaborate an understanding of individual, collective, and group being as situated in webs of connection that are "practice-inert" as well as discursive, embodied as well as cognitive. Grosz's elaboration of a "corporeal feminism" (1989, 1994) provides perhaps the most sustained attempt to articulate such a "theory of the flesh."[7] She builds on Irigaray's understanding of difference as being always inscribed upon the experiences of the sexed body:

> I want to go back to the natural material which makes up our bodies, in which our lives and environment are grounded . . . a latent materiality which our so-called human theories . . . move away from [and] progress through . . . with a language which forgets the matter it designates and through which it speaks.
>
> (Irigaray, 1986, quoted in Grosz, 1989: 172)

Here, the body is considered not as the passive container of social being but as a living assemblage of biological materials and processes which both register and orient our senses of the world. Although always configured through particular social orders of meaning, technology, and practice, these corporeal properties are no less conditional of the very capacities of cognition and communication that mark the abstracted ideals of individual autonomy and human distinctiveness. As Grosz goes on to suggest in her more recent work (1994) such an understanding of the body undermines the political myth of self-authorship and the privileged ethical status of humans as cognitive, communicative subjects.

A second theme in feminist ethics that is particularly pertinent to the elaboration of an intersubjective conception of the situated self is the praxis of care. This builds on the contention that feminisms can only move beyond "the impasse of (in)difference" (Probyn, 1993) by simultaneously articulating questions of "who am I?" with those of "who is she?" This ethical incarnation of "difference in relation" derives from a number of impulses in feminist work other than philosophy, particularly from psychoanalytic feminism (Meyers, 1994). A major stimulus was the empirical work of psychologist Gilligan (1982), who reported a marked tendency for women to articulate more relational senses of self and stronger senses of responsibility for connected others than do men—what she called a "different ethical voice" from that institutionalized in conventional justice. The recognition and enacting of these relational senses of self and responsibility constitute what has become known as the "feminist care ethic." Although much debated, it centers on a concern with ethical praxis and the practical connectivities which secure the well-being of those least mobile and most vulnerable, *not* as discursive subject positions, but as mortal others-in-relation such as the hungry, the sick, and the abused (Lovibond, 1994).

This understanding of ethical agency and community recognizes a bodily intentionality to human existence and social life that knits together multiple and apparently fragmentary collective identities, each of which is itself the outcome of a multiplicity of prior and present praxes (Kruks, 1995, page 15).[8] Although such an understanding certainly helps to substantiate an appreciation of the ineluctable embodiment of intersubjective being, it is restricted purely to *human* being disembedded from webs of connection with other life-forms and processes. It is here that environmental ethics promises to make an important contribution.

ENVIRONMENTAL ETHICS: (RE)CONSIDERING OTHERS

> the multiplicity of living organisms retain, ultimately, their peculiar, if ephemeral, characters and identities but they are systemically integrated and mutually defining.
>
> (Callicott, 1989: 111)

In contrast to much feminist work, environmentalists have invested considerable energies in trying to extend the ethical domain of the autonomous self, as a bearer of social rights, beyond the human.[9] This has taken shape in one of two ways. The first, which might be termed *moral extensionism* and is associated with long-standing concerns over animal rights, transports the liberal figure of the rights-bearing individual wholesale to a range of nonhuman creatures. These extensions are made either on the criterion of intelligence in the form of reasoning and linguistic capacities, which is usually restricted to primates and cetaceous mammals, or of sentience, a more inclusive criterion centered on the capacity to suffer or experience pain, which covers all mammals with a central nervous system. Informed by new perspectives in animal biology and psychology, particularly primate cognition, this approach culminates in the proposal of a "subject-of-life" criterion for extending ethical standing to all animate beings (Regan and Singer, 1989). Such approaches build on mainstream utilitarian or Kantian ethical arguments and are open to the critiques of liberal individualism rehearsed above (see Benton, 1993).

A second approach, broadly aligned with deep ecological perspectives and informed by Gaian organicism, has involved the elaboration of various notions of *expanded human consciousness* to encompass a recognition of our embeddedness in constitutive relations with the nonhuman world. These efforts do not restrict the extension of ethical standing to animate organisms but include vegetal and inanimate elements and processes under the collective term of *earth others* (Bigwood, 1993). This wider ethical compass frequently relies upon the evocation of a spiritual dimension to "being in the world" which resonates uneasily with the intellectual register of the academy. Prominent examples of this approach include Mathews's concept of the "ecological self" (1991), Naess's notion of "self-actualisation" (1989), and Fox's idea of the "transpersonal self" (1990). In a sustained critique of these approaches, Plumwood has identified such concepts with what she calls the "imperialism of the self" (1993). As attempts to construct an intersubjective conception of ethical agency, they are flawed by a colonizing humanism which subsumes the ethical considerability of nonhuman organisms into the conception of human being, denying them subject status in their own right. This highlights a key dilemma for environmental ethics. Feminist difficulties with the privileged status of cognitive and linguistic competences in fashioning the ethical subject are amplified for environmentalists whose constituency consists of subjects without (intelligible) voices, a constituency of nonpersons more resolutely excluded from the status of ethical subjects than any human.

This dilemma has stimulated an important development in recent work on environmental ethics. Picking up Kruk's insistence on a materially situated, practically engaged self as the embodiment of an intersubjective understanding of ethical agency, this work has begun (re)exploring a dialogical understanding of relations between the self and the world centered on the corporeal immersion of humans in the biosphere. This conceptualization of intersubjectivity recognizes humans as

> beings thoroughly entwined with an extralinguistic world . . . [and that] to deny this entwinement is to bind ourselves to a quest for an abstract and empty sovereignty that destroys the world and is self-defeating.
>
> (Coles, 1993: 231)

Like feminist evocations of a "theory of the flesh," some of these explorations draw inspiration from traditions of dialectical reasoning, such as that of Adorno (Coles, 1993), and of phenomenology, particularly that of Merleau-Ponty (Abram, 1988). They simultaneously emphasize the corporeal embeddedness of cognitive processes in the visceral dynamics of brain, eye, skin, etc., and the configuration of human well-being and interdependence with that of other living beings. Arguably it has been feminist environmentalists, particularly those writing from postcolonial perspectives (Mohanty *et al.*, 1991), who have done most to transform these ideas into an ethical praxis in the form of a "politicized ecological care ethic" (Donovan, 1993). This translates the recognition of webs of connectivity between the livelihood practices and cultural values of particularly situated human actors (collective and individual) and the life-habits and relationships of other biotic agents into acknowledged responsibilities, both in the sense of caring *about* "generalized others" and caring *for* "concrete others" (Curtin, 1991; King, 1991). A good example is the global actor network DAWN (Development with Women working for a New Era) which, since 1984, has sought to articulate material connectivities between environmental, livelihood, and health issues and the centrality of "third world" women in this nexus (Braidotti *et al.*, 1994).

The feminist and environmentalist approaches outlined in this section are each ongoing and contested discourses which inform, and are informed by, a wide variety of political practices. My treatment of them has been necessarily highly selective. The main contributions which I would attribute to the particular threads of feminist and environmentalist ethics I have traced are their various attempts to substantiate a corporeal conception of the situatedness of ethical agency and the extralinguistic connectivities of ethical community. Moreover, they are suggestive of spatial imaginaries of ethical community which do not replicate the bimodal geographies of public–private morality. Equally, however, these approaches share shortcomings which are important in terms of my broader argument. Even amidst the talk of intersubjectivity, embodiment, and embeddedness the categories "human" and "nonhuman" remain unproblematic both in themselves and as an encoding of society and nature as discrete, if subsequently reconnected, terrains. Moreover, although the distinction between general and concrete others is an heuristic device which has no necessary spatial predisposition, feminist and environmentalist care ethics have tended in practice to map it simplistically onto the geographical binaries of distance–proximity, global–local, and outside–inside, for example in the praxis of "bioregionalism" (Cheney, 1989) and "communities of place" (Friedman, 1989). In the next section of the paper I turn to consider the concept of *hybridity* as a means of disrupting the polarization of "society" and "nature" and to begin to explore some alternative cartographies for a relational ethics.

HYBRID CARTOGRAPHIES OF ETHICAL COMMUNITY

> Evidence is building of a need for a theory of "difference" whose geometries, paradigms and logics break out of binaries . . . and nature/culture modes of any kind.
>
> (Haraway, 1991: 129)

Bringing ideas of *difference in relation*, both in the discursive and in the corporeal sense, to bear on the question of political community has been most extensively explored in the work of Haraway and Latour in their elaboration of concepts of hybridity. Haraway's argument is that we "cannot not want" something called humanity because nobody is self-made, least of all humans (1992a: 64). But in order to recuperate a progressive commitment to humanity as a moral community the dualisms associated with *humanism* have to be jettisoned. This requires a hybrid concept of community which disrupts the purification of culture and nature into distinct ontological zones, onto which the binary of "human"–"nonhuman" is then mapped. Haraway's cyborg metaphor articulates a political vision which appreciates the instability of boundaries between human, animal, and machine and their discursive and technological malleability, particularly in the hands of corporate science (1985). Political agency and community emerge from this vision through "webs of connection" between situated and partial knowing selves fashioned through "shared conversations," and what she calls "semiotic-material technologies" which link meanings and bodies (Haraway, 1991: 192). Ethical agency and community likewise emerge as the performance of multiple lived worlds, weaving threads of meaning and matter through and between these "webs of connection."

As with so many of Haraway's provocative ideas, what she means by semiotic-material technologies is hard to fix. Her favorite examples are prosthetics, genetics, and organ transplants in which particular codified knowledges become stabilized as technological artifacts which, in turn, are grafted into and mobilized by living beings. These examples tend to site the dilemmas of hybrid subjectivity, and the cyborg figure used to signify them, within an individuated being—"a hybrid creature composed of organism and machine" (Haraway, 1991: 1). There is a tension, then, in Haraway's account of the status and configuration of her hybrid subject the cyborg. It is not clear whether, as Kruks asks, these hybrid subjects stitch their own parts together, in which case they become more cohesive than Haraway wants to admit, or whether this "stitching together" is better understood as an operation taking place from without (Haraway, 1985: 9). If the first, then Haraway's hybrid subject falls back on an account of political and ethical agency which privileges cognitive and discursive faculties in the constitution of "knowing selves" (however partial or unfinished the project of self-fabrication). If the second, then it is not clear from Haraway's account just what it is that connects diverse knowing selves together other than the capacity for linguistic communication evoked in her notion of "shared conversations." In short, although Haraway's account of hybridity successfully disrupts the purification of nature and society and the relegation of "nonhumans" to a world of objects, it is less helpful in trying to "flesh out" the "material" dimensions of the practices and technologies of connectivity that make the communicability of experience across difference, and hence the constitution of ethical community, possible. These dimensions require a closer scrutiny of overlapping life-practices and corporeal processes, for example those mediated by food, energy, disease, birth, and death, than Haraway has so far admitted.

In this context, I find Latour's account of hybridity, through the metaphor of the "hybrid network," more suggestive for elaborating a relational understanding of ethical agency and community. The network metaphor places greater emphasis on the multiple agency of hybridity—the mobilization of animate, mechanical, and discursive modalities of being within and between differently configured actants. Such networks not only connect pregiven individual entities but shape the possibilities for individuality. Moreover, Latour is explicit about the implications of this interpretation of hybridity for the reordering of ethical community. Hybrid networks, he argues, force us to

> take into account the objects that are no more the arbitrary stakes of [human] desire alone than they are the simple receptacle of our mental categories. (Latour, 1993: 117)

The intersubjective understanding of hybridity articulated in the metaphor of networks disrupts the opposition between objects and subjects prescribed by an ethics centered on instrumental reason and its encoding in the purified domains of "Nature" and "Society." Instead, a multitude of mediators, what Latour calls "nature–culture collectives," are exposed, built with raw materials made out of "poor humans and humble nonhumans" (1993: 115). It is these collectives which constitute the topography of political and ethical community, communities which are ever lengthening as larger and larger numbers of nonhumans are enlisted by the technologies of science, governance, and market into networks that are increasingly global in reach. But Latour insists that such networks are by no means comprehensive or systematic. They are "connected lines, not surfaces, points of view on networks that are by nature neither local nor global" (Latour, 1993: 120). Instead, hybrid networks are conceived as occupying narrow lines of force that allow us to pass with continuity from the local to the global, from the human to the nonhuman, through partial and unstable orderings of numerous practices, instruments, documents, and bodies.

Though by no means unproblematic, Latour's notion of hybridity as networks of nature–culture collectives seems to me to breach the impasse of individualist ethics at a number of key points. First, it releases "nature" and nonhuman beings from their relegation to the status of objects with no ethical standing in the human pursuit of individual self-interest, *without* resorting to the extension of this liberal conception of ethical agency to other animals. Second, it substantiates an intersubjective understanding of ethical agency and community by which the corporeal connectivities between differently constituted actants can be traced in particular material circumstances and specified cases. And finally, it liberates the geographical imaginary of ethical community from the territorialized spaces of the embodied individual, the local neighborhood, and the nation-state, to trace the threads of ethical considerability through more dynamic, unstable, and performed spatial orderings of flow, mobility, and synthesis (see Shields, 1992).

I want to illustrate these themes briefly through the example of food, which represents one of the most pervasive corporeal mediators of hybrid communities spanning differently situated people, artifacts, biotic complexes, and practices (Lupton, 1996). As Atkinson has remarked, "Food is a liminal substance . . . bridging . . . nature and culture, the human and the natural, the outside and the inside" (1983: 11). The transformation of human food-production and food-consumption processes has involved the proliferation of hybrids, through the genetic engineering of plants and animals, and the pollution of biotic networks, through the release of synthetic chemical waste and the absorption of hormonal and chemical additives into the bodies and organs of producers and consumers of agrofood goods. The material and discursive economies of these hybrid networks connect the life-practices of human food-consumers and food-producers with those of other animals, plants, and environments over considerable distances. The ethical connectivities between actants at one location in the network and those at other locations are no less intimate or immediate for the physical distance or lack of proximate knowledge involved. Figure 7.1 traces in a simplified way the corporeal contours of ethical community for one hybrid network constituted through the fluid geographies of milk.

The figure illustrates the transfigurations of milk in animal (including human) bodies, variously inscribed by hormonal, genetic, and chemical treatments, and in biophysical spaces, such as in the form of nitrate runoff into river catchment areas. It highlights the myriad ways in which the connectivities between people, variously situated in the social organization of milk production and consumption, are fashioned in and through animals, habitats, and technologies, whose *presence* is integral to recognizing ethical community. Such a recognition informs numerous ethical practices, for example those manifested in alternative food networks which enact more equitable relations between producers and consumers, based on the principles of "fair trade" and more sustainable

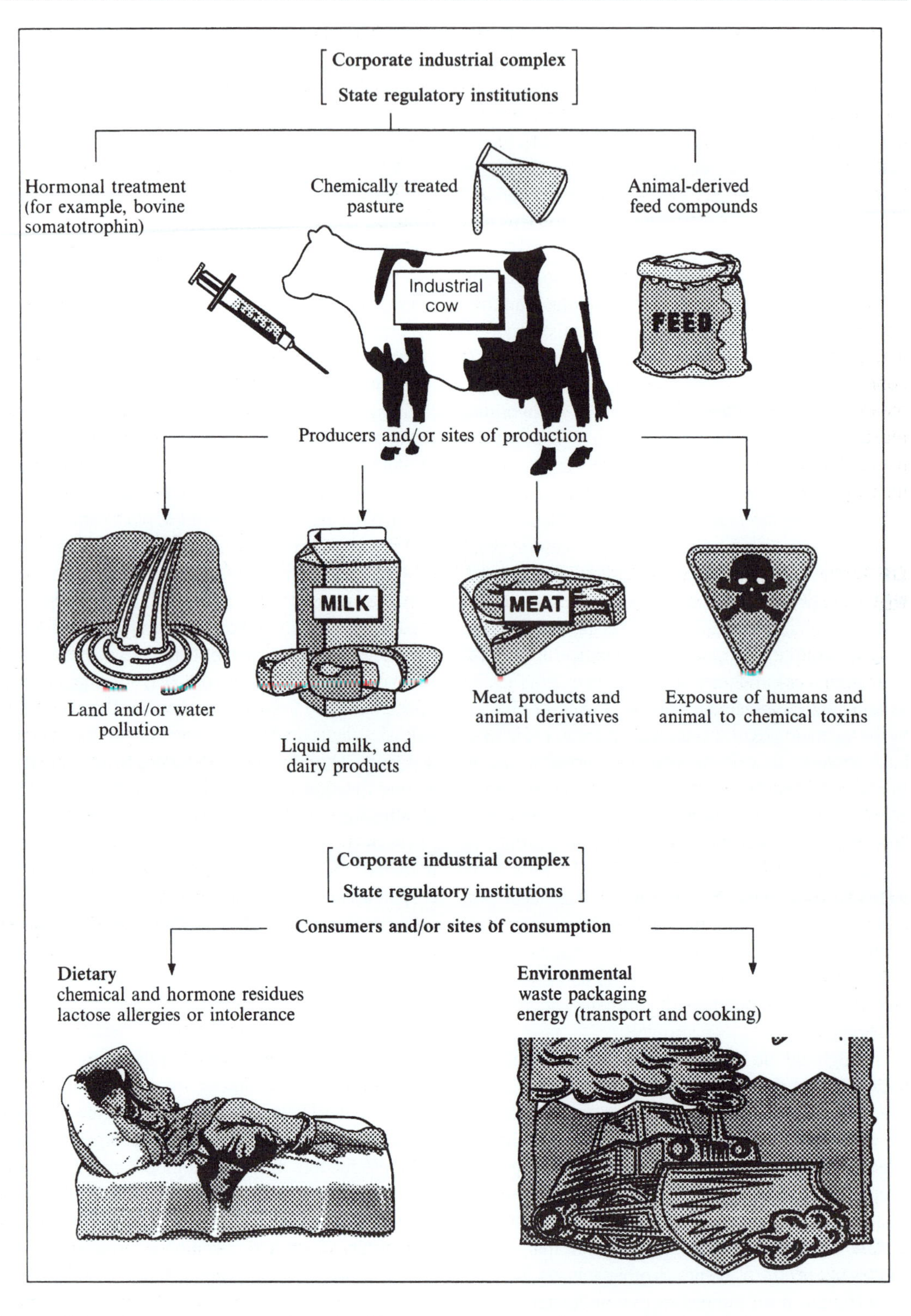

Figure 7.1 Corporeal geographies. Hybrid networks of embodiment and embeddedness: an agrofood example.

relations with other living constituents of the network through the adoption of "organic" farming methods.

The example raised earlier of BSE represents another such hybrid network, centered this time on a "prion" disease which has passed from sheep to cattle through infected animal feed and from cattle to humans through infected meat products (Lacey, 1994). The ethical (and political) implications of the BSE epidemic in Britain center precisely on recognizing the material properties of the BSE prion as a mobile constituent of a hybrid collective; an intricate network of corporeal relations between humans, animals, and technologies. In both cases, intimate ethical connections between people and places, bodies and meanings, sometimes over considerable distances, make sense only through an acknowledgement of the material properties of nature–culture hybrids such as milk and BSE.

GEOGRAPHICAL DIRECTIONS FOR A RELATIONAL ETHICS

> Modernity is changing the locus of belonging: our language of attachments limps suspiciously behind, doubting that our needs could ever find larger attachments.
>
> (M. Ignatieff, quoted in Corbridge, 1993: 449)

In an effort to articulate an intersubjective conception of ethical agency and a relational understanding of ethical considerability I have identified corporeality and hybridity as key modalities for reconfiguring the cartographies of ethical community. As Keller has noted, "it is precisely in embodiment that the many are becoming one and the outer becoming the inner" (1990: 236). Critically engaging with feminist and environmentalist critiques of traditional ethical discourse I have highlighted three issues which need simultaneously to be considered in pursuing this project:

1 extending the body politic beyond the human subject;
2 grounding cognitive processes and rationalities as specifically embodied and practiced; and
3 displacing the fixed and bounded contours of ethical community.

This understanding of ethical community is relational in concept, insisting on the situatedness of individual and collective efforts to realize new ethical connections and codes and their emergence through the political process rather than some ideal, rational, abstraction. As Cornell has noted,

> the opportunity to participate in . . . political life requires more than liberalism's formal recognition of each of us as abstract subjects equal before the law. It depends on the achievement of the material and cultural conditions for participation.
>
> (Cornell, 1985: 368)

This insistence on the situatedness of ethical practice and discourse recognizes the entrenched contours of the sovereign individual and the nation-state as sites of material struggle and resistance rather than narrative play. For although state sovereignty affirms that we have our primary political identity as participants in a particular civic community, we retain a potential connection with "humanity" through participation in broader international institutions (encoding both civic and human rights) (Walker, 1991: 256). Similarly, Haraway's figure of the cyborg suggests new possibilities for mapping the individual in an era of the "postorganic" embodiment of self in which knowledge projects such as genetic codification and recombinant DNA technologies breach the categorical cordons erected to distinguish humans from other animals, as well as between animal and plant species (Haraway, 1992b). No longer "a physical place to which one can go" (Haraway, 1992a: 66), "nature" emerges through notions of corporeality and hybridity as a staple figure of a relational ethics understood as a "sphere of judgement regarding the possibility and actuality of connections, arrangements, lineages, machines" (Grosz, 1994: 197).

Recognizing ethical communities as practically constructed and corporeally embedded points towards a world of fragile heterogeneous networks in which equality (in the sense of an equivalence of being rather than a universal rational ideal) remains the common premise of emancipatory subjects or movements (Thrift, 1996; Whitt and Slack, 1994). The practical and discursive stability of such networks can only be realized through a rethinking of the language of attachment and the locus of belonging in ways which breach the implicit spatial encoding of ethical consciousness and performance to proximate "others." As Harvey has argued, the issues of spatial and temporal scale are central to the question of building new ethical communities because the political power to act, to decide upon socioecological projects and to regulate their unintended consequences has always to be defined in relation to institutionalized scales (1993: 41). Theoretical and practical efforts in this direction are most visibly underway in the field of "development studies" (Corbridge, 1993) and in the political arenas of international governance. Nowhere is the urgency of such a rethinking more apparent than in the politics of so-called global environmental management in which the fate of the "distant poor" and "nonhumans" is cast in the shadow of the social institutions and practices of capital accumulation and material consumption led by advanced industrial countries concentrated in the "north" (Cooper and Palmer, 1995). As Visvanathan (1991) has observed, in such a Cartesian consumerist world it is self-restraint rather than self-authorship that would seem to promise more viable spaces for new forms of ethical community.

ACKNOWLEDGMENTS

This paper was conceived and largely written under the auspices of a Research Fellowship funded by the Economic and Social Research Council's Global Environmental Change Programme (L32027 3073). An early version of the paper was presented in the "Metatheory and ethics" session at the meetings of the Association of American Geographers in San Francisco in April 1993. I should like to thank the organizers and participants for their stimulation in working these issues through. In revising the paper for publication, I am particularly indebted to Andrew Sayer for his patience and detailed comments; Les Hepple, Doreen Massey, and Nigel Thrift for their encouragement; and two anonymous referees for valuable suggestions for improvement.

NOTES

1 Notable reworkings of the natural law tradition include those of Acquinas and Grotius, but Locke's work best epitomizes early modern tensions between notions of "common good" and "individual good."

2 Contemporary writers in this Kantian tradition have modified their reliance on the impartiality of justice by recognizing that competent moral agents are contracted on unequal terms; a theme pursued most influentially by Rawls (1971) in his "difference principle," and by Kymlicka (1991) in his notion of the "pluralist contract."

3 Persons in law can be nonindividuals, for example states, corporations, unions, etc. McHugh has argued that if the concept of "security of individual" (central to human rights law) were extended from persons to human beings, this would contribute towards the realization of substantive equality (that is, in terms of the material prerequisites for participating as equal members of a polity) (1992: 460).

4 It is no coincidence that the language of early women's struggles for political rights, notably in the writings of Mary Wollstonecraft, should borrow from those for the abolition of slavery in likening the status of wives to that of slaves (see Ferguson, 1992).

5 This is *not* to suggest that these are the only responses (for example, Habermasian critical theory is also notable) but rather that they have been the most influential in the sense of being translated into discourses beyond the academy.

6 For example, see contributions by Massey, Mouffe, and Natter to a recent edition of this journal [13(3), 1995] deriving from a session at the 1994 meetings of the Association of American Geographers entitled "Post-Marxism, democracy, and identity." Interestingly, Mouffe points to similar problems to those raised here with what she calls "a certain type of extreme postmodern fragmentation of the social" (1995: 262)—but without identifying any alleged "extremists."

7 See also Diprose's notion of "corporeal schema" which takes up Merleau-Ponty's idea of the body's directional activity or "intentional arc" (1994: 106).

8 Thrift makes a similar point, drawing on the Heideggarian notion of "comportment," in the marvelous introductory chapter to his book *Spatial Formations* (1996).

9 The ethical standing of animals has been a matter of dispute in moral philosophy, well in advance of contemporary environmentalism. Particularly influential contributions include the Thomist legacy of Thomas Aquinas in the natural law tradition, and the utilitarian legacy of Jeremy Bentham in the social contract tradition.

REFERENCES

Abram, D. 1988 "Merleau-Ponty and the voice of the earth," *Environmental Ethics* 10(3) 110–125.

Anderson, A. 1992 "Cryptonormativism and double gestures: the politics of post-structuralism," *Cultural Critique* 13(1): 63–95.

Atkinson, A. 1983 *Principles of Political Economy* (Belhaven, London).
Bauman, Z. 1992 *Mortality, Immortality and Other Life Strategies* (Polity Press, Cambridge).
Beck, U. 1992 *Risk Society* (Sage, London).
Benhabib, S. 1987 "General and concrete others," in *Feminism as Critique*, eds S. Benhabib and D. Cornell (Polity Press, Cambridge), pp. 77–96.
Benton, T. 1993 *Natural Relations: Ecology, Animal Rights and Social Justice* (Verso, London).
Bigwood, C. 1993 *Earth Muse: Feminism, Nature and Art* (Temple University Press, Philadelphia, PA).
Braidotti, R. 1992 "On the female feminist subject, or: from 'she-self to 'she-other'," in *Beyond Equality and Difference*, eds G. Block and S. James (Routledge, London), pp. 177–192.
Braidotti, R., Charkiewicz, E., Hausler, S. and Wieringa, S. 1994 *Women, the Environment and Sustainable Development. Towards a Theoretical Synthesis* (Zed Books, London).
Buckle, S. 1991 "Natural law," in *A Companion to Ethics*, ed. P. Singer (Blackwell, Oxford), pp. 161–174.
Callicott, J. B. 1979 "Elements of an environmental ethic: moral considerability and the biotic community," *Environmental Ethics* 171–81.
—— 1989 *In Defence of the Land Ethic: Essays on Environmental Philosophy* (State University of New York Press, Albany, NY).
Caraway, N. 1992 "The cunning of history: empire, identity and feminist theory in the flesh," *Women and Politics* 12(2): 1–18.
Cheney, J. 1989 "Postmodern environmental ethics: ethics as bioregional narrative," *Environmental Ethics* 11(2): 117–134.
Coles, R. 1993 "Eco-tones and environmental ethics," in *In the Nature of Things*, eds J. Bennett and W. Chaloupka (Minnesota University Press, Minneapolis, MN), pp. 226–249.
Cooper, D. and Palmer, J. (eds) 1995 *Just Environments* (Routledge, London).
Corbridge, S. 1993 "Marxisms, modernities, and moralities: development praxis and the claims of distant strangers," *Environment and Planning D: Society and Space* 11: 449–472.
Cornell, D. 1985 "Towards a post-modern reconstruction of ethics," *University of Pennsylvania Law Review* 133: 291–380.
Curtin, D. 1991 "Towards an ecological ethic of care," *Hypatia* 6(1): 60–74.
Diprose, R. 1994 *The Bodies of Women: Ethics, Embodiment and Sexual Difference* (Routledge, London).
Donovan, J. 1993 "Animal rights and feminist theory," in *Ecofeminism. Women, Animals, Nature*, ed. G. Gaard (Temple University Press, Philadelphia, PA), pp. 167–194.
Ebert, T. 1991 "The (body) politics of feminist theory," *Phoebe* 3(2): 56–65.
Ferguson, M. 1992 "Mary Wollstonecraft and the problematic of slavery," *Feminist Review* 82: 102–124.
Fox, W. 1990 *Towards a Transpersonal Ecology* (Shambhala Publications, London).
Friedman, M. 1989 "Feminism and modern friendship: dislocating the community," *Ethics* 99: 275–290.
Giddens, A. 1991 *Modernity and Self-identity* (Polity Press, Cambridge).
Gilligan, C. 1982 *In a Different Voice* (Harvard University Press, Cambridge, MA).
Gobetti, D. 1992 *Private and Public: Individuals, Households and Body Politic in Locke and Hutcheson* (Routledge, London).
Gore, A. 1992 *Earth in Balance* (Houghton Mifflin, Boston, MA).
Gross, E. 1986 "Philosophy, subjectivity and the body. Kristeva and Irigaray," in *Feminist Challenges*, eds C. Pateman and E. Gross (Allen and Unwin, Sydney), pp. 125–143.
Grosz, E. 1989 *Sexual Subversions* (Allen and Unwin, Sydney).
—— 1994 *Volatile Bodies: Toward a Corporeal Feminism* (Indiana University Press, Bloomington, IN).
Haraway, D. 1985 "Manifesto for cyborgs: science, technology and socialist feminism in the 1980s," *Socialist Review* 80: 65–108.
—— 1991 *Simians, Cyborgs, and Women. The Reinvention of Nature* (Free Association Books, London).
—— 1992a "Otherworldly conversations; terran topics; local terms," *Science as Culture* 3(1): 64–98.
—— 1992b "The promises of Monsters: a regenerative politics for Inappropriate/d Others," in *Cultural Studies*, eds L. Grossberg, C. Nelson and P. Treichler (Routledge, London), pp. 295–337.
Harvey, D. 1993 "The nature of environment. Dialectics of social and environmental change," *Socialist Register* 30: 1–51.
Hebdige, D. 1987 "The bottom line on planet one: squaring up to the face," *Ten-8* 19: 40-49.

Hennessy, R. 1993 *Materialist Feminism and the Politics of Discourse* (Routledge, London).

hooks, b. 1990 "Third world diva girls," in *Yearning: Race, Gender and Cultural Politics*, ed. b hooks (Turnaround, London), pp. 89–102.

Keller, C. 1990 "Secrets of god, nature and life," *History of the Human Sciences* 3(2): 229–242.

King, R. 1991 "Caring about nature: feminist ethics and the environment," *Hypatia* 6(1): 75–89.

Kruks, S. 1995 "Identity politics and dialectical reason: beyond an epistemology of provenance," *Hypatia* 10(2): 1–22.

Kymlicka, W. 1991 "The social contract tradition," in *A Companion to Ethics*, ed. P. Singer (Blackwell, Oxford), pp. 186–196.

Lacey, R. 1994 *Mad Cow Disease: The History of BSE in Britain* (Cypsela, St Helier, Jersey).

Laclau, E. and Mouffe, C. 1985 *Hegemony and Socialist Strategy. Towards a Radical Democratic Politics* (Verso, London).

Latour, B. 1993 *We Have Never Been Modern* (Harvester Wheatsheaf, Brighton).

Levin, D. 1985 "The body politic: political economy and the human body," *Human Studies* 8: 235–278.

Lovibond, S. 1994 "Maternalist ethics: a feminist reassessment," *The South Atlantic Quarterly* 93: 779–802.

Lupton, D. 1996 *Food, the Body and the Self* (Sage, London).

McHugh, J. 1992 "What is the difference between a 'person' and a 'human being' within the law?," *The Review of Politics* 54: 445–461.

Massey, D. 1995 "Thinking radical democracy spatially," *Environment and Planning D: Society and Space* 13: 283–288.

Mathews, F. 1991 *The Ecological Self* (Routledge, London).

Meyers, D. 1994 *Subjection and Subjectivity. Psychoanalytic Feminism and Moral Philosophy* (Routledge, London).

Mohanty, C., Russo, A. and Torres, L. (eds.) 1991 *Third World Women and the Politics of Feminism* (Indiana University Press, Bloomington, IN).

Moraga, C. and Anzaldua, G. 1981 *This Bridge Called My Back* (Persephone, Watertown, MA).

Mouffe, C. 1995 "Post-Marxism: democracy and identity," *Environment and Planning D: Society and Space* 13: 259–265.

Naess, A. 1989 *Ecology, Community and Lifestyle* (Cambridge University Press, Cambridge).

Natter, W. 1995 "Radical democracy: hegemony, reason, time and space," *Environment and Planning D: Society and Space* 13: 267–274.

O'Neill, J. 1985 *Five Bodies* (Cornell University Press, Ithaca, NY).

Pateman, C. 1989 *The Disorder of Women* (Stanford University Press, Stanford, CA).

Pile, S. and Thrift, N. 1995 "Introduction," in *Mapping the Subject. Geographies of Cultural Transformation*, eds S. Pile and N. Thrift (Routledge, London), pp. 13–56.

Plumwood, V. 1993 *Feminism and the Mastery of Nature* (Routledge, London).

Poole, R. 1991 *Morality and Modernity* (Routledge, London).

Porter, E.1991 *Women and Moral Identity* (Allen and Unwin, Sydney).

Probyn, E. 1993 "Technologizing the self," in *Cultural Studies*, eds L. Grossberg, C. Nelson and P. Treichler (Routledge, London), pp. 501–511.

Rawls, J. 1971 *A Theory of Justice* (Oxford University Press, Oxford).

Regan, T. and Singer, P. (eds) 1989 *Animal Rights and Human Obligations*, 2nd edition (Prentice-Hall, Englewood Cliffs, NJ).

Sandel, M. 1982 *Liberalism and the Limits of Justice* (Cambridge University Press, Cambridge).

Shields, R. 1992 "A truant proximity: presence and absence in the space of modernity," *Environment and Planning D: Society and Space* 10: 181–198.

Squires, J. (ed.), 1993 *Principled Positions: Postmodernism and the Rediscovery of Value* (Lawrence and Wishart, London).

Thrift, N. 1996 *Spatial Formations* (Sage, London).

Visvanathan, S. 1991 "Mrs Brundtland's disenchanted world," *Alternatives* 16: 377–384.

Walker, R. 1991 "On the spatio-temporal conditions of democratic practice," *Alternatives* 16: 243–262.

Whitt, L. and Slack, J. 1994 "Communities, environments and cultural studies," *Cultural Studies* 8(1): 5–31.

Wolch, J. and Emel, J. 1995 "Bringing the animals back in," *Environment and Planning D: Society and Space* 13: 632–636.

Young, I. 1989 "Polity and group difference: a critique of the ideal of universal citizenship," *Ethics* 99: 250–274.

PART 3

Practicing politicized geographic thought

Introduction

If it is understood that geographic scholarship is situated in the ways detailed above (i.e. as always political, and potentially to be useful for "progressive" political purposes), then we proceed to a discussion of different sorts of agency with which geographic scholarship is to be vested. In particular, this section orients geographic thought toward people's ability to represent the world, participate in the world, and change the world. These are neither "politically correct" orientations nor fringe geographical ideas. Knowledge always-already stands in relation to the relative presence and absence of these social "abilities" to know critically and act purposively with some success. In this section, therefore, we want to explore some of the implications of these orientations for the practices of geographic thought and scholarship.

Earlier, we referred to the notion of situated knowledge, and the embedded nature of knowledge production; knowledge is always produced in relational ways by embodied actors in particular times and places. In framing geographic thought as we have thus far, it should be clear that a sensitive and reflective stance is necessary for considering the power relations ever-present in the practices and products of our work. It is important to remain cognizant of the conditions in which scholarship is co-produced, the ways in which the products of scholarship circulate, and the possibility that these knowledges can be appropriated and re-presented by others for purposes which we might never intend. This does not mean that all such circumstances can be anticipated, predicted, or controlled, but a commitment to progressive scholarship entails an obligation to be as alert as possible to these contextual issues.

A few of these concerns have already been signaled in the papers above. In this section we want to highlight three particular elements of scholarly work, all of which will be further examined in the final parts of the book: (1) the products of our work as geographers; (2) our relationships with others in the processes of knowledge production; and (3) sites of progressive practices.

To examine the multiple issues connected to the products of our work as geographers, we draw upon the work of the late J. Brian Harley (1932–1991), who produced a provocative array of scholarship on that most basic and ubiquitous element of geographic knowledge: the map. Harley's project concerning the history of cartography was intimately bound up with other developments in social theory, and particularly with the nature of knowledge production and its connections with the social contexts of its production. For Harley maps are never innocent products that simply and faithfully reproduce or represent artifactual places and spaces. They are always multiply layered insights into the conceptions held by the map-maker, as well as the conditions that produce those conceptions. They are always made from a point of view, and are always deployed for a set of purposes (some of which may be unconscious from the perspective of the person making the map). As such, maps (as texts to be read and interpreted) are essential parts of the Foucauldian notion of "discourse." They are both drawn from and help to constitute systems of meaning. Such discourses are also always saturated with relations of power. One essential task for geographers, as Harley argues, is to excavate the meanings contained in maps as cultural products, and to understand the ways in which particular maps (as well as the enterprise of map-making itself) are connected to systems of power. Another is for geographers to be ever alert to these effects of maps as cultural products. We would like the reader to keep in mind that the map is used here as an exemplar (though a particularly apt one) of all products of geographic and other scholarship.

In Chapter 8, Harley is concerned with three major aspects of maps and map-making. First is his view of maps as a kind of language. Through this theoretical lens, Harley is able to consider both the specific content of maps as well as their rhetorical effects. Maps are produced in a particular "vocabulary" and "grammar" (i.e. the conventions of map-making) that themselves vary over time and space. An understanding of this language, Harley argues, provides important insights into the ideological dimensions that are always at play in map-making. Casting the production and reception of maps in the frame of language also prompts Harley to inquire about the nature and condition of authors and authorship, of readers and readership, and the development of meaning as an intersubjective relationship between/among them. Important questions that Harley raises concern the position (subject as well as geographic and historic) of the map-maker, his/her motivations in making a particular map, the cartographic choices embodied in the product, and their connections to the map's intended audience and desired effects.

A second perspective derives from the concept of iconology (or the interpretation of symbols and symbolic meanings) in painting and other forms of art. Here Harley is interested in applying this form of interpretation to dig beneath the surface features, as presented, in order to uncover deeper meanings, and particularly the nature and content of the political power represented in maps.

Finally, drawing upon Michel Foucault and Anthony Giddens, Harley explores map knowledge as a social product and as always connected to forms of power. The bulk of the paper elaborates this point and, through careful analysis, identifies the mechanisms employed in map-making to instantiate and reinforce (though also sometimes to subvert) particular moments of political power. Along the way, Harley points to the long-term complicity of geographic scholarship, particularly as "cartographic science," with the political projects of nation-states and imperialism, with property rights and capital régimes, and with the vital inventory functions necessary for primitive accumulation (a process more recently described by Harvey [2003] as accumulation through dispossession). In all of these contexts, Harley points to the power of maps and the processes of map-making, which draw upon all of the legitimating power of "science," to both represent and naturalize political relations "on the ground." Maps not only show how the world is divided up at any particular time and place (a putatively scientific and descriptive task), but also that such divisions are proper, legitimate, and to be taken for granted (a thoroughly normative task).

Harley's paper also points to the ways in which such tasks as map-making and other forms of scholarly knowledge production can often proceed without the producer being consciously aware of the ways in which his/her efforts work to reproduce or reinforce existing relations of power (what Harvey, in Chapter 1, calls *status quo* or counter-revolutionary scholarship). Through careful attention to the conventions, languages, discourses, and symbologies of map-making, Harley alerts us to the constant need to be mindful of the powerful ideologies that always underlie scholarship, and to be attentive to the effects that our products and practices produce in the world.

The next two papers, each with a different emphasis, explore an issue first expressed in this volume in the paper by David M. Smith (Chapter 6). One of Smith's concerns as he examines professional ethics, and in the questions he poses about good geographers, is the ethical aspects of "the treatment and representation of research subjects . . . [and] the importance of a communicative ethics in participatory research . . . the legitimacy of persons writing about a group . . . of which they are not members . . . [and] the conduct of cross-cultural research involving encounters with alternative views of the world" (see p. 104). These are thorny issues indeed.

The kinds of relational identity and connectivity that we discussed earlier, and that are raised by most of the papers thus far, obligate progressive geographers to think very carefully about knowledge production as a social enterprise. Who we propose to represent, how we represent those involved in our research, and the respective (and respectful) roles for researchers and the "researched" are all critical matters in progressive scholarship.

BOX 2: WORKING IN THE WORLD: PROGRESSIVE CHANGE AND MULTIVOCAL SCHOLARSHIP

An early and perceptive voice in this conversation was that of Chandra Mohanty (currently a Professor of Women's Studies and Humanities at Syracuse University). In her 1988 paper, "Under Western Eyes: Feminist Scholarship and Colonial Discourses," Mohanty uses the trope of the "Third World woman" to open up vexing questions of identity, essentialism, representation, and the inevitable politics of scholarly practices. The paper also works through a very practical example of what Young (Chapter 4) characterizes as cultural imperialism and what Blaut (Chapter 2) describes as "ethnoscience."

Mohanty begins by placing "Western feminist scholarship," itself a potentially essentialized category (though she is careful to avoid this problem), into a wider, global geopolitical-economic context in which the privileged power status of Western scholars becomes both legible and salient. Because Mohanty eschews the possibility of apolitical scholarship (as we also argue on p. 4), it is crucial to situate such scholarship or knowledge production in its necessary and continuous juxtapositions with power. For Mohanty, in this paper, a critical element of this is the "First World/Third World" tension (our use of quotation marks here follows Mohanty's own skepticism and caution in using these fraught terms), and her recognition of the domination of the West over "the Rest." Here she is resonating quite closely with Young's conception of cultural imperialism, its ability to impose dominant ideas as the norm, and the effects this kind of domination is capable of producing.

In her analysis of specific exemplars of Western, feminist scholarship, Mohanty first registers the multiple mechanisms through which oppressive stereotypes are both produced and reinforced to construct the "average Third World woman," and then goes on to consider the methodological, conceptual, and political difficulties that such constructs effect. Methodologically, Mohanty faults this work for its failure to pursue grounded, situated cases in sufficient detail to lay out the overlapping relationships through which "real women" are constructed in great complexity. The conceptual problem that this produces is a conception of a set of essentialized, stable categories like "woman" or "Third World woman" that ignore the kinds of nuanced difference that are produced when gender is cut through by class, race, age, ability, and other markers of identity.

For Mohanty all of this contributes to a set of debilitating political constraints. First, she finds this analysis reductive and leaves power relations defined as a stable binary, as she says "people who have it (read: men), and people who do not (read: women)." Furthermore, if we fail to identify the multiple ways in which women are oppressed, options for resisting oppression go unidentified or are mis-specified. Finally, if we fail to recognize the important differences among women, creativity is diverted from identifying effective bases for political coalition organization. Ultimately, Mohanty finds the binaries that emerge from this form of analysis politically crippling. The analyses fail to come to grips with more relationally defined notions of power, and end up perpetuating, she argues, exactly those oppressive systems that must be challenged if justice is to prevail.

Mohanty concludes her paper by characterizing this type of Western, feminist scholarship as itself a microcosm of the colonial and neo-colonial impulses of Western humanism more broadly. Both projects represent, for her, vigorous, sustained efforts to develop and perpetuate a set of hierarchical arrangements that attempt to justify the political and economic interests of privileged, dominant élites in the West. Mohanty's paper brings a sharp, specific focus on a set of cases that allow us to see how the concepts of ethnoscience and cultural imperialism play out in constructing, maintaining, stabilizing, and naturalizing oppression. Her analysis also points to some of the essential elements of strategic, progressive scholarship aimed at destabilizing and bringing down such edifices of injustice.

The paper by Richa Nagar (Chapter 9) (a geographer and Associate Professor of Gender, Women and Sexuality Studies at the University of Minnesota) and her collaborators (without seeking to minimize the contributions of Nagar's consultants, but rather to avoid cumbersome constructions in the discussion in this section, we will simply refer to the authors of this paper as Nagar) picks up on Mohanty's call for more carefully calibrated on-the-ground studies, and the complexities inherent in such sensitive, collaborative scholarship.

In this reflection on her own ongoing encounters with research collaborators across a variety of borders, Nagar is assessing the kinds of barriers that inhibit collective scholarship, and that go beyond the by now typical ruminations about how ultimately to share credit and representation in the outputs of the work. Although Nagar makes clear that these issues of representation and outputs are far from settled matters with important and complex problems, she finds that there are much more fundamental impediments to producing truly effective and collaborative work. For Nagar, the problems reside in the most elemental mismatches between Western, academic theorizations and political orientations, and the exigent priorities of the individuals and groups with whom such academics seek to work and collaborate. For Nagar this is fundamentally a failure of relevance, i.e. a failure to demonstrate to non-academic co-workers what is useful, non-obvious or helpful about academic theories, methods, and insights.

Nagar's analysis parallels, in many ways, our earlier discussion of both Young's (Chapter 4) and Fraser's (Chapter 5) papers and their conceptions of justice. Nagar recognizes the kinds of power hierarchies in knowledge production that Mohanty describes, and argues that the most effective remedies lie not in a distributive solution (i.e. the equitable sharing of credit and representation in the outcomes of research, important as these may be), but much more in a productive form of justice (i.e. Young's notions of control over decision processes and Fraser's concept of parity of participation). For Nagar, this kind of productive justice entails an equitable sharing of the construction of the theoretical and political frameworks that orient the research process at the outset. It is in these structural dimensions of knowledge production, she contends, that privileged hierarchies of power inhere, and constitute the sites where constructive engagement with collaborators must take place.

Nagar maintains that this kind of collaboration must include all elements of the research process. Problem formulation and the setting of priorities, theoretical frameworks, methodological considerations, as well as outputs (products, political tactics and strategies, audiences, venues) must all be co-produced. The imposition of any of these elements by privileged (i.e. Northern, Western, colonial, patriarchal, etc.) participants serves to undermine the kind of deep, collaborative endeavor Nagar is after. Exactly how this is to be done remains a question, but Nagar's alerting us to the necessity and appropriateness of the effort is itself quite valuable. (In fact, in her most recent forays into this complicated territory, Nagar continues to find such issues particularly vexing; see Benson and Nagar 2006.)

The final two papers in this section, again with different inflections, draw our attention to the sphere of pedagogy as a site of progressive geographic practice. Rich Heyman (Chapter 10) (a geographer, currently at the University of Texas, Austin), in an ardent promotion of the role of critical pedagogy, and a lament at its neglect by most academic geographers, situates his argument within an evocation of the contested nature of both universities and knowledge itself. As Heyman traces the evolution of the modern U.S. research university from its German predecessors, he demonstrates that the trajectory taken was neither predictable nor inevitable. The university, as the producer of "useful" (or instrumental) knowledge was not envisioned, Heyman insists, by the founders of the University of Berlin, but a later imposition by Daniel Coit Gilman when he established the Johns Hopkins University in Baltimore. Heyman is not putting forward a particular normative assertion about the proper role of a university here, but rather is making the more general case that the articulation of such a proper role has long been a matter of contestation and struggle.

Though the U.S. model of university structure and function has been quite compatible with the needs of a capitalist state (as a purveyor of apposite cultural norms, and the training of an appropriate workforce), Heyman's evocation of the historical contingency of that model allows him to conclude that both the nature and purposes of the modern university are still worth struggling over. He therefore urges critical scholars, including radical geographers, not to abandon this battle.

For Heyman, a key element in this contest, and a key site for engagement, is the classroom and what goes on there. Heyman's conceptions of knowledge production, and thus his views of pedagogy, resonate closely with

Nagar's perspectives on the research process (presumably the source of much of the "knowledge" that fuels teaching). For him, this is an interpersonal relationship, in which the "results" of research (whether a written text or some other form) become the beginning point for a negotiated co-production of meaning between teachers and students. Drawing upon the work of Paulo Freire (and other radical educators), Heyman challenges the typical hierarchical model of pedagogy as information transmission as both oppressive and contributing to the perpetuation of *status quo* relations of power.

Taking up Heyman's suggestions means that the nature of knowledge, of knowledge production, the respective roles of teachers and students, and the purposes of teaching all must be given the same careful, reflexive appraisal as other elements of scholarship, as well as the intersections among these elements (e.g. the connections between teaching and research). Viewed this way, the classroom becomes an important potential site for truly critical thinking and appraisal, and helps to explain the vehemence with which reactionary élite actors try to police the activities that go on there.

It is quite fitting to end this section with the paper by Andy Merrifield (Chapter 11) (currently a geographer and independent scholar living in France). By using as a touchstone the geographic expedition (see the discussion of the Horvath paper in Box 1 on p. 10), Merrifield's analysis helps us to link the concerns expressed by the radical geographers in the late 1960s with many of the strands of related scholarship that have emerged in the several succeeding decades. His paper also connects quite closely with the elaborations of justice and ethics that our discussions and the preceding papers have offered.

Merrifield takes up a theme that runs through much of our dialogue thus far, situated knowledge, and emphasizes particularly the political implication that situatedness (i.e. that knowledge is always produced somewhere, sometime, by relationally constructed someones) "implies that an understanding of reality is accountable and responsible for an enabling political practice" (see p. 174). He then uses this notion of "accountable responsibility" to negotiate between the need for a kind of universal ethical justification for progressive scholarship and an inescapable recognition of the role of difference in the world. This should be reminiscent of the arguments above by Young, Fraser and Smith (Chapters 4–6).

Merrifield then invokes the expedition idea first proposed by William Bunge in the 1960s as a particularly apt mechanism for producing situated knowledge, especially from the "standpoint" of those in dominated, subjugated positions in society. (In this regard, Merrifield [following Hartsock, 1989/90, and Marx, 1975], argues that such subjugated positions "present a truer and more adequate account of reality" (see p. 175). This is not a viewpoint that is held universally among advocates of standpoint epistemology. Haraway [1991], for example, would not accord such privilege to particular positions whether subjugated or dominant.) Merrifield's description of the geographical expedition ties it both to issues of social justice and to the kind of radical pedagogy just advocated by Heyman. Along the way he provides us with a brief biography of Bunge, and a contextualization of his motivations growing out of the heated atmosphere of the times. The expeditions (which were self-consciously termed so by Bunge to challenge geography's long complicity with colonialism and imperialism) were ardently anti-capitalist and anti-racist, and were designed to produce knowledge democratically and with a position (both geographic and political) explicitly staked out.

As Merrifield "resituates" the geographical expedition he brings it into engagement with the multiple issues we have touched upon: a commitment to justice and engaged geographic practice to uncover institutionalized mechanisms of oppression, alienation, and exploitation; concerns with gender, race, class, and other aspects of (relational) identity; the sensitive use of mapping and other geographic methods to elucidate the power structures that saturate everyday life; attention to an ethical component to space and landscape; and finally, in his examination of the role of personal biography in scholarship, a reprise of the concerns of Mohanty and Nagar with issues of representation and collaborative knowledge production.

The concluding arguments of the paper engage directly with Heyman's call for a reconsideration of pedagogy as a site for radical, progressive, geographic practice. For Merrifield, the geographical expedition (in both its original deployment, as well as in the resuscitated form he calls for here) emblematizes a valuable exemplar of just such engaged pedagogy, and the kinds of co-produced knowledge envisioned in this kind of model. He also raises a few cautionary points that hark back to Harvey's admonition, in Chapter 1, regarding the need to seek "real" as opposed to merely "liberal" commitment to change. As Bunge and others involved with the expeditions found, much was at stake for them, both personally and professionally, in pursuing radical scholarship. Merrifield situates such considerations in the present climate of speed-up in academic work, and puts forward Bunge's suggestion that committed radical geographers position themselves in the cracks "between the academy and broader social life." As the papers and discussions in the subsequent parts of the book demonstrate, this has been the ambition and vision for numerous, progressive geographic scholars for several decades now.

8

Maps, knowledge, and power*

J. Brian Harley

from D. Cosgrove and S. Daniels, eds, *The Iconography of Landscape*. Cambridge: Cambridge University Press, 1988, pp. 277–312

> Give me a map; then let me see how much
> Is left for me to conquer all the world, . . .
> Here I began to march towards Persia,
> Along Armenia and the Caspian Sea,
> And thence unto Bithynia, where I took
> The Turk and his great empress prisoners.
> Then marched I into Egypt and Arabia,
> And here, not far from Alexandria
> Whereas the Terrene and the Red Sea meet,
> Being distant less than full a hundred leagues
> I meant to cut a channel to them both
> That men might quickly sail to India.
> From thence to Nubia near Borno lake,
> And so along the Ethiopian sea,
> Cutting the tropic line of Capricorn,
> I conquered all as far as Zanzibar.
>
> (Christopher Marlowe, *Tamburlaine*, Part II, V.iii. 123–39)

A book about geographical imagery which did not encompass the map[1] would be like *Hamlet* without the Prince. Yet although maps have long been central to the discourse of geography they are seldom read as "thick" texts or as a socially constructed form of knowledge. "Map interpretation" usually implies a search for "geographical features" depicted on maps without conveying how as a manipulated form of knowledge maps have helped to fashion those features.[2] It is true that in political geography and the history of geographical thought the link is increasingly being made between maps and power—especially in periods of colonial history[3]—but the particular role of maps, as images with historically specific codes, remains largely undifferentiated from the wider geographical discourse in which they are often embedded. What is lacking is a sense of what Carl Sauer understood as the eloquence of maps.[4] How then can we make maps "speak" about the social worlds of the past?

THEORETICAL PERSPECTIVES

My aim here is to explore the discourse of maps in the context of political power, and my approach is broadly iconological. Maps will be regarded as part of the broader family of value-laden images.[5] Maps cease to be understood primarily as inert records of morphological landscapes or passive reflections of the world of objects, but are regarded as refracted images contributing to dialogue in a socially constructed world. We thus move the reading of maps away from the canons of traditional cartographical criticism with its string of binary oppositions between maps that are "true and false," "accurate and inaccurate," "objective and subjective," "literal and symbolic," or that are based on "scientific integrity" as opposed to "ideological distortion." Maps are never value-free images; except in the narrowest Euclidean sense they are not in themselves either true or false. Both in the selectivity of their content and in their signs and styles of representation maps are a way of conceiving, articulating, and structuring the human world which is biased towards, promoted by, and exerts influence upon particular sets of social relations.[6] By accepting such premises it becomes easier to see how appropriate they are to manipulation by the powerful in society.

Across this broad conceptual landscape I shall pinpoint three eminences from which to trace some of the more specific ideological contours of maps. From the first I view maps as a kind of language[7] (whether this is taken metaphorically or literally is not vital to the argument).[8] The idea of a cartographic language is also preferred to an approach derived directly from semiotics which, while having attracted some cartographers,[9] is too blunt a tool for specific historical enquiry. The notion of language more easily translates into historical practice. It not only helps us to see maps as reciprocal images used to mediate different views of the world but it also prompts a search for evidence about aspects such as the codes and context of cartography as well as its content in a traditional sense. A language—or perhaps more aptly a "literature" of maps—similarly urges us to pursue questions about changing readerships for maps, about levels of carto-literacy, conditions of authorship, aspects of secrecy and censorship, and also about the nature of the political statements which are made by maps.

In addition, literary criticism can help us to identify the particular form of cartographic "discourse" which lies at the heart of this essay. Discourse has been defined as concerning "those aspects of a text which are appraisive, evaluative, persuasive, or rhetorical, as opposed to those which simply name, locate, and recount."[10] While it will be shown that "simply" naming or locating a feature on a map is often of political significance, it nevertheless can be accepted that a similar cleavage exists within maps. They are a class of rhetorical images and are bound by rules which govern their codes and modes of social production, exchange, and use just as surely as any other discursive form. This, in turn can lead us to a better appreciation of the mechanisms by which maps—like books—became a political force in society.[11]

A second theoretical vantage point is derived from Panofsky's formulation of iconology.[12] Attempts have already been made to equate Panofsky's levels of interpretation in painting with similar levels discernible in maps.[13] For maps, iconology can be used to identify not only a "surface" or literal level of meaning but also a "deeper" level, usually associated with the symbolic dimension in the act of sending or receiving a message. A map can carry in its image such symbolism as may be associated with the particular area, geographical feature, city, or place which it represents.[14] It is often on this symbolic level that political power is most effectively reproduced, communicated, and experienced through maps.

The third perspective is gained from the sociology of knowledge. It has already been proposed that map knowledge is a social product,[15] and it is to clarify this proposition that two sets of ideas have been brought to bear upon the empirical examples in this essay. The first set is derived from Michel Foucault who, while his observations on geography and maps were cursory,[16] nevertheless provides a useful model for the history of map knowledge in his critique of historiography: "the quest for truth was not an objective and neutral activity but was intimately related to the 'will to power' of the truth-seeker. Knowledge was thus a form of power, a way of presenting one's own values in the guise of scientific disinterestedness."[17]

Cartography, too, can be "a form of knowledge and a form of power." Just as "the historian paints the landscape of the past in the colors of the present"[18] so the surveyor, whether consciously or otherwise, replicates not just the "environment" in some abstract sense but equally the territorial imperatives of a particular political system. Whether a map is produced under the banner of cartographic science—as most official maps have been—or whether it is an overt propaganda exercise, it cannot escape involvement in the processes by which power is deployed. Some of the practical implications of maps may also fall into the category of what Foucault has defined as acts of "surveillance,"[19] notably those connected with warfare, political propaganda, boundary making, or the preservation of law and order.

Foucault is not alone in making the connection between power and knowledge. Anthony Giddens, too, in theorizing about how social systems have become "embedded" in time and space (while not mentioning maps explicitly) refers to "authoritative resources" (as distinguished from material resources) controlled by the state: "storage of authoritative resources involves above all the retention and control of information or knowledge. There can be no doubt that the decisive development here is the invention of writing and notation."[20] Maps were a similar invention in the control of space and facilitated the geographical expansion of social systems, "an undergirding medium of state power." As a means of surveillance they involve both "the collation of information relevant to state control of the conduct of its subject population" and "the direct supervision of that conduct."[21] In modern times the greater the administrative complexity of the state—and the more pervasive its territorial and social ambitions—then the greater its appetite for maps.

What is useful about these ideas is that they help us to envisage cartographic images in terms of their political influence in society. The mere fact that for centuries maps have been projected as "scientific" images—and are still placed by philosophers and semioticians in that category[22]—makes this task more difficult. Dialectical relationships between image and power cannot be excavated with the procedures used to recover the "hard" topographical knowledge in maps and there is no litmus test of their ideological tendencies.[23] Maps as "knowledge as power" are explored here under three headings: the universality of political contexts in the history of mapping; the way in which the exercise of power structures the content of maps; and how cartographic communication at a symbolic level can reinforce that exercise through map knowledge.

POLITICAL CONTEXTS FOR MAPS

> *Tsar.* My son, what so engrosses you? What's this?
> *Fyodor.* A map of Muscovy; our royal kingdom
> From end to end. Look, father,
> Moscow's here
> Here Novgorod, there Astrakhan.
> The sea there,
> Here is the virgin forestland of Perm,
> And there Siberia.
> *Tsar.* And what may this be,
> A winding pattern tracing?
> *Fyodor.* It's the Volga.
> *Tsar.* How splendid! The delicious fruit of learning!
> Thus at a glance as from a cloud to scan
> Our whole domain: its boundaries, towns, rivers.
> (Alexander Pushkin, *Boris Godunov*)

In any iconological study it is only through context that meaning and influence can properly be unraveled. Such contexts may be defined as the circumstances in which maps were made and used. They are analogous to the "speech situation" in linguistic study[24] and involve reconstructions of the physical and social settings for the production and consumption of maps, the events leading up to these actions, the identity of map-makers and map-users, and their perceptions of the act of making and using maps in a socially constructed world. Such details can tell us not only about the motives behind cartographic events but also what effect maps may have had and the significance of the information they communicate in human terms.

Even a cursory inspection of the history of mapping will reveal the extent to which political, religious, or social power produce the context of cartography. This has become clear, for example, from a detailed study of cartography in prehistoric, ancient and medieval Europe, and the Mediterranean. Throughout the period, "mapmaking was one of the specialized intellectual weapons by which power could be gained, administered, given legitimacy, and codified."[25] Moreover, this knowledge was concentrated in relatively few hands and "maps were associated with the religious élite of dynastic Egypt and of Christian medieval Europe; with the intellectual élite of Greece and Rome; and with the mercantile élite of the city-states of the Mediterranean world during the late Middle Ages."[26]

Nor was the world of ancient and medieval Europe exceptional in these respects. Cartography, whatever other cultural significance may have been attached to it, was always a "science of princes." In the Islamic world, it was the caliphs in the period of classical Arab geography, thc Sultans in the Ottoman Empire, and the Mogul emperors in India who are known to have patronized map-making and to have used maps for military, political, religious, and propaganda purposes.[27] In ancient China, detailed terrestrial maps were likewise made expressly in accordance with the policies of the rulers of successive dynasties and served as bureaucratic and military tools and as spatial emblems of imperial destiny.[28] In early modern Europe, from Italy to the Netherlands and from Scandinavia to Portugal, absolute monarchs and statesmen were everywhere aware of the value of maps in defense and warfare, in internal administration linked to the growth of centralized government, and as territorial propaganda in the legitimation of national identities. Writers such as Castiglione, Elyot, and Machiavelli advocated the use of maps by generals and statesmen.[29] With national topographic surveys in Europe from the eighteenth century onwards, cartography's role in the transaction of power relations usually favored social élites.

The specific functions of maps in the exercise of power also confirm the ubiquity of these political contexts on a continuum of geographical scales. These range from global empire building, to the preservation of the nation state, to the local assertion of individual property rights. In each of these contexts the dimensions of polity and territory were fused in images which—just as surely as legal charters and patents—were part of the intellectual apparatus of power.

MAPS AND EMPIRE

As much as guns and warships, maps have been the weapons of imperialism (see Figure 8.1). Insofar as maps were used in colonial promotion, and lands claimed on paper before they were effectively occupied, maps anticipated empire. Surveyors marched alongside soldiers, initially mapping for reconnaissance, then for general information, and eventually as a tool of pacification, civilization, and exploitation in the defined colonies. But there is more to this than the drawing of boundaries for the practical political or military containment of subject populations. Maps were used to legitimize the reality of conquest and empire. They helped create myths which would assist in the maintenance of the territorial *status quo*. As communicators of an imperial message, they have been used as an aggressive complement to the rhetoric of speeches, newspapers, and written texts, or to the histories and popular songs extolling the virtues of empire.[30]

In these imperial contexts, maps regularly supported the direct execution of territorial power. The grids laid out by the Roman *agrimensores*, made functional in centuriation, were an expression of power "rolled out relentlessly in all directions . . . homogenizing everything in its path,"[31] just as the United States rectangular land survey created "Order upon the Land" in more senses than merely the replication of a classical design.[32] The rediscovery of the Ptolemaic system of co-ordinate geometry in the fifteenth century was a critical cartographic event privileging a "Euclidean syntax" which structured European territorial control.[33] Indeed, the graphic nature of the map gave its imperial users an arbitrary power that was easily divorced from the social responsibilities and consequences of its exercise. The world could be carved up on paper. Pope Alexander VI thus demarcated the Spanish and Portuguese possessions in the New World.[34] In the partitioning of North America, itself "part of a vast European process and experiment, an ongoing development of worldwide imperialism," the

> very lines on the map exhibited this imperial power and process because they had been imposed on the continent with little reference to indigenous peoples, and indeed in many places with little reference to the land itself. The invaders parceled the continent among themselves in designs reflective of their own complex rivalries and relative power.[35]

In the nineteenth century, as maps became further institutionalized and linked to the growth of geography as a discipline, their power effects are again manifest in the continuing tide of European imperialism. The scramble for Africa, in which the European powers fragmented the identity of indigenous territorial organization, has become almost a textbook example of these effects.[36] And in our own century, in the British partition of India in 1947, we can see how the stroke of a pen across a map could determine the lives and deaths of millions of people.[37] There are innumerable contexts in which maps became the currency of political "bargains," leases, partitions, sales, and treaties struck over colonial territory and, once made permanent in the image, these maps more than often acquired the force of law in the landscape.

MAPS AND THE NATION STATE

The history of the map is inextricably linked to the rise of the nation state in the modern world. Many of the printed maps of Europe emphasized the estates, waterways, and political boundaries that constituted the politico-economic dimensions of European geography.[38] Early political theorists commended maps to statesmen, who in turn were among their first systematic collectors.[39] The state became—and has remained—a principal patron of cartographic activity in many countries.[40]

Yet while the state was prepared to finance mapping, either directly through its exchequer or indirectly through commercial privilege, it often insisted that such knowledge was privileged. In western Europe the history of cartographic secrecy, albeit often ineffective, can be traced back to the sixteenth-century Spanish and Portuguese policy of *siglio*.[41] It was the practice to monopolize knowledge, "to use geographic

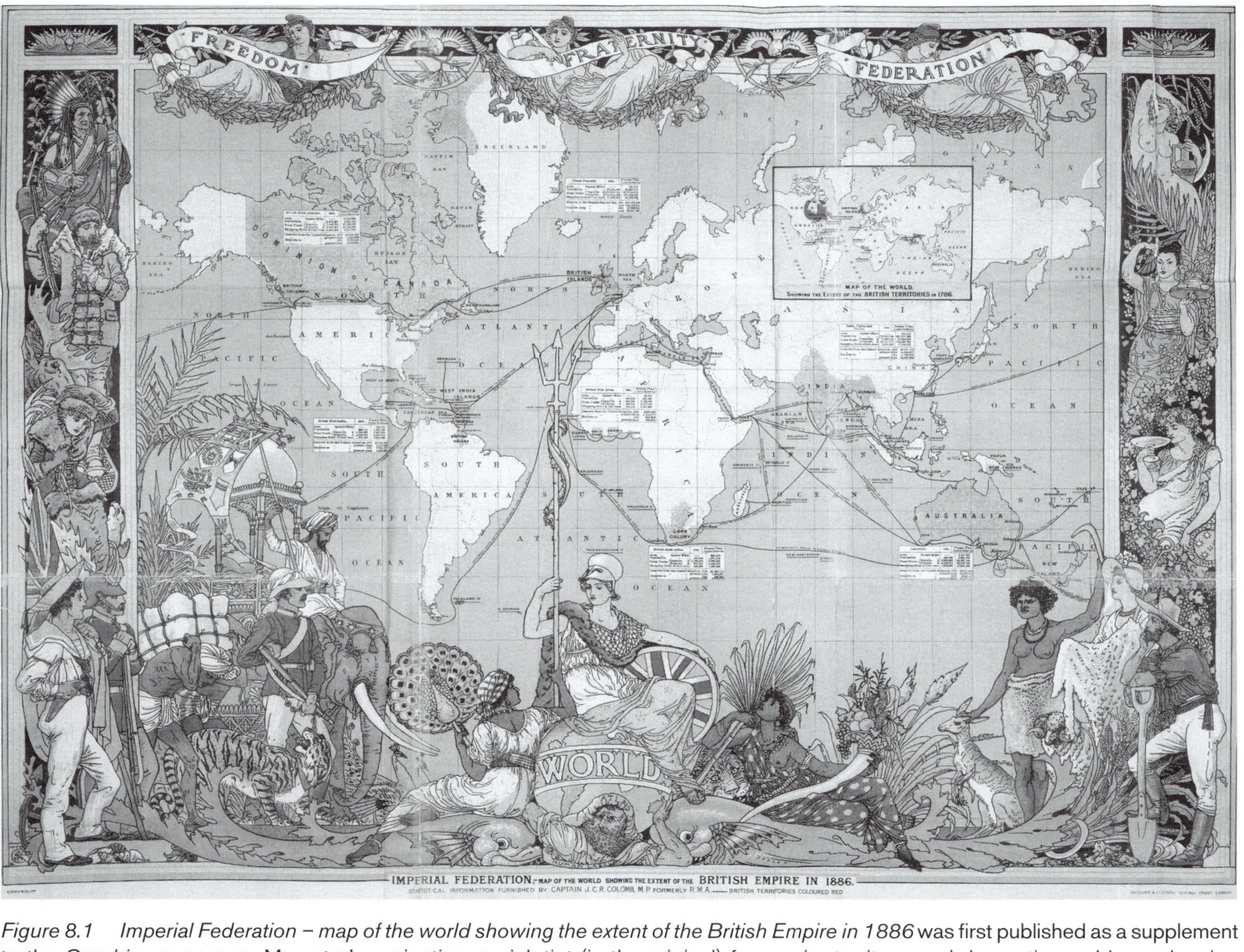

Figure 8.1 *Imperial Federation – map of the world showing the extent of the British Empire in 1886* was first published as a supplement to the *Graphic* newspaper. Mercator's projection, a pink tint (in the original) for empire territory, and decorative emblems showing Britannia seated on the world are used to articulate the message of the "New Imperialism." By coutesy of the Mansell Collection.

documents as an economic resource, much as craft mysteries were secreted and used."[42]

A major example of the interaction between maps and state polity is found in the history of military technology. In military eyes, maps have always been regarded as a sensitive sort of knowledge and policies of secrecy and censorship abound as much today in the "hidden" specifications of defense and official map-making agencies as in the campaign headquarters of the past.[43] At a practical level, military maps are a small but vital cog in the technical infrastructure of the army in the field. As the techniques of warfare were transformed from siege tactics to more mobile strategies, especially from the eighteenth century onwards, so too were the maps associated with them transformed.[44] Even in these active contexts, however, there were subtler historical processes at work. Map knowledge allows the conduct of warfare by remote control so that, we may speculate, killing is that much more easily contemplated.[45] Military maps not only facilitate the technical conduct of warfare, but also palliate the sense of guilt which arises from its conduct: the silent lines of the paper landscape foster the notion of socially empty space.

Not all military maps are silent; many stridently proclaim military victory. Just as there were military parades, songs, and poems, so too, at least from the fifteenth century onwards in Europe, there have been battle plans designed to commemorate the sacred places of national glory.[46]

MAPS AND PROPERTY RIGHTS

Cadastral or estate maps showing the ownership of property reveal the role of mapping in the history of agrarian class relations. Here the map may be regarded as a means by which either the state or individual landlords could more effectively control a tenant or peasant population.[47] In Roman society the codified practices of the *agrimensores* may be interpreted not just as technical manuals of land division in a theoretical sense but also as a social apparatus for legally regulating appropriated lands and for exacting taxation.[48] The maps themselves, whether cast in bronze or chipped in stone, were designed to make more permanent a social order in which there were freemen and slaves and for which the territorial division of land was the basis of status.[49] In early modern Europe, too, though the sociological context of mapping was different, some of the same forces were at work. The extent to which the mapping of local rural areas was locked into the process of litigation can leave us in no doubt about its socio-legal context and as a means by which conflict between lords and peasants over private rights in land could be more effectively pursued.[50] Maps fitted as easily into the culture of landed society as they had into the courtly diplomacies and the military maneuvers of European nation states in the Renaissance.

In similar terms maps can be seen to be embedded in some of the long-term structural changes of the transition from feudalism to capitalism. The world economy and its new geographical division of labor were produced with the aid of geographical documents including maps.[51] Accurate, large-scale plans were a means by which land could be more efficiently exploited, by which rent rolls could be increased, and by which legal obligations could be enforced or tenures modified. Supplementing older, written surveys, the map served as a graphic inventory, a codification of information about ownership, tenancy, rentable values, cropping practice, and agricultural potential, enabling capitalist landowners to see their estates as a whole and better to control them.[52] Seeing was believing in relation to the territorial hierarchies expressed in maps. Whether in the general history of agricultural improvement, of enclosure, of the draining or embankment of fens and marshes, or of the reclamation of hill and moor, the surveyor ever more frequently walks at the side of the landlord in spreading capitalist forms of agriculture.[53]

Maps impinged invisibly on the daily lives of ordinary people. Just as the clock, as a graphic symbol of centralized political authority, brought "time discipline" into the rhythms of industrial workers,[54] so too the lines on maps, dictators of a new agrarian topography, introduced a dimension of "space discipline." In European peasant societies, former commons were now subdivided and allotted, with the help of maps, and in the "wilderness" of former Indian lands in North America, boundary lines on the map were a medium of appropriation which those unlearned in geometrical survey methods found impossible to challenge. Maps entered the law, were attached to ordinances, acquired an aureole of science, and helped create an ethic and virtue of ever more precise definition. Tracings on maps excluded as much as they enclosed. They fixed territorial relativities according to the lottery of birth, the accidents of discovery or, increasingly, the mechanism of the world market.

MAP CONTENT IN THE TRANSACTION OF POWER

> "Is that the same map?" Jincey asked. She pointed to the large map of the world that hung, rolled up for the summer, above the blackboard behind Miss Dove. "Is China still orange?" "It is a new map," Miss Dove said. "China is purple." "I liked the old map," Jincey said. "I like the old world." "Cartography is a fluid art," said Miss Dove.
>
> (Frances Gray Patton, *Good Morning, Miss Dove*)

Cartographers and map historians have long been aware of tendencies in the content of their maps that they call "bias," "distortion" "deviance," or the "abuse" of sound cartographic principles. But little space in cartographic literature is devoted to the political implications of these terms and what they represent, and even less to their social consequences. Such "bias" or "distortion" is generally measured against a yard-stick of "objectivity," itself derived from cartographic procedure. Only in deliberately distorted maps, for example in advertising or propaganda, are the consequences discussed.[55] "Professional" cartography of the Ordnance Survey, the USGS [United States Geological Survey], Bartholomew or Rand McNally or their predecessors would be regarded as largely free from such politically polluted imagery. That maps can produce a truly "scientific" image of the world, in which factual information is represented without favor, is a view well embedded in our cultural mythology. To acknowledge that all cartography is "an intricate, controlled fiction"[56] does not prevent our retaining a distinction between those presentations of map content which are deliberately induced by cartographic artifice and those in which the structuring content of the image is unexamined.

Deliberate distortions of map content

Deliberate distortions of map content for political purposes can be traced throughout the history of maps, and the cartographer has never been an independent artist, craftsman, or technician. Behind the map-maker lies a set of power relations, creating its own specification. Whether imposed by an individual patron, by state bureaucracy, or the market, these rules can be reconstructed both from the content of maps and from the mode of cartographic representation. By adapting individual projections, by manipulating scale, by over-enlarging or moving signs or typography, or by using emotive colors, makers of propaganda maps have generally been the advocates of a one-sided view of geopolitical relationships. Such maps have been part of the currency of international psychological warfare long before their use by Nazi geopoliticians. The religious wars of seventeenth-century Europe and the Cold War of the twentieth century have been fought as much in the contents of propaganda maps as through any other medium.[57]

Apparently objective maps are also characterized by persistent manipulation of content. "Cartographic censorship" implies deliberate misrepresentation designed to mislead potential users of the map, usually those regarded as opponents of the territorial *status quo*. We should not confuse this with deletions or additions resulting from technical error or incompetence or made necessary by scale or function. Cartographic censorship removes from maps features which, other things being equal, we might expect to find on them. Naturally this is less noticeable than blatant distortion. It is justified on grounds of "national security," "political expediency," or "commercial necessity" and is still widely practiced. The censored image marks the boundaries of permissible discourse and deliberate omissions discourage "the clarification of social alternatives," making it "difficult for the dispossessed to locate the source of their unease, let alone to remedy it."[58]

The commonest justification for cartographic censorship has probably always been military. In its most wholesale form it has involved prohibiting the publication of surveys.[59] On the other hand settlement details on eighteenth-century maps were left unrevised by Frederick the Great to deceive a potential enemy, just as it has been inferred that the towns on some Russian maps were deliberately relocated in incorrect positions in the 1960s to prevent strategic measurements being taken from them by enemy powers.[60] Since the nineteenth century, too, it has been almost universal practice to "cleanse" systematically evidence of sensitive military installations from official series of topographical maps.[61] The practice now extends to other features where their inclusion would be potentially embarrassing to the government of the day; for example, nuclear waste dumps are omitted from official USGS topographical maps.

Deliberate falsification of map content has been associated with political considerations other than the purely military. Boundaries on maps have been subject to graphic gerrymandering. This arises both from attempts to assert historical claims to national territory,[62] and from the predictive art of using maps to project and to legitimate future territorial ambitions.[63] For example, disputed boundaries, whether shown on official maps, in atlases, or in more ephemeral images such as postage stamps, have been either included or suppressed according to the current political preference.[64] Nor do these practices apply solely to political boundaries on maps. It is well documented how the geographies of language, "race," and religion have been portrayed to accord with dominant beliefs.[65] There are the numerous cases where indigenous place-names of minority groups are suppressed on topographical maps in favor of the standard toponymy of the controlling group.[66]

"Unconscious" distortions of map content

Of equal interest to the student of cartographic iconology is the subtle process by which the content of maps is influenced by the values of the map-producing society. Any social history of maps must be concerned with these hidden rules of cartographic imagery and with their accidental consequences.[67] Three aspects of these hidden structures—relating to map geometry, to "silences" in the content of maps, and to hierarchical tendencies in cartographic representation will be discussed.

SUBLIMINAL GEOMETRY

The geometrical structure of maps—their graphic design in relation to the location on which they are centered or to the projection which determines their transformational relationship to the earth[68]—is an element which can magnify the political impact of an image even where no conscious distortion is intended. A universal feature of early world maps, for example, is the way they have been persistently centered on the "navel of the world," as this has been perceived by different societies. This "omphalos syndrome,"[69] where a people believe themselves to be divinely appointed to the center of the universe, can be traced in maps widely separated in time and space, such as those from ancient Mesopotamia with Babylon at its center, maps of the Chinese universe centered on China, Greek maps centered on Delphi, Islamic maps centered on Mecca, and those Christian world maps in which Jerusalem is placed as the "true" center of the world.[70] The effect of such "positional enhancing"[71] geometry on the social consciousness of space is difficult to gauge and it would be wrong to suggest that common design features necessarily contributed to identical world views. At the very least, however, such maps tend to focus the viewer's attention upon the center, and thus to promote the development of "exclusive, inward-directed worldviews, each with its separate cult center safely buffered within territories populated only by true believers."[72]

A similarly ethno-centric view may have been induced by some of the formal map projections of the European Renaissance. In this case, too, a map "structures the geography it depicts according to a set of beliefs about the way the world should be, and presents this construction as truth."[73] In the well-known example of Mercator's projection it is doubtful if Mercator himself—who designed the map with navigators in mind to show true compass directions—would have been aware of the extent to which his map would eventually come to project an image so strongly reinforcing the Europeans' view of their own world hegemony. Yet the simple fact that Europe is at the center of the world on this projection, and that the area of the land masses is so distorted that two-thirds of the earth's surface appears to lie in high latitudes, must have contributed much to a European sense of superiority. Indeed, insofar as the "white colonialist states" appear on the map relatively larger than they are while "the colonies" inhabited by colored peoples are shown "too small" suggests how it can be read and acted upon as a geopolitical prophecy.[74]

THE SILENCE ON MAPS

The notion of "silences" on maps is central to any argument about the influence of their hidden political messages (see Figure 8.2). It is asserted here that maps—just as much as examples of literature or the spoken word—exert a social influence through their omissions as much as by the features they depict and emphasize.

So forceful are the political undercurrents in these silences that it is sometimes difficult to explain them

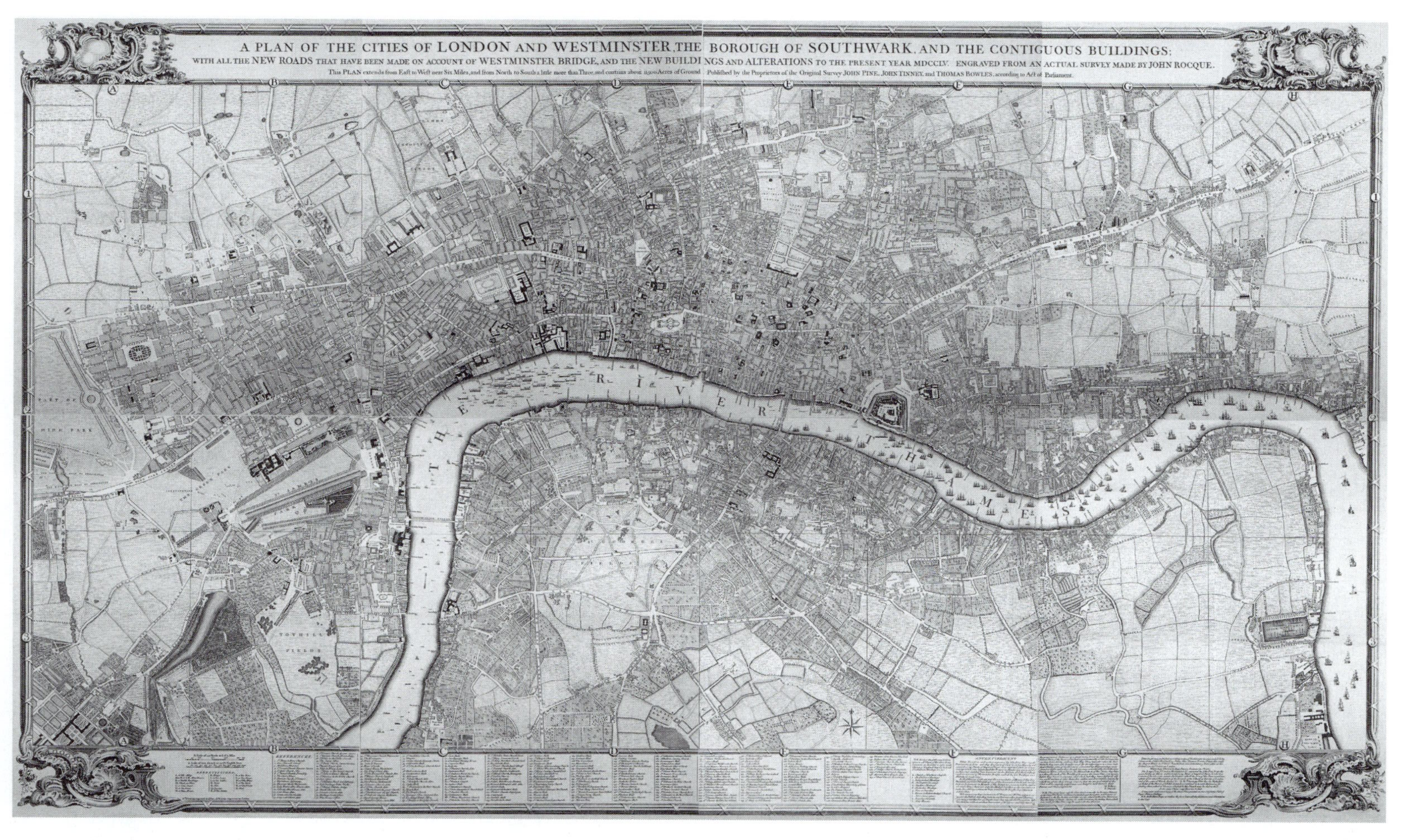

Figure 8.2 Silences on maps: part of John Rocque's "Plan of the Cities of London and Westminster . . ." (1755) showing the built-up area west of the City of London and the prestigious new green field developments of Bloomsbury. While districts to the north of Covent Garden and around Broad Street and St Giles were rapidly becoming slums, the cartographer has produced an idealized view of the city which emphasizes the gracious rurality of the main squares but fails to convey urban squalor. By permission of the British Library.

solely by recourse to other historical or technical factors. In seventeenth-century Ireland, for example, the fact that surveyors working for English proprietors sometimes excluded the cabins of the native Irish from their otherwise "accurate" maps is a question not just of scale and of the topographical prominence of such houses, but rather of the religious tensions and class relations in the Irish countryside.[75] Much the same could be said about omissions on printed county surveys of eighteenth-century England: the exclusion of smaller rural cottages may be a response as much to the ideal world of the map-makers' landed clients as to the dictates of cartographic scale.[76] On many early town plans a map-maker may have unconsciously ignored the alleys and courtyards of the poor in deference to the principal thoroughfares, public buildings and residences of the merchant class in his conscious promotion of civic pride or vaunting commercial success.[77] Such ideological filtering is a universal process. In colonial mapping, as in eighteenth-century North America, silences on maps may also be regarded as discrimination against native peoples. A map such as Fry and Jefferson's of Virginia (1751) suggests that the Europeans had always lived there: where "Indian nations" are depicted on it, it is more as a signpost to future colonial expansion than as a recognition of their ethnic integrity.[78] In this way, throughout the long age of exploration, European maps gave a one-sided view of ethnic encounters and supported Europe's God-given right to territorial appropriation. European atlases, too, while codifying a much wider range of geographical knowledge, also promoted a Eurocentric, imperialist vision, including as they did a bias towards domestic space which sharpened Europeans' perception of their cultural superiority in the world system.[79] Silences on maps—often becoming part of wider cultural stereotypes—thus came to enshrine self-fulfilling prophecies about the geography of power.

REPRESENTATIONAL HIERARCHIES

The role of the map as a form of social proclamation is further strengthened by the systems of classification and modes of representation—the so-called "conventional" or cartographic signs[80]—which have been adopted for landscape features. It has long been one of the map-maker's rules that the signs for towns and villages—whether depicted iconically or by abstract devices—are shown proportionally to the rank of the places concerned. Yet the resulting visual hierarchy of signs in early modern maps is often a replica of the legal, feudal, and ecclesiastical stratifications. Indeed, the concept of a tiered territorial society was by no means lost on contemporary map-makers. Mercator, for example, had hoped in his 1595 atlas to show "an exact enumeration and designation of the seats of princes and nobles."[81] Like other map-makers before him, he designed a set of settlement signs which, just as truly as the grids which have already been discussed, reify an ordering of the space represented on the map by making it visible. On other maps, towns occupy spaces on the map—even allowing for cartographic convention—far in excess of their sizes on the ground.[82] Castle signs, too, signifying feudal rank and military might, are sometimes larger than signs for villages, despite the lesser area they occupied on the ground. Coats of arms—badges of territorial possession—were used to locate the *caput* of a lordship while the tenurially dependent settlements within the feudal order were allocated inferior signs irrespective of their population or areal size. This was particularly common on maps of German territory formerly within the Holy Roman Empire. Such maps pay considerable attention to the geography of ecclesiastic power. The primary message was often that of the ubiquity of the church. Whether in "infidel" territory held by the Turk, inlands under the sway of the Papacy, in areas dominated by protestants in general, or by particular sects such as the Hussites, maps communicated the extensiveness of the temporal estate within the spiritual landscape. As a secondary message, not only do these maps heighten the perception of the power of the church as an institution within society as a whole, but they also record the spatial hierarchies and conflicting denominations within the church itself. On the former point, we may note that on Boazio's map of Ireland (1599), an exaggerated pictorial sign for "a Bishopes towne" is placed at the head of its key,[83] just as on the regional maps of Reformation England the signs for church towers and spires often rose far above the requirement of a notional vertical scale. On the matter of hierarchy, individual signs for archbishoprics and bishoprics, in arrays of single or double crosses, or crosiers, miters, and variations in ecclesiastical headgear, testify to the social organization of religion.[84] Here again, the selective magnifications of cartographic signs were closely linked to the shifting allegiances of opposing faiths. They survive as expressions of the religious battlegrounds of early modern Europe.

But if map signs sometimes reacted to changing religious circumstances they also tended to favor the *status quo*, legitimizing the hierarchies established on earlier maps. They were a socially conservative vocabulary. In France, for example, map-makers, as servants of the crown, inscribed images as a form of state propaganda, emphasizing the administrative mechanisms of its centralized bureaucracy and depicting aspects of the legal code of the *Ancien Régime*.[85] In 1721, when Bouchotte codified the signs to be used on regional maps (*cartes particulières*), for the territories which gave holders their titles, no less than seven of these are listed (*Duché Paine, Principauté, Duché, Marquisat, Comté, Vîcomté, Baronnie*) as well as five ecclesiastical ranks (archbishopric, bishopric, abbey, priory, *commanderie*).[86]

THE CARTOGRAPHIC SYMBOLISM OF POWER

> The earth is a place on which England is found,
> And you find it however you twirl the globe round;
> For the spots are all red and the rest is all grey,
> And that is the meaning of Empire Day.
>
> (G. K. Chesterton, "Songs of Education: 11 Geography," *The Collected Poems of G. K. Chesterton*)

In the articulation of power the symbolic level is often paramount in cartographic communication and it is in this mode that maps are at their most rhetorical and persuasive. We may consider the symbolic significance of the group of maps found within paintings, where maps are embedded in the discourse of the painting. Alternatively we may assess how artistic emblems—which may not be cartographic in character but whose meaning can be iconographically identified from a wider repertoire of images within a culture—function as signs in decorative maps where they are embedded in the discourse of the map. Having linked the meaning of particular emblems with the territory represented on the map, we may consider how non-decorative maps may equally symbolize cultural and political values.

MAPS IN PAINTING

The use by artists of globes and maps as emblems with their own specific symbolism can be traced back to the classical world. As a politically laden sign the globe or orb has frequently symbolized sovereignty over the world.[87] From Roman times onwards—on coins and in manuscripts—a globe or orb was held in the hand of an emperor or king. In the Christian era, now surmounted by a cross, the orb became one of the insignia of the Holy Roman Emperors and, in religious painting, it was frequently depicted held by Christ as *Salvator Mundi*, or by God the Father as *Creator Mundi*.[88] Such meanings were carried forward in the arts of the Renaissance. By the sixteenth century, globes which like maps had become more commonplace in a print culture,[89] were now shown as part of the regalia of authority in portraits of kings, ambassadors, statesmen, and nobles. But now they were primarily intended to convey the extent of the territorial powers, ambitions, and enterprises of their bearers. These paintings proclaimed the divine right of political control, the emblem of the globe indicating the world-wide scale on which it could be exercised and for which it was desired.[90]

Maps in painting have functioned as territorial symbols. The map mural cycles of the Italian Renaissance, for example, may be interpreted as visual summa of contemporary knowledge, power, and prestige, some of it religious but most of it secular.[91] In portraits of emperors, monarchs, statesmen, generals, and popes, maps also appear as a graphic shorthand for the social and territorial power they were expected to wield. It is apt that Elizabeth I stands on a map of sixteenth-century England; that Louis XIV is portrayed being presented with a map of his kingdom by Cassini;[92] that Pope Pius IV views the survey and draining of the Pontine marshes;[93] and that Napoleon is frequently shown with maps in his possession, whether on horseback, when campaigning, or seated and discussing proposed or achieved conquest.[94] Even when the medium changes from paint to photography and film the potent symbolism of the map remains, as the makers of films about Napoleon or Hitler readily grasped.[95] In newspapers, on television screens, and in innumerable political cartoons, military leaders are frequently shown in front of maps to confirm or reassure their viewers about the writ of power over the territory in the map. Map motifs continue to be accepted as geopolitical signs in contemporary society.

THE IDEOLOGY OF CARTOGRAPHIC DECORATION

Since the Renaissance, map images have rarely stood alone as discrete geographical statements, but have been accompanied by a wide range of decorative emblems.[96] From Jonathan Swift onwards these elements have been dismissed as largely incidental to the purposes of cartographic communication.[97] Decorative title pages, lettering, cartouches, vignettes, dedications, compass roses, and borders, all of which may incorporate motifs from the wider vocabulary of artistic expression, helped to strengthen and focus the political meanings of the maps on which they appeared (see Figure 8.3). Viewed thus, the notion of cartographic decoration as a marginal exercise in aesthetics is superannuated.

Such a symbolic role for decoration can be traced through much of the history of European cartography. The frontispieces and title pages of many atlases, for example, explicitly define by means of widely understood emblems both the ideological significance and the practical scope of the maps they contain.[98] Monumental arches are an expression of power; the globe and the armillary sphere are associated with royal dedications; portraits of kings and queens and depictions of royal coats of arms are incorporated into the design; royal emblems such as the *fleur de lys* or the imperial eagle also triggered political as well as more mundane geographical thoughts about the space mapped. The figures most frequently personified are those of nobles, bishops, wealthy merchants, and gentry. On English estate maps, microcosmic symbols of landed wealth, it is the coats of arms, the country house, and the hunting activity of the proprietors which are represented.[99] To own the map was to own the land.

In atlases and wall maps decoration serves to symbolize the acquisition of overseas territory. European navigators—portrayed with their cartographic trade

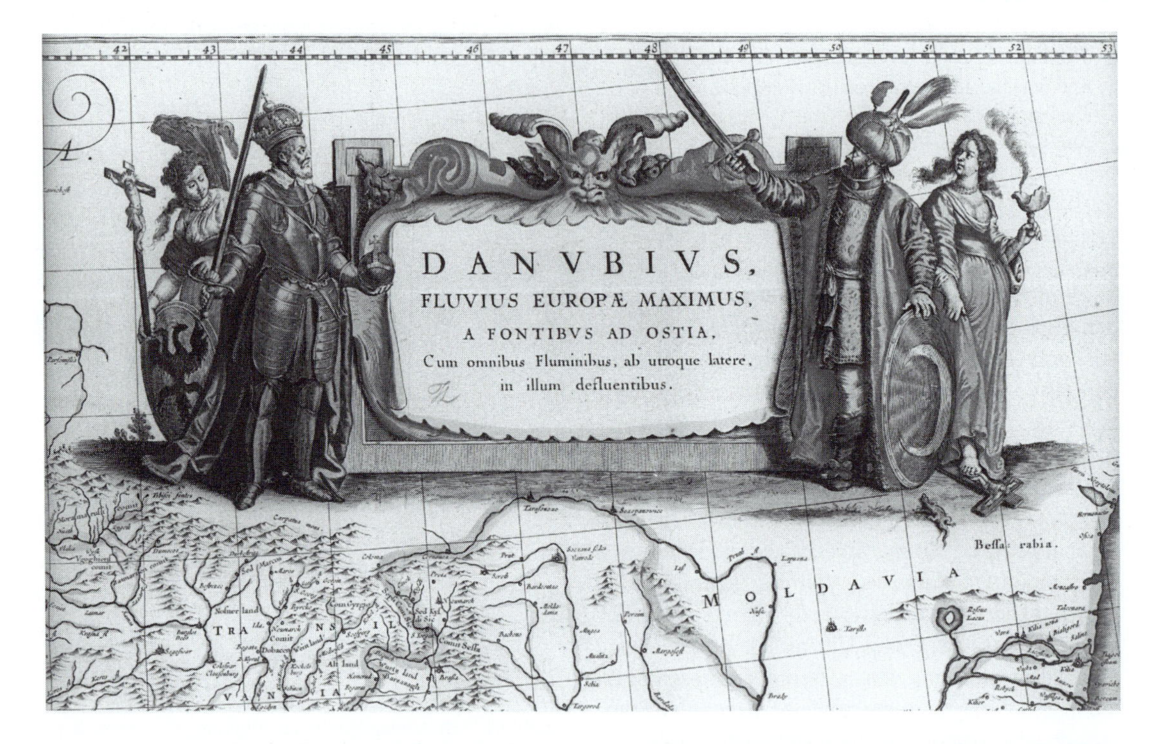

Figure 8.3 Religious and territorial conflict is epitomised in the cartouche to the map of the Danube in *Mayor o Geographia Blaviana*, vol. 3: *Alemania* (Amsterdam, 1662). Here, the Holy Roman Emperor (left), vested with emblems of power and the Christian faith, confronts the infidel Sultan, enemy of Christendom and spoiler of the cross. By courtesy of the American Geographical Society Collection, University of Wisconsin-Milwaukee.

symbol of compasses or dividers in hand[100]—pore earnestly over *terrae incognitae* as if already grasping them before their acts of "discovery," conquest, exploration, and exploitation have begun. Indeed, it is on the maps of these overseas empires that we find some of the most striking examples of ideological reinforcement through decoration. Whether we are looking at the French explorer's maps of South America in the sixteenth century[101] or nineteenth-century British maps of African territories decoration plays a part in attaching a series of racial stereotypes and prejudices to the areas being represented. This is manifestly so with Africa. The decoration on maps produced in Europe disseminated the image of the Dark Continent. Some of the motifs employed suggest that Europeans found it hard to accept that African humanity *was* different. Thus, in the margins of many maps African faces stare out with European features. African men were given "ideal" physiques and poses found in the iconography of figures in classical Greece and Rome; and African rulers —in obedience to the assumption that the political systems of Europe were universal—were usually depicted on maps as "kings."

In other cases the symbols of "otherness" assumed the form of a bizarre racism. Natives are shown riding an ostrich or a crocodile, engaged in cannibal practices, located in captions as "wild men," or, as on one French map of the eighteenth century, include "a race of men and women with tails." Female sexuality in depictions of African women and allegories for America and the other continents is often explicit for the benefit of male-dominated European societies.[102] Nor are the symbols of European power ever far from African space. European ships, castles, forts, and soldierly figures in European uniforms are deployed on maps in coastal regions; African "kings" are subject to European authority; and allegorical angels, the Bible, or the cross bring to the "barbarous" Africans the benefits of Christianity as part of a colonial package of enlightenment. Sometimes, too, cartouches and vignettes symbolize the colonial authority of individual nations: on a French map of 1708, black Africans are shown with a lion below the arms of France.[103]

CARTOGRAPHIC "FACT" AS SYMBOL

It is a short step to move back from these examples of artistic expression to consider another aspect of "real" maps. Having viewed maps in metaphorical contexts, it is easier to realize how a map which lacks any decorative features, or even caption and explanation, can nevertheless stand on its own as a symbol of political authority. Such maps are characterized by a "symbolic realism," so that what appears at first sight to be cartographic "fact" may also be a cartographic symbol. It is this duality of the map which encompasses much cartographic discourse and is a principal reason why maps so often constitute a political act or statement.

Once the ubiquity of symbolism is acknowledged, the traditional discontinuity accepted by map historians, between a "decorative" phase and a "scientific" phase of mapping, can be recognized as a myth.[104] Far from being incompatible with symbolic power, more precise measurement intensified it. Accuracy became a new talisman of authority. For example, an accurate outline map of a nation, such as Cassini provided for Louis XIV, was no less a patriotic allegory than an inaccurate one, while the "plain" maps of the Holy Land included in Protestant Bibles in the sixteenth century, in part to validate the literal truth of the text, were as much an essay in sacred symbolism as were more pictorial representations of the region.[105]

These are not exceptional examples of the historical role of measured maps in the making of myth and tradition.[106] Estate maps, though derived from instrumental survey, symbolized a social structure based on landed property; county and regional maps, though founded on triangulation, articulated local values and rights; maps of nation states, though constructed along arcs of the meridian, were still a symbolic shorthand for a complex of nationalist ideas; world maps, though increasingly drawn on mathematically defined projections, nevertheless gave a spiraling twist to the manifest destiny of European overseas conquest and colonization. Even celestial maps, though observed with ever more powerful telescopes, contained images of constellations which sensed the religious wars and the political dynasties of the terrestrial world.[107] It is premature to suggest that within almost every map there is a political symbol but at least there appears to be a *prima facie* case for such a generalization.

CONCLUSION: CARTOGRAPHIC DISCOURSE AND IDEOLOGY

I have sought to show how a history of maps, in common with that of other culture symbols, may be interpreted as a form of discourse. While theoretical insights may be derived, for example, from literary criticism, art history, and sociology, we still have to grapple with maps as unique systems of signs, whose codes may be at once iconic, linguistic, numerical, and temporal, and as a spatial form of knowledge. It has not proved difficult to make a general case for the mediating role of maps in political thought and action nor to glimpse their power effects. Through both their content and their modes of representation, the making and using of maps have been pervaded by ideology. Yet these mechanisms can only be understood in specific historical situations. The concluding generalizations must accordingly be read as preliminary ideas for a wider investigation.

The way in which maps have become part of a wider political sign-system has been largely directed by their associations with élite or powerful groups and individuals and this has promoted an uneven dialogue through maps. The ideological arrows have tended to fly largely in one direction, from the powerful to the weaker in society (see Figure 8.4). The social history of maps, unlike that of literature, art, or music, appears to have few genuinely popular, alternative, or subversive modes of expression. Maps are preeminently a language of power, not of protest. Though we have entered the age of mass communication by maps, the means of cartographic production, whether commercial or official, is still largely controlled by dominant groups. Indeed, computer technology has increased this concentration of media power. Cartography remains a teleological discourse, reifying power, reinforcing the *status quo*, and freezing social interaction within charted lines.[108]

The cartographic processes by which power is enforced, reproduced, reinforced, and stereotyped consist of both deliberate and "practical" acts of surveillance and less conscious cognitive adjustments by map-makers and map-users to dominant values and beliefs. The practical actions undertaken with maps: warfare, boundary making, propaganda, or the preservation of law and order, are documented throughout the history of maps. On the other hand, the undeclared processes of domination through maps are more subtle and elusive. These provide the "hidden rules" of cartographic discourse whose contours can be traced in the subliminal geometries, the silences, and the representational hierarchies of maps. The influence of the map is channeled as much through its representational force as a symbol as through its overt representations. The iconology of the map in the symbolic treatment of power is a neglected aspect of cartographic history. In grasping its importance we move away from a history of maps as a record of the cartographer's intention and technical acts to one which locates the cartographic image in a social world.

Maps as an impersonal type of knowledge tend to "desocialize" the territory they represent. They foster the notion of a socially empty space. The abstract quality of the map, embodied as much in the lines of a fifteenth-century Ptolemaic projection as in the contemporary images of computer cartography, lessens the burden of conscience about people in the landscape. Decisions about the exercise of power are removed from the realm of immediate face-to-face contacts.

These ideas remain to be explored in specific historical contexts. Like the historian, the map-maker has always played a rhetorical role in defining the configurations of power in society as well as recording their manifestations in the visible landscape. Any cartographic history which ignores the political significance of representation relegates itself to an "ahistorical" history.[109]

NOTES

* Harley's original figures 1, 4, 8, and 10 appear here as figures 8.1, 8.2, 8.3, and 8.4, respectively. The four figures deemed most important for conveying Harley's arguments were selected.

1 Geographical maps are but one aspect of the wider discourse of maps which extends to embrace other genres such as cosmological and celestial representations and maps of fictional areas.

2 Historians are also primarily concerned with the extent to which the evidence of maps can be evaluated as a "true" record of the facts of discovery, colonization, exploration, or other events in space.

3 On this view Margarita Bowen, *Empiricism and geographical thought from Francis Bacon to Alexander von Humboldt* (Cambridge, 1981); and D. R. Stoddard (ed.), *Geography, ideology and social concern* (Oxford, 1981), esp. pp. 11, 58–60.

4 Carl O. Sauer, "The education of a geographer," *Annals of the Association of American Geographers*, 46 (1956), pp. 287–99, esp. p. 289.

5 W. J. T. Mitchell, *Iconology: image, text, ideology* (Chicago, 1986), pp. 9–14.

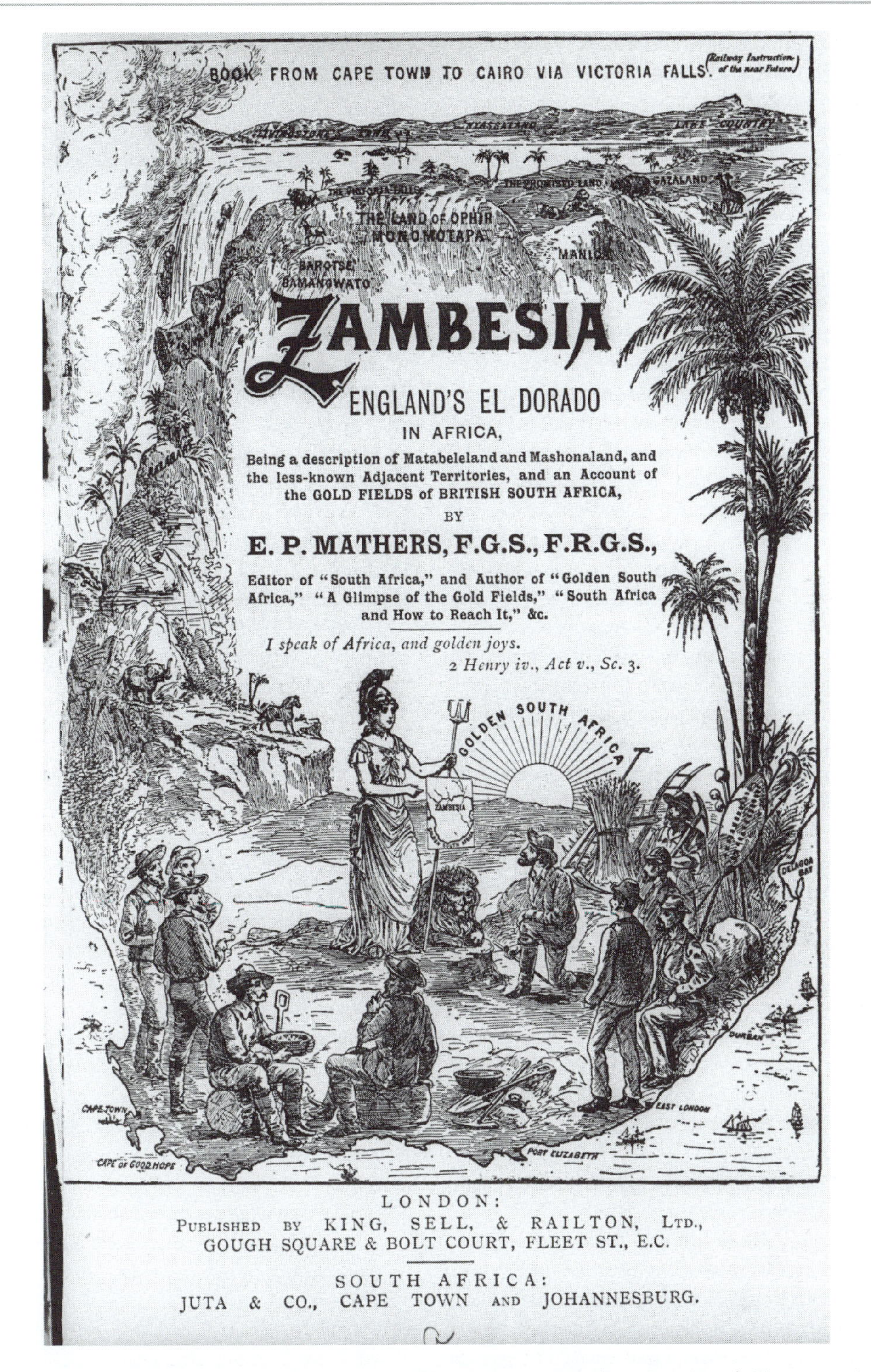

ZAMBESIA

ENGLAND'S EL DORADO

IN AFRICA,

Being a description of Matabeleland and Mashonaland, and the less-known Adjacent Territories, and an Account of the GOLD FIELDS of BRITISH SOUTH AFRICA,

BY

E. P. MATHERS, F.G.S., F.R.G.S.,

Editor of "South Africa," and Author of "Golden South Africa," "A Glimpse of the Gold Fields," "South Africa and How to Reach It," &c.

I speak of Africa, and golden joys.
2 *Henry iv., Act v., Sc.* 3.

LONDON:
PUBLISHED BY KING, SELL, & RAILTON, LTD.,
GOUGH SQUARE & BOLT COURT, FLEET ST., E.C.

SOUTH AFRICA:
JUTA & CO., CAPE TOWN AND JOHANNESBURG.

Figure 8.4 Title page from *Zambesia, England's El Dorado in Africa* (London, 1891). The scene is set on an outline map of southern Africa. Britannia, displaying a map of Zambesia, entices white colonists to take advantage of the economic wealth of the country while the indigenous African population is excluded from the stage. By courtesy of the American Geographical Society Collection, University of Wisconsin-Milwaukee.

6 Cf. the analysis of art in "Art as ideology," in Janet Wolff, *The social production of art* (London, 1981), p. 49.

7 How widely this is accepted across disciplines is demonstrated in W. J. T. Mitchell (ed.), *The language of images* (Chicago, 1980).

8 Arthur H. Robinson and Barbara Bartz Petchenik, *The nature of maps: essays toward understanding maps and mapping* (Chicago, 1976), discuss the analogy at length. It is rejected by J. S. Keates, *Understanding maps* (London, 1982), p. 86, although he continues to employ it as a metaphor for the ways maps "can be studied as ordered structures": Another recent discussion is C. Grant Head, "The map as natural language: a paradigm for understanding," in Christopher Board (ed.), *New insights in cartographic communication*, Monograph 31, *Cartographica*, 21, 1 (1984), pp. 1–32, and Hansgeorg Schlichtmann's "Discussion" of the Head article, *ibid.*, pp. 33–6.

9 Jacques Berlin, *Semiology of graphics: diagrams, networks, maps*, transl. William J. Berg (Madison, 1983); see also Hansgeorg Schlichtmann, "Codes in map communication," *Canadian Cartographer* 16 (1979), pp. 81–97; also Hansgeorg Schlichtmann, "Characteristic traits of the semiotic system 'Map Symbolizm,'" *Cartographic Journal*, 22 (1985), pp. 23–30. A humanistic application of semiology to maps is found in Denis Wood and John Pels, "Designs on signs: myth and meaning in maps," *Cartographica*, 23, 3 (1986), pp. 54–103.

10 Robert Scholes, *Semiotics and interpretation* (New Haven, 1982), p. 144.

11 In accepting that maps can be regarded as an agent of change in history we can draw on the ideas of Lucien Febvre and Henri-Jean Martin, *The coming of the book: the impact of printing 1450–1800*, transl. David Gerard (London, 1976); see also Kenneth E. Carpenter (ed.), *Books and society in history: papers of the Association of College and Research Libraries Rare Books and Manuscripts Preconference 24–28 June 1980, Boston, Massachusetts* (New York, 1983).

12 Erwin Panofsky, *Studies in iconology: humanistic themes in the art of the Renaissance* (Oxford, 1939).

13 A preliminary discussion is in M. J. Blakemore and J. B. Harley, *Concepts in the history of cartography. A review and perspective*, Monograph 26, *Cartographica*, 17, 4 (1980), pp. 76–86, and in J. B. Harley, "The iconology of early maps," *Imago et mensura mundi: atti del IX Congresso internazionale di Storia delta Cartographia*, ed. Carla Marzoli, 2 vols (Rome, 1985), 1, pp. 29–38. A narrower context is found in J. B. Harley, "Meaning and ambiguity in Tudor cartography," in Sarah Tyacke (ed.), *English map-making 1500–1650: historical essays* (London, 1983), pp. 22–45. For another application see Patricia Gilmartin, "The Austral continent on 16th century maps; an iconological interpretation," *Cartographica*, 21, 4 (1984), pp. 85–90. See also Brian S. Robinson, "Elizabethan society and its named places," *Geographical Review*, 63 (1973), pp. 322–33.

14 W. H. Stahl, "Representation of the earth's surface as an artistic motif," in *Encyclopedia of world art* (New York, 1960), 3, cols 851–4.

15 Mitchell, *Iconology*, p. 38.

16 See "Questions on geography," in Colin Gordon (ed.), *Power/knowledge: selected interviews and other writings 1972–1977 of Michel Foucault*, transl. Colin Gordon, Leo Marshall, John Mepham, and Kate Soper (Brighton, 1980), pp. 63–77, esp. pp. 74–5.

17 Mark Poster, "Foucault and history," *Social Research*, 49 (1982), pp. 116–42, esp. pp. 118–19.

18 *Ibid.*

19 M. Foucault, *Discipline and punish*, transl. Alan Sheridan (London, 1977), esp. pp. 195–228.

20 Anthony Giddens, *The contemporary critique of historical materialism: power, property and the state* (London, 1981), p. 94 (emphasis added).

21 *Ibid.*, p. 5.

22 See, for example, Nelson Goodman, *Languages of art: an approach to a theory of symbols* (Indianapolis and New York, 1968), pp. 170–3.

23 These arguments will be more fully developed in J. B. Harley, *The map as ideology: knowledge and power in the history of cartography* (London, forthcoming).

24 Oswald Ducrot and Tzvetan Todorov, *Encyclopedic dictionary of the sciences of language*, transl. Catherine Porter (Oxford, 1981), pp. 333–8.

25 J. B. Harley and David Woodward, "Concluding remarks," in J. B. Harley and David Woodward (eds), *The history of cartography*, vol. 1: *Cartography in prehistoric, ancient, and medieval Europe and the Mediterranean* (Chicago, 1987), p. 506.

26 *Ibid.*

27 Islamic cartography is most authoritatively described in E. van Donzel, B. Lewis, and Ch. Pellat (eds), *Encyclopaedia of Islam* (Leiden, 1978), vol. 4, pp. 1077–83.

28 Joseph Needham, *Science and civilization in China*, vol. 3, sec. 22 (Cambridge, 1959).

29 B. Castiglione, *The courtier* [1528], transl. George Bull (Harmondsworth, 1967), p. 97; Thomas Elyot, *The boke named the gouernour*, ed. from the first edn of 1531 by H. H. S. Croft, 2 vols (London, 1880), vol. 1, pp. 45, 77–8;

Machiavelli, *Arte della guerra* [1521], ed. S. Bertelli (Milan, 1961), pp. 457–8.

30 For the classical empires see O. A. W. Dilke, *Greek and Roman maps* (London, 1985), pp. 41–53 (on Agrippa's map) and pp. 169–70 (on the world map of Theodosius II). Maps of the British Empire became popular during the Victorian era: see Margaret Drabble, *For Queen and country; Britain in the Victorian age* (London, 1978), where the map by Maclure & Co., London, 1886, is reproduced. The geopolitical message of such maps and globes is unequivocably conveyed by G. K. Chesterton, "Songs of Education: II Geography," quoted on p. 139 above.

31 Samuel Y. Edgerton, Jr., "From mental matrix to *mappamundi* to Christian empire: the heritage of Ptolemaic cartography in the Renaissance," in David Woodward (ed.), *Art and cartography* (Chicago, 1987), p. 22.

32 Hildegard Binder Johnson, *Order upon the land. The U.S. rectangular land survey and the upper Mississippi country* (New York, 1976).

33 Claude Raffestin, *Pour une Géographie du pouvoir* (Paris, 1980), p. 131.

34 Alexander's bull regarding the demarcation line is given in Anne Fremantle (ed.), *The papal encyclicals in their historical context* (New York, 1956), pp. 77–81.

35 D. W. Meinig, *The shaping of America: a geographical perspective on 500 years of history*, vol. 1: *Atlantic America, 1492–1800* (New Haven, 1986), p. 232. A similar point is made by Robert David Sack, *Human territoriality: its theory and history* (Cambridge, 1986), p. 11.

36 See P. A. Penfold (ed.), *Maps and plans In the Public Record Office*, vol. 3: *Africa* (London, 1982), *passim*; J. Stengers, "King Leopold's imperialism," in Roger Owen and Bob Sutcliffe (eds), *Studies in the theory of imperialism* (London, 1972), pp. 248–76.

37 For a vivid reconstruction of Radcliffe's partition of India employing relatively small-scale maps see Larry Collins and Dominique Lapierre, *Freedom at midnight* (London, 1982), pp. 245–8.

38 Chandra Mukerji, *From graven images: patterns of modern materialism* (New York, 1983), p. 83. See also Giuseppe Dematteis, *Le metafore della terra: la geografia umana tra mito e scienzia* (Milan, 1985), pp. 54–9.

39 On early map collections see R. A. Skelton, *Maps: a historical survey of their study and collecting* (Chicago, 1972), pp. 26–61; Harley, "The map and the development of the history of cartography," in Harley and Woodward (eds), *History of cartography*, pp. 6–12.

40 For early examples of state involvement in topographical mapping see Lloyd A. Brown, *The story of maps* (Boston, 1949), esp. pp. 241–71.

41 Daniel J. Boorstin, *The discoverers* (New York, 1983), pp. 267–9; on the Dutch East India Company's policy see Gunter Schilder, "Organization and evolution of the Dutch East India Company's hydrographic office in the seventeenth century," *Imago Mundi*, 28 (1976), pp. 61–78; for an English example, Helen Wallis, "The cartography of Drake's voyage," in Norman J. W. Thrower (ed.), *Sir Francis Drake and the famous voyage, 1577–1580* (Los Angeles and London, 1985), pp. 133–7.

42 Mukerji, *From graven images*, p. 91; see also Chandra Mukerji, "Visual language in science and the exercise of power: the case of cartography in early modern Europe," *Studies in Visual Communications*, 10, 3 (1984), pp. 30–45.

43 Official map-making agencies, usually under the cloak of "national security," have been traditionally reticent about publishing details about what rules govern the information they exclude especially where this involves military installations or other politically sensitive sites.

44 Christopher Duffy, *Siege warfare. The fortress in the early modern world 1494–1660* (London, 1979), esp. p. 81; and *The fortress in the age of Vauban and Frederick the Great 1660–1789* (London, 1985), esp. pp. 29, 72, 142. On the effect of cartography on more mobile warfare see R. A. Skelton, "The military surveyor's contribution to British cartography in the 16th century," *Imago Mundi*, 24 (1970), pp. 77–83.

45 Phillip C. Muehrcke, *Map use: reading, analysis, and interpretation* (Madison, WI, 1978), pp. 299–301.

46 Probably the majority of published battle plans and campaign maps issued "after the event" in Europe down to the end of the eighteenth century fall either into this category or illustrated histories justifying the conduct of warfare:

47 A comparison can be made here with written documents; see, for example, M. T. Clanchy, *From memory to written record: England 1066–1307* (London, 1979), esp. pp. 149–265.

48 O. A. W. Dilke, *The Roman land surveyors. An introduction to the Agrimensores* (Newton Abbot, 1971).

49 P. Anderson, *Passages from antiquity to feudalism* (London, 1974), esp. pp. 147–53, 185, 188–9, 207–8.

50 P. D. A. Harvey, *The history of topographical maps: symbols, pictures and surveys* (London, 1980), *passim*.

51 Mukerji, *From graven images*, p. 84; Immanuel Wallerstein, *The modern world-system*, vol. 2: *Mercantilism and the consolidation of the European world economy,*

1600–1750 (New York, 1980), offers many clues to this process. Appropriately enough, the frontispiece to the volume is a world map by Jan Blaeu (1638).

52 J. R. Hale, *Renaissance Europe 1480–1520* (London, 1971), pp. 52–3.

53 F. M. L. Thompson, *Chartered surveyors: the growth of a profession* (London, 1968).

54 David S. Landes, *Clocks and the making of the modern world* (Cambridge, Mass., 1983), pp. xix, 2, 25, 228–30, 285–6; and Stephen Kern, *The culture of time and space* (London, 1983), pp. 10–35.

55 There is an extensive literature on maps in the pre-war German school of geopolitics. See, for example, Hans Speir, "Magic geography," *Social Research*, 8 (1941), pp. 310–30; Louis O. Quam, "The use of maps in propaganda," *Journal of Geography*, 42 (1943), pp. 21–32; Louis B. Thomas, "Maps as instruments of propaganda," *Surveying and Mapping*, 9 (1949), pp. 75–81; and John Ager, "Maps and propaganda," Society of University Cartographers, *Bulletin*, 11 (1977), pp. 1–14.

56 Muehrcke, *Map use: reading, analysis, and interpretation*, p. 295.

57 Geoffrey Parker, *The Thirty Years' War* (London, 1984), plates 10, 13.

58 T. J. Jackson Lears, "The concept of cultural hegemony: problems and possibilities," *American Historical Review*, 90 (1985), pp. 567–93.

59 Harry Margary, *The old series Ordnance Survey maps*, vol. 3 (Lympne Castle, 1981), p. xxxiv.

60 Speir, "Magic geography," p. 320; F. J. Ormeling, Jr., "Cartographic consequences of a planned economy —50 years of Soviet cartography," *The American Cartographer*, 1, 1 (1974), pp. 48–9; "Soviet cartographic falsifications," *The Military Engineer*, 62, 410 (1970), pp. 389–91.

61 For "security" reasons not even the existence of these practices is reported, although in Britain, for example, in recent years they have been unearthed by investigative journalism: see *New Statesman*, 27 May 1983, p. 6, which reported that "Moles within the Ordnance Survey have sent us a most interesting secret manual which lists and defines the places in Britain which do not officially exist, and therefore cannot appear on maps."

62 For example, in West Germany, the publishers of atlases have been obliged to obey a set of detailed ministerial regulations relating to political boundaries for maps that are to be used in schools. These did not receive approval for publication unless they showed the 1937 boundaries of Germany as well as those of today: K. A. Sinnhuber, "The representation of disputed political boundaries in general atlases," *The Cartographic Journal*, 1, 2 (1964), pp. 20–8.

63 Numerous examples occur in the eighteenth-century British and French maps of North America: Percy G. Adams, *Travelers and travel liars 1660–1800* (New York, 1980), pp. 64–79, who, however, misses the ideological significance of the cartographic falsification he describes. See also J. B. Harley, "The bankruptcy of Thomas Jefferys: an episode in the economic history of eighteenth century map-making," *Imago Mundi*, 20 (1966), pp. 28–48, esp. pp. 33–40. For a nineteenth-century example see Charles E. Nowell, *The rose-coloured map: Portugal's attempt to build an African empire from the Atlantic to the Indian Ocean* (Lisbon, 1982).

64 For political aspects of carto-philately see Bruce Davis, "Maps on postage stamps as propaganda," *Cartographic Journal*, 22, 2 (1985), pp. 125–30.

65 H. R. Wilkinson, *Maps and politics. A review of the ethnographic cartography of Macedonia* (Liverpool, 1951).

66 F. J. Ormeling, *Minority toponyms on maps: the rendering of linguistic minority toponyms on topographic maps of western Europe* (Utrecht, 1983).

67 The idea of the hidden rules of cartography comes from Michel Foucault, *The order of things, an archaeology of the human sciences* (London, 1966; repr. 1970).

68 These geometrical elements also include the manipulation of scale and orientation and the use of cartographic grids to organize space. On the wider social significance of these geometries see Robert Sack, *Conceptions of space in social thought: a geographic perspective* (London, 1980), *passim*.

69 The phrase is that of Edgerton, "From mental matrix to *mappamundi*," p. 26.

70 On European examples see Harley and Woodward, *The history of cartography*, vol. 1; on Chinese maps, Needham, *Science and civilization in China*, vol. 3; and on Islamic maps, *Encyclopaedia of Islam*, vol. 4.

71 The concept is E. H. Gombrich's *The sense of order* (Ithaca, 1979), pp. 155–6.

72 Edgerton, "From mental matrix to *mappamundi*," p. 27. For potential insights into how maps could have contributed to the infrastructure of social cosmologies, see Michael Harbsmeier, "On travel accounts and cosmological strategies: some models in comparative xenology," *Ethnos*, 50, 3–4 (1985), pp. 273–312.

73 Denis E. Cosgrove, *Social formation and symbolic landscape* (London, 1984), p. 8.

74 Arno Peters, *The new cartography* (New York, 1983), p. 63; see also Terry Cook, "A reconstruction of the

world: George R. Parkin's British Empire map of 1893," *Cartographica*, 21, 4 (1984), pp. 53–65, for the deliberate use of Mercator's projection in a map promoting the "New Imperialism" of the pan-Britannic world of the late nineteenth century. The recent reaction of cartographers towards the "unscientific" nature of the alternative "Peters' projection," which adjusts some of these distortions in favor of the Third World, provides a contemporary gloss on the entrenched scientism among map-makers which still gives credibility to the mathematically constructed map while ignoring the possibility of the social and political effects of its imagery. For example, see the comments by John Loxton, "The Peters' phenomenon," *The Cartographic Journal*, 22, 2 (1985), pp. 106–8, which attempt to discredit Peters as a "Marxist" and "Socialist." "The so-called Peters' projection," in *ibid.*, pp. 108–10, which is presented as the considered view of the German Cartographical Society is in some respects more polemical than Peters in its "defense of truthfulness and pure scientific discussion." See also A. H. Robinson, "Arno Peters and his new cartography," *American Cartographer*, 12 (1985), pp. 103–11, and Phil Porter and Phil Voxland, "Distortion in maps: the Peters' projection and other devilments," *Focus*, 36 (1986), pp. 22–30.

75 J. H. Andrews, *Plantation acres: an historical study of the Irish land surveyor and his maps* (Belfast, 1985), pp. 157–8.

76 J. B. Harley, "The re-mapping of England 1750–1800", *Imago Mundi*, 19 (1965), pp. 56–67; Paul Laxton, "The geodetic and topographical evaluation of English county maps, 1740–1840," *The Cartographic Journal*, 13, 1 (1976), pp. 37–54.

77 Cf. Juergen Schulz, "Jacopo de' Barbari's view of Venice: map making, city views and moralized geography before the year 1500," *Art Bulletin*, 60 (1978), pp. 425–74; J. B. Harley, "Meaning and ambiguity in Tudor cartography," pp. 28–32.

78 For the development of this argument see J. B. Harley, "Society, ideology, and the English geographical atlas in the eighteenth century," in John A. Wolter (ed.), *Images of the world: the atlas through history* (Washington, D.C., forthcoming).

79 James R. Akerman, "National geographical consciousness and the structure of early world atlases," Paper presented at the Eleventh International Conference on the History of Cartography, Ottawa, Canada, July 1985.

80 I am indebted to Catherine Delano Smith for discussion and the sight of a draft manuscript on "Cartographic signs in the Renaissance," to be published in J. B. Harley and David Woodward (eds), *The history of cartography*, vol. 3: *Cartography in the age of Renaissance and discovery* (Chicago, forthcoming).

81 Catherine Delano Smith, "Cartographic signs on European maps and their explanation before 1700," *Imago Mundi*, 37 (1985), pp. 9–29, where Mercator's *Advice for the use of maps: atlas sive cosmographicae. Meditationes de fabrica mundi et fabricati figura* (1595) is quoted, pp. 25–6.

82 See Christian Sgrothen's maps of the Netherlands (1573) where towns such as Bruges, Brussels, and Ghent are depicted in high oblique in such a way—and with so large a sign—as to ensure ample scope for the detailed display of the attributes of their commercial success and civic pride

83 Edward Lynam, "Boazio's map of Ireland," *British Museum Quarterly*, 11 (1937), pp. 92–5.

84 François de Dainville, *Le Langage des géographes: termes, signes, couleurs des cartes anciennes, 1500–1800* (Paris, 1964), pp. 236–44.

85 François de Dainville, "Le Signe de 'justice' dans les cartes anciennes," *Revue historique de droit français et étranger*, 4th ser., 34 (1956), pp. 111–14. For a broader context see Yi Fu Tuan, *Landscapes of fear* (Oxford, 1980).

86 Buchotte, *Les Règles du dessin et du lavis* (Paris, 1721), plate facing p. 124.

87 Helen Wallis, "Globes in England up to 1660," *The Geographical Magazine*, 35 (1962–3), pp. 267–79.

88 David Woodward, "Medieval *mappaemundi*," in Harley and Woodward (eds), *The history of cartography*, vol. 1, pp. 334–42.

89 Victor Morgan, "The literary image of globes and maps in early modern England," in Tyacke (ed.), *English map-making 1500–1650*, pp. 46–56.

90 For other meanings of the globe see James Hall, *Dictionary of subjects and symbols in art* (London, 1974), p. 139; and J. E. Cirlot, *A dictionary of symbols*, 2nd edn, transl. Jack Sage (London, 1971), pp. 118–19.

91 Juergen Schulz, "The map mural cycles of the Renaissance," in Woodward (ed.), *Art and cartography*, pp. 97–120.

92 Reproduced in *Arte e scienza per il disegno del mondo* (Milan, 1983), p. 57; see also the plate on p. 56.

93 Roberto Almagià, *Monumenta cartographica vaticana*, 4 vols. (Vatican City, 1952), vol. 3: *Le pitture murali della galleria delle carte geografiche*, pp. 7, 12.

94 *Cartes et figures de la terre* (Paris, 1980), p. 354; *A la Découverte de la terre. Dix siècles de cartographie* (Paris, 1979), facing p. 57.

95 Abel Gance, *Napoleon* (France, 1927); *The Great Dictator* (US, 1940). On the Gance film see Peter Pappas, "The superimposition of vision: *Napoleon* and the meaning of Fascist art," *Cineaste. A Magazine on the Art and Politics of the Cinema* (1983), pp. 5–13.

96 A. G. Hodgkiss, *Understanding maps: a systematic history of their use and development* (Folkestone, 1981), pp. 184–98; MacDonald Gill, "Decorative maps," *The Studio*, 128 (1944), pp. 161–9.

97 So Geographers in Afric-Maps
With Savage-Pictures fill their Gaps;
And o'er unhabitable Downs
(Place Elephants for want of Towns.
(Jonathan Swift, *On poetry: a rhapsody*)

"Savage-Pictures," "Elephants," and a "want of Towns" (towns being one of the hallmarks of European civilization) suggest that a stereotype of African geography, promoted by maps, was already in existence. On present-day attitudes towards decoration, see R. A. Skelton, *Decorative printed maps of the 15th to 18th centuries* (London, 1952), p. 1.

98 These have been treated as decorative ephemera for collectors: R. V. Tooley, *Title pages from 16th to 19th century* (London, 1975). Historians of cartography still have to attempt the depth of iconographic analysis revealed in M. Corbett and R. Lightbown, *The comely frontispiece: the emblematic title-page in England 1550–1660* (London, 1979), or F. A. Yates, *Astraea: the imperial theme in the sixteenth century* (London, 1975), p. 63.

99 Harley, "Meaning and ambiguity in Tudor cartography," pp. 37–8; Hilda Marchant, "A 'Memento Mori' or 'Vanitas' emblem on an estate map of 1612," *Mapline*, 44 (1986), pp. 1–4.

100 In different contexts compasses have other meanings: see Hall, *Dictionary of . . . Symbols*, p. 73.

101 H. Wallis, *The boke of idrography of Jean Rotz* (Oxford, 1982), esp. pp. 67–72; Bernadette Bucher, *Icon and conquest: a structural analysis of the illustrations of de Bry's Great Voyages* (Chicago, 1981).

102 On the female personifications for America see Hugh Honour, *The new golden land: European images of America from the discoveries to the present time* (New York, 1975), pp. 85–117, and Clare Le Corbeiller, "Miss America and her sisters: personifications of the four parts of the world," Metropolitan Museum of Art, *Bulletin*, 19, New Series (1961), pp. 209–23. I owe these two references to Howard Deller.

103 Oscar I. Norwich, *Maps of Africa: an illustrated and annotated carto-bibliography* (Johannesburg, 1983). For comparison see Leonard Bell, "Artists and empire: Victorian representations of subject people," *Art History*, 5, 1 (1982), pp. 73–86.

104 R. Rees, "Historical links between cartography and art," *Geographical Review*, 70 (1980), pp. 60–78; David Woodward, "Introduction," in Woodward (ed.), *Art and cartography*, vol. 2.

105 The continued symbolic significance of the map is indicated by Louis XIV's dismay in the thought that his kingdom had shrunk as a result of more accurate survey. Brown, *Story of maps*, facing p. 246. On biblical maps see the prefatory "epistle" to the 1559 Geneva Bible of Nicolas Barbier and Thomas Courteau where the usefulness of the maps in interpreting the scriptures is explained: I owe this reference to Catherine Delano Smith.

106 Göran Therborn's argument in *The ideology of power and the power of ideology* (London, 1980), pp. 81–4, about "affirmative symbolism or ritual" is relevant to maps; see also Eric Hobsbawm and Terence Ranger (eds), *The invention of tradition* (Cambridge, 1983), esp. pp. 1–100, 211–62.

107 Deborah J. Warner, *The sky explored: celestial cartography 1500–1800* (New York and Amsterdam, 1979), pp. xi–xii, discusses the iconographies of constellations produced by astronomers supporting the Reformation and the Counter Reformation respectively.

108 There is a parallel here to some of the tendencies identified by Robert David Sack, "Human territoriality: a theory," *Annals of the Association of American Geographers*, 73, 1 (1983), pp. 55–74; the ideas are more fully developed in Sack, *Human territoriality: its theory and history*.

109 This paper was given in a preliminary form at a meeting of the "Visual Documentation Group" of the History Workshop Center for Social History, held at Ruskin College, Oxford, in February 1984. It has subsequently been presented in seminars at the Department of Art History and Theory in the University of Essex and at the Department of Geography at the University of Wisconsin at Madison. I am grateful for the constructive suggestions received on those occasions and, for helpful comments, to John Andrews, Peter Barber, Mark Blacksell, Mark Cleary, Catherine Delano Smith, Ann Godlewska, Derek Gregory, Nicola Gregson, Roger Kain, Richard Oliver, Raphael Samuel, and David Woodward.

9

Collaboration across borders

Moving beyond positionality

Richa Nagar, in consultation with Farah Ali[1] and *Sangatin* women's collective, Sitapur, Uttar Pradesh, India

from *Singapore Journal of Tropical Geography*, 2003, 24(3): 356–372

INTERROGATING 'RELEVANCE' WITH BORDER-CROSSINGS

In September 2002, Ellen Messer-Davidow (2002a), at a talk about her book *Disciplining Feminism*, cited an incident where Donna Shalala, the former United States (US) Secretary of Health and Human Services, had maintained that academic research was useless to the Clinton administration when it was reforming welfare policy because it was too slow in coming out, produced conflicting results, used impenetrable jargon and failed to address questions that concerned policymakers. Shalala was not bad-mouthing welfare scholars, argued Messer-Davidow (2002b), she was simply calling attention to what the academy expects all scholars to do:

> it expects us to complexify, theorise and debate problems that have been constituted by our disciplines . . . [S]uch fields as feminist, cultural and GLBT [gay, lesbian, bisexual and transgender] studies use highly politicised rhetorics and espouse social-change objectives but produce knowledge that has little impact on real-world politics other than igniting backlashes.
>
> (Messer-Davidow, 2002b)

From another part of the world, Jean Dreze (2002: 817) echoed similar sentiments after a sustained involvement with two people's movements (*Mazdoor Kisan Shakti Sangathan* and *Akal Sangharsh Samiti*) in the Indian state of Rajasthan. Even after 15 years of researching hunger and famines, and 'perhaps entitled to feel like an expert of sorts on these matters' (especially after collaborating with Amartya Sen, the Nobel Prize winner for Economics), Dreze had not always found himself better equipped to understand the practical issues that arose in those groups. In fact, he had often felt 'embarrassingly ignorant' compared to village folk, who had little formal education but a 'sharp understanding of the real world', and for whom the main insights of his research delivered no more than a 'fairly obvious' message. Underscoring an urgent need to produce more accessible and relevant social scientific knowledges, Dreze states, rather provocatively:

> social scientists are chiefly engaged in arguing with each other about issues and theories that often bear little relation to the real world. It is in this foggy environment that common sense ideas have a cutting edge. Their power, such as it is, springs not so much from great originality or profundity as from their ability to bring some basic clarity to the confused world of academia. It is no wonder that these common sense ideas often fail to capture the imagination of people who are not exposed to that confusion in the first place.
>
> (Dreze, 2002: 817)

Scholars who have undertaken or theorized border-crossings have long struggled with such tensions and contradictions between the academic and non-academic realms by highlighting the problems of voice, authority and representation (Spivak, 1988; Ortner, 1995). At the same time, very few have grappled explicitly with what Visweswaran (1990:32) identifies as a main challenge for postcolonial feminist ethnography: 'If we have learned anything about anthropology's encounter with colonialism, the question is not really whether anthropologists can represent people better, but whether we can be accountable to people's own struggles for representation and self-determination.' In this article, I engage with this issue by focusing on two heightened concerns: (1) that the gulf between the theories produced in Northern academic institutions and the priorities of Southern intellectuals, activists and communities continues to widen; and (2) that very few Anglophone feminist and/or postcolonial geographers are explicitly engaged with the challenge of co/producing knowledges that 'speak' the theoretical and political languages of communities beyond the academy (Frisch, 1990; Larner, 1995; Alatas, 2001; Peake and Trotz, 2001; Raju, 2002). These general problems seem particularly acute when they involve partnerships between researchers located in Northern academic institutions and their postcolonial subjects in tropical and subtropical locations.

Of course, it is widely accepted that scholars must produce different kinds of products to reach different audiences in the multiple worlds they inhabit and research. There is also a partially shared understanding that we can guard against betraying people's sociopolitical interests by disseminating the views of marginalized actors and by transferring skills and legitimacy from professional to community researchers (Abu-Lughod, 1993; Ong, 1995; Red Thread, 2000). And there are the cautionary reminders that we must interrogate a rhetoric that valorizes these crossings too readily lest they mimic and supplement the language of the increasingly corporate university establishment (Pratt, 2000). But when it comes to addressing the reasons behind our limited ability to excite the imagination of our 'subjects' – subaltern or otherwise – located in those 'Other' worlds, or shift the forms, boundaries and languages of what is regarded as meaningful academic discourse, there is very little out there to grasp as a tool for charting new possibilities for postcolonial geographies and transnational feminisms (Red Thread, 2000; Peake and Trotz, 2001; Dreze, 2002).

Here, my aim is neither to rehash a critical analysis of previously attempted or problematized border-crossings, nor to perpetuate an uncomfortable romancing of collaboration across borders. It is, rather simply, to share some evolving thoughts triggered by my repeated encounters, similar to those described by Jean Dreze, of working with individuals and groups who simply failed to see academic insights into power, space, identity or representation as anything more than what was fairly obvious to them, or as anything that could usefully contribute to their own struggles around these issues.[2] But at the same time, rather than a disdain, mistrust or indifference towards academic knowledge, I have found these actors to be quite sophisticated when it comes to determining the parameters of their relationships with 'western'/diasporic researchers. Despite (or perhaps because of) being acutely aware of the turbulent politics of location and positionality that mould these relationships, these actors often had a strong sense of the relative privileges (e.g. mobility and resources) that 'overseas' academics had access to, and of the role that successful dialogues and collaborative efforts could play in furthering the personal, organizational, political and/or intellectual agendas of all involved parties.

What I have then done since 1996 is actively identified specific groups and individuals who are interested in building collaborative relationships with me, and reflected with them on the conditions, goals and processes that could give a concrete form and language to our evolving dialogues and collaborative agendas. My efforts have emanated from the belief that discussions surrounding the politics of representation – and of reflexivity, positionality and identity as a way to address those politics – have reached an impasse (Nagar and Geiger, 2000; Nagar, 2002a). It is only in and through such moments – successful and failed – of dialogue and collaboration that we can hope to move beyond the impasse and find new possibilities for postcolonial *and* transnational feminist geographical knowledges that can be simultaneously theorized, accessed, used, critiqued and revised across national, institutional and socioeconomic borders.

FROM PARTIAL KNOWLEDGES TO COLLABORATIVE BORDER-CROSSINGS

It is probably not an exaggeration to say that the idea of 'border-crossings' has now become a trendy prerequisite – at least in the US academy – for any critical social scientific scholarship to be regarded as 'cutting-edge'. The enthusiasm for such cutting-edge theories and accounts has undoubtedly encouraged an active, healthy and desperately needed interrogation of almost every conceivable border – borders of disciplines, methods, nations and social categories. At the same time, however, relatively little concern has been expressed for the manner in which the products of such crossings can/should become socially or politically relevant – or the means and languages by which they are rendered irrelevant or exclusionary – across the boundaries of the Northern academy, especially in tropical and subtropical locations.

In feminist geography, the discussion that has come closest to addressing this question of relevance across borders has focused on the politics of representation and reflexivity (Radcliffe, 1994; Rose, 1997; Pratt, 2000; Nagar, 2002a). Sarah Radcliffe (1994: 26), for instance, discusses the connections between authorial representation and political representation and asks: how can 'Western First World geographers write about Third World women in their teaching/productions, without at the same time (perhaps by the same means) claiming to represent these women politically?' Gillian Rose (1997) suggests that this problematic of representation can only be addressed by moving away from the notion of a 'transparent reflexivity', in which any attempt at self-positioning by the author only serves 'the purpose of stabilizing interpretation and removing bias in order to uncover the truth' and thereby reproduces the idea of a detached, universalizing gaze (Pratt, 2000: 641). This imperative of transparent reflexivity is problematic because it depends on certain notions of agency (as conscious) and power (as context), and also assumes that both are knowable. Rose (1997: 311) argues that there is an inherent contradiction when 'a researcher situates both herself and her research subjects in the same landscape of power, which is the context of the research project in question'. It is contradictory because 'the identity to be situated does not exist in isolation but only through mutually constitutive social relations, and it is the implications of this relational understanding of position that makes the vision of a transparently knowable self and world impossible' (Rose, 1997 312). Geraldine Pratt responds to Rose's call to explore how the researcher herself is reconstituted through the research process within a fissured space of fragile and fluid networks of connections and gaps. Through an interrogation and problematization of her own 'research performances' undertaken at/with the Philippine Women Centre in Vancouver, Pratt (2000: 642) presents a reflexive account in which the researcher, instead of being firmly located, is marked by 'absences, fallibilities, and moments that require translation'.

These writings have contributed to a rich discussion of the concepts of reflexivity and positionality in geographical research, but their primary focus has remained on textual and representational strategies rather than on the theoretical, empirical and political content of the stories that geographers seek to tell (exceptions include Red Thread, 2000; Peake and Trotz, 2001). This kind of focus on dismantling or interrogating power hierarchies through representational and textual strategies has often resulted in an unintended widening of the gulf between the theories produced by Northern academics and the priorities of their Southern subjects (see example in Nagar, 2002a). In making this observation, I do not want to diminish the importance of acknowledging the partialities of the knowledge(s) 'we' produce and of the ways in which these are, indeed, ridden with gaps and fissures. Nor am I suggesting that the politics and strategies of representation should cease to be our concern, for academic writing – especially when it crosses politicized borders of any kind – necessarily implies struggle(s) as well as strategic choices around representation. However, I do believe that if 'our' acknowledgement of partial and fissured landscapes of knowledge production does not also go beyond textual performances, it runs the danger of reproducing an unbridgeable gap created by our own practice, a gap not very different from the one that Messer-Davidow writes about:

> The problem was a gap I couldn't seem to bridge when I wrote about academic feminism as a change project. The change I had grasped from all those years of doing activism I couldn't reformulate in scholarly terms, and the change I knew from reading scholarship I couldn't deploy in activism. Eventually I realised that practice created the problem. The activist me had acquired know-how by planning, escalating and modifying direct action, and the academic me had acquired knowledge by analysing, refuting and reframing esoteric propositions. These very different sets of practice didn't provide two perspectives on the same thing; rather,

> they constituted change as two divergent things. Tactical practices engendered changes that were orchestrated, whereas intellectual practices generated schemas that were debated . . . How could I bridge the divides between intellectual and tactical practices, academic and societal arenas, discourses and dollars?
>
> (Messer-Davidow, 2002b)

If our goal is to transform the power hierarchies embedded in knowledge production, it is clearly not going to happen merely through a discussion of how we represent others and ourselves. What we need is a revamping of the theoretical frameworks so that the stories and struggles we write about do not become completely inaccessible and/or meaningless to the people whose sociopolitical agendas we want to support or advance. This need to shift our theoretical frameworks is not embedded in a romantic or presumptuous idea that our work could always be relevant to the subjects of our research. Rather, I am suggesting that the analyses we produce remain theoretically and politically impoverished in the absence of close scrutinies and critiques by those postcolonial subjects whose interests we want to advance, or whose histories and geographies we want to (re)write. Such a rethinking and extending of our theoretical and political frameworks is only possible in/through spaces of collaborative knowledge production – spaces in which academic agendas and frameworks can get interrogated and recast, and where we can generate new transformative possibilities in the fissures, gaps, absences and fallibilities of our critical frameworks whose cutting-edge status we have taken for granted.

This article, then, argues for an urgent need to develop postcolonial and transnational feminist praxes that focus explicitly and deliberately on (1) conceptualizing and implementing collaborative efforts that insist on crossing multiple and difficult borders; (2) the sites, strategies and skills deployed to produce such collaborations; and (3) the specific processes through which such collaborations can find their form, content and meaning. To ground this discussion, I draw upon two recent initiatives I have begun, both in the state of Uttar Pradesh (UP), North India: the first, with a woman known as Farah Ali in Kanpur; the second, with *Sangatin*, a collective of rural women activists working through a state-funded women's organization (and in the process of establishing a new organization independent from the state) in Sitapur. Although the sociopolitical and spatial processes and interrelationships that are at work in each case are quite different, both collaborations deploy personal narratives revolving around multiple forms of violence in gendered, classed and communalized lives, and the struggles around that violence. Instead of seeking to 'uncover' the processes that constitute these experiences of violence and struggle, my analysis aims to highlight strategies that are available for producing new collaborative geographies; for exploring the ways in which these geographies are/can be simultaneously embedded in and speak to multiple sites and landscapes of struggle and survival; and for imagining the processes by which we might begin to re-evaluate and reclaim previously colonized and appropriated knowledges.

BORDER-CROSSINGS IN TRANSLATION

First border-crossing: speaking 'with' Farah

> Do you know what my fight is about, Richa? I'm fighting to speak my way so that no family member, no community, no organisation, no researcher, no media person gets to distort my story to sensationalise my life! . . . I am speaking to you, seeking you out, building a relationship with you so that I can help you by telling you what you want to know. But I do so with an understanding that you are committed to helping me out when I need you, whether you are here in Kanpur or in the US.
>
> (Farah Ali, interview, 27 March 2002)

In these four bold and straightforward sentences, Farah Ali powerfully summarizes her own struggle as well as the nature of my partnership with her. I met Farah in 2002 through *Sahara*,[1] a nongovernmental organization (NGO) that for the last 16 years has served as a legal counselling cell and support centre in Kanpur for women of all classes and religious groups on issues of domestic and dowry-related violence and troubled marital relationships. As such, they work with not only women and their male partners, but also with key members of their extended families who often play a critical role in the creation and escalation of their 'marital problems'. Although I had known of and sometimes participated in *Sahara* activities since my college days, it was only in 2000–1 that my focus on women's NGOs and their relationships with globalization and communalism (religious extremism) brought me to *Sahara* as a researcher interested in exploring the possibility of a long-term collaboration with the organization. Although my relationship and (limited) collaboration with *Sahara* are not a theme of focus here, a brief background is necessary to understand and contextualize the story of Farah.

Sahara officers wanted me to help them document, analyse and collectively reflect on their work and, initially, I was excited about the potential embedded in such a collaboration. After working with the NGO over a period of four months, however, I discovered that there was little openness among their leaders to internal or external criticism, especially in relation to their strong organizational hierarchy and a problematic underplaying (at times, negation) of class- and religion-based differences. These factors affected not only their internal structure but also the manner in which *Sahara* reached out to and intervened in the lives of the women who sought its help. The coordinator of the organization was aware of my reservations about their approach, and we often had long, sometimes uncomfortable, discussions on the subject.

One of the questions that interested me during my work with women's NGOs in India was the interrelationship between communal violence and domestic violence. For example, how are the rise of Hindu nationalism and the state-sponsored instances of anti-Muslim violence shaping the manner in which questions surrounding domestic violence are being addressed, recast or stifled inside/across familial and communal borders? Whenever this question came up in discussions at *Sahara*, one name that was repeatedly mentioned was that of Farah Ali, a 37-year-old Muslim woman who had filed but subsequently withdrawn her case with the NGO because she refused to adopt any of the steps that their counsellors advised her to take. One counsellor described Farah as 'a sophisticated, US-returned Muslim woman' who was uncomfortable with the organization because she wanted her matter to remain private, whereas *Sahara* believed in politicizing domestic violence issues by making them public. The counsellor gave me Farah's number but also warned me not expect a positive response from her.

As it turned out, however, Farah was living just a few blocks from my parents' house, with her parents and her brother's family, and very eager to talk to me – not on the phone, but at a neighbourhood restaurant. We met at a street corner a few blocks from our homes and rode there together on a loud *tempo* (three-wheeler). As we began to sense and share fragments of our histories and geographies, Farah and I recognized some striking similarities in our social locations that neither of us had encountered before: our upbringing in lower-middle-class families (hers Muslim, mine Hindu) in the same city; our 'unexpected' journeys to the US; and our shared status as mothers with a very young daughter, living with our parents and brother's family – as well as the deep contradictions, joys and pains embedded in that reality. There is much to be noted and analysed along these lines about the telling, recording and retelling of Farah's story, but for the purposes of this article, I want to summarize the complex strands of Farah's struggle and return to the question of collaboration.

Why no one can give Farah a voice

Let me summarize the pieces that contribute to making Farah's story sensational and exotic in the eyes of 'outsiders' – not just the outsiders who can gaze at her from the west, but also the multiple gazes that stifle Farah's voice in her 'own' home, city and nation. Farah, a well-educated social worker from a liberal, middle-class, Sunni family, married Aamir in 1994. The marriage was arranged through their families, but she and Aamir spent ten months getting to know each other during the period of engagement, and both consented happily to the marriage. In 1995, Aamir got an opportunity to work as a scientist at a top US university, and she joined him after spending two months at his parents' home in Meerut. Farah had deep reservations about how his family treated her, but she chose not to discuss her feelings with Aamir and focused her energies on building a healthy partnership with him once she reached the US. Despite her suspicions and discomfort about Aamir's growing pull towards extremist interpretations of Islam, Farah mostly remembers herself as a happy, content wife and mother in New Jersey – until December 1998, when everything turned upside-down on a trip back to India.

By March 1999, Farah found herself abandoned with her five-month-old daughter Juhie in her in-laws' home in Meerut because Aamir had taken possession of her immigration documents and returned to New Jersey. In April 2000, he divorced Farah from the US – on the grounds that she had failed to fulfil her duties as a Muslim wife and woman. Farah refused to accept the divorce but the Muslim Personal Law Board

(MPLB) of India declared it legal. She had wanted to fight this, but then came 11 September 2001, followed in India by the re-escalation of the Hindu fanaticism over building a Ram Temple in Ayodhya and the senseless massacre of Muslims in Gujarat. Says Farah:

> To tell you the truth, my voice has been snatched. From my brothers, their wives and my parents to the rest of my community . . . and from the folks at Sahara and the Muslim Law Board to the white guys in the US Embassy [in New Delhi] . . . I feel like everyone's hands are pressing against my mouth to silence me . . . All I have to do is just let out one word . . . and the media and the people will just find one more reason to dehumanise Muslims.
>
> (Farah Ali, interview, 27 March 2002)

Farah is correct. She is suspicious of anyone who wants to speak on her behalf, convinced that this would only serve their sociopolitical or careerist agendas while undermining her own objectives. In extremely delicate political times in North India, when the MPLB and her own family are asking her not to talk about her issues in public, *Sahara* wanted Farah to challenge Aamir by shaming him and his family in the mainstream media. Embarrassing his family in public, according to *Sahara*, would force Aamir to reconsider – or perhaps withdraw – the divorce statement. It came as no surprise, then, that one well-intentioned *Sahara* worker proceeded to leak Farah's story to a producer at Z-TV, who then approached Farah for an interview, with a promise of 'tremendous publicity' that would eventually help her win a parliamentary election!

Farah does not believe that any of these people can give her 'voice'. She considers *Sahara*'s thinking to be too localized and parochial to understand her 'case'. She hates the guts of the Z-TV producer, who she sees as no different from those who caricatured Khomeini in the 1980s and are demonizing Osama bin Laden today. Farah is incensed by the stance of the MPLB but appreciates why this is not the time to publicly criticize them. She also recognizes how her family's hands are tied, why they have to ask her to be silent about Aamir in these times of state-sponsored repression of Indian Muslims, but she also feels that she and her daughter are increasingly becoming unwanted burdens in her natal home.

In these circumstances, Farah believes that the only tool she has left to regain her voice and fight for justice is through gaining entry into the US, where she can confront Aamir through the US law – not because it is inherently more just or sensitive than the Indian law, but because the US courts will not recognize the *lalaqnama* (Muslim deed of divorce) and/or would require Aamir to provide adequate maintenance for her and their daughter. Farah, whose parents- and sister-in-law have effectively prevented her from having any direct communication with Aamir since March 2000, also wonders if meeting Aamir face-to-face would make him realize the implications of what (she thinks) he has done under his family's pressure. Her final reason for regarding the US as her best option is familiarity; she has lived and worked there before so it seems to be the easiest place for her to start a new life as a single mother and give her daughter the environment that she needs to blossom and to have a bright future. But Farah also fears that 9/11 (and its aftermath) has irrevocably injured her relationship with the US Immigration and Naturalization Service (INS), the Embassy in New Delhi, and perhaps even with the very place where she had hoped to find a hope.

Reading/retelling Farah's story

Farah's story has many rich and complicated strands. There are multiple actors embedded in multiple locations, of which I will name just three. There is Farah, who angrily – and rather perceptively – states that her fate is straddling Kanpur (her natal home) and Meerut (her conjugal household), family and community, the US and India, and the INS and the MPLB. There is *Sahara* – an NGO committed to a particular strategy of politicizing violence against women at the local level – which fails to appreciate Farah as a transnational subject, and which she dismisses as too parochial and as lacking subtlety in tactics. And there is me, a US-based researcher working 'back home', trying to build a complex alliance with Farah while also remaining committed to certain ethical and political stances.

In terms of the currently existing postcolonial and poststructuralist frameworks that can be deployed to make a 'cutting-edge' theoretical intervention based on this story, the possibilities are tremendous. I could choose to enact a highly innovative textual performance; I could theorize the multiple border zones and border-crossings that are at work in this story; I could problematize existing theorizations of communalism, secularism and the postcolonial state; and I could revisit the famous trope of colonial feminism about brown women being saved from brown men by white men. But I must accept that none of these approaches will have much worth for Farah, for women like her who are battling with similar forms of violence in similar locations, or for organizations like *Sahara* that are struggling to find new conceptual frameworks that could enable them to better understand and address the kind of violence and silencing that Farah faces.

Here I return to the partnership that Farah described between her and me at the outset of this section. Farah wants me to help her return to the US by discussing her situation with individuals working with specific South Asian women's NGOs in New York and San Francisco who can advise on how she can approach the INS to reclaim her green card status, and how she can sue Aamir through the Indian courts. Farah has needed my assistance in tracking down Aamir and, after placing her confidence and trust in me so generously, expects me to be there for her as an ally and friend. To her, these are the most important parts of our collaboration.

For me – as for many other feminist scholars – the kinds of commitments and obligations described above come with any research that involves close relationships between a researcher and her 'subjects'. Generally speaking, there is nothing wrong in believing or acting upon this idea. However, we lose a critical opportunity to interrogate and extend our theoretical frameworks when we reduce such visions/expectations of partnership articulated by our research subjects to the status of commitments and obligations that are either post-fieldwork or independent of theory/academic production. What we need to do instead is engage in a serious and honest examination of why the existing possibilities of framing and analysing Farah's story contribute little or nothing towards advancing the struggles that concern Farah or *Sahara*. Why is it that the most sophisticated and complex theories – when translated into an accessible language – fail to deliver anything beyond a fairly obvious message to Farah and her family, and to *Sahara*! And what possibilities for extending or revamping those theoretical frameworks emerge when creating relevant knowledge for actors such as Farah and *Sahara* becomes my main academic goal?

The next step of this collaboration between Farah Ali and me seeks to explore the ways in which current feminist work on transnational citizenship and violence can speak to Farah's experiences and to the organizations in New Jersey, New Delhi and her natal city that cannot at this moment interfere in Farah's case or advance her cause. In this process, Farah also wants to build bridges with specific Muslim activists who are making feminist interventions in the politics of communalism and gendered violence in India in the aftermath of the Gujarat massacre. Hence, the real test of the relevance of this analysis, and the extent to which it can do justice to the enmeshing of local, global and transnational subjectivities, power relationships and citizenships cannot be based merely on my ability to provide another twist to the existing academic debates on these subjects. It hinges, instead, on Farah's ability to draw sustenance, hope, direction and a sense of fulfilment from this collaboration, and from my ability to deploy insights from transnational feminist theories to help reach that goal.

Second border-crossing: producing a methodology to 'speak with' Sangatin

Manju: The Chamar and Yadav[3] in my village are at each other's throats and everyone blames me . . . It all started on 13 March when Hari [a Yadav] and Kishan [Manju's brother-in-law] broke into my home and beat me mercilessly . . . 1 went to the police station and said, 'I dare any man in this village to touch me or humiliate me again for the rest of my life' . . . Kishan screamed, 'This woman is evil. She keeps three men.' I said, 'Yes, I have three men. I will keep two more. Why are my men his responsibility?' . . . But for some reason, Kishan got released and Hari was arrested under the Harijan Act.[3] Now it has become a big caste war.

Eighteen rural women workers of the *Mahila Samakhya* programme in Sitapur (MSS) discuss Manju's intervention and the complicated political situation it has created in her village. Rita and Sunita reflect on how caste and family politics enmesh to shape Manju's current circumstances. Rohini and Gauri draw connections between Manju's mismatched marriage arranged by her more prosperous (and therefore more influential) younger sister, the physical violence inflicted upon her by that sister and the sister's husband, and Manju's intimacy with one of her husband's cousins. Vineeta argues that the humiliation Manju suffers is closely linked to the manner in which agricultural land is divided between her husband and his brothers. Manju agrees with some of these statements and modifies or responds

> to the others. She fears that the caste politics in the village and accusations hurled against her will result in her murder – just as her friend Noor was killed last month. There are tears. There is concern. The 18 women sitting in the circle know that Manju's fear is grounded in something too real and familiar. The group decides to hold a public meeting in Manju's village in a week.
>
> (notes of MSS *mahasangh* or general assembly in Sitapur, 25 March 2002; all personal names are pseudonyms)

In June 1996, I had the rare opportunity to join Richa Singh, the coordinator of the newly launched MSS, when she and her co-workers had just begun training eight local women as mobilizers. Each was responsible for mobilizing women in ten villages, mostly in the vicinity of her natal and conjugal villages. The idea was to give birth to a new model of education and literacy in these villages that allowed the poorest women from the 'scheduled castes' and 'other backward classes' to collectively understand, address and change the processes and structures responsible for their own marginalization. Another goal was to enable the women to build their own grassroots organization that would replace MSS at the end of the initial period of its activity funded by the governments of India and the Netherlands, and the World Bank. In 1999, the eight mobilizers, along with Richa Singh, registered as co-founders of a new collective called *Sangatin*, which would continue the work of MSS when the funding from its current donors stops.

In the seven years since 1996, MSS activists have become well known in UP, especially for their sustained efforts to challenge and modify specific festivals and rituals that sanction violence against girls and women. On a somewhat smaller scale, these women have also addressed the ways in which violence inflicted on the poorest women's bodies is intricately connected with their access to land and wages, and with local religious and caste-based politics. It is not surprising, then, that Manju's narration of her conflict with Hari and Kishan developed into a detailed, insightful and multilayered discussion among MSS women, where they explored the material, metaphoric and political connections between landlessness, untouchability, poverty, morality and sexualized violence in Manju's life and in their own lives.

Unlike the heavily researched work on some other similar women's organizations in India, most of the accomplishments of MSS and *Sangatin* have remained undocumented partly because of the desire of the rural women to be centrally involved as researchers in any documentation and analysis of their work. This factor, combined with my previous work with *Manila Samakhya* programs in Uttar Pradesh (see Nagar, 2000, 2002a, 2002b), led Richa Singh to contact me in March 2002 with an explicit request to visit Sitapur and explore with key MSS activists the possibility of undertaking and planning a collaborative research project. Between March and December, I interacted with approximately 60 MSS workers (face-to-face and through detailed letters) to collectively determine the goals and processes that would define such a collaboration. Three central decisions were made. First, since *Sangatin* will continue the work of MSS, the collaboration must focus on giving a vision and direction to *Sangatin* for its future work in Sitapur. Second, to determine their future goals, strategies and political stances as a collective, it is necessary for women to engage in an in-depth reflection and analysis of their past achievements and failures *through* the life-histories of key grassroots activists in their own midst, whose work around gender and caste-based violence they have found to be the most challenging and inspiring. At a time when rural activists are experiencing a deep disillusionment with changing structures and agendas of government funded NGOs, *Sangatin* wants to reflect carefully on the organizational limitations that frustrate and paralyse them. Third, women whose life-histories are collected and analysed for this project must simultaneously acquire training as community researchers so that they can continue the work of documentation, reflection and analysis of their own work on an ongoing basis without any reliance on the expertise and agendas of outside researchers.

In December 2002, women who had worked with MSS in more than 80 villages of Sitapur collectively chose the eight founding members *of Sangatin* as women whose life-histories they considered most central for understanding and documenting their history of struggles and accomplishments as a collective. These eight women formally invited me to work on their collective history project as a part of their research team. Together, we spent ten days and nights jointly laying out the methodology, budget, process and rules to be followed in the production, sharing and dissemination of the eight life-stories, and the analyses emerging from them. These ten days were marked by moments in which all of us wrote our autobiographical journals in the same space, laughed and cried together as we shared our accounts, confronted each other with difficult questions and produced new dynamics, where some people learned to suppress their voices while others found the voices and words whose presence they had never realized.[4]

This border-crossing-in-progress with *Sangatin* seeks to make an intervention in the theory and praxis of 'North–South' collaborations in four critical ways. First, it focuses on how rural activists theorize, strategize, prioritize and act upon their own understandings of development, globalization, violence and empowerment. Second, it allows me to use my analytical and linguistic skills, mobility and access to resources to help the activists meet their own goals, while also gaining new insights into ways that collaborative theories and methodologies on questions of development and empowerment can be produced across borders. Third, it prioritizes activists' own articulations of how they want their understandings to be recorded, written, disseminated and deployed, and the kind of role they want the academic researcher to play in these processes. Finally, both feminist social scientists and NGOs have come to regard life-histories as an exceptionally rich tool for understanding personal experiences, identities and social relations, and how individual biographies intersect with social processes (although their efforts and agendas have mostly remained separate). This last collaboration extends the methodological discussions in both NGOs and academia by interrogating the dualisms of theory/praxis, expert/non-expert and academic/community-based, and by confronting the questions of voice, authority and representation at each step of this project's conceptualization and implementation.

IMAGINING COLLABORATIVE FEMINIST POSTCOLONIAL GEOGRAPHIES

The idea that postcolonial researchers should produce diverse knowledges to reach different audiences in the multiple worlds they straddle has gained increased currency across disciplines. Important differences remain, however, among those who hold this position. While some argue that academics can protect people's interests by disseminating the views of the marginalized, others remain highly sceptical of the degree to which the agendas of academics and grassroots workers can be harmonized. These divergent positions emanate, in part, from a lack of systematic research that explores the reasons behind our limited ability to excite the imagination of those whose struggles we study in the 'South', or to shift the forms, boundaries and languages of what is regarded as meaningful academic discourse.

Peake, Trotz and Kobayashi are among the few who have explicitly grappled with the question of how Third World and First World women can work together 'in ways that are authorized by dialogue with [Third World subjects] and not just First World audiences' (Peake and Trotz, 1999: 28). Reflexive questioning of ourselves and of the techniques we use to develop multivocality, they remind us, must be accompanied by a continued interrogation of how our supposedly 'improved' representational strategies might constitute new silences (Peake and Trotz, 1999: 35). Such an interrogation requires that we challenge the divide between politics on the ground and research as an academic practice *through* a geography of engagement that taps into the tremendous potential of activism and produces critical analyses based on local feminist praxis, and the ways that these connect with broader relations of domination and subordination (Peake and Kobayashi, 2002; Peake and Trotz, 2001).

It is in the context of these broader struggles of domination and subordination under globalization that these feminist geographies of engagement become explicitly postcolonial and of critical relevance to the theory and praxis of social sciences in the tropics and subtropics. As Spivak (2000) observes, the expansion since 1989 of a full-scale globalized capitalism regulated by the World Trade Organization, World Bank and International Monetary Fund has been accompanied by a complex politics of state and international civil society. International platforms such as the United Nations (UN) are dominated by a 'global feminist' agenda rooted in problematic assumptions such as a sex-gender system, an unacknowledged biological determination of behaviour and an 'object-choice scenario that defines female life' (Spivak, 2000: 321) in terms of choosing between children or public life, population control or 'development'. In this political scenario, the interventions made by powerful NGOs often end up serving the interests of global capital, despite being feminist in their professed interest in gender.

These processes, Spivak argues, demand both a revision of feminist theory and a rethinking of the 'subaltern' within the feminist mode. The genetically reproductive body as a site of production questions feminist theories based only on the ownership of the phenomenal body as means of reproduction as well as feminist psychological theories reactive to reproductive genital penetration as normality. Politically, this new understanding of subalternity leads to global social movements supported by a Marxist analysis of exploitation, calling for an undoing of the systemic–anti-systemic binary opposition and requiring an engagement with global feminism (Spivak 2000: 321). Thus:

> If the dominant is represented by the centreless centre of electronic finance capital, the subaltern woman is the target of credit-baiting without infrastructural involvement, thus opening a huge untapped market to the international commercial sector. Here a genuinely feminist politics would be a monitoring one, that forbids the ideological appropriation of much older self-employed women's undertakings, and further, requires and implements infrastructural change rather than practises cultural coercion in the name of feminism.
>
> (Spivak, 2000: 322)

Instead of invoking strategic use of essentialism, then, Spivak (2000: 327) emphasizes a need to underscore how oppositions are being generated in dominant discursive formations of global feminisms and a process of 'learning to learn from below'.

The border-crossings that I have initiated here with the help of Farah and *Sahara*, and MSS/*Sangatin* can be seen as an effort to further imagine and enact postcolonial and transnational geographies of engagement through collaboration. Such collaborative processes provide concrete spaces to 'learn from below' and co-determine the specific ways in which we can be accountable to people's struggles for self-representation and self-determination (Visweswaran, 1990). Thus, collaboration becomes a tool to understand how women themselves conceptualize and represent their subalternity in complex ways that challenge the problematic assumptions made by a UN-style 'global feminist' agenda. It, furthermore, becomes a vehicle for the collaborators to imagine new ways in which they can resist processes that make the subaltern woman 'the target of credit-baiting without infrastuctural involvement' (Spivak, 2000: 322).

At the same time, the words, commitments and obligations shared between these women/organizations and me do not serve a predetermined agenda (theirs or mine). Rather, our exchanges continue to take place in the spirit of listening, sharing and collaborative decision-making about where these stories should speak, for whom, in what languages and with what purpose. These collaborations have the potential to fruitfully extend existing academic frameworks and yield more 'relevant' insights across national and institutional borders on how familial structures, socioeconomic processes, spatial (im)mobility and politicized religion intersect to shape the multiple forms of violence in the lives of North Indian women, the resources and strategies that women create to resist this violence, as well as the contradictions that remain buried in their efforts to overcome their silences. In so doing, these collaborations allow us to exploit the political possibilities created by discursive materialities of global capitalism and international civil society. They permit us to complicate assumptions of élite theory about modernity in postcolonial societies and allow us to appreciate the dilemmas as well as the possibilities of Dalit/women's struggles (John, 1996).

But what about authorship? Why are Farah and members of *Sangatin* identified as 'consultants' and not co-authors of this article? A simple answer is that neither wants to be, because the broader issue of what constitutes a postcolonial geographical methodology is not one that they find particularly relevant to their concerns. While they are interested in the specific representation of their own struggles and of our collaborative process, neither is (re)defining geography central to their struggles, nor are they interested in becoming token co-authors.

A more complex discussion of this subject, however, demands an in-depth interrogation of more traditional forms of collaboration where research agenda – and theoretical and methodological underpinnings – are determined (fully or largely) by the Northern academic researcher and her institutional context, and the names of the non-academic actors she worked with appear as a way to denote shared power and authority; or where two academic researchers from different institutional and sociopolitical locations co-produce an academic text. In either case, the collaboration is represented narrowly in terms of formal co-authorship, with the names of the authors appearing below the title of the academic text. The practice of crediting only the formal author(s) of a text is itself a faulty one that gives undue credit to authorship of a *text*, downplaying issues of collaboration in the processes of defining and addressing the research problems themselves. The expectation that our collaborators would always want to be co-authors, furthermore, assumes that speaking to academic audiences is a priority for all involved and that, like Northern academics, their non-academic collaborators in the tropics and subtropics are also invested in securing intellectual property rights and/or recognition by academic audiences.

A more radical and more complex idea of collaboration must problematize these assumptions. If the intellectual agenda, research questions and approaches evolve as part of a collaboration between actors in different institutional, sociopolitical and geographical locations, then the collaborators must also understand that, as long as they maintain their commitment to the shared intellectual and political agenda, they might be required to produce knowledges and theories for different audiences, with different goals and strategies. This implies that the specific products emerging from collaboration will sometimes be written jointly and sometimes by an individual or sub-group in consultation with others. Nonetheless, the knowledges produced, as well as the purposes for which they are deployed, remain inherently and deeply collaborative, irrespective of the formal co-authorship of the actual texts that get produced and circulated.

The challenge for postcolonial and feminist geographers, then, is to conceptualize border-crossings that are committed to forming collaborative partnerships with academic and non-academic actors in 'other' worlds, in every sense of the term – partnerships in which the questions around how power and authority would be shared cannot be answered beforehand, but are imagined, struggled over and resolved through the collaborative process itself. Since the issues I raise here defy conclusion by their very nature, I offer as a *non-conclusion* to these thoughts-in-progress another semi-translated border-crossing – or the painful reality of a collaboration that could not happen.

CHUPPI: *THE COLLABORATION THAT DID NOT HAPPEN*

Grahwal ke ek nukkad par
Hum dono intezaar kar rahe the
Ek hi bas ka.
On a little street corner of
Garhwal
She and I waited
For the same bus.
Aur woh aurat baithi thi
Dukan ki seedhion par
Apni 8 maheenon ki bitiya ko Chhatiyon se zabran alag kiye –
Mano koshish kar rahi ho sari bheed se
Apna sookhajism chhipane ki.
She sat
On the steps of a shop
Holding her eight-month-old
daughter
Away from her breasts,
As if to shield from the eyes of the
crowd
Her drained, empty body.
Baar-baar bitiya
Uske badan se
Apna moonh ragadti
Aur baar-bar woh thel deti us
Ziddi moonh mein
Nipple lagi ek pani ki botal
Over and over again,
The child's mouth, struggled to
Rub itself against her, and
Each time she
Stuffed in that persistent mouth
A rubber nipple attached to a water
bottle.
Lekinwo
Nanhi-dubli taaqatwarjaan
Botal hata-kar har-baar Maa ki
Chhati se chipak jati.
Uska bhookha-ziddi moonh pahle unhe
Chichodta, phir bhookh se beqaboo hokar
Betarah cheekhkarrota.
But that tiny, thin-boned bundle of
energy
pushed the bottle away each time,
and clung to her mother's breasts.
Her thirsty, stubborn mouth
searched those breasts desperately,
and then
screamed crazily
with an uncontrolled hunger.
Do hi qadam door main chupchap
sab khadi dekhti rahi gunahgar bankar.
I stood
just two steps away
watching all this

like a criminal.
Gunahgar
is baat ki nahin ki
apni 18 maheenon ki
aulad ko main 'kaam' ki wajah se
Amreeka chhod aayi thi.
Balki gunahgar us
kadwi ghair-barabari ki -
jisne us aurat ke bajae
meri qameez doodh se bhigo dali . . .
Jisne lakh chahne par bhi
rok liya mujhe us
bachchi ko apne bheetar
Samet lene se . . .
Wohi ghair-barabari
jo dheeth si
chattan bankar
khadi ho gayi mere aur us aurat ke
darmiyan . . .
Aur hamare beech guzre is adhoore
kathin samvaad ko jisne, ek lambi
bojhil chuppi ke age
badhne hi nahin diya . . .
A criminal –
not of the 'crime' that had forced
me
to leave my 18-month-old daughter
in the USA
(so that 1 could 'study' women's
struggles in India)
But the crime of that
bitter inequality, which
in stead of wetting that woman, had
drenched my shirt with milk . . .
Which
despite my intense pain
prevented me from
holding that child tightly
against my chest . . .
That very same
inequality
which stood between me and that
woman like a
stubborn cliff
determined to prevent this
difficult,
incomplete dialogue between us
from becoming
anything more than a
long, burdened silence . . .

ACKNOWLEDGEMENTS

This work was supported by a McKnight Land-Grant Professorship, a McKnight Presidential Fellowship and a Grant-in-Aid from the University of Minnesota. I am grateful to David Faust for his sustained feedback and support, to Abidin Kusno for his inspiring words, and to Derek Gregory for helping me see more clearly why it was necessary to undergo the discomfort caused by disrupting certain borders. To Farah, Richa, Shashi Dighia, Sheela, Reshma, Vibha, Anupam, Shashi Vaishya and Surbala, I express my *dili shukriya* for the countless hours of sharing which made it possible for us to dream and cry together.

NOTES

1 Farah is the pseudonym chosen by the narrator of the first border-crossing, and likewise, to avoid certain risks, the pseudonyms *Sahara*, Aamir and Juhie, for the primary actors that appear in her story. Kanpur, Meerut and New Jersey are real places, but selected because they most resemble the ones where the actual events in her life took place.

2 Although I have had an opportunity to interact with activists and workers in NGOs from various parts of South Asia, my most sustained and in-depth encounters have been with women activists in (my 'home' state) Uttar Pradesh.

3 Chamar are officially classified as among the Scheduled Castes (formerly 'untouchables'; also Harijan/Dalit) and Yadav as Other Backward Castes (OBCs). The Harijan Act punishes discrimination against the Scheduled Castes.

4 To summarize briefly, our methodology involves an organized schedule of tasks in the collection of each life-history that simultaneously focuses on developing each member's skills as a community researcher. The activities include writing, sharing and collective reflection on journal entries over a period of four months, followed by recording, transcribing and editing the life-history interviews of each member. Subsequently, Sangatin members and I will (1) analyse and evaluate the nine life-histories; (2) reflect on the collaborative process and lessons learned by the team; and (3) share the life-histories in the open meetings of MSS, where women will discuss (in small and large groups) what they have learned from each life-history individually and from all life-histories collectively, and the ways in which the life-histories should

shape the goals, plans and strategies for their future organizing. The team will compile and organize the life-histories and draft a book (in Hindi) that can be read and used by women and organizations in the region.

REFERENCES

Abu-Lughod, L. (1993) *Writing Women's Worlds*, Berkeley: University of California Press.

Alatas, S. F. (2001) 'The study of the social sciences in developing societies: Towards an adequate conceptualization of relevance', *Current Sociology*, 49(2), 1–19.

Dreze, J. (2002) 'On research and action', *Economic and Political Weekly*, 2 March, 37(9), 817.

Frisch, M. (1990) *A Shared Authority: Essays on the Craft and Meaning of Oral and Public History*, Albany: SUNY Press.

John, M. E. (1996) 'Dalit women in western ethnography', *Economic and Political Weekly*, 24 February, 463–64.

Larner, W. (1995) 'Theorising "difference" in Aotearoa/New Zealand', *Gender, Place and Culture*, 2(2), 177–90.

Messer-Davidow, E. (2002a) *Disciplining Feminisms: From Social Activism to Academic Discussion*, Durham: Duke University Press.

—— (2002b) 'Feminist studies and social activism', paper presented at the Feminist Studies Colloquium Series, Department of Women's Studies, University of Minnesota, Minneapolis, 30 September.

Nagar, R. (2000) '*Mujhe Jawab Do!* (Answer me!): Feminist grassroots activism and social spaces in Chitrakoot (India)', *Gender, Place and Culture*, 7(4), 341–62.

—— (2002a) 'Footloose researchers, traveling theories and the politics of transnational feminist praxis', *Gender, Place and Culture*, 9(2), 179–86.

—— (2002b) 'Shodh: Mujhe jawab do/Chitrakoot ke gavon mein Vanangna ki pukar', *Bharat Rang*, 2, 76–88.

Nagar, R. and Geiger, S. (2000) 'Reflexivity, positionality and identity in feminist fieldwork: Beyond the impasse', unpublished manuscript.

Ong, A. (1995) 'Women out of China: Traveling tales and traveling theories in postcolonial feminism', in R. Behar and D. Gordon (eds), *Women Writing Culture*, Berkeley: University of California Press, 350–72.

Ortner, S. (1995) 'Resistance and the problem of ethnographic refusal', *Comparative Studies in Society and History*, 37(1), 173–93.

Peake, L. and Trotz, D. A. (1999) *Gender, Ethnicity and Place: Women and Identity in Guyana*, New York: Routledge.

—— (2001) 'Feminism and feminist issues in the South', in V. Desai and R. B. Potter (eds), *The Companion to Development Studies*, London: Arnold, 34–37.

Peake, L. and Kobayashi, A. (2002) 'Policies and practices for an anti-racist geography at the millennium', *The Professional Geographer*, 54(1), 50–61.

Pratt, G. (2000) 'Research performances', *Environment and Planning D: Society and Space*, 18(5), 639–51.

Radcliffe, S. A. (1994) '(Representing) post-colonial women: Authority, difference and feminisms', *Area*, 26(1), 25–32.

Raju, S. (2002) 'We are different, but can we talk?', *Gender, Place and Culture*, 9(2), 173–77.

Red Thread (2000) *Women Researching Women: Study on Issues of Reproductive and Sex Health and of Domestic Violence Against Women in Guyana*, report of the Inter-American Development Bank (IDB) Project TC-97-07-40-9-GY conducted by Red Thread Women's Development Programme, Georgetown, Guyana, in conjunction with Dr. Linda Peake, available at www.sdnp.org.gy/hands/wom_surv. htm.

Rose, G. (1997) 'Situating knowledges: Positionality, reflexivities and other tactics', *Progress in Human Geography*, 21(3), 305–20.

Spivak, G. C. (1988) 'Can the subaltern speak?', in C. Nelson and L. Grossberg (eds), *Marxism and the Interpretation of Culture*, Urbana: University of Illinois Press, 271–313.

—— (2000) 'Discussion: An afterword on the new subaltern', in P. Chatterjee and P. Jeganathan (eds), *Subaltern Studies XI: Community, Gender and Violence*, New Delhi: Permanent Black, 305–34.

Visweswaran, K. (1990) *Fictions of Feminist Ethnography*, Minneapolis: University of Minnesota Press.

10

Research, pedagogy, and instrumental geography

Rich Heyman

from *Antipode*, 2000, 32(3): 292–307

INTRODUCTION

> It may be ruled out that immediate economic crises of themselves produce fundamental historical events; they can simply create a terrain more favourable to the dissemination of certain modes of thought, and certain ways of posing and resolving questions.
>
> (Antonio Gramsci, *Selections from the Prison Notebooks*, 1971: 184)

Recently many academic geographers have felt a subtle, yet powerful, institutional pressure to become academic "entrepreneurs." In an era of decreasing direct support for universities by government, this has meant a heightened emphasis on grant writing and a shift towards self-funded research projects. For radical geographers, the most disturbing outcome of this situation has been the increasing presence of corporations in the academy and the move by some geographers to embrace this post-Fordist "corporatized" institution. The vision of Lay James Gibson, President of the Western Regional Science Association, encapsulates one such attempt to articulate academic geography with industry:

> [W]e need to be prepared to proactively work with a different corporate culture. And we need to remind ourselves that there are multiple markets for research . . . If we are not shy about claiming new territory, we will probably become increasingly familiar with those working downstream [in the marketing and TQM (total quality management) stages of "commercialization models"]. And this will have substantial benefits for those in the academy as we increasingly find ourselves chasing funding into the latter stages of the research enterprise.
>
> (Gibson, 1998: 464–5)

However, Gibson's vision of a "client-driven" "full-service" regional science infused with the language of TQM and post-Fordist manufacturing processes represents only the most brazen embrace of the corporatized university (1998: 464–5). While commodified research has a long history in the academy,[1] David Noble's recent wake-up call, "Digital diploma mills," has alerted us to the advent of the commoditized classroom—the move by companies into a relatively untapped arena of potential profit realization aided by new information technologies (Noble, 1998).

I would like to make two points regarding the possibilities for challenging such a "corporatization" of the university. First, I would caution against too hastily concluding that we are witnessing the penetration of market principles into the academy. In order to fully understand what is happening today we need to view the current "corporatization" in the historical context of an institution that has been contested since its inception in the early nineteenth century, and we need to see the founding of the modern research university in the context of changing modes of intellectual inquiry and the advent of the disciplines. Second, I argue that the classroom, as a site of political praxis, has been neglected in mainstream geography and is a crucial place in which this "corporatization" can be challenged.

What ties together Gibson's vision of a client-driven geography and the threatened classroom is the common conception of knowledge as *information* that is unproblematically transmissible, as a commodity that can be readily exchanged for the price of a book, a consulting fee, or university tuition. As Noble has revealed, the conception of knowledge as information has fueled a boom in the commodification of the classroom through new information technologies. To use an example from my own institutional context, in a recent speech Wallace Loh, chief educational advisor to Washington State Governor Gary Locke, called for the development and implementation of new education "delivery systems" (Loh, 1998). Published on the editorial page of the *Seattle Times*, Loh's comments were generated in response to Governor Locke's creation of an advisory commission to "re-think and re-invent higher education from the ground up" (Locke, 1998). Such "reinvented" institutions, ones that treat education as a series of "delivery systems," proceed on the understanding of knowledge as instrument. One significant aspect of this attitude, as Readings (1996) has rightly pointed out, is the clientization of students that has become so prevalent in current discourses of universities and has been the justification for many of the reforms that we read as "corporatizing" the university. For example, in pursuing more "accountability" in the state of Washington's universities and making them more "relevant" to the state's high-tech industries, the Washington State Higher Education Coordinating Board is currently developing a battery of "outcomes testing" that redefines university education by quantifying narrowly defined technical proficiency in selected fields such as writing, mathematics, and computer science.

While branching out to embrace new conceptions of a critical science, geography has devoted precious little space to theorizing a pedagogy capable of contesting these trends in hope of maintaining a place for critical thought in the university of the twenty-first century. Geographers have expended much energy on working out new methods of research and analysis, but have not adequately addressed the link between knowledge production and pedagogy. In this paper I hope to bring questions of radical pedagogic practices into mainstream discussions in geography. To do this, in the next section I briefly show how knowledge came to be viewed primarily in instrumentalist terms during the nineteenth century. In the last section I then discuss how recent challenges to positivism can open the door to more sophisticated discussions of the classroom, ones that move us beyond talking about students as empty vessels to be filled with knowledge for the price of their tuition fees. By doing this, we will be better equipped to defend our classrooms and more able to promote teaching that matters to radical geographers: social justice, critical citizenship, and participatory democracy.

INSTRUMENTAL KNOWLEDGE AND THE CONTESTED UNIVERSITY

At first glance, Gibson's desire to make science useful to corporations strikes more radical geographers as a poor political choice. We might say, why not put geography to some better use? This has been the progressive social science response for much of this century. In the past thirty years, radical geography and feminist geography have reframed the questions "useful knowledge for whom?" and "towards what social ends?" Together with critical theory, this has led geography to begin to question not only how geographical knowledge is used but on what bases it is made. Along these lines, we might ask Gibson what assumptions underlie his regional science methodologies—exposing them to be just as ideologically driven as the use to which he wants them put—and could then continue to question his motives. However, we wouldn't normally think twice about his desire to produce useful knowledge. That, after all, is the point of doing research and the point of having research universities.

Such an attitude would strike the group of German Idealists led by Wilhelm von Humboldt, with whom we credit the idea of the modern research university and its dual mission of research and teaching, as a near-total failure of the institution with which their names are associated. During the first decade of the nineteenth century, when the University of Berlin—recognized as the first modern research university—was founded, Humboldt and his companions were engaged in a lengthy intellectual struggle to define the proper mode of inquiry in the pursuit of knowledge. However, by the time geographer Daniel Coit Oilman established the first research university in the US in 1876 at Johns Hopkins, which set the standards for such American institutions, the primary impetus for such an institution was its possible usefulness to a rapidly expanding American industry and empire. In this section of the paper, I trace this crucial shift to show how instrumentalism came to dominate the university system between Humboldt's time and Gilman's.[2] My point is not to argue that universities have been exclusively concerned with producing instrumental knowledge. Rather, it is to show that the conception of knowledge as "useful" is historically contingent and has been continually contested. Because universities are products of such a history they are pluralistic places, full of competing, often contradictory, processes and practices that persist into the present despite pressures towards capitalist homogenization. The goal for radical geography is to create greater, not less, space in the future for those practices that challenge the current hegemony.

Kant's last book, *The Conflict of the Faculties* (originally published in 1798), described universities as divided between "true" and "useful" knowledge (Kant, 1979). The former resided in the Faculty of Philosophy, encompassing the disciplines that we would now call Arts and Sciences; and the latter was to be found in the Faculties of Law, Medicine, and Religion, where civil servants were trained. Emphasizing speculation as the superior mode of inquiry, Kant hoped a philosophical science could function as an opposition party in the parliament of knowledge, keeping the training of the young on the road to "truth." At the same time, others who followed Bacon, Locke, and Newton promoted the utility of certain scientific knowledges derived from empirical observation and experiment. The conflict between speculative and empirical modes of inquiry quickly came to be aligned with Kant's division between "true" and "useful" knowledge, and by the time Humboldt presided over the founding of the University of Berlin science and philosophy had begun to be seen by some as fundamentally different.

Along with his brother Alexander and other thinkers of the day, Humboldt held a more holistic vision of science than that inherent in this growing dualism. As he explained in a 1797–8 essay on Goethe, "There are those who appeal to actual observation, others who appeal to philosophical analysis . . . Insofar as an individual accords more emphasis to one or the other of these two basic activities of the mind, he departs from the course of true experience, either in a too empirical or a too speculative manner" (Humboldt, 1963: 107–8). He argued that joining research and teaching would aid in bridging the growing breach between truth and use: "a mind which has been trained in this way will spontaneously aspire to science and scholarship [*Wissenschaft*]" (Humboldt, 1970: 247; 1903: 256).[3] *Wissenschaft* is what we would today call critical thinking, with this crucial difference: that it was *not* solely the business of the humanities or departments of philosophy, because "science" as we now conceive of it had not yet precipitated out from the general field of knowledge and shed its claims to philosophical truth.[4]

Humboldt's concept of knowledge failed to take hold, however, and the natural sciences became increasingly differentiated from the human sciences during the nineteenth century, partly as a result of the pressures of state demands in the imperialist era and partly due to the related growth of capitalist emphasis on instrumental knowledge. There developed what social science historian Wolf Lepenies calls "a functional division of labor between production of knowledge and achievement of orientation" (1989: 57)—that is, between practical knowledge and questions of value. In Germany, as the disciplines became redefined, questions of value retreated from other fields until they were left primarily to the discipline of philosophy, so that the university came to expound "an unwavering faith in the power of science and philosophy to generate knowledge and culture" (Lepenies, 1988: 59), that is, for science to generate useful knowledge and for philosophy to generate culture.[5] By the time Daniel Coit Gilman transferred the model of the research university to the US in 1876, Humboldt's *Wissenschaft* had come to designate natural science rather than "the principles of the cosmos" (McClelland, 1980: 133).

The German university quickly developed into an institution that we would recognize today: one that relied heavily on the state as well as on industry for its funding, pursued specialized research for utility rather than pure *Wissenschaft*, and produced technical specialists and excellently efficient administrators to run the increasingly industrialized and highly bureaucratized German society.[6] The process only accelerated after 1870 with the founding of the German Empire and the Reich, and the university became a crucial instrument in the rise of Germany industrially and as a world power on the imperial stage. The "integration of the universities with the administrative and economic structures of imperial Germany" was in large part responsible for its power and prestige within Germany and with Anglo-American educators, many of whom were trained in this system (McClelland, 1980: 314). Mitchell Ash maintains that "it was precisely the mixture of state and private industry funding that made the German universities and basic research so productive under the German Empire" (1997: xvi).

The modern American research university followed its German elder by integrating its structure with that of industry and the nation-state. As historian Burton Bledstein argues, "Not only has higher education brought coherence and uniformity to the training of individuals for careers, it has structured and formalized the instrumental techniques Americans employ in thinking about every level of existence" (1978: 289–90). While the humanities concerned themselves with the study and stewardship of "culture," Bledstein maintains, "[i]t was the primary function of American universities to render universal scientific standards credible to the public" (1978: 326). Just as it had in the industrializing and imperially aspiring Germany, the university became a powerful social institution in the US precisely because it could successfully contain value questions in the "cultural" disciplines and open up science to the pursuit of instrumental reason. By the turn of the century, the new American universities and the newly restructured colleges met the bureaucratic and governmental needs of an expanding empire in the same way that they were used by Bismarck in the Germany of the 1870s: they not only produced bureaucratic subjects to administer an empire, but also produced the technological know-how to win such an empire industrially and militarily. And Daniel Coit Gilman played a crucial role in shaping and cementing that relationship.[7]

When Gilman traveled to Germany in 1875 to study German universities as possible models for Johns Hopkins, he found a system of research and education fundamentally tied to the imperial aspirations of the German Reich. At this time his own discipline, geography, was experiencing an "explosive institutionalization" in Germany due to its "practical relevance" to the imperial project (Sander and Rossler, 1994: 115–19). In fact, it was in Germany in 1874 that "[g]eography as a field of advanced study taught by professionally qualified individuals first appeared" (Martin and James, 1993: 133).

When Gilman began to establish the hegemony of the research university in the US, he used the relationship between geography and the German Reich as his model. He found theoretical justification for his version of knowledge production in geographical thought: he based his idea of the social role of the university on the instrumentalist theories of his teacher and mentor Arnold Guyot, who argued in his book *The Earth and Man* (1850) that geography provided "scientific" justification for the Euro-American domination of the world. Guyot concludes this book by looking forward to

> the elaboration of the material wealth of the tropical regions, for the benefit of the whole world. The nations of the lower races, associated like brothers with the civilized man of the ancient Christian societies, and directed by his intelligent activity, will be the chief instruments. The whole world, so turned to use by man, will fulfil its destiny.
>
> The three northern continents, however, seem made to be the leaders; the three southern, the aids. The people of the temperate continents will always be the men of intelligence, of activity, the brain of humanity, if I may venture to say so; the people of the tropical continents will always be the hands, the workmen, the sons of toil.
>
> (Guyot, 1850: 331)

Instead of a geographical inquiry (and all forms of *Wissenschaft*) with the goal of producing critical thinkers, Guyot's geography bolstered a strictly instrumentalist approach in which man[8] stood apart from nature: nature existed so that (European) man could conquer it, and science would prove to be a central tool in that project.

In a speech entitled "Books and Politics" in which he invoked Guyot's geography, delivered at Princeton at the conclusion of the Spanish–American War, Gilman explained the central importance of a research university to the new imperial project that the US had undertaken. He believed that the "highest service" that universities could "render to the community in which they are placed" would be to produce "the man behind the gun" of American imperialism, a man "disciplined in accuracy, coolness, memory, ingenuity, judgement, and intellectual strength" (Gilman, 1906: 198–9). It was the "intellectual strength" of the modern research university that allowed Gilman to make it an institution indispensable to the project of American industrial imperialism. Throughout the nineteenth century, this intellectual strength did not so much replace physical or military might in Germany and the US as it became a crucial component in those nations' development of industrial nationalism in the late imperial era. Through the work of Gilman and others, the university became a crucial site for the production of that might and for the entrance of Germany and the US onto the global imperial stage at the end of the nineteenth century, a position it has retained and strengthened since the beginning of World War II, particularly since the explosion of government-funded research during the Cold War.[9]

That Gilman's instrumentalism beat out Humboldt's during the nineteenth century to emerge as the hegemonic model for knowledge production in the American university does not mean that the university is "in ruins," nor does it mean that the university should be abandoned by critical thinkers. Indeed, political scientist Clyde Barrow argues that "[w]hat radical scholars must therefore rediscover is not merely that intellectuals play a significant role in the reproduction of capitalism and the capitalist state, but that education has been and remains every bit as much a contested terrain as the shop floor, the party caucus, and the halls of legislative assemblies" (1990: 9). The contested history of the university means that these institutions remain crucial sites for political engagement. Just as geographers and geography have been central to historical conceptions of the modern research university—through, for example, Humboldt's theory of the "cosmos" or Guyot's racist imperial dreams —so too can radical geography play a vital role in reimagining a progressive role in the contested university today.

THEORY, PRAXIS, PEDAGOGY

The last thirty years has seen a productive, if unfinished from the perspective of radical geography, rethinking of the relationship between researcher and subject and of the epistemological basis of geographical research in general. However, little space has been devoted to discussing the implications of new epistemologies for classroom practice and pedagogical theory. Despite all the post-positivist reflexivity on the research process, academic knowledge is still overwhelmingly treated instrumentally: that is, once written, it is conceived of as *information* that is unproblematically transmissible, as a commodity that can be readily exchanged for the price of a book, a consulting fee, or university tuition.

For example, Rubin and Rubin's book-length treatment of qualitative interviewing ends with a chapter on "Sharing the Results" in which they claim that "[t]he last step in the research is to put this information into a report that is convincing, thought-provoking, absorbing, vivid, and fresh . . . If people from the research arena and others who are familiar with that arena say, 'Yes, this is the way it is,' then your research is finally complete. Your work has passed its last test" (1995: 257–74). Rubin and Rubin believe the research process achieves closure the moment writing is completed. According to this theory, once "produced," geographical knowledge is viewed outside the wider social context in which it exists. Once written, this knowledge loses the dynamic qualities that theory has given it and becomes objectified, ready for quick and easy dissemination as a piece of writing, whether to policy élites or to other academics. Academic geographers may have become aware of the social relations that affect the research process, but we have failed to extend that understanding to the wider social context in which the product of research exists—the world of teaching and learning that is the research university.

I want to resist the kind of closure that Rubin and Rubin bring to the research process and argue that the work of knowledge production does *not* end with a written text. Keeping open the problematics of knowing beyond the end of the writing and extending to work *inside* the academy the lessons learned from contemplating post-positivist methods in geographical knowledge production means developing more sophisticated approaches to pedagogy that do not reduce knowledge to information that is easily transmissible. Such an approach will itself be an engaged form of radical politics. It also suggests ways to talk about the student–teacher relationship that move us beyond the current discourse of student-as-consumer, which perpetuates the instrumentalization and commodification of knowledge and contributes to what radical education theorist Henry Giroux refers to as the "objectification of thought itself" that is part of a "culture of positivism" in the academy (1997: 25).

American philosopher and educator John Dewey describes how the process of learning is itself an active engagement with the world that cannot be reduced to the transfer of information:

> Normally every activity engaged in for its own sake reaches out beyond its immediate self. It does not wait for information to be bestowed which will increase its meaning; it seeks it out . . . It is the business of educators to supply an environment so that this reaching out of an experience may be fruitfully rewarded and kept continuously active.
>
> (Dewey, 1925: 245)

Post-positivist models of research imply the dynamic, open-ended, and interactive qualities of knowledge that education theorists have posited as central to learning. Radical geographers need a clearly articulated pedagogy that refuses the kind of closure Rubin and Rubin bring to research and reclaims the classroom from the quantifiers of technical proficiency.

Theorists of "critical pedagogy" argue that classroom practices—the ways in which teachers formally approach their students in attempts to convey the "content" of their knowledge—are structured by ideological assumptions, just as much as is knowledge itself. Without a pedagogy theoretically equal to the production of its knowledge, radical geography runs the risk of losing the current battle over the classroom. Radical educator Paulo Freire describes how objectification of knowledge within the classroom—one that treats knowledge as a thing to be unproblematically transferred from teacher to student—serves as an instrument of domination and oppression, despite its "content" (Freire, 1990; see also hooks, 1994). For Freire, such a pedagogic approach treats students as empty vessels for teachers to fill with knowledge, thus disempowering them and devaluing their own experiences and powers of critical thought.[10] Ultimately, such objectification presents the world as a static and fixed structure to which students must conform. Even radical "messages" delivered in this manner contribute to the perpetuation of existing power relations.[11] Extending post-positivist theory to the classroom admits the participatory nature of knowledge and invites an active and critical engagement with the world through which students are empowered to transform their world; it is, bell hooks says, "to teach in a way that liberates, that expands consciousness, that awakens, . . . [and] challenge[s] domination at its very core" (1994: 75). Giroux describes the political practices implied by a post-positivist pedagogy:

> Critical pedagogy needs to be informed by a public philosophy dedicated to returning schools to their primary task: furnishing places of critical education that serve to create a public sphere of citizens who are able to exercise power over their own lives and especially over the conditions of knowledge production and acquisition. This is a critical pedagogy defined, in part, by the attempt to create the lived experience of empowerment for the vast majority. In other words, the language of critical pedagogy needs to construct schools as democratic public spheres. In part, this means that educators need to develop a critical pedagogy in which the knowledge, habits, and skills of critical rather than simply good citizenship are taught and practiced. This means providing students with the opportunity to develop the critical capacity to challenge and transform existing social and political forms, rather than simply adapt to them.
>
> (Giroux, 1997: 218)

Deobjectifying knowledge in the classroom as well as in "the field" exposes the classroom as a site of practical political engagement and disrupts the boundary between theory and praxis. As much as the field, the classroom is a place where we come together to make meaning and knowledge about the world(s) we inhabit. Knowledge is not merely an object to be used as an instrument of technocratic rationality with which to better manage the world. Rather, it is itself a dynamic pedagogical encounter. Viewed this way, it has the potential to empower rather than dominate.

While the presence of periodicals like the *Journal of Geography in Higher Education* (*JGHE*) and the *Journal of Geography* bring much-needed pedagogical discussions to the discipline, such discussions remain on the periphery of geography. Even Alan Jenkins, cofounder and longtime editor of *JGHE*, argued in his 1997 retrospective that teaching is less valued now in the discipline than it was in 1977, when the journal was first published, and that the current economic shifts in universities exacerbate the situation: "the pressures of budget cuts and underfunding put greater pressures on [junior faculty's] time and attention. These pressures make it harder for them and their institutions to 'value' teaching effectively" (1997: 13). Furthermore, much of the writing about pedagogy in geography, when not attempting to implement new information "delivery" systems in the classroom (Towse and Garside, 1998: 386; but see Ó Tuathail and McCormack, 1998a, 1998b), stresses "effective" or "efficient" methods of teaching without addressing the political or ideological issues that concern radical geographers. A notable exception is the recent *JGHE* Symposium on teaching sexualities (March 1999 issue), especially the article by J. K. Gibson-Graham, which takes as one of its premises that the classroom itself is a space for political struggle and an important site for the project of "queer(y)ing" capitalism (*JGHE* Symposium, 1999; Gibson-Graham, 1999). Accepting the dynamic, interactive, and political nature of the pedagogical encounter means defining the classroom as a vital public space that needs to be defended against the forces of commoditization that would reduce it to a mere medium of transmission.

A radical pedagogy that resists the closure of knowledge also resists the clientization of students discussed by Readings (1996). If knowledge is seen as a cooperative project, rather than an object or commodity, then students must be regarded as partners in education, rather than consumers of it. By moving beyond an instrumentalist notion of the student–teacher relationship, we may yet fulfill Derek Gregory's "plea" that critical human geography "restores human beings to their own worlds and enables them to take part in the collective transformation of their own human geographies" (1978: 172). The classroom is one vital place in which we enact this transformation.

What implications does this have for new pedagogies appropriate to a radical *geography*? Dewey says,

> Geography and history supply subject matter which gives background and outlook, intellectual perspective, to what might otherwise be narrow personal actions or mere forms of technical skill. With every increase of ability to place our own doings in their time and space connections, our doings gain in significant content . . . Thus our ordinary daily experiences cease to be things of the moment and gain enduring substance.
>
> (Dewey 1925: 244)

The kind of "intellectual perspective" that Dewey describes here differs greatly from the objectification of knowledge driven by dominant educational practices, which prepare students to be efficient and expert users of information in a world dominated by technocratic rationality. As Giroux and Freire make clear, such intellectual growth demands classroom practices that go beyond mere "delivery" of information (Giroux, 1994, 1997; Freire, 1990). As Dewey suggests, geography and radical geographers are uniquely equipped to make an important contribution to such a pedagogical and political project. For instance, in a reversal of the logic driving current university restructuring, Matthew Sparke (1999) has argued for a radical geography that instills in students a "geographic accountability," a type of critical cosmopolitanism that seeks to draw students, who are soon to become active members in a new global order, into a critical understanding of the ethical dilemmas of this new geographical connectivity that we call globalization (1999: 95–6; see also Castree and Sparke, 2000). The goal of this mindset is not merely to describe a new world order, but to empower students to reshape it according to principles of social justice, critical citizenship, and participatory democracy.

The intellectual project of empowerment must be pursued by deobjectifying knowledge in the classroom, thus breaching the walls that have been artificially erected between theory and practice. Only by fully engaging in the pedagogical mission of the university can critical thought have an influence on the social production of a better world and not be further marginalized in the university. In an era in which the main function of university teaching is to produce efficient workers for the information sector of corporate capital, critical human geographers can play an important pedagogical role in instilling in the corporate bureaucrats of tomorrow—the managers of our so-called "information economy"—a sense of geographical accountability, an understanding of our situatedness in the contemporary world.[12] Through a full account of the new international division of labor, critical human geography can show how our lives and the lives of our students are also products of the labor of others. It can help make clear what William Cronin called capitalism's "landscape of obscured connections" (1991: 340). Critical human geography can help students reflect on how the work that they will do in life has the potential to participate in the reproduction of the world and the perpetuation of uneven development—or in the transformation of the world.

In this context, J. Hillis Miller's fear that the university is becoming "an increasingly less important site" is unfounded (Miller, 1996: 7). Rather, the university's central role in social reproduction makes it a primary site for struggle and engagement. With the advent of critical theory and a new recognition that all knowledge is socially constructed, critical human geographers can play a part in stopping the instrumentalization of knowledge and the increasing industrial domination of the world and can help shape a better future. Such a project comes out of the history of our own discipline. Nearly 150 years ago, Alexander von Humboldt insisted that "the character of the landscape, and of every imposing scene in nature, depends so materially upon the mutual relation of the ideas and sentiments simultaneously excited in the mind of the observer" (1849, 1: 6). Critical human geography must make radical pedagogy a central concern—and must recognize the classroom as a site of practical political engagement.

NOTES

1 In fact, Gibson draws on the long-standing "external funding models followed by colleagues in the hard sciences, engineering, and medicine" (1998: 463).
2 The history that I outline here is specific to American universities, which followed a different trajectory from British institutions.
3 Humboldt's plan for the University of Berlin, "Über die innere und aussere Organisation der höheren wissenschaftlichen Anstalten in Berlin" ("On the Inner and Outer Organization of Institutions of Higher Learning in Berlin"), appears in volume X of the 1903 *Gesammelte Schriften* on pp. 250–60. I have also consulted three translations: Marianne Cowan's in *Humanist without Portfolio* (1963), the one which appeared in the journal *Minerva* (1970), and that by Clifton Fadiman in *The Great Ideas Today* (1969). References to those editions are indicated by year of publication and page number. I have modified the translation of this piece throughout. Humboldt's now-famous "plan," which is assumed to be the foundational document for the German model university, was unknown until around 1900. A fragment of a memo that Humboldt may or may not have sent to the king, the document was found by historian Bruno Gephardt at the end of the nineteenth century in a drawer in the Prussian ministry and subsequently published (Fallon, 1997: 149; vom Bruch, 1997: 12). It is the basis of much of the contemporary discussion of thc "Humboldtian University."
4 The social sciences did not separate out from the humanities and natural sciences until the end of the nineteenth century. See Lepenies (1988). For discussion of the emergence of *Geisteswissenschaften* and *Naturwissenschaften* and the emergence of the modern conception of aesthetic culture, see Gadamer (1989) and Iggers (1983). Through the influence of Hegel, Dilthy, and others, *Geisteswissenschaft*, with a focus on "spirit" and speculation, emerged as the accepted philosophical model of inquiry, while *Naturwissenschaften* became defined primarily in positivist terms. Kant's aesthetics gave rise to a philosophical tradition in which the "doctrine of taste and genius" (Gadamer, 1989: 42) was equated with culture.
5 In Britain and the US, however, this "cultural" function came to be housed in the humanities in general and literature in particular. See Readings (1996) and Graff (1987).
6 For the history of German universities in the nineteenth century, see McClelland (1980), vom Bruch (1997), and Fallon (1997).
7 For more on Gilman and his influence, see Flexner (1946), Franklin (1910), French (1946), Hawkins (1960), Heyman (1998), and Ryan (1939), as well as Gilman's own reflections (1891, 1906).
8 There is no space in this essay to adequately address the important issue of Gilman's (and Guyot's and Humboldt's) masculinism; let me, therefore, merely note it here.
9 For discussion of university funding during and after the Cold War, see Lewontin (1997). For a general discussion of the relationship between the university and the state after World War I, see Barrow (1990).
10 For a discussion of the problematics of importing Freire into a North American context, see Stygall (1988) and Giroux (1994).
11 This is of course something that marginalized groups within the academy have long recognized (see hooks, 1994).
12 This sense of "accountability" carries with it an implicit imperative towards Harvey's notion of social justice (Harvey, 1996).

REFERENCES

2020 (1998) "Commission on the Future of Higher Education Charge." 12 August, http://www.washington.edu/faculty/facsenate/2020/2020charge.html.
Ash, M. G. (ed.) (1997) "Introduction." *German Universities Past and Future*. Providence, RI: Berghahn Books, pp. 3–27.
Barrow, C. W. (1990) *Universities and the Capitalist State*. Madison: Wisconsin University Press.
Bledstein, B. J. (1978) *The Culture of Professionalism: The Middle Class and the Development of Higher Education in America*. New York: W. W. Norton.
Bowen, M. (1981) *Empiricism and Geographical Thought from Francis Bacon to Alexander von Humboldt*. Cambridge: Cambridge University Press.
Brubacher, J. S., and W. Rudy (1968) *Higher Education in Transition: A History of American Colleges and Universities, 1636–1968*. New York: Harper and Row.
Castree, N., and M. Sparke (2000) "Introduction: Professional geography and the corporatization of

the university: Experiences, evaluations, and engagements." *Antipode* 32(3).

Cowley, W. H., and D. Williams (1991) *International and Historical Roots of American Higher Education*. New York: Garland.

Cronin, W. (1991) *Nature's Metropolis*. New York: Norton.

Dewey, J. (1925) *Democracy and Education*. New York: Macmillan.

Fallon, D. (1997) "Interpreting Humboldt for the twenty-first century." In M. G. Ash (ed.) *German Universities Past and Future*. Providence, RI: Berghahn Books, pp. 149–57.

Flexner, A. (1946) *Daniel Coit Oilman: Creator of the American Type of University*. New York: Harcourt, Brace and Co.

Franklin, F. (1910) *The Life of Daniel Coit Gilman*. New York: Dodd, Mead and Co.

Freire, P. (1990) *The Pedagogy of the Oppressed*. New York: Seabury Press.

French, J. C. (1946) *A History of the University Founded by Johns Hopkins*. Baltimore: Johns Hopkins Press.

Gadamer, H.-G. (1989) *Truth and Method*. Second edition. New York: Continuum.

Gibson, L. J. (1998) "Institutionalizing regional science." *The Annals of Regional Science* 32: 459–467.

Gibson-Graham, J. K. (1999) "Queer(y)ing capitalism in and out of the classroom." *Journal of Geography in Higher Education* 23: 80–5.

Gilman, D. C. (1860) "Humboldt, Ritter, and the new geography." *The New Englander* 18(70): 277–306.

—— (1891) "The Johns Hopkins University (1876–1891)." In *Studies in Historical and Political Science*. Ninth series, no. 3–4. Baltimore: Johns Hopkins Press.

—— (1906) *The Launching of a University*. New York: Dodd, Mead and Co.

Giroux, H. A. (1994) *Disturbing Pleasures*. London: Routledge.

—— (1997) *Pedagogy and the Politics of Hope*. Boulder, CO: Westview Press.

Graff, G. (1987) *Professing Literature: An Institutional History*. Chicago: University of Chicago Press.

Gramsci, A. (1971) *Selections from the Prison Notebooks*. New York: International Publishers.

Gregory, D. (1978) *Ideology, Science and Human Geography*. London: Hutchinson.

—— (1994) *Geographical Imaginations*. Cambridge, MA: Blackwell Publishers.

Guyot, A. (1850) *The Earth and Man*. Boston: Gould, Kendall and Lincoln.

Haraway, D. (1996) "Situated knowledges: The science question in feminism and the privilege of partial perspective." In J. Agnew, D. N. Livingstone, and A. Rogers (eds) *Human Geography: An Essential Anthology*. Oxford: Blackwell Publishers, pp. 109–28.

Harvey, D. (1989) *The Condition of Postmodernity*. Cambridge, MA: Blackwell Publishers.

—— (1996) *Justice, Nature, and the Geography of Difference*. Cambridge, MA: Blackwell Publishers.

Hawkins, H. (1960) *Pioneer: A History of the Johns Hopkins University, 1874–1889*. Ithaca, NY: Cornell University Press.

Heyman, R. (1998) "Geographical thought, ideology, and the university: The Humboldt brothers and Daniel Coit Gilman." Unpublished master's thesis. University of Washington.

hooks, b. (1994) "Toward a revolutionary feminist pedagogy." In D. H. Richter (ed.) *Falling into Theory*. Boston: Bedford, pp. 74–9.

Horkheimer, M., and T. W. Adomo (1990) *Dialectic of Enlightenment*. New York: Continuum.

Humboldt, A. von (1849) Cosmos: *A Sketch of a Physical Description of the Universe*. 2 vols. London: H. G. Bonn.

Humboldt, W. von (1903) "Über die innere und aussere Organisation der höheren wissenschaftlichen Anstalten in Berlin." In *Gesammelte Schriften*, vol. X. Berlin: B. Behr's Verlag, pp. 250–60.

—— (1963) *Humanist without Portfolio: An Anthology of the Writings of Wilhelm von Humboldt*. Translated by Marianne Cowan. Detroit: Wayne State University Press.

—— (1967) "On the historian's task." *History and Theory* 6(1): 57–71.

—— (1969) "On the organization of institutions of higher learning in Berlin." Translated by Clifton Fadiman. In R. M. Hutchins and M. J. Adler (eds) *The Great Ideas Today 1969*. New York: Encyclopaedia Britannica, pp. 348–55.

—— (1970) "On the spirit and the organizational framework of intellectual institutions in Berlin." *Minerva* 8(2): 242–9.

Iggers, G. G. (1983) *The German Conception of History*. Middletown: Wesleyan University Press.

JGHE Symposium (1999) "Teaching sexualities in geography." *Journal of Geography in Higher Education* 23(1): 77–124.

Jaspers, K. (1959) *The Idea of the University*. Boston: Beacon Press.

Jenkins, A. (1997) "Twenty-one volumes on: Is teaching valued in geography in higher education?" *Journal of Geography in Higher Education* 21(1): 5–15.

Judy, R. A. T. (1993) *(Dis)Forming the American Canon*. Minneapolis: Minnesota University Press.

Kant, I. (1979) *The Conflict of the Faculties*. Translated by Mary J. Gregor. New York: Abaris. Originally published in 1798.

Kaplan, E. A., and G. Levine (eds.) (1997) *The Politics of Research*. New Brunswick: Rutgers University Press.

Kerr, C. (1969) "The pluralistic university in the pluralistic society." In R. M. Hutchins and M. J. Adler (eds) *The Great Ideas Today 1969*. New York: Encyclopaedia Britanica, pp. 4–29.

—— (1972) *The Uses of the University*. Cambridge, MA: Harvard University Press.

Knoll, J. H., and H. Siebert (1967) *Wilhelm von Humboldt: Politician and Educationist*. Bad Godesberg: Inter Nationes.

Lepenies, W. (1988) *Between Literature and Science: The Rise of Sociology*. Cambridge: Cambridge University Press.

—— (1989) "The direction of the disciplines." In E. S. Shaffer (ed.) *Comparative Criticism: An Annual Journal. II: The Future of the Disciplines*. Cambridge: Cambridge University Press.

Lewontin, R. C. (1997) "The Cold War and the transformation of the academy." In N. Chomsky *et al. The Cold War and the University*. New York: The New Press, pp. 1–34.

Locke, G. (1998) "Remarks by Governor Locke: 2020 Commission Announcement." 4 February, http://www.govemor.wa.gov/press/98020401.htm.

Loh, W. (1998) "Future of higher education: New realities for universities." *Seattle Times*, 14 June, http://www.seattletimes.com; search archives.

Lucas, C. J. (1994) *American Higher Education: A History*. New York: St. Martin's.

Lyotard, J.-F. (1988) *The Postmodern Condition: A Report on Knowledge*. Minneapolis: Minnesota University Press.

McClelland, C. E. (1980) *State, Society, and University in Germany 1700–1914*. Cambridge: Cambridge University Press.

Martin, G. J., and P. E. James (1993) *All Possible Worlds: A History of Geographical Ideas*. Third edition. New York: John Wiley & Sons.

Miller, J. H. (1996) "Literary study in the transnational university." *Profession* 1996: 6–14.

Nelson, C. (1995a) "Lessons from the job wars: Late capitalism arrives on campus." *Social Text* 13(3): 119–34.

—— (1995b) "Lessons from the job wars: What is to be done?" *Academe* 81(6): 18–25.

Noble, D. (1998) "Digital diploma mills." Distributed via email. Available online at http://www.firstmonday.dk/issues/issue3_l/noble/index.html.

Ó Tuathail, G., and D. McCormack (1998a) "Global conflicts online: Technoliteracy and developing an internet-based conflict archive." *Journal of Geography* 97: 1–11.

—— (1998b) "The technoliteracy challenge: Teaching globalization using the internet." *Journal of Geography in Higher Education* 22(3): 347–61.

Pelikan, J. (1992) *The Idea of a University*. New Haven: Yale University Press.

Readings, B. (1996) *The University in Ruins*. Cambridge, MA: Harvard University Press.

Reill, P. H. (1994) "Science and the construction of the cultural sciences in late enlightenment Germany: The case of Wilhelm von Humboldt." *History and Theory* 33(3): 345–66.

Rubin, H. J., and I. S. Rubin (1995) *Qualitative Interviewing*. Thousand Oaks: Sage.

Ryan, W. C. (1939) *Studies in Early Graduate Education*. New York. Carnegie Foundation for the Advancement of Teaching.

Sander, G., and M. Rossler (1994) "Geography and empire in Germany, 1871–1945." In A. Godlewska and N. Smith (eds) *Geography and Empire*. Oxford: Blackwell Publishers, pp. 115–27.

Smith, P. (1990) *Killing the Spirit: Higher Education in America*. New York: Viking.

Soley, L. C. (1995) *Leasing the Ivory Tower: The Corporate Takeover of Academia*. Boston: South End Press.

Sparke, M. (1999) "Teaching geographic accountability with newspapers." In S. Knowlton and B. Barefoot (eds) *Using National Newspapers in the College Classroom*. University of South Carolina: First Year Experiences Center Monograph Series #28, pp. 95–6.

Stygall, G. (1988) "Teaching Freire in North America." *Journal of Teaching Writing* 1(1): 113–25.

Sweet, P. R. (1978) *Wilhelm von Humboldt: A Biography*. Two vols. Columbus: Ohio State University Press.

Towse, R. J., and P. Garside (1998) "Integration and evaluation of CAL courseware and automated assessment in the delivery of geography module." *Journal of Geography in Higher Education* 22(3): 385–96.

Veysey, L. (1965) *The Emergence of the American University*. Chicago: University of Chicago Press.

vom Bruch, R. (1997) "A slow farewell to Humboldt? Stages in the history of German universities, 1810–1945." In M. G. Ash (ed.) *German Universities Past and Future*. Providence, RI: Berghahn Books, pp. 3–27.

11

Situated knowledge through exploration

Reflections on Bunge's 'Geographical Expeditions'

Andy Merrifield

from *Antipode*, 1995, 27(1): 49–70

> The need to lend a voice to suffering is a condition of all truth.
>
> (Theodor Adorno, 1973)

> [T]o seek to know before we know is as absurd as the wise resolution of Scholasticus, not to venture into the water until he had learned to swim.
>
> (G. W. F. Hegel, 1931)

> It's a strange world. Some people get rich and others eat shit and die.
>
> (Hunter Thompson, 1988)

INTRODUCTION

Over the last decade or so notions of "situated knowledge," "standpoint theory," and "positionality" have received an enormous amount of attention in radical scientific and social scientific circles and in the humanities. At its most simplistic, this state of affairs can be taken as something of a reaction to, and critical engagement with, postmodernist and post-structuralist modes of thought which have gained increasing credence within the academic left. Within feminist academic and activist inquiry particularly, the concept of "situated knowledge" has called into question the epistemological basis of the western Enlightenment philosophical tradition and scientific practice (see, e.g., Harding, 1986; Hartsock, 1987, 1989/90; Nicholson, 1990; and Haraway, 1991). At the heart of the issue lies a fundamental insistence on the *contextualized* nature of *all* forms of knowledge, meaning, and behavior. There is a further recognition of the partial and partisan edge to inquiry, theory construction, and scholarly (re)presentation, as well as an explicit acknowledgement of the importance of the author's biography in this creative process. These challenging insights have in recent years percolated down to the discipline of human geography (see Christopherson, 1989; McDowell, 1991, 1992; Jackson, 1993). In so doing, they have sparked intense debate and ushered in something of a reappraisal and realignment of the left critical geographical program (cf. Massey, 1991; Harvey, 1992).

This engagement has had both constructive and destructive ramifications, though it is not my intention to discuss these further here. Instead, I will argue that versions of situated knowledge and standpoint theory have had a long and rich legacy within radical academic, intellectual, and political endeavor. In this paper, I propose to explore in more depth just one strand of this tradition within geography: the "geographical expedition" program, an initiative that was linked with the development of radical approaches to urban phenomena in the late 1960s and early 1970s. I want to argue, firstly, that a critical engagement with the legacy of geographical expeditions can contribute toward the methodological and epistemological discussions now echoing within left geography. Secondly, a contemporary redefinition of expeditions can inform the development of a more sophisticated critical urban geography, one that acknowledges the progressive aspects of postmodernist theory, yet does not jettison the hard-fought insights gained from the expeditions' modernist tradition.[1] Finally, I contend that within the expedition concept there lurks an undeveloped *pedagogic* component that can be teased out and built upon as part of a progressive radicalization strategy for teaching and for localized political activity. In what follows I shall confront these themes in turn. The account, however, will be prefaced with some clarifying remarks about situated knowledge arguments.

SITUATED KNOWLEDGES

Epistemologies of situated knowledges offer adherents a conceptual platform from which to call into question all privileged knowledge claims. A contextualized basis for knowledge production and scientific accomplishment implies the denial of meta-narratives and totalizing perspectives, especially those that speak in the name of objectivity and neutrality. Rejection of universal truth claims thus gives voice to marginalized "Others," those traditionally oppressed and excluded from Western (white male) intellectual and political practices. Situated knowledge, according to feminist historian of science Haraway (1991), is nothing more than a shorthand term that provides intellectual and political space for hitherto silenced voices to be heard. Hence, situated knowledges are "particularly powerful tools to produce maps of consciousness for people who have been inscribed within the marked categories of race and sex that have been so exuberantly produced in the histories of masculinist, racist and colonialist dominations." Haraway (1991: 191) argues that this hegemonic Western cultural narrative produces disembodied, detached, unbeatable, and *irresponsible* knowledge claims. For Haraway, furthermore, irresponsibility means "unable to be called into account" because it purports to "see everything from nowhere." It follows from this that a spurious doctrine of scientific objectivity provides an ideological veil—a ruse Haraway calls a "god-trick"—simultaneously beclouding and reinforcing existing and unequal power relations.

Situated knowledges appeal to the social constructivist argument since, as Nicholson stresses, "[d]escriptions of social reality bear a curious relation to the reality that they are about; in part such descriptions help constitute the reality" (cited in McDowell, 1992: 60). Haraway's (1992) research into primatology illustrates that the detached eye of objective science is an ideological fiction and so it is legitimate to criticize the natural sciences on the level of "values" as well as that of "facts." The upshot is an invocation "for a politics and epistemologies of location, positioning, and situating, where partiality and not universality is the condition of being heard to make rational knowledge claims" (Haraway, 1991: 195). Under such circumstances, knowledge is always embedded in a particular time and space; it doesn't see everything from nowhere but rather *sees somethings from somewhere* (cf. Hartsock, 1989/90: 29). A situated understanding, therefore, provides a position from which to organize, conceptualize, and judge the world. Yet this is always partial, never finished nor whole; it is always woven imperfectly and holds no justifiable claims to absolute privileged knowledge. There are always different and contrasting ways of knowing the world, equally partial and equally contestable.

But this doesn't mean any viewpoint will suffice. Again Haraway is helpful: "'equality' of positioning is a denial of responsibility and critical enquiry. Relativism is the perfect mirror twin of totalization in the ideologies of objectivity." Relativism and absolutism present themselves as commensurate "god-tricks:" both deny the stakes in location, embodiment, and partial perspective, both "make it impossible to see well" (Haraway, 1991: 191). To this extent a committed and situated knowledge offers a corrective to the god-tricks of positivism and some postmodernism: situatedness implies that an understanding of reality is accountable and responsible for an *enabling* political practice. Ultimately, then, the realm of politics conditions what may count as *true* knowledge.

This fundamental insistence is instructive for my desire to resituate the geographical expeditions phenomenon within an epistemology of situated knowledges and attempt to circumvent the current paralysis within "strong" postmodern critical theory. [. . .] The situated knowledge conceptualization, framed within the expedition ideal, permits a theoretical and political alternative bold enough not to relinquish some sort of universal, ethical anchoring to scholarly endeavor, yet acknowledges "otherness" and difference, and recognizes that a partial and partisan perspective is preferable precisely because it can be held accountable. Such an understanding, too, places at a premium knowledge produced from the standpoint of the subjugated; and, as Hartsock (1989/90) tells us, knowledge constructed from the standpoint of the dominated and marginalized holds a claim to present a truer and more adequate account of reality.[2]

Geographical expeditions were implemented precisely to acquire such a subjugated standpoint: Bunge's theoretical and practical invocation to set up "base camp" in the inner city was, *inter alia*, a search for a situated knowledge. Base camp offered quite literally a critical bivouac from where a partisan, responsible, and accountable vision of urban society could be constructed in such a way as to inform action. From there geographers had the opportunity "to learn how to see faithfully from another's point of view."[3] Expeditions permitted critical vision consequent upon a "critical positioning" in urban space. Later, I will explore whether geographers' ability to understand the situation of "the Other" is restricted by their own individual biographies. In the interim, let me outline more directly the geographical expedition tradition.

SITUATEDNESS THROUGH EXPLORATION

The initial impetus to the geographical expedition "movement"—if it's possible to call it that—rested of course in the turbulent social and political atmosphere of the 1960s: protests against the imperialist war in Vietnam, the May 1968 insurrection in Paris, and the proliferation of civil right demonstrations and large-scale rioting on the part of oppressed and impoverished urban populations in the United States. These widescale and cataclysmic events provided a *Zeitgeist* in which many geographers were compelled to reconsider the conceptual and practical basis of their discipline (see Harvey and Smith, 1984). For those most radicalized by such a state of affairs, this was to be a heart-wrenching process of reevaluation which prompted a necessity for greater social relevancy in their geography, as well as a concomitant rejection of an erstwhile "nice" or *status quo* geography (Harvey, 1973: chap. 4). This was to involve a deep conviction to an intellectual and political project intent on changing society and fashioning a critique of dominant values, ideology, and scholarly practices. These concerns were at odds with the inherently capitalistic and imperialistic geographical establishment, and many radical geographers, like Bunge, were forced out of their jobs, marginalized and ostracized from academic geographical circles (see Horvath, 1971). (Bunge, from his Quebec "exile," continues to remain something of a geographical *persona non grata* today.) At the time, the pursuit of intellectual and societal transformation was no place for the career-orientated.[4]

Hitherto, Bunge's research was, like that of many who later turned toward radical reinterpretations of spatial and social phenomena (notably Harvey), rooted in the positivistic tradition. His *Theoretical Geography* (1962) (which he dedicated to Walter Christaller) employed mathematical modeling and mapping techniques to attack, *inter alia*, Hartshorne's insistence on locational uniqueness. This commitment to mapping, it should be noted, remained a dominant motif of Bunge's radical geography. But what had changed was the target and scale of Bunge's focus. In "Perspective on theoretical geography" (1979a: 170), for example, he passionately describes the way in which a stay in a black ghetto hotel in Chicago for the 1966 Martin Luther King demonstrations taught him "how you have to 'get ready to kill the world' to walk across the street to get a corned beef sandwich; that is, I could make it on 'the mean streets'—an indispensable skill for urban exploration in antagonistic systems." In Detroit, moreover, a young black woman, Gwendolyn Warren (who became Director of the Detroit Geographical Expedition), taught Bunge a further lesson on urban reality: children were starving to death and being killed by automobiles in front of their own homes. For the neophyte Bunge, this experience had a powerful effect on his geography; and as he recounts (1979a: 170), Warren and other ghetto dwellers were "furiously interpreting the world all around me that I could not see because my life had been spent buried in books . . . [This] caused me to reverse my scale and I wrote a book about one square mile in the middle of black industrial Detroit."

Yet, the said book, *Fitzgerald: Geography of a Revolution* (Bunge, 1971), didn't start out as a geography text. Nevertheless, the collection of highly charged maps and evocative photographs—in a large, atlas-size format—emphasized the usefulness of academic geography while convincing Bunge of the urgent political necessity to "bring global problems down to earth, to the scale of people's normal lives" (1979a: 170).[5] That, for Bunge, had to be a geographer's *raison d'être*: in blunt terms, one had to work at being useful. Unsurprisingly, this fundamental insistence necessitated a categoric rejection of contented "campus geography" which, according to Bunge, tends to sever theory from practice and prioritizes citing as opposed to *sighting*.[6] Crucial here was the practice of exploration: the construction of a critical vantage point that wasn't an exotic quest of geographical plundering or escapism, but rather a "contributive" expedition. Indeed, Bunge (n.d.: 48–49) claimed that the seven mile journey from rich suburban Detroit to its poor inner city is a trip half-way around the world in terms of infant mortality rates. And as he deduces, "If half-way around the world is compressed into just seven miles, only a micro-mapping of it could show even the massive features of its geography."

Such a geography tended "to shock because it includes the full range of human experience on the earth's surface; not just the recreation land, but the blighted land; not just the affluent, but the poor; not just the beautiful, but the ugly. In America, since most of the humans live in cities, it implies the exploration of these cities" (Bunge, 1977a: 35). Geographers, Bunge insisted, had to take their geographical knowledge to the poor, and local people were to be incorporated in expeditions as both students and professors. Ironically, Bunge retained the label "expedition" in an attempt to subvert the exploration practices of the nineteenth century. Now, an expedition was to realize its full potential by helping—rather than destroying—the human species, and hence ensure the collective survival of humankind in a machine age which Bunge saw as intent on threatening itself with annihilation.

For Bunge, *survival* became the fragile thread binding logic, ethics, and politics (Bunge, 1973a, 1973b). And he did not pull any punches about where his own political loyalties lay: they were and remain virulently anti-capitalist and anti-racist. Thus, Bunge's geography was informed by a deep commitment to socialism:

> It is an illustration of the nature of the mental labor called geography. Geography has been the overwhelming force in leading me to such a deep 'political' position. Having lived and struggled in this neighborhood of Detroit called Fitzgerald, from which I write, how could I avoid directing my attention to this region? And in the dialectics of work, the commerce between the labor and the worker, how else could my work not help shape what I think—and, therefore, as a geographer, shape me?
>
> (Bunge, 1973a: 320)

There is a certain affinity here with the situated, partial, located and responsible standpoint that Haraway asserts as a means of gaining "objective knowledge" of the world. To be sure, much of Bunge's (1973a) essay on "Ethics and logic in geography" anticipates and cuts an epistemological swath for Haraway's more recent radical conviction within the history of science "to see faithfully from the standpoint of the subjugated." From such a situated and responsible perspective, the expedition would strive to be a democratic rather than an élitist pursuit. Consequently, the points of view of local people themselves were given a relative priority. Professional geographers worked in unison with "folk geographers:" practically informed lay persons such as members of residents' associations, community activists, socially-responsible citizens of all stripes, as well as taxi drivers who, says Bunge, possess an invaluable and sensitive knowledge of urban environments that should be tapped.[7] An important proviso, however, was that the "power of the expedition itself, who hires and fires, who writes checks and so forth must be in the hands of the people being explored, risky as that sounds to academics" (Bunge, 1977a: 39).

The prototypical Detroit Geographical Expedition and Institute (D.G.E.I.), established in the summer of 1969, incorporated these prerequisites and went on to implement a program of community research and education for the black residents of Detroit. Horvath (1971: 73–74) writes that the main purpose of D.G.E.I. is to "find a way in which geographers could make available educational and planning services to inner city Blacks; it represents an attempt by the black community and some professional geographers to build an institution that would link the university to the needs of the disadvantaged Blacks in the city of Detroit." The interconnection between research and education was therefore fundamental to the operation of the expedition. In terms of the educational component, professional geographers (explorers) set up free university extension/outreach programs on cartography and geographical aspects of urban planning (in conjunction with the University of Michigan) with the aim that any black person could walk off the street and take 45 hours of university credit courses, and if they attained a C grade or better could transfer with sophomore status to any Michigan university (Horvath, 1971: 73–74). All campus teachers were volunteers, the use of practical case studies was the major teaching mode, and the local community participated in any decisions over structure and content of the courses. Such circumstances enabled a productive commerce between campus explorers and local people: "[b]eyond learning the technical skills of the academics, these folk geographers learn to generalize their experiences to a larger world. In return, the campus explorers gain valuable knowledge and insights into the community" (Stephenson, 1974: 99).

Field research was equally crucial to the expedition concept. Each participant in the expedition was involved in research around issues such as political districting, interregional money flows, transportation problems, cartographic skills, and the geography of child death (see Antipodean Staff Reporters, 1969). This information gave considerable grist to the local community's mill insofar as it enabled them to organize themselves around the issues investigated. In 1971, for example, the D.G.E.I. was active in lobbying against the encroachment of Wayne State University into the Trumbull community (particularly into Mattaei Playfield; see *Field Notes* (4), 1972). Field work data gave sustenance to any political program because it could be presented to city politicians/planners and reports made political lobbying much more effective (Stephenson, 1974). The D.G.E.I. produced *A Report to the Parents of Detroit on School Decentralization* (see *Field Notes* II, 1970) that highlighted the very real difficulties low-income blacks faced just showing up for free classes: some would come hungry and others couldn't afford bus fares. Geographers and local community participants responded with a series of maps indicating a more suitable and socially just geographical allocation of educational resources, which gave black community leaders a technical study so effective that Detroit Board of Education was compelled to respond (see Horvath, 1971).

Elsewhere, field work investigated the relationship between children and machines in the context of inner city community spaces (or lack of them) allocated for children's play. Expedition research emphasized how low-income high-rise environments, with their dearth of play space, force children onto the streets where they are vulnerable to speeding traffic. Understanding the complex issue of machine versus human space—especially how it varied between rich suburban kids and their impoverished downtown counterparts—became a pivotal concern of the expedition program, as did the antagonism between community and non-community land use. General environmental quality of urban landscapes was also brought under intense scrutiny: hidden landscapes, private landscapes, landscapes of the powerless, toyless landscapes, rat-bitten baby landscapes were all explored with considerable élan.

Many innovative and imaginative ideas were later deepened and sharpened when Bunge moved from Detroit to Toronto, where he helped establish the Canadian–American Geographical Expedition (CAGE) in October 1972 (the results of which were published in *The Canadian Alternative* (Bunge and Bordessa, 1975; see, too, Stephenson, 1974). Here, five geographical scales were charted: (1) One square mile of Toronto (the base camp neighborhood of Christie Pits); (2) Toronto itself; (3) Canada; (4) North America; and (5) the world. Exploration focused on the way in which each scale impinged upon three different types of spaces: human-kind, machine-kind, and nature. These five different "scales of survival" in their mutual interaction reveal the relationship between the unique and the general, especially as it unfolds, concretizes, and impacts upon the daily life of low-income urban populations.

So in both Detroit and Toronto theory and practice were galvanized, and expedition "manuals" and field data reports (like the *Field Notes* series) were compiled to promote community activism and enhance local empowerment (see Colenutt, 1971). Expedition teams in both cities spent time in local people's homes and professional geographers were taught lessons seldom discussed on university campuses. Many of the ideas propounded by Bunge went beyond the received geographical literature of the time. Impoverished black people brought their desperate and often lurid experiences to the campus classroom, where academic geographers would listen, attempt to understand, and henceforward incorporate these insights into intellectual and political endeavor.

Researchers grappled to gain trust and respect in the base camp community (Bunge, 1977a: 39). The geographer studied the area from the point of view of the people that live there, investigating in the process "what is geographically out of whack" (1977a: 37). They did so, Bunge says, by "getting a 'feel' of the region. By talking, listening, arguing, befriending, and by making enemies of the humans in the region." The geographers' fate was the fate of the locals. Accordingly, geographers could be held responsible and accountable simply because they were "expected to live with the mess they help create." Such situatedness and positionality meant that the geographer was able to be called into account. This excursion beyond the cloisters of the academy was the route whereby geographers became both rigorous and useful in their inexorable quest for knowledge. But it meant, too, a redefinition of the research problematic and intellectual commitment of the researcher away from a smug campus career to one incorporating a dedicated community perspective which pivots around what Howe (1954) in another context called a "spirit of iconoclasm."

RESITUATING GEOGRAPHICAL EXPEDITIONS

A particularly illuminating insight of the expeditions was the explicit acknowledgement of the point of reproduction—the so-called "second front" of struggle (Bunge, 1977b)—in the organization and perpetuation of capitalism. Workplace struggles and practices had heretofore assumed a relative privilege in left research agendas. Yet class, gender, and race dynamics outside the factory—in the home, the community, and the neighborhood—are crucial in conditioning workplace social relations. For Bunge (1977b: 60), the "geography of the working class is overwhelmingly at the point of reproduction not the point of production." It followed that achieving working class unity required a dual concentration at the point of production and reproduction; the "full power" of the working class cannot be mobilized unless this unified struggle is achieved (Bunge, 1977b).

Bunge's thought also exhibited a geographical sensitivity to the way in which women—particularly poor black women—struggle to look after their children under the oppressive structure of patriarchal capitalism. The "hidden landscape" of the home became a legitimate domain of geographical inquiry, as did the worlds of child care and child's play. Exploring these hidden landscapes, noted Bunge, was "an uncovering of furtive and underground groups relative to public and assertive groups" (*Field Manual* (5), n.d.: 3). And, he added, "[finding these groups and establishing their geography, their perception of the space, helps them establish their rightful claim to their turf" (n.d.: 5). Bunge readily admits that much of this awareness of the often desperate plight of women at the point of reproduction was gained from the experiences with the black women folk geographers in Fitzgerald.

Another path-breaking aspect of the expedition tradition was an insistence on deepening our understanding of alienation, oppression, and exploitation within the practices of everyday life. Curiously enough, this painstaking desire for the everyday—which is to say, the desire to bring seemingly abstract global problems down to the scale of people's lives—bore all the hallmarks of the libertarian, humanist Marxist principles espoused in continental Europe by Henri Lefebvre (see, e.g., Merrifield, 1993a).[8] Of further interest, too, is that both Bunge's and Lefebvre's sensitivity to lived experience and city space have been deeply affected by periodic stints as taxi drivers (in Toronto and Paris respectively). Thus Bunge, like Lefebvre, berates *remote* sensing and appeals instead for what he calls an *intimate* sensing (Bunge, n.d.: 41). A passage from Bunge splendidly characterizes this micro-sensibility of everyday, lived urban spaces; it could have easily been lifted from Lefebvre's *Critique of Everyday Life* (1991; cf. pp. 57, 97, 134):

> "if you sit on a front step and listen to people talk they often talk about the urban geography of the city but seldom about anything that the census is measuring: A child was almost hit at the corner by a speeding teenager, a landlord refused to fix the plumbing, the park is becoming a hangout for teenage narcotic users, a lady had her purse snatched on the nearby commercial street.
>
> (*Field Manual* (8), n.d.: 7–8)[9]

Bunge, then, was for a geography, just as Claes Oldenburg (1990: 728) was for an art, "that embroils itself with the everyday crap and still comes out on top."

Yet mapping became a crucial strategic tool used by the geographer Bunge to represent the complex dialectical relationship between the thing/process and form/flow nature of reality. In other words, mapping the form of everyday life enabled the instantiation of the misery and alienation that abstract money and capital flows produced when they grounded—or did not ground—themselves in specific places. As Bunge (1974b: 86, 1979a: 172) maintained, the "simplicity" of descriptive maps makes for better propaganda and agitation. However, Bunge saw that in practice the human misery that money transfers—particularly the "spatial injustice" conditioned by capital and revenue flowing between the inner city and the suburbs—is difficult to depict. So he tried to resolve this problem by showing both kinds of maps: abstract money flows and the concrete descriptive human misery that these flows beget. Here, Bunge had recourse to his own brand of mathematical-theoretical geography; and, on the face of it, Livingstone (1992: 330) is correct to spell out that Bunge's "*analytical* apparatus remained firmly that of the spatial science brigade."

But it is equally clear that expeditions represented a strategic and politically-charged quantitative mapping exercise. Though this doesn't mean Bunge can be fully exculpated from positivistic overtones. Nevertheless, Bunge was wary that the façade of "logic" and "objectivity" in mathematical geography could be used to divert attention away from value judgements, or be deployed as a means to reinforce domination (Bunge, 1973a: 325). As a result, Bunge's maps weren't meant to be disembodied, ostensibly objective representations; they were situated, partial and accountable representations, visualizations constructed from the standpoint of the disenfranchised (cf. Wood, 1993: 186–88). So while, for instance, Pile and Rose (1992: 132) are certainly right to assert that the "development of cartographic skills went hand in hand with colonial expansion," they surely go too far in adding that "it is the powerful bourgeois male gaze that constructs maps. By implicitly employing a rhetoric of neutral description, maps institutionalize a certain understanding of space, certain claims to objective worldviews. The act of mapping assumes a totally transparent society and denies not only difference but also different kinds of difference." For me, this is somewhat simplistic; it also asphyxiates committed geographical enquiry and reverts to the kind of essentialist logic Pile and Rose are trying to challenge in the first place (cf. Gregory, 1994: 6–7). Maps can, as Bunge affectionately underscored, sometimes be enabling for "Others" to assert and represent their differences.

True, there are big troubles with mapping and "computer-print-out geography," especially the "kinds of information being fed into the computer, or better, the lack of certain needed information" (Bunge, 1973c: l). It follows that the dilemma for Bunge was scale of scrutiny and geographers' positionality in the exploration of the human environment. Seen in this light, Bunge (1973a, 1979a) believed that certain forms of mathematical geography could be employed in the struggle for a genuine and liberating humanism.[10] He argues, for example, that a computer mapping program was implemented in Detroit to investigate the election situation, unemployment data, and those School Boards most sympathetic to children (Bunge, 1971, 1973a). Furthermore, mathematical geographers were instrumental in resisting the building of an expressway in Philadelphia during the early 1970s (Bunge, 1973a: 324).

The philosophical and methodological trajectory of Bunge's project followed Schaefer's earlier contention that spatial relations rather than mere description are the subject matter of a rigorous geography (see Bunge, 1979b). In Bunge's hands, however, the search for generality and a nomothetic understanding of location served not only an intellectual rejoinder to Hartshorne's neo-Kantian idiographic geography, but also offered considerable scope for political maneuver: the "generality of locations and humans is the essence of the methodological fight necessary to break into a scientific geography and scientific socialism" (Bunge, 1979a: 173). Yet Bunge's search for the "generalizability of the story" (as he puts it in the foreword to *Fitzgerald*) was situated and partial; one Bunge called a "disciplined objectivity:" in contradistinction to Hartshorne, locations for Bunge are both unique and general at the same time, namely, they are "sort of unique" in much the same vein as humans are as similar yet as different as snowflakes (Bunge and Bordessa, 1975: 286).[11]

In saying this, certain assumptions implicit in the expedition concept nevertheless spark a potentially bothersome question: To what degree can the individual biography of the geographer involved in the expedition invalidate the ability to empathize and situate oneself authentically in an impoverished community? This concern about authenticity, to be sure, has a strikingly familiar ring about it, and parallels—if not anticipates—contemporary debates within human geography, anthropology, and urban ethnography about whether it is possible to speak meaningfully about "the Other" or the subaltern (Clifford and Markus, 1984; Jackson, 1985; Harvey, 1992; Katz, 1992; see also Spivak, 1988). Accordingly, the thorny issue of political representation has to be confronted. Bunge (1977a) isn't oblivious to the problematical nature of a researcher investigating, joining, and maybe even representing what he calls a "foreign-to-his-childhood group." Indeed, as he affirms (1977a: 37), "[b]ig important gaps will exist" because it is "not possible to totally undo one's past . . . [and] no matter how empathetic, [the researcher] cannot entirely do it." Thus, biographical baggage is unquestionably influential and can, unless it is recognized and engaged with, compromise the ability of the geographer to learn *with* and comprehend the way "the Other" exists in an impoverished inner city community.

Certainly, the task will involve emotional difficulty and an honest political and intellectual commitment to the expedition; nonetheless this can, Bunge (1977a) asserts, be achieved through patience (not patronage) and with "dogged determination" on the part of the geographer. A sensitive dialogue can, therefore, overcome what Harvey (1992) has recently called "vulgar conceptions of situatedness." These, Harvey (1992: 303) claims, "dwell almost entirely on the relevance of individual biographies for the situatedness of knowledge. In so doing they dwell on the separateness and non-compatibility of language games, discourses, and experiential domains, and treat these diversities as biographically and sometimes even institutionally, socially and geographically determined."

For Harvey, this is a non-dialectical and debilitating rendition of situatedness: it anesthetizes and renders powerless any critical and empathetic impulse because it "proceeds as if each of us exists as autonomous atoms coursing through history and as if none of us is ever capable of throwing off even some of the shackles of that history or of internalizing what the condition of being 'the other' is all about" (1992: 303).[12]

Spivak, too, in her complex essay "Can the subaltern speak?" (1988), expresses a similar oppositional stance to a problematic essentialist act of foreclosing. While Spivak is concerned with whether it's possible to speak of—or for—the post-colonial other (notably the subaltern women), her insistence that the colonized subaltern subject is "irretrievably heterogeneous" (1988: 284) suggests that the thesis can (and must) be durable and broad enough to encompass all categories of subaltern groups. Invariably, she cautions, the subaltern cannot speak. Spivak propounds a positionality that "accentuates the fact that calling the place of the investigator into question remains a meaningless piety" (1988: 271). Thus intellectuals always have an "institutional responsibility" which behooves them to be critical of a complete privileging of subaltern consciousness (1988: 284). Indeed, as Spivak (1988: 285) reiterates, for the "'true' subaltern group, whose identity is its difference, there is no unrepresentable subaltern subject that can know and speak itself; the *intellectual's solution is not to abstain from representation*" (emphasis added).[13]

Bunge's own situatedness and positionality within the expedition program have themselves been made available through a critical engagement with his own past. His experiences in Fitzgerald have certainly been formative in throwing off the ideological shackles of what Bunge candidly admits was a privileged childhood. "Being raised bourgeois," he avers, "I always knew my class were thieves. It was the explicitness of the misery this produced, not the process, which I had to discover" (1979a: 172). According to Bunge, his father, William Bunge Sr., was head of the fifth largest mortgage bank in the United States and so had a hand in redlining black neighborhoods in many cities. In *Fitzgerald*, Bunge Jr. admitted a certain existential duality that has been forged out of a "history of generational tensions. Bunge generations alternate between money-making and cause-serving" (1971: 135). "From this vantage point," the book recounts, "Bunge [Jr.] has had good opportunities to explore the attitudes of the rich and poor towards each other;" all of which seemingly corroborates how subjectivity is inevitably fragmented: the Other is internalized within the self in a manner that complements Spivak's and Harvey's dialectical approach. Because, then, it doesn't shy away from themes of situatedness and representation, the geographical expedition principle embodies within its intellectual and organizational ethos a practical and theoretical route that can perhaps deepen ongoing debates within left geography. Let me explore this contention more closely.

TOWARDS A SITUATED PEDAGOGY AND PRAXIS?

Learning "to see faithfully" from the subjugated standpoint of the oppressed is maybe the most challenging insight to regain from the expedition program. By confronting the dialectical relationship between a researcher subject (geographer) and a researched object, the expedition concept seemingly strives for a genuine "pedagogy of the oppressed" of the sort that radical Brazilian educationalist Paulo Freire (1972) steadfastly invokes.[14] The geographer herein gains a platform to look faithfully at the world through a *dialogical encounter* with others. Within this interaction, the academic geographer has critical faculties to offer in the exploration of cities, particularly in recognizing the shortcomings of framing things solely in terms of the "concrete experience" of the oppressed or in a quasi-empiricist manner of "what actually happens" (cf. Spivak, 1988).[15] As Bunge was keen to observe, "academic geographers have a sense of scale but no sense, while members of communities have sense but no sense of scale" (interview, 1994). A commerce between local people and campus geographers is thus implied, rather than domination of one group by the other (Bunge, n.d.: 76).

The political practice emerging from this dialogue is formed *with* oppressed urban groups. Such is the nub of Freire's argument. For the geographer involved in an expedition, positionality is simultaneously that of teacher and student; and knowledge of particular urban realities is attained through "common reflection and action" as researcher and researched alike discover themselves dialectically through "intentional practice" (Freire, 1972: 44). To the degree that they attempt to negate the formalized and contradictory relationship between the one who teaches and the one who is taught, expeditions could become a powerful teaching instrument—much the way they did in both Detroit and Toronto during the late 1960s and 1970s. This, too, would be consistent with Bunge's behest that geographers should on occasion be put under the tutelage of folk geographers; both of whom henceforward give a coherence and an awareness of function to a social group and thereby represent something akin to Gramsci's (1971: 5–6) *organic intellectual.*

Via this activist route, academic geographers can articulate the "collective will of a people" (Gramsci's phrase) by gaining access and speaking to power élites, or by giving evidence at public inquiries and the like. Here, geographers participating in the expedition can use their research skills and written pamphlets as vital weapons of resistance for oppressed groups. Stephenson (1974: 101), who was active in the Toronto Geographical Expedition, notes that when "[a]rmed with a comprehensive research report, political lobbying is much more effective, but even failure to implement proposals has some value. People begin to realize their position in society, which may in turn lead to more active agitation for change." The authenticity of the geographer, then, simply lies in the ability to do committed and accountable urban geography.

This dialogue between oppressed communities and professional geographers is, therefore, an indispensable instrument in the process which Freire (1972) calls "conscientization".[16] Under such conditions, expeditions become more than an attempt to learn *about* the impoverished: they become an effort to learn *with* them the oppressive reality that confronts ordinary people in their daily lives.[17] From this standpoint, the academic geographers involved in geographical expeditions come to know through a dialogue of mutual recognition both oppressed people's objective situation in the city and their awareness of that situation (cf. Freire, 1972: 68). Furthermore, the researchers can come to have a recognition of themselves as part of this synergy. It might be possible to press this point further if we interpolate Kojève's (1980) highly influential reading of Hegel which asserts that the self-consciousness of an individual subject can only be achieved through the *recognition* (as an equal) of the Other. Such a dialogical interaction could provide the opportunity both to discover, following Haraway, the significant differences ushered in by global systems of domination and to stimulate awareness (for the geographer as well as oppressed subjects) of these restrictive mechanisms, since they cannot be rendered intelligible solely at the level of "concrete experience."

That is why radical geographers have a vital contributing role to play. Through expeditions it is incumbent upon the geographer to become a person of action, a radical problem-raiser, a responsible critical analyst participating *with* the oppressed. That said, the ambit of the geographer's responsibility is always ambiguous. And while expeditions in the 1960s and 1970s were intended to be a mutual learning and consciousness-raising program for researcher and researched alike, academic geographers were not, as Bunge insisted, to organize the local community.[18] Academic geographers can stifle community mobilization or centralize the expedition's organizational structure. According to Bunge, the Vancouver Geographical Expedition was a failure because it lacked a true community base and was never self-critical about its democratic failings (Bunge, n.d.: 81). So, there is always an immanent hazard that the voice heard in the supposed symbiosis between academic geographer and folk geographer is skewed toward the overzealous—though well-meaning—academic geographer. As the voice of the oppressed is muted, the expedition program degenerates into a paternalism reminiscent of nineteenth-century Western missionaries and settlement houses.

In both Detroit and Toronto, however, this enfeebling impulse was avoided through *organic* interaction between academics and the local community. In each case, academic geographers persistently asked the local community about their own priorities. Bunge's *Fitzgerald*, for instance, impresses by the sensitivity expressed in the text and through the emotive photographs. Therein, via the medium of the Detroit Geographical Expedition, it was unquestionably black voices documented in an honest and non-patronizing fashion; *Fitzgerald* evoked a representational (not represented) experience of oppressed black people in an American city. In Detroit, Bunge consolidated the radical intellectual ideal that Marshall Berman (forthcoming) articulates as the conjoining of the "stacks in the library with the signs in the street."

CONCLUDING REMARKS

It's been some twenty years since expeditions were initiated. Difficulties plague any assessment of their efficacy, either theoretically or practically. Yet, critical assessment would appear to be in order. Bunge recently—and rather elusively—confirmed that it was too early to tell whether the Detroit expedition and the accompanying book, *Fitzgerald*, were successful (interview, 1994). Nevertheless, expeditions did for a brief sparkling moment threaten what Harvey (1973: 147–152) labeled *status quo* understandings of capitalist society. Presumably this was why they came under assault from dominant class and social forces within the academy. Bunge claims he was driven out of his university teaching posts (interview), and Horvath (1971: 84) described how expedition radicals were fired, denied promotions, and refused admittance to graduate school, grimly concluding: "[d]ealing with the poor and powerless transforms the advocate into a marginal man."

What of the reinstigation of a similar radical venture today? Aside from all else, one factor precluding a reassertion of the expedition principle may be lack of time: with burgeoning teaching and administrative workloads and the competitive stresses of "publish or perish" in an ever more marketized academic world, finding the time to begin living, working and getting to know a potential base camp locality as an insider is extremely difficult. Such pressures aren't denied by Bunge. Though he suggests that they can be partly circumvented by geographers implanting themselves within what he calls the "cracks" between the academy and broader social life (interview, 1994). That way, a potentially creative tension might ensue as the scholar restlessly gravitates between formal academic and community-based duties, whereby one simultaneously informs and enriches the other. The prospect for the insertion of some kind of expedition (nominally or otherwise) might thus blossom within the contact zones of these conflictual realities.

To summarize: I've argued that there is much that is instructive about Bunge's geographical expedition program for radical debates resonating today, especially over situatedness, positionality, representation and the political role of left academics. In this paper, I have tried to sketch out the numerous ways expeditions previously acknowledged that these issues were vital aspects of critical scholarship and knowledge production. While it would be hasty—and foolhardy—to think Bunge had definitive answers to such problematical themes, his expeditions did at least show geographers a possible way *into* these dilemmas; that they did so in such a palpable manner makes them all the more suggestive and radical today. At any rate, exploring more deeply the tradition's successes and failures might illuminate the pursuit of the genuinely accountable and responsible situated knowledge that Haraway *et al.* now invoke. At a time when left geography is in grave danger of being rendered anodyne through a heady prioritization of discourse and textual politics, there is much to learn from the legacy of practical expeditions into the world of the exploited and oppressed outside the academy: it might at least ensure that critical theory is truly critical.

ACKNOWLEDGEMENTS

I am grateful to the Advanced Studies Committee at Southampton University, to Bob Colenutt, David Harvey, James Sidaway, Dick Walker, and Jane Wills for their helpful remarks on earlier versions of this paper, and to Bill and Donia Bunge for their hospitality. Responsibility for the paper's content is mine.

NOTES

1 I take expeditions to be part and parcel of the modernist tradition of opposition that Berman (1982) so brilliantly identifies: It is a tradition, for example, that offers a "celebration of urban vitality, diversity and fullness of life" (1982: 316), and is intimately related to a "shout in the street."

2 Marx's critique of political economy was a powerful version of this thesis, accepting that there are epistemological and ontological distinctions between Marx and Haraway and Hartsock in terms of their

notions of truth and objectivity, inasmuch as Marx's scathing analysis of modern capitalism was established from the subjugated standpoint of the working class. For Marx, *true* knowledge could only be produced *within* the confines of capitalist power relations (see Harvey, 1989) and pretending to be "outside of" or "beyond" a position in the world through an appeal to any notion of objective neutrality is either intellectually shoddy—because it fails to confront the *political nature* of knowledge production—or outrightly dishonest. "Objective truth," according to Marx in his *Theses on Feuerbach* (1975, III), "is not a theoretical but a practical question. It is in praxis that humans must prove the truth." However, I think it is also important to bear in mind that my discussion below recognizes that Bunge's expedition concept relied on a Marxist notion of truth and science, and as such differs somewhat from the manner in which Haraway and Hartsock deploy the situated knowledge stance.

3 It is also worthwhile here to underscore that this understanding holds a certain similarity with the work of anthropologist Clifford Geertz, notably his *Local Knowledge* (1983) (see especially the chapter on "From the Native's Point of View").

4 This said, it is perhaps ironic to look at the respectability of much radical geography and the established academic reputation of radical geographers today, many of whom hold senior ranking positions in Anglo-American geography departments. (For a fierce, though marred, polemic on the contradictions between tenure politics and radical politics, see Jacoby, 1987.) This scholarly respectability of geography can, of course, be witnessed by *Antipode*'s own decision to professionalize in 1986 and prosper from the academic credibility provided by publisher Basil Blackwell (cf. Jacoby's comments, 1987:181–82).

5 See, for example, Lewis's (1973) and Ley's (1973) review of *Fitzgerald*, and Bunge's (1974a) trenchant rejoinder to Lewis.

6 "Armchair geographers of the world arise, you have nothing to lose but your middle-aged flab," was a flamboyant Bunge clarion call at the time (see Bunge, 1977a). "The academic geographer," Bunge colorfully adds, "needs to get off their camp-ass." This opinion, it should be added, is one Bunge reinforces today (interview, 1994).

7 Interview, 1994.

8 Bunge's project likewise parallels some of the ideas expounded by the Situationists in continental Europe between 1957 and 1972. The strategies formulated by this group focused on the creation of *situations*, creatively constructed encounters and directly lived experiences that could subvert and transform alienated everyday life within urban settings (see Knabb, 1981). Herein, the active production of situations lay at the core of the Situationists' manifesto of an integrated urbanism (so-called unitary urbanism) inasmuch as these situations provided a critical vantage point from which proponents could understand, contest and agitate against the sterility and oppressive nature of market-driven urban landscapes and practices of everyday life.

9 As I read it, this standpoint also closely resembles Marshall Berman's special notion of modernism and its connections with the "signs in the street": both Bunge's and Berman's theorizations are thus passionate and partisan ones that are deeply embedded in the everyday life of ordinary people. Berman's (1984) vignette on the desolation and struggles involved in New York daily life, for example, evoked as a response to what he sees as the "remoteness" of Perry Anderson's vision of modernity, is a telling recognition of this concern.

10 In the foreword to *Fitzgerald*, the book is ambiguously described as "science: its data are maps, graphics, photographs, and the words of people. But the book also makes a value judgment—the desirability of human survival—and thus transforms itself into a steel-hard hammer of humanism." And later on: "The end product [of *Fitzgerald*], like science and art, hopes to be more real than facts alone." For a brief discussion of the dissolution of art, science, and humanism in Bunge's *Fitzgerald*, see Meinig (1983). Certain situated knowledges, such as those drawing upon the anti-humanism of Foucault's post-structuralism—and it is uncertain as to whether Haraway falls into this camp or not—would doubtless want to distance themselves from Bunge's humanist predilections.

11 Bunge's early explorations here were already keenly sensitive to the problem of geographical scale, a topic that writers such as Smith (1992) and Harvey (1993) have recently sought to address more directly. To this extent, Bunge's expedition project, in Toronto especially, recognized, as Neil Smith has more recently, that capitalism operated in some sort of "nested hierarchical space." While Bunge didn't, of course, have any definitive answers to this dilemma, he did at least pinpoint the theoretical and practical importance of arbitrating and translating between different spatial scales.

12 Buber (1987: 88) provides succinct confirmation here when he points out that "no [hu]man is pure person and no [hu]man is pure individuality." And Sennett's (1970) far-sighted debunking of the desire for a "purified identity" in city life likewise reiterates Buber's concern, though less mystically. "In order to sense the Other,"

Sennett (1990: 148) has written more recently, "one must do the work of accepting oneself as incomplete."

13 Hartsock (1989/90) makes a similar claim: "[t]here is a role for intellectuals in making these [situated knowledges] clear [and] in explaining a group to itself, in articulating taken-for-granted understandings." Hartsock's postulation, of course, has close affinities with Gramsci's (1971) category of "organic intellectual." For Gramsci, these comprise intellectuals who are *organically bonded* to a place and to a people, to the degree that they feel "the elementary passions of the people, understanding them and therefore explaining and justifying them in the particular historical [and geographical] situation and connecting them dialectically to the laws of history" (1971: 418).

14 Since the paper was first drafted, it has come to my attention that Gregory (1978: 161–64) had already made allusions with respect to this possibility. Although Gregory's main purpose was to emphasize the parallels between Freire's "theory of dialogical action" and Habermas's "theory of communicative competence" for furthering "committed explanation in geography," Gregory stresses that "when the practical lessons which they [geographical expeditions] contained were translated into theoretical terms the language was Freire's" (1978: 162).

15 The problematical nature of "concrete experience" or "common sense" understandings of social reality was of course emphasized by Gramsci. Indeed, he wrote (1971: 422) that common sense is "a chaotic aggregate of disparate conceptions, and one can find there anything that one likes."

16 Conscientization refers, for instance, "to learning to perceive social, political, and economic contradictions and to take action against the oppressive element of society" (Freire, 1972: 15).

17 As a shorthand definition, by "ordinary people" I mean those people who may have the capabilities to intellectualize but don't, as Gramsci identified, have the capacity to *function* as an intellectual.

18 A further point of qualification might also be useful here: I accept, as did Bunge, that the nature of representation *within* respective oppressed communities is problematical; and I am aware that local community leaders playing a "vanguard" role is both unavoidable and frequently divisive. Moreover, while I likewise accept Young's (1990) favoring of a "politics of difference" for checking romantic—and potentially reactionary—notions of community, I also believe it possible, then as now, to speak of "community action" comprising a group of people bonded by *commonality* and organizing around a common grievance in a way that isn't simply a NIMBY ordeal (see Merrifield, 1993b). And as with any collective action there are leaders and spokepersons. A community's mobilization here is, furthermore, likely to involve a "militant particularist" component and activate internal as well as external controversy. This might be as much about the *form* of resistance as it is about the actual grievance itself.

REFERENCES

Adorno, T. (1973) *Negative Dialectics*. London: Routledge.

Antipodean Staff Reporters (1969) "Where it's at: Two programs in search of geographers II: The Detroit Geographical Expedition." *Antipode* 1: 45–46.

Berman, M. (1982) *All that Is Solid Melts into Air*. London: Verso.

—— (1984) "The signs in the street: A response to Perry Anderson." *New Left Review* 144: 114–23.

—— (forthcoming) "Justice/Just-Us: Rap and social justice in America." In A. Merrifield and E. Swyngedouw (eds) *The Urbanization of Injustice*. London: Lawrence & Wishart.

Buber, M. (1987) *I and Thou*. Edinburgh: T. & T. Clark.

Bunge, W. (1962) *Theoretical Geography*. Stockholm: Lund Studies in Geography.

—— (1971) *Fitzgerald: Geography of a Revolution*. Cambridge, Mass.: Schenkman.

—— (1973a) "Ethics and logic in geography." In R. J. Chorley (ed.) *Directions in Geography*. London: Methuen, 317–31.

—— (1973b) "The geography of human survival." *Annals of the Association of American Geographers* 63: 275–295.

—— (1973c) "Urban stations—the tradition of geographic base camp urbanized," unpublished paper.

—— (1974a) "Fitzgerald from a distance." *Annals of Association of American Geographers* 64:485–489.

—— (1974b) "Simplicity again." *Geographical Analysis* 6: 85–89.

—— (1977a) "The first years of the Detroit Geographical Expedition: A personal report." In R. Peet (ed.) *Radical Geography* London: Methuen, 31–39.

—— (1977b) "The point of reproduction: A second front." *Antipode* 9: 60–76.

—— (1979a) "Perspective on theoretical geography." *Annals of the Association of American Geographers* 69: 169–174.

—— (1979b) "Fred K. Schaefer and the science of geography." *Annals of Association of American Geographers* 69: 128–32.

—— (no date) "The methodology of exploring cities," unpublished manuscript.

Bunge, W. and R. Bordessa (1975) *The Canadian Alternative—Survival, Expeditions and Urban Change.* Ontario: York University.

Christopherson, S. (1989) "On being outside 'the project.'" *Antipode* 21: 83–89.

Clifford, J. and G. Markus (1984) (eds.) *Writing Culture: The Poetics and Politics of Ethnography.* Berkeley: University of California Press.

Colenutt, R. (1971) "Postscript on the Detroit Geographical Expedition." *Antipode* 3: 85.

Field Manual (no date) No. 5: *Hidden Landscapes.* Detroit: Detroit Geographical Expedition.

—— (no date) No. 8: *Toronto—Detroit: A Tale of Two Countries.* Toronto: Toronto Geographical Expedition.

Field Notes (1970) No. 2: *A Report to the Parents of Detroit on School Decentralization.* Detroit: Detroit Geographical Expedition & Institute.

—— (1972) No. 4: *The Trumbull Community.* Detroit: Detroit Geographical Expedition.

Freire, P. (1972) *Pedagogy of the Oppressed.* Harmondsworth: Penguin.

Geertz, C. (1983) *Local Knowledge: Further Essays in Interpretative Anthropology.* New York: Basic Books.

Gramsci, A. (1971) *Selections from Prison Notebooks.* London: Lawrence & Wishart.

Gregory, D. (1978) *Ideology, Science and Human Geography.* London: Hutchinson.

—— (1994) *Geographical Imaginations.* Oxford: Basil Blackwell.

Haraway, D. (1991) "Situated knowledges: The science question in feminism and the privilege of partial perspective." In D. Haraway *Simians, Cyborgs and Women.* London: Free Association Press, 183–201.

—— (1992) *Primate Visions: Gender, Race and Nature in the World of Modern Science.* London: Verso.

Harding, S. (1986) *The Science Question in Feminism.* Milton Keynes: Open University Press.

Hartsock, N. (1987) "Rethinking modernism: Minority versus majority theories." *Cultural Critique* 7: 187–206.

—— (1989/90) "Postmodernism and political change: Issues for feminist theory." *Cultural Critique* 14: 15–33.

Harvey, D. (1973) *Social Justice and the City.* London: Edward Arnold.

—— (1989) "From models to Marx: Notes on the project to 'remodel contemporary geography.'" In W. MacMillan (ed.) *Remodelling Geography.* Oxford: Basil Blackwell, 211–216.

—— (1992) "Postmodern morality plays." *Antipode* 24: 300–26.

—— (1993) "The nature of the environment: The dialectics of social and environmental change." In L. Panitch and R. Miliband (eds) *Socialist Register 1993.* London: Merlin Press, 1–51.

Harvey, D. and N. Smith (1984) "Geography: From capitals to capital." In B. Oilman and E. Vernoff (eds) *The Left Academy: Marxist Scholarship on American Campuses Volume II.* New York: Praeger, 99–121.

Hegel, G. W. F. (1931) *The Logic of Hegel (Part 1 of The Encyclopedia of the Philosophical Sciences).* London: Oxford University Press.

Horvath, R. (1971) "The 'Detroit Geographical Expedition and Institute' experience." *Antipode* 3: 73–85.

Howe, I. (1954) "This age of conformity." *Partisan Review* 21: 7–33.

Jackson, P. (1985) "Urban ethnography." *Progress in Human Geography* 9: 157–76.

—— (1993) "Editorial: Visible and voice." *Environment & Planning D: Society & Space* 11: 123–26.

Jacoby, R. (1987) *The Last Intellectuals—American Culture in the Age of Academe.* New York: Noonday Press.

Katz, C. (1992) "All the world is staged: Intellectuals and the projects of ethnography." *Environment & Planning D: Society and Space* 10: 495–510.

Knabb, K. (1981) *Situationist Anthology.* Berkeley: Bureau of Public Secrets.

Kojève, A. (1980) *Introduction to the Reading of Hegel: Lectures on the Phenomenology of Spirit.* Ithaca: Cornell University Press.

Lefebvre, H. (1991) *Critique of Everyday Life—Volume One.* London: Verso.

Lewis, J. (1973) "Review of William Bunge's 'Fitzgerald.'" *Annals of the Association of American Geographers* 63: 131–32.

Ley, D. (1973) "Review of William Bunge's 'Fitzgerald.'" *Annals of Association of American Geographers* 63: 133–35.

Livingstone, D. (1992) *The Geographical Tradition*. Oxford: Basil Blackwell.

McDowell, L. (1991) "The baby and the bathwater: Deconstruction and feminist theory in geography." *Geoforum* 22: 123–33.

—— (1992) "Multiple voices: Speaking from inside and outside 'the Project.'" *Antipode* 24: 56–71.

Marx, K. (1975) "Theses on Feuerbach." In *Early Writings*. Harmondsworth: Penguin, 421–423.

Massey, D. (1991) "Flexible sexism." *Environment & Planning D: Society & Space* 9: 31–57.

Meinig, D. (1983) "Geography as an art." *Transactions of the Institute of British Geographers* 8: 314–28.

Merrifield, A. (1993a) "Place and space: A Lefebvrian reconciliation." *Transactions of the Institute of British Geographers* N.S. 18: 516–31.

—— (1993b) "The struggle over place: Redeveloping American Can in southeast Baltimore." *Transactions of Institute of British Geographers* N.S. 18: 102–21.

Nicholson, L. (ed.) (1990) *Feminism/Postmodernism*. London: Routledge.

Oldenburg, C. (1990) "I am for an Art . . ." In C. Harrison and P. Wood (eds) *Art in Theory 1900–1990*. Oxford: Basil Blackwell, 727–730.

Pile, S. and G. Rose (1992) "All or nothing: Politics and critique in the modernism–postmodernism debate." *Environment & Planning D: Society & Space* 10: 123–36.

Sennett, R. (1970) *The Uses of Disorder. Personal Identity and City Life*. Harmondsworth: Penguin.

—— (1990) *The Conscience of the Eye—The Design and Social Life of Cities*. London: Faber & Faber.

Smith, N. (1992) "Geography, difference and the politics of scale." In J. Doherty, E. Graham and M. Malek (eds) *Postmodernism and the Social Sciences*. London: Macmillan.

Spivak, G. (1988) "Can the subaltern speak?" In C. Nelson and L. Grossberg (eds) *Marxism and the Interpretation of Culture*. London: Macmillan, 271–313.

Stephenson, D. (1974) "The Toronto Geographical Expedition." *Antipode* 6: 98–101.

Thompson, H. (1988) *Generation of Sivine*. London: Picador.

Wood, D. (1993) *The Power of Maps*. London: Routledge.

Young, I. M. (1990) *Justice and the Politics of Difference*. Princeton, New Jersey: Princeton University Press.

SECTION 3

Goals and arenas of struggle

What is to be gained and how?

Introduction

The embeddedness of intentions, tactics, and strategies in rights-, justice-, and ethics-based worldviews

In this last section of the book we continue to track the ways in which critical scholarship in geography has been linked to progressive social change. The difference is that now we assemble a collection of works that addresses specific, grounded instances of social and political struggle. Moreover, we want to notice how and why specific ideas, concepts, and theoretical perspectives are actually mobilized and put to use within struggles and to notice how these struggles then refresh or alter conceptual understandings of the world that are of interest to geographers. We want to draw readers' attentions to the fact that social and political struggles, whatever else they might be, always involve a struggle over how to know and comprehend the world. And, pushing this a bit further, we are interested in the ways that social and political struggles, however locally particular they might be, are often embedded in or framed within broader worldviews or discourses. (This is the idea behind the rights-, justice-, and ethics-based worldviews that we alluded to early on and that are the substance of the essay you have now begun reading.) In deciding which essays to anthologize we chose ones that contain a review of the events that comprise a particular struggle and an analysis or interpretation of the struggle, our motivation being to find works that offer interesting examples of the roles scholarship can have when in alliance (sometimes more implied than explicit) with progressive struggles and movements.

Before continuing we offer a note on method, for we have overlaid this section of the book with some interpretive/organizing themes of our own. Even the most casual observer will realize there is a staggering diversity of social and political struggles around the world. This means that an organizational format or categorizing scheme of some kind can be useful. Such schemes reduce variety into a manageable simplicity but, more importantly, serve as a medium through which to compare and contrast diverse phenomena. Not surprisingly there is no shortage of schemes. One can organize struggles by chronology (e.g. "old" versus "new" social movements) or by ideology (e.g. feminist, nationalist, environmentalist) or by geographical locale (e.g. South Africa, US, Mexico, India) or by manner of connectivity (e.g. transnational/cross-border activisms, "rainbow" coalitions, religion–labor alliances). These are all ways of placing different struggles on the same plane so they don't have to be understood in isolation and so that instructive lessons can be learned or at least provocative questions can be asked. If one proceeds with care and does not insist that different struggles are perfectly equivalent or analogous to each other—class or gender oppression is not exactly the same as racial oppression—then it is possible to learn from comparative analysis (a point made with stunning precision by Janet Jakobsen [1998]). As we reviewed different organizational formats, we came to feel that any of them might work, though we knew we would not be able to offer anything like encyclopedic coverage, no matter the choice. But as we deliberated our options and poured over the scholarship on various movements and struggles, none of the prevailing schemes seemed to us to help with a question that continually haunted us: "*On what grounds does a given struggle seem to claim its legitimacy*?" We came to be specifically curious about the approaches used in social and political struggles to understand the kinds of problems or oppressions faced *and* the bases upon which their appeals for remedy are asserted. We call these approaches "worldviews" and differentiate among them as follows: **rights-based**, **justice-based**, and **ethics-based** worldviews. It is according to this heuristic that we anthologize selected examples of recent critical geographic scholarship, with a few selections from other fields. This means that each part of Section 3 of this book, though it represents diverse struggles and places, coheres around a particular worldview of problems and solutions. We must caution that by "cohere" we mean a reasoned decision on our part to represent a given essay as consistent with a particular worldview of problems and solutions. Sometimes this seemed easy enough to do, as some of the essays self-select. Other times, we reasoned as best we could. So, even though the assembled readings cut across specific arenas of struggle, and even though the authors of the essays treat their subject matter with varying methods and intentions, we hope the framework we have imposed offers a constructive reading of what social and political struggles can be about.

But why only three worldviews? These worldviews are in one sense a heuristic and like any heuristic the aim is to simply offer a way into a complicated subject matter. At the same time any heuristic has complexities of its own, and these can be useful to think about. For example, we do not suggest that these worldviews are mutually exclusive. On the contrary, they speak to actually existing tactics and strategies that have very often been brought together. ("Rights," "justice," and "ethics" are therefore more than just a heuristic.) What we wish to do is to make a (minimal) set of useful distinctions among them so that overlaps, blending, and synergies will be made explicit, visible, and very much the subject of progressive politics. (What counts specifically as progress, we leave up to the authors of these works.) These worldviews are, in short, and as Chela Sandoval skillfully documents in Chapter 20, flexible forms of oppositional consciousness and practice that serve simultaneously (or sequentially) to legitimate struggle and define its goal. How then do we distinguish these worldviews? Here is a brief synopsis, to be followed by an extended discussion and guide to the readings.

By **rights-based goals** we mean those struggles whose purpose (not necessarily sole purpose) is to gain access to a right or entitlement, to create some new right or entitlement, or to otherwise expand capacities to act and "become" in the world. What we hope to capture in the discussion of rights-based goals and the accompanying readings is the manifold struggle to expand the forms of identity and ways of living that enhance people's capacity to be at home in their bodies, to work and play, to belong to a place, to be mobile, to gain critical knowledge, to form loving bonds with other beings, to exercise "citizenship" at multiple scales, and the like. Critics of rights often see rights as merely endorsing or imposing existing forms of "being," that is, forms that are conducive to the perpetuation of power as currently constituted (see Chapter 13; cf. Esteva and Prakash 1998). But perhaps the specific promise of rights is that they will secure our capacities to become *in ways that are reproducible and repeatable (i.e. not a one-time gain) but also in a revisable form*. The view of rights we adhere to, perhaps unconventionally in the eyes of some scholars, is that rights can be consistent with a politics of "becoming," while the politics of becoming cannot be exclusively a rights-based one—at a minimum justice and ethics are involved as well. Indeed, rights-based struggles can rigidify or reify the process of becoming into circumscribed forms of "being"—that is, into identities and practices into which groups and individuals may be locked. But rights need not be synonymous with rigidity.

By **justice-based** goals we refer to struggles for equal participation in productive and distributive issues, for the redress of wrongs, and for the "just" resolution of conflicting rights, whether these struggles be aimed at reforming dominant social practices and structures or at revolutionizing them. There is no hard and fast distinction to be made between rights-based and justice-based worldviews. Arguably at least some, thin conception of rights is essential for determining whether justice is served (see Chapter 16). In any event, struggles for justice refer particularly to the struggle to secure a processual arena in which domination, oppression, or wrongs of some sort can be fought, as well as to the actual "fighting." Justice-based goals do not demand such an arena be formalized in a juridical or governmental sense, merely that some space of encounter be produced or appropriated. Arguably, justice has no endpoint; as with rights-based struggles it is about the struggle to keep possibilities for becoming perpetually open to new possibilities for becoming. Yet it is our view that *a politics of becoming begets justice-based goals in a characteristic way*. This has to do with the struggle to adjudicate the assertion of rights (thickly or thinly conceived and practiced) when those assertions come into conflict. Justice and rights and the struggles for them are necessarily caught in an ontological embrace. It is not simply that a conception of rights enables us to identify whether justice is served, then. Indeed, struggles for justice may lead to new or revised conceptions of rights and the "identities" to whom they belong, a point borne out repeatedly in Noriko Ishiyama's Chapter 17 (see also Pulido [1996] on the invention of the "people of color" identity in the U.S. Southwest).

Struggles, we argue, are productive of new or revised forms and modes of being and therefore new political imaginaries of becoming.

The **ethics-based** worldview is concerned with the production (as such) of revised or new political imaginaries of becoming. This includes analyses of how certain notions of justice or rights become a means of domination and oppression. (We might recall, for example, Marx's argument in the first volume of *Capital* that in bourgeois society the right to equal participation in exchange was defined in such a way as to be nonexistent inside the factory gate [Marx 1967].) Ethics, for us, and as divined in the readings collected for Part 3 of this section of the book, is about subjecting received and accepted notions of the good, the just, the right to a radical critique so that their limits and boundaries become clear and their warrant "tested" against the "outsides" they might produce or desiderata yet to be practiced. In our view, ethics is a practical orientation toward new concepts that potentially open up new political practices, and new or revised alliances and identities for those politics. At a minimum, the ethical stance is a skepticism toward well-trodden, well-known political trajectories and practices. In more positive terms, Gibson-Graham's contribution (Chapter 21) specifically advises, *ethics directly concerns the field of immanence*, that is, it discovers the potential openings for further action, the possible connections toward others, and the new self-fashionings and identities that inhere in (i.e. are immanent to) social and political struggles (also see Box 2 on pp. 271–72 focusing on the work of David Featherstone). The ethical imagination fuels, if you will, struggles for rights and for justice and hopes to prevent them from ossifying.

As we have mentioned previously this book is interested in more than progressive politics and social change. It is vested in the idea that scholarly work and critical reflection have a role to play and, as we have taken some care in asserting, that such a role has become one of the most noticeable aspects of geography in recent decades. The discussion of rights-, justice-, and ethics-based worldviews that follows therefore poses two sets of questions to the essays collected in this final section of the anthology:

- What is the relationship between scholarship and social struggle in these pieces? What sort of work is scholarship doing here?
- What forms of social-geographical knowledge emerge? What specific concepts are struggled over and why?

In fact there are many possible kinds of relationship between academic scholarship and progressive social struggles. For each article we discuss we identify in the heading preceding the discussion what we think is the major role that research is playing, however modest this may seem. We encourage readers to develop their own reflections along these lines, too, and ask only that these pieces be read in a spirit of generosity.

PART 1

Rights-based goals

Introduction

A wide variety of struggles in many different parts of the world have called for rights as a desideratum, though not necessarily as a sole, simple solution to social change. The pieces collected here consider very specific struggles over rights in Canada, the U.S., East Timor, the African continent, and Mexico. They concern rights to mobility, sovereignty, development, and claims to national identity. Readers will learn something of the debates over the problems with rights as a concept and as a political goal. Perhaps more than justice and ethics, the very concept of rights and the utility and disutility of that concept are hotly contested. For a number of critics rights, *qua* rights, are seen as an imposition of the West on the rest of the world. For others they allow for perverse forms of multiculturalism (e.g. "white" rights), while for others rights legitimate the state as their sole guarantor, when the state itself is viewed as a problem. The readings we have assembled convey some sense of these debates. For us, and we think for the authors represented here, rights are a site of struggle, including struggles over the meaning of rights. As a site of contestation, the hope is that these meanings can be kept open, revisable, and subject to equitable and democratic deliberation. How they are to be kept open is an important question. As one of the contributors argues, rights need to be grounded in discourses and practices beyond rights themselves. Readers can bear this in mind as they read through the justice and ethics sections of the anthology. To have gained a right, though, is to have gained an entitlement of some kind that one expects to be able to retain. In a non-trivial sense, struggles over rights are struggles to not have to begin again from the same place when social and political struggle is an ongoing process.

What is the place of scholarship in the rights-based worldview? We draw attention to the following: Scholarship can play a basic and therefore important role in examining the place of rights in struggles for social and political change. It can explore what makes a struggle for rights a distinctly progressive struggle (or not). It can suggest how academic social theory can be used to augment a struggle for rights. And scholarship can be used to explain why a particular movement achieved the success it did.

EXAMINING THE PLACE OF RIGHTS AS A GOAL (AMONG OTHERS) IN STRUGGLES FOR EMPOWERMENT

Should we accept *a priori* that securing rights—human rights, civil rights, free speech rights, rights to free association, to sexual freedom, to unionize, etc.—should be the, or even *a*, goal of social struggle? What can geographical research tell us about this question? The first two essays in this section argue that a struggle for rights *per se* should not be a taken for granted goal. The essays do make a case, however, for the specific utility of a quest for rights in certain situations. It might seem odd of course to readers who live in the liberal democracies of the West to call the desirability of rights into question. This seems utterly counterintuitive. Yet, as discussed in Chapter 12, "Mobility, empowerment and the rights revolution," by Nicholas K. Blomley (a geographer at Simon Fraser University in British Columbia), rights have been viewed with considerable skepticism. One argument, going back to at least Karl Marx, is that rights are configured so as to never challenge fundamental economic arrangements, as when the rights of the individual, under capitalism, serve to uphold ownership of private property, especially the private ownership of the means of production. For Marx, the virtual monopoly ownership of the means of production by the bourgeoisie ensures that non-property owners must sell their labor power to survive and forfeit the surplus value they produce. Even a fight for the right to higher wages or a shorter work week still accepts (works within) a working-class identity. Rights are thus devised to uphold the capitalist mode of production, not change it. For other skeptics rights are simply too ambiguous. An individual's right to liberty or equal treatment can be exceedingly tricky, because there is no necessary agreement over what "liberty" or "equality" means. They are as open to conservative interpretation as they are to progressive interpretation and may depend on the particular political cast of the authorities charged with upholding some right. Any interpreter who happens to be empowered may claim that rights are being upheld, when by another reckoning they are not. As Blomley puts it, the basic problem of ambiguity is that "interpretations must, of necessity, appeal to conceptions of social and political life *external* to rights-discourse itself" (see p. 202, emphasis added). Another form of skepticism argues that a rights-based politics oversimplifies the complexity of social struggle. Here the problem is that a claim to one sort of right may be yoked to another right that is not desired. Think here of lesbian or gay litigants who seek insurance or retirement benefits for their partners. They find themselves having to deploy a "family" right "despite their suspicion of patriarchal familial ideology" (see p. 203). Rights skeptics, one might say, are liable to argue that rights-based struggles neither cut deep enough through dominant social arrangements, nor challenge sufficiently how social identities are defined. Critiques of rights take yet another turn when legal, and not only moral, rights are sought. Legal rights must be recognized *and* enforced to be effective. Since this typically requires an authoritative (i.e. law-enforcing) mechanism, the appeal is most often made to some element of the state, which means an appeal to an important guarantor of the *status quo*. While such appeals may be useful and successful (i.e. the right is recognized, accorded, and enforced), the same act further legitimates the authority of the state and, by extension, its role in maintaining the *status quo*.

And, yet, as Blomley recounts, rights remain a powerful and continuing basis of struggles for social change. The title of his contribution speaks of a "rights revolution," which, begun long ago, shows no signs of abating. In this context he finds compelling recent attempts to *reconceptualize* rights. Not least in this effort is the "minority critique" aimed at rights skeptics. The minority critique points to the significance that gaining rights has had in minority communities, as well as to the continuing deficit of rights in those communities. To skeptics who view rights as too ambiguous and indeterminate, the minority critique responds, "what else could a right be other than an abstraction for someone who has never had their abstractions taken away or denied?" (R. Williams, quoted by Blomley; see p. 204). The rights discourse and specific rights claims have been invaluable in focusing and solidifying minority struggle. Moreover, to struggle for a right is not just to struggle for a right, it is to struggle to be political *per se*, to have access to an abstraction!

Rethinking rights has been of interest to many on the Left, too, who argue that the diversity of social struggles (whether around race, sexuality, gender, environment, etc.) has just not been taken seriously enough by the Marxist tradition. More, this diversity cannot be simply channeled into a single politics of class. To make this point, Blomley draws in particular upon the arguments of Ernesto Laclau and Chantal Mouffe, as presented in their influential and contentious treatise *Hegemony and Socialist Strategy* (1985). Laclau and Mouffe argue that the diversity of 20th-century social struggle should be accepted for what it is—evidence of an explosion of sites of struggle which ought not to be assimilated into a single struggle (say, the class struggle in some versions of Left politics). If these struggles are not to be assimilated into a single struggle, they are nonetheless linked along a "chain of equivalences." What this means is that the diverse struggles for self-determination, for civil rights, for free speech, for sexual liberation, for women's rights, for environmental justice, for workplace safety, and so on, are roughly equivalent with respect to the democratic revolution begun during the Enlightenment; that is, they are extensions of the democratic political practices and agitations inaugurated during that period. Put another way, they are immanent to the democratic revolution, a socially and politically logical extension of its promise of increasing liberty. The task of progressive politics is to continue to widen the set of freedoms immanent to democratic revolution: "The expansionary logic of rights-discourse must be extended to ever wider social relations, Laclau and Mouffe argue, and an alternative model of the liberal citizen must be constructed, premised not on possessive individualism but on participation and collective action" (see p. 205).

Readers will note that Blomley steers a pragmatic course between skepticism and optimism. For him the litmus test of how useful rights-based struggles are is "the location within which rights are to be put to work." As the title of his article suggests, he stakes out a pragmatic middle course in an examination of the struggle for "mobility" rights in Canada. In so doing, he not only assesses what role this right plays in a particular social struggle, he also revivifies and thickens the meaning of the concept of mobility itself.

The emphasis above on the European and American domains of the Enlightenment may have alerted readers to another major source of skepticism toward rights as a goal of social struggle: the concern that the West universalizes *its* ideas of what rights are and whose rights count. With a focus on rights-based "development" in Africa, Giles Mohan and Jeremy Holland address this concern in "Human rights and development in Africa: moral intrusion or empowering opportunity?" (Chapter 13) (Giles Mohan is a geographer in the Development Policy and Practice Department, at the Open University, Milton Keynes. Jeremy Holland is a social development consultant and expert on participatory approaches to research, based with Oxford Policy Management, Ltd.) Here it is shown in some detail that expanding rights to ever wider social relations is no simple task (neither Blomley nor Laclau and Mouffe suggest that it is). The larger geopolitical context and the international political-economic conditions within which a rights-based struggle for development occurs have to be taken into account. In particular, Mohan and Holland document how the recent emergence of rights-based development politics has been given a neoliberal spin (where neoliberalism means the extension of free trade, privatization of community resources, lean-and-mean government, and *de facto* Euro-American hegemony). The threat exists that human rights, if codified in a certain way, can be a means through which global capital insinuates itself, on the one hand, while entrusting enforcement of rights to local authoritarian régimes, on the other hand.

We hope that readers will find especially useful Mohan and Holland's account of how the historical geography of human rights and of development discourse and practice converged in the late 20th century. This involved the concept of development becoming more socially complex (turning away from top-down technocratic approaches and moving toward an emphasis on local participation), and the idea of universal human rights being adopted as a strategy by development activists. The strength of Mohan and Holland's discussion is, in part, its emphasis on the contingent nature of the convergence of rights and development. That is, there is nothing that prevents rights-based development from becoming the handmaiden of neoliberal policies (of the World Bank, for example) but nor is there anything determining it will. A certain vigilance is called for to keep rights-based development from marching in step with neoliberal practice. A central issue is the distinction between universal versus local notions of human rights. Mohan and Holland argue that the claim of universality is in fact an extension of Euro-American ideas (see Chapter 2 by James Blaut), which when claimed as universal makes it very difficult to see as legitimate any claim that such ideas may not be locally desirable. Yet a virtue of the Mohan and Holland contribution is their desire to give up on neither universal nor particular notions of, and struggles for, rights. In particular they endorse economic and developmental rights (a right to a decent standard of living, for example) but argue that these rights must be locally meaningful and based upon direct participation.

WHAT MAKES A GIVEN STRUGGLE FOR RIGHTS PROGRESSIVE OR NOT? IS IT POSSIBLE OR EVEN DESIRABLE TO PLACE "PROGRESSIVE" ON ONE SIDE OF A BINARY OPPOSITION?

In the "Staking claims" section (Section 2) of the anthology we presented some ideas and works that examined how people become aware that a situation is oppressive or that a "right" has been abrogated. But is every struggle to right a wrong a struggle you would want to endorse? One contribution that scholarship can make is to take up that question. It is a virtue of Carolyn Gallaher and Oliver Froehling's comparative study of the Patriot Movement in the U.S. and the Zapatista movement in Mexico that they describe some analytical tools with which to compare and contrast these two struggles and their emancipatory potential (Chapter 14).

For Gallaher, a geographer at the American University, Washington, D.C., and Froehling, a geographer at the Universidad de la Tierra en Oaxaca and Centro de Encuentros y Diálogos Interculturales, Oaxaca, Mexico, the Zapatistas and U.S. patriots are "linked" in a couple of different ways. First, each is a response to neoliberal reforms that "weakened the ability of the state to provide social guarantees [while not diminishing] its traditional control of the means of violence" (see p. 234). During the 1980s neoliberal reforms were common in numerous parts of the world—hence the "new world order" alluded to in the title of the essay. As examples of how these reforms were received, the authors contrast the political activisms of hard-pressed peasants in the southern Mexican state of Chiapas and economically strapped farmers and factory workers in the U.S. These peasants and workers were each drawn into political movements, and in many instances faced very strong reactions from their respective national governments. Second, and more important to the authors, though, both movements claim an identity as "nation." This might seem innocuous. But in doing so movement participants and leaders force the issues of what the nation means, what it ostensibly promises, and what scales it might operate through other than that codified by the national-level government, or "state". Each movement maintains a distinction between "nation" and "state," the first providing an identity, the second being an apparatus for governance. Their protest is self-avowedly on behalf of "the nation," is meant to appeal to others who are concerned with the status of the nation, but also offers a different version of what that term is all about. In short, these movements each lay claim to a distinct national identity; they wish to have some measure of autonomy within the state; and they push for a state whose powers would be restructured in such a way as to admit greater local autonomy. The issue then is how scholarship might be used to evaluate the two movements.

Again, the work of Laclau and Mouffe is salient in the article. Recall that for Laclau and Mouffe rights have an expansionary logic along a chain of equivalence. In the expansion of that chain there is no guarantee that political movements, which very often butt up against each other, will develop "agonistically" (living with difference) as opposed to "antagonistically" (erasing difference). The central problem is that any identity, every identity, is posited around or outside an "other." Identities are in a sense defined by their outsides, by what they exclude. This makes their exclusions part of what they are. And while this is, in generic terms, a constant, it is variable whether a given identification develops agonistically or antagonistically with respect to its outsides, its others. As Gallaher and Froehling then put it,

> [W]e must ask how social identification may be constructed in such a way that "others" may co-exist peacefully and fairly under extant forms of governance. Such a proposition may be considered radical because it represents a break with standard liberal politics that seek to eradicate differences, as illustrated, for example, in liberal calls to create a "color blind society." Rather, these scholars [Laclau and Mouffe] argue that the goal for radical democracy should be to root out antagonistic forms of identification, where the "other" is considered dangerous and in need of extermination, and actively work to "agonize" them . . . creating a context where articulating difference(s) is seen as crucial to rather than dangerous for the political whole.
>
> (see p. 236; also see the essay by Chantal Mouffe, Chapter 19)

Readers will discover that Gallaher and Froehling see the Zapatistas as a much more agonistic movement than the U.S. patriot movement, although it is not exclusively so. They refer, for example, to the group of women within the movement who seek to ensure that gender equality remains an enduring goal of the Zapatistas.

By the same token, the authors do not condemn the *concerns* of the much more antagonistic politics of the U.S. patriot movement. The concerns of people in the movement are very real, trapped as they have been by farm crisis on the one side and deindustrialization on the other, and abandoned by an economically weakened state that could otherwise provide support. Moreover, the authors refuse to will away struggles for a right to "national" identity. They assume that the "nation" will be an enduring site of political identity and struggle; conversely, individual movements will likely come and go. The question for those seeking an evaluative stance is to take notice of the relative antagonist versus agonistic mode through which movements do their work. Arguably, Mohan and Holland adopt a similar evaluative framework to Gallaher and Froehling, even if they do not employ the same terminology. The former's call for rights-based development is surely in an agonistic mode. It sides with a universal rights framework that will not be blind to the need for local translation and responsibility for rights enforcement. But it also sides with a local rights framework that must be accountable to an international community of rights activists. In their own ways Gallaher and Froehling and Mohan and Holland remind us that the local as a site of politics, and as a site of ethical knowledge, is not something we should romanticize. Ethical authority does not reside at any one scale; just because something is local does not make it good (see Box 3, highlighting the work of David Slater, on pp. 327–28).

BRINGING SOCIAL (FEMINIST) THEORY TO THE STRUGGLE FOR RIGHTS

The connections between theory and practice, between ideas and their development or deployment in social struggle, are not predictable. Certainly, the utility of a given theoretical development to some social struggle is not determined by theory itself; connections need to be actively imagined and forged. In Jack Kloppenburg's essay on the de/reconstruction of agricultural science (Chapter 15) readers will find just such an imagination at work (cf. Ingram 2007). Kloppenburg's purpose is to outline some of the successes of the alternative agriculture movement and to present a detailed case for how feminist interventions in science studies can propel the movement forward. Readers will note that the essay's disciplinary concern is with rural sociology, agricultural science, and what these can learn from feminist theories of science—Kloppenburg is a rural sociologist at the University of Wisconsin, Madison. The seeming lack of "geography" should not be a worry. Geographers have had a longstanding interest in food systems, local knowledge, and scientific practice, both separately and as they intersect. Moreover, that Kloppenburg finds inspiration in feminist science studies jives nicely with the uptake of this field in geography in recent years (e.g. Whatmore 2004). It is a real merit of the essay that it conveys so well the core arguments of the feminist science studies literature and then shows exactly how they might be brought to bear on the need for an alternative knowledge production system in U.S. agriculture. These are enduring themes that inhabit but also transcend disciplinary quarters.

The successes so far of the movement for an alternative agriculture in the U.S. have been encouraging, even if profoundly compromised (Guthman 2004). That the National Academy of Sciences would endorse the project of an alternative agriculture is, for Kloppenburg, a sign of how far the movement has come and a sign that knowledge production is indeed a legitimate terrain of struggle for the movement. "Social theory and the de/reconstruction of agricultural science: local knowledge for an alternative agriculture" speaks directly to the greater possibility for a continuing interchange between progressive activism and theories of knowledge. In the struggle for a healthier, more locally vibrant food system, it matters whose knowledge counts and who counts as a producer of knowledge. Like many people in movements for an alternative agriculture, and an alternative food system in general, Kloppenburg is concerned that the dominant agricultural science, with its close connection to industrialized food production, "may fail to respect the exigencies and needs of a specific locality" (see p. 254). Yet it is not simply the connection to industrialized agriculture that accounts for this possibility. Dominant scientific practice in agriculture, with its emphasis on producing universalizable knowledge (or "immutable mobiles" in Bruno Latour's memorable phrase) bears some responsibility, Kloppenburg argues. The movement for an alternative agriculture, he suggests, is at once a struggle to gain a place for local knowledges and locally viable farm practices and an effort to reconstruct agricultural science itself.

Readers of Kloppenburg's essay will find that the struggle for an alternative agricultural science is at base a struggle for a right to incorporate the contingencies of place into the food people eat. If you like, it is a struggle for a right to sustenance that expresses and embodies locality. But this has a number of related meanings. It means first that there are multiple routes to agricultural knowledge, including those emerging from the practice of agriculture in different environments, undertaken by farmers themselves. (Kloppenburg is interested in the harm that comes to non-humans and humans alike when local variability in the food ecosystem is elided.) It means that sustainable practices must be linked to time-proven but also adaptable understandings of locally variable environments. It means accepting that the binary between expert and non-expert does not do justice to the practice-derived knowledge that places farmers in a knowledge-holding position. It means, as suggested in the essays in earlier sections of this volume by Andy Merrifield and Janice Monk and Susan Hanson, rethinking the very notions of science and objectivity and instituting the concept of partiality as a characteristic of all knowledge. By implication, it means democratizing the processes through which agricultural science is built. Readers may wish to reflect back on the agonistic model of politics described in Gallaher and Froehling's work. From that perspective Kloppenburg is interested in how feminist approaches to knowledge production can serve as a resource to remake agricultural science and create "a context where articulating difference(s) is seen as crucial to rather than dangerous for the political whole" (see Gallaher and Froehling: p. 236).

BOX 1: AFFORDABLE HOUSING AND ACCOMMODATION OF DIFFERENCE: THE ONTOLOGICAL EMBRACE OF RIGHTS AND JUSTICE

In the Jack Kloppenburg essay on agricultural science that you just read about (Chapter 15), not only food is at stake. He joins the struggle for multiple ways of knowing to be recognized and put into practice. Along the way he disturbs conventional boundaries between expert and lay understandings. A specific goal is that society should not become trapped by default into conventional technoscience. Why is this important? His essay is concerned with the question of who gets to determine whether needs are being met; he asks whether the right needs are even being recognized. His analysis broaches what some call the "politics of difference." Instead of concerning itself with the fair social distribution of existing needs, the politics of difference concerns the struggle to define new needs (or to define old needs in new ways). This text box discusses the work of geographer and disability rights researcher Flora Gathorne-Hardy in her article "Accommodating Difference: Social Justice, Disability and the Design of Affordable Housing" (Gathorne-Hardy 1999). The article joins together two developments, the influential theoretical work of Iris Marion Young, especially her concept of "politics of difference," for which she is well known (Young 1990), and the movement for affordable housing for physically disabled people.

The "politics of difference" represents a departure from mainstream notions of fairness, which have tended to rely upon distributive or procedural criteria. Let's think back for a moment: as explained in the essay that introduces Section 2 of this anthology, "distributive justice" is said to be achieved when the available pool of resources, inclusive of goods, privileges, and duties, at society's disposal is equitably divided. (There are, of course, all kinds of debates about what counts as equitable. Especially divisive, for example, is the issue of whether achieving equality of opportunity is sufficient or whether equality of outcomes needs also to be achieved.) "Procedural justice" is said to be achieved when all enfranchised persons have access to the institutions that decide distributional matters, that is, when persons have been given due consideration by those institutions (see Chapter 16 for further discussion; cf. Smith 1994.) But neither distributive nor procedural justice questions what sorts of things, powers, resources, and so on actually get distributed and how adequate these are for various groups of people. Neither mode challenges the social norms regarding what counts as a need and who is authorized to say so (see Chapter 5). The "politics of difference," however, does pose such a challenge. It "argues that ideas of distributive justice function ideologically, representing the institutional context in which they arise as natural or necessary and forestalling criticism or debate about alternative social arrangements" (Gathorne-Hardy, 1999: 242). Readers should note where Young (and Gathorne-Hardy) locates the impetus for this argument. The impetus comes from political struggles themselves, from social actors who struggle against the strong compulsion toward assimilation as the path to justice. Their struggle identifies that compulsion as *itself* unjust. Young seizes on this dynamic, arguing that social movements are capable of generating important new social knowledges. For Gathorne-Hardy fairness demands the social invention of "ways to ensure that socially and culturally differentiated groups are able to participate in collective, democratic processes of decision-making about issues that affect their lives" (Gathorne-Hardy, 1999: 242).

Where to begin? Gathorne-Hardy chooses one window, what she calls "just design" of affordable housing, to get at what a politics of difference entails. At stake in this particular struggle are the social and spatial arrangements that would allow a group of physically disabled people to have both affordable housing and housing whose design suits their "differences." At the same time a critical approach to difference is called for: "A central question . . . is how to achieve a prior political recognition that people with physical impairments have 'different' design needs without creating crude dichotomies between able-bodiedness and physical disability" (Gathorne-Hardy, 1999: 243).

Gathorne-Hardy notes that the peculiarity of housing itself plays a role in the production of disability. For example, housing is expensive by its very nature and more expensive still if it is to be adapted for use by specific groups. Much of it is out of reach for groups who already tend to have low incomes, which in the case of the disabled may result from discriminatory hiring practices. A sort of vicious circle results in which "disabled people are often left living in ill-designed accommodation. Poor design results not only in unnecessary practical problems, but also has an adverse impact on people's dignity, privacy and opportunity for self-determination" (Gathorne-Hardy, 1999: 243). That legislative change has been inadequate or very slow in coming exacerbates the experience of impairment, as do the prevailing models of urban planning and housing design that entrust these activities to professionals and experts. In short, disability does not merely come prior to the quest for housing; it is also an *effect* of housing as social institution. For these reasons Gathorne-Hardy took an interest in an innovative, affordable housing design and construction project that involved collaboration between a group of disabled people (the future residents) and a group of design professionals. Gathorne-Hardy's purpose was to assess the success of this collaboration from the perspective of a politics of difference. In addition to the goal of simply gaining affordable living accommodations, the design was to answer specific needs, while anticipating future ones. (After all, the domicile as a fixed space must house animate beings whose needs change over the life course.)

The results were mixed, as many difficulties arose along the way, including problems with inadequate funding and residents not having been consulted at every stage of the design process. It proved extremely difficult not to reproduce, at least in some measure, the abled/disabled binary. But Gathorne-Hardy does not judge the project an outright failure. Indeed, most residents felt some measure of satisfaction with the outcome, something that Gathorne-Hardy attributes to the incredible tenacity of the struggle. At the same time, one of the residents point to a crucial insight:

> I would have no qualms about doing it again. But we would know [about the pitfalls] next time, and that's why I feel quite strongly that people should know what has happened here. Because this would be ideal for other disabled people. It would give them independence. It would keep them out of community care. They could have their careers in their own homes. I think it's brilliant. The concept is brilliant. But then, like all concepts, they sound brilliant in theory—then you put them into practice and it doesn't work.
>
> (Gathorne-Hardy, 1999: 250)

It "doesn't work" and yet there are "no qualms about doing it again." This is a testament to the iterative nature, and intertwining, of theory and practice that is the substance of political activity. It runs in strong parallel with Young's account of the very origins of a "politics of difference," which proposes not a once and for all solution but an enfranchising, deliberative process which can only be secured through persistent struggle.

What Gathorne-Hardy narrates is a two-pronged struggle for rights and justice, a right of recognition and a claim to just, democratic deliberation among all claimants. In her words:

> Far greater political discretion, financial resources and time must be available to resident groups to enable them to draw upon the services of housing providers, community designers, advocates and

> other relevant professionals in order to ensure that housing produced is sensitive to their particular needs. Such a strategy has to be accompanied by a commitment to universal entitlement to secure affordable housing provision, as well as a politicization of housing design and what are presented as normal images of the home and home life.
>
> (Gathorne-Hardy, 1999: 253)

The politics of difference to some degree reimagines rights, placing them not in the sphere of assimilation but in the sphere of democratic deliberation. As we enter the discussion of "justice-based goals" we may need to be mindful that justice and rights perhaps ground each other—their order is not hierarchical but horizontal.

12

Mobility, empowerment and the rights revolution[1]

Nicholas K. Blomley

from *Political Geography*, 1994, 13(5) (September): 407–422

At first sight, a concern with rights and empowerment might seem removed from the traditional domain of the human geographer. However, this is to ignore an undercurrent of geographic research that has long placed such issues at its centre. Such an interest can be traced to several sources, including the nuanced and often overlooked writings of Peter Kropotkin (1885), who both wrote on the relation among law, space and rights and appealed to a broader conception of human rights and social justice. Marxist and anarchist geographers of the 1970s also placed human rights and social justice at the centre of their moral and political vision (Harvey, 1973), whilst a concern with distributional rights underlay the work of those such as David Smith (1977). Significantly, a concern with rights has recently been revived and extended. Discussions have centered on the link between locality and rights (Smith, 1989), emancipation and public and private spaces (Rose, 1990), local conflicts between different conceptions of rights (Clark, 1990; Mitchell, 1992), citizenship (Fyfe, 1993) and the tensions between universal principles of social justice and the post-modern critique (Harvey, 1992).

These writings are interesting at two levels. First, they beg several questions concerning the emancipatory potential of rights. For example, whilst there seems to be a general assumption that rights offer considerable potential in effecting social and political change, these accounts often fail to explore the mechanisms by which this might occur. Sceptics might also suggest that some offer an overly optimistic account of the redemptive power of rights. Second, and more specifically, they raise the issue of the *geography* of rights. Is there anything peculiarly geographic concerning rights and rights-discourse? Are there contributions that the geographer is uniquely equipped to provide?

This paper seeks to address both these questions. First, I hope to address the emancipatory potential of rights by briefly outlining a debate within Left and legal theory concerning rights. This debate, I shall suggest, is instructive for geographers in its clarification of the potential of rights-claims for empowerment. However, it is also worth visiting given its warning of the real risks involved in such rights-based strategies. In this, it directs us to the need to specify carefully the social and political spaces within which rights are deployed. Put simply, I shall argue that in certain community-based settings, rights claims can be powerful and progressive weapons, both as critique and mobilizer. In more regulated, juridical settings, however, the invocation of rights claims can backfire.

I continue to explore this question in my discussion of the geography of rights. It is notable that geographic writings have tended to invoke rights that have geographies, such as equality (cf. Smith, 1977); justice (Harvey, 1992); or democracy (Rose, 1990). However, it is remarkable that little attention has been given to mobility, a right that is geographical in a more immediate sense. I consider the progressive potential of mobility rights, arguing that, like other political and social rights, it has considerable potency. However, I go on, drawing on the Canadian experience, to question the wisdom of any political programme that would invoke it in the courtroom.

THE LEFT AND RIGHTS

Are rights empowering? This question has long been a controversial one on the Left. If I value, defend and seek to advance my rights, does this speak of my essential dignity, my aspiration to respect and my desire for flexible and responsive forms of political association, or is it an indication of my fundamentally self-serving or alienated condition? As progressives, can we materially advance our goals if we engage in 'rights-talk', or are we drawn into a sophisticated game of smoke and mirrors which not only blunts the force of our claims, but may even deradicalize and contain our opposition? Can rights provide the vocabulary by which marginalized groups can, at last, articulate their historically silenced demands or do they negate the experiences of the disenfranchised? Is rights discourse flexible enough to allow for a progressive extension and expansion, or is it irredeemably circumscribed by its liberal provenance?

This debate has sharpened in recent years, given a 'rights revolution' that is both political and intellectual in nature. For many years, the received wisdom on the Left has been that rights-struggle is, at best, diversionary; at worst, counter-progressive. This 'rights scepticism' has appeared, to many, increasingly untenable, given the persistent and pervasive hold of rights-consciousness upon the political imaginary. Some progressive intellectuals have, as a result, sought to reoccupy a field, long vacated by the Left, insisting on the potential of rights to mobilization, empowerment and political critique.

Despite their differences, however, both rights sceptics and optimists tend to agree on one thing: the open-textured nature of rights. That the meaning of rights is indeterminate and open is, to the sceptics, cause for alarm. Meaning can only be provided by appeal to an overarching theory of social life and political association which, they argue, is all too frequently hostile to the progressive project. For the optimists, however, this same fluidity is cause for celebration, in that it opens rights-discourse to progressive co-option. The semantic slipperiness of rights, they argue, gives them a subversive and explosive quality, allowing them to be radically extended so as to encompass progressive possibilities.

The 'so-called rights of man'

For Karl Marx, the 'so-called rights of man' were simply so much 'obsolete verbal rubbish'. Emancipation that was 'merely' political, he insisted, was not 'the final form of human emancipation, but [only] . . . the final form of human emancipation within the framework of the prevailing social order' (1978: 35). This has been the received wisdom of the Left for many years. Not only are rights compromised by their cynical deployment by the bourgeoisie, it is argued, but they are themselves beyond redemption, imbued with the regressive lexicon of individualism (Campbell, 1983; Kennedy, 1971).

A similar rights-scepticism can be found within recent legal writing, although here the supposed ambiguities of rights receive greater attention. For many legal scholars, the specific meaning to any right – such as liberty and equality – is not inherent to the linguistic category itself, but is formed by the interpretative actions of social institutions – most importantly, the judiciary. These interpretations must, of necessity, appeal to conceptions of social and political life external to rights-discourse itself. Despite the sincerity and apparent objectivity by which this proceeds, the critics suggest that these grounding norms are usually deeply conservative. For Tushnet (1984), for example, rights are usually defined exclusively in negative terms, serving to protect the hypostatized 'free' individual from the predations of the collective. Rights thus 'becomes a loaded gun', suggests Roberto Unger (1983), that the 'rightholder can shoot at will in his corner of town' (p. 597) in order to protect an autonomous zone of individual agency. The substantive meaning of individual rights, moreover, can become legally fixed in problematic ways as Alan Freeman (1990) notes in a discussion of US anti-discrimination law. The 'plasticity of legal characterization', he argues, has provided space for the US Supreme Court to circumscribe equality with reference to restrictive notions of formal rights, under which racism is cast as individualized 'prejudice'. The effect, ironically, is to 'celebrate inequality, while compelling those who fail to "make it" to internalize a despairing sense of self-worth' (p. 143).

The danger, moreover, is that a recourse to rights can signal an enervating 'diversion from *true* political language, political modes of communication about the nature of reality' (Gabel and Kennedy, 1984: 33, emphasis added; see also Gabel, 1984). Rights-discourse, the critics argue, offers a very thin gruel compared with 'real' political consciousness and struggle, which has an immediacy, reality and vibrancy that cannot (and should not) be confined within the desiccated categories of rights-discourse (Tushnet, 1984).

Not only does rights-talk misrepresent the urgency and complexity of political life, the critics argue, but it can also be politically disempowering. Any 'rights victory' by a progressive group, it is argued, is likely to be a Pyrrhic one, as political movements find their vocabularies refracted and distorted through the lens of rights. Battles over abortion, for example – which can variously be described as struggles over life and death, freedom and violence, feminism and fundamentalism – are reduced to the arid oppositions of 'choice' or 'life'. North American First Nations find their concerns, drawing from a long history of cultural liberty and oppression, freedom and resistance, and articulated with a complex mix of humour, anger, passion, hope and despair, reduced to the arid legal plea for recognition of 'rights of ownership' over territory (Monet and Skanu'u, 1992). More seriously, perhaps, the terms of reference which a progressive movement is forced to adopt in its legal challenge can serve, ultimately, to inscribe the very power relations that it opposes. For example, feminist groups engaged in sexual violence litigation may find themselves using the consent/coercion dichotomy, despite their claim that such a choice is meaningless in a patriarchal society (Fudge, 1989). Lesbian litigants, seeking to ensure employee benefits for their same-sex partners, may find themselves arguing that as a 'family' they enjoy the same rights as heterosexuals, despite their suspicion of patriarchal familial ideology (Herman, 1990).

The rights revolution

If so, why is it 'that concepts which appear as "hopelessly confused" when examined in abstraction . . . become such a powerful mobilizing force in everyday life' (Harvey, 1992: 396)? Why does the traditional pantheon of liberal rights – democracy, equality, justice – retain such an obstinate hold on the popular imagination? The 'dream-like narrative' dismissed by Left legal scholars is both catalyst and touchstone for political struggles throughout the world. This 'rights revolution' has been central to political struggle in the liberal heartland, often crystallizing around constitutional documents. Twelve years ago, for example, Canada entrenched a binding Charter that ushered in a profound transformation of the political and legal landscape. In Britain, left-liberals have recently argued for a similar 'Bill of Rights' (Rustin, 1992). In the United States, of course, constitutional rights have long been a volatile and hotly debated issue. Recent years, moreover, have seen an extension of rights-discourse to emergent and evolving social and political relations. As a result, many issues – such as racist hate literature or the plight of the terminally ill – have been recast in terms of rights, such as those of liberty, choice or life.

Rights-struggle has been expansive in a second, geographic sense. The popular appeal and transformative power of rights appear to have been pivotal in political struggles beyond Western Europe and North America. The New World Order, to some, is premised less on the embrace of western capitalism than on the deployment of the corrosive potential of western rights in the struggle for social justice across the world (Bowles and Gintis, 1987: ix).

Rights have not gone away. As such, the dismissal of rights-based struggle as incoherent or counter-progressive seems condescending. It is in response to this that we can understand a recent reconceptualization of rights. Rather than abandoning them to the Right, the attempt, broadly, is to reclaim rights for the Left. In so doing, theorists have radically rethought and extended the meaning and progressive possibilities of rights.

As a result, the rights-critique of legal scholars has, itself, become subject to challenge; notably from those active in the feminist and anti-racist movements. The so-called 'minority critique' has, according to Bartholomew and Hunt (1990), several lines of attack: it draws attention to the historic significance of rights-struggle for oppressed groups, highlights the manner in which rights can serve to protect the disenfranchised, and identifies the complexity and difference of rights experiences. Again, however, it is the relative openness of rights-discourse – the fact that rights can be read in various ways – that is central to the argument. The significant context against which rights are to be read now, however, is not the court-room but the concrete experiences of the oppressed and disenfranchised.

In one powerfully 'righteous' essay, for example, the African-American legal thinker Patricia Williams describes the process of apartment hunting with critical legal scholar Peter Gabel (who has cast rights as 'hallucinations'). Whilst Gabel obtains an apartment with nothing but a handshake, Williams, in her 'rush to show good faith and trustworthiness ... signed a detailed, lengthily negotiated, finely printed lease' (1991:147). This difference, to Williams, is suggestive. Gabel's reaction, she argues, embodies not only his intellectual refusal to legalize a social relation, ideally based on informality and sodality, but also expresses a social expectation – that the apartment will be forthcoming, that the deposit will be returned – that is far removed from her experience, centred on white racism and legal invisibility. Gabel's treatment of rights as abstractions and encumbrances, she argues, is far removed from the Black experience, in which the concrete *denial* of rights is the everyday experience and expectation. Rights, as she puts it, may only be an appropriate place from which to jump for those who have already attained the 'Olympus of rights discourse'. To Robert Williams (1987: 25), 'what else could a right be other than an abstraction for someone who has never had their abstractions taken away or denied?'

Arguing against rights-discourse, from this perspective, denies the specificity of the Black experience, as well as that of other oppressed people (cf. Rhode, 1990). Not only does it negate the violence attached to a systematic denial of rights, it also ignores the historic centrality of rights-discourse to political struggles such as the civil rights movement, or feminism. These struggles, however, have not accepted the thin and desiccated definitions of the orthodoxy; rather, they have exploded the progressive possibilities of rights: 'this was not the dry process of reification ... but its opposite. This was the resurrection of life from ashes four hundred years old' (P. J. Williams, 1991: 163). The mere invocation of rights, of course, was and is insufficient. However, without the vocabulary of rights to give form and focus to this struggle it would have become seen as 'unrealistic' or 'other-worldly'. The importance of rights, Patricia Williams argues, rests less with the material gains that they brought than in their mobilizing power; especially significant for a people historically excluded from the magic circle of enfranchisement:

> 'Rights' feels new in the mouths of most black people. It is still deliciously empowering to say. It is the magic wand of visibility and invisibility, of inclusion and exclusion, of power and no power. The concept of rights, both positive and negative, is the marker of our citizenship, our relation to others.
>
> (P. J. Williams, 1991: 164)

Kimberlé Williams Crenshaw (1988) also takes the 'critique of rights' to task for its failure to comprehend the specific forms of oppression that Black Americans experience. If racism served to make the racialized invisible, then political struggle in and through rights discourse not only offers political possibility but perhaps presents itself as *the* only option to those experiencing racism:

> Because rights that other Americans took for granted were routinely denied to Black Americans, Blacks' assertion of their 'rights' constituted a serious ideological challenge to white supremacy. Their demand was not just for a place in the front of the bus, but for inclusion in the American political imagination. In asserting rights, Blacks defied a system which had long determined that Blacks were not and should not have been included.
>
> (Crenshaw, 1988: 1365)

She readily admits to the critical argument that such strategies can entail co-option. However, not only is the 'power of legal ideology to counter some of the most repressive aspects of racial domination' (p. 1376) undeniable, but a dismissal of such strategies fails to acknowledge 'the limited range of options presented to Blacks in a context where they were deemed "other"' (p. 1385). Rights-struggles, then, should not be dismissed as forms of false consciousness, but seen as 'intensely powerful and calculated political acts' (p. 1382).

The reassertion of rights has not only gone in legal theory: Left political theory also bears witness to the rights revolution. Ernesto Laclau and Chantal Mouffe (1985) have made one of the more influential arguments in this regard, connecting, in several significant ways, with that of the 'minority critique'. One immediate point of connection turns on what Laclau and Mouffe term the 'rejection of privileged points of rupture' (p. 152). The political terrain of classical Marxism, they argue, fails to capture the plurality of subject positions around which political struggle increasingly occurs. As the 'minority critique' demonstrates, the political terrain has become fragmented, as historically silenced groups organized around anti-sexist, anti-racist, urban or ecological antagonisms have found political voice.

For Laclau and Mouffe, these struggles can be 'sutured' (or connected), not through the assertion of certain privileged positions such as class, but through the construction of a 'chain of equivalences'. It is to the liberal tradition that Laclau and Mouffe turn for the links in this chain. What unites these new social movements, they argue, is their common deployment of rights and the language of the Enlightenment. This assertion rests on their claim that the language of rights, or the 'democratic revolution' of the Enlightenment, signalled a profound and subversive transformation in the representation of social relations. For the first time, the vocabulary of rights provided the means by which relations of *subordination* could be recast as relations of *oppression* and *domination*. In other words, the political yardstick of rights allowed power relations previously understood as organic and natural to be reconceived as unnatural and social. As such, rights offer a 'discursive "exterior" from which the discourse of subordination can be interrupted' (Laclau and Moffe, 1985: 154).

The subversive possibilities of rights are also explored within the 'minority critique'. To Crenshaw (1988: 1366), for example, 'engaging in rights rhetoric can be an attempt to turn society's "institutional logic" against itself – to redeem some of the rhetorical practices and the self-congratulations that seem to thrive in American political discourse'. The political promise of citizenship, for example, can be used to powerful effect. The civil rights movement, she suggests, 'proceeded as if American citizenship were real, and demanded to exercise the "rights" that citizenship entailed' (1988: 1368). Interestingly, Chantal Mouffe (1991) has also recently invoked the concept of citizenship as a means by which a 'chain of equivalences' between democratic struggles can be achieved.

However, it is not only the Left that can reconfigure rights, as Laclau and Mouffe warn. Recognizing their explosive logic, conservative theorists such as Nozick, Hayek and Friedman have spearheaded an 'anti-democratic' offensive, centred on negative liberty, property and the individual, which attempts to recover an early-modern vision of rights. This hegemonic remapping, in turn, has provided ideological justification for the welfare roll-backs, punitive policing and institutionalized racism, sexism and homophobia of the New Right (cf. Hall, 1983). Laclau and Mouffe's project, then, claims an added urgency. In the face of the discursive onslaught of the Right, the Left must re-enter the field it has so long vacated: '*the task for the Left . . . cannot be to renounce liberal-democratic ideology, but on the contrary, to deepen and extend it in the direction of a radical and plural democracy*' (1985:176, emphasis in original). The expansionary logic of rights-discourse must be extended to ever wider social relations, they argue, and an alternative model of the liberal citizen must be constructed, premised not on possessive individualism but on participation and collective action.

LAW AND RIGHTS

To summarize, although both rights-sceptics and optimists accept the plasticity of rights, the consequences of that openness are treated very differently. For many critical scholars, it is the conservative logic of liberalism that fills the empty space, making rights not only ill-suited as a vehicle for progressive change, but positively harmful. For the rights optimists, conversely, the semantic possibilities of rights provide room for progressive mobilization. Rights-discourse is not only often the only space available to the historically disenfranchised but it has, given its flexibility, a real subversive potential, allowing a means by which the rhetoric and 'self-congratulatory' promises of the liberal state can be called to account.

What are we to make of this? I am torn between both positions. Whilst sympathizing with the 'minority critique' and recognizing the corrosive logic of rights, it is undeniable that progressive movements that rely on rights have all too often been disappointed. Perhaps, then, we need to 'retreat from an either/or position', as Didi Herman (1990: 809) suggests, and from the internal struggles on the Left. Perhaps, moreover, this task has never been more urgent: the real enemies of rights are neither the rights-sceptics nor rights-optimists, but the New Right.

But I want to argue that the two positions are not as far apart as they seem. Indeed, the divide is an illusory one. Both see rights, in the abstract, as open to a number of possible meanings. Where they differ, however, is in the social spaces in which they locate the production of these meanings. For optimists, rights acquire meaning and progressive potency when deployed in *community* settings as mobilizers and political yardsticks; for the pessimists, it is the circulation of rights within the *juridical* domain that ensures that their meaning is counter-progressive. Seen this way, both arguments could be correct. If so, rights claims must be used tactically and carefully. If, as Laclau and Mouffe suggest, rights have an explosive logic, it is also one that can detonate in the face of the bearer. When making rights-based arguments, geographers, like activists, need to be careful in clarifying the location within which rights are to be put to work.

MOBILITY

Mention of geography brings me to my second related point. Are there specific geographies of rights? Given the discussion above, how do such geographic rights contribute to empowerment and disempowerment? I propose to explore these questions in an examination of one right – that of mobility. I choose mobility rights for two broad reasons. First, and most immediately, mobility is intrinsically geographic. As such, it perhaps offers one means by which we can examine the uses to which spaces are put in political life and political relations. This is not to say, of course, that other rights do not also allow for such interrogations. However, the limited attention given to mobility rights perhaps offers new possibilities. Second, it offers an interesting test case of the arguments outlined above. Rights and entitlements attached to mobility have long had a hallowed place within the liberal pantheon and, as such, mobility is part of the 'democratic revolution' of Laclau and Mouffe. Moreover, its meaning seems hard to pin down; this ambiguity gives it an 'explosive' and 'subversive' potential. It can be empowering within numerous political struggles. However, it has a judicial meaning that seems, as the sceptics would argue, semantically restrictive and politically disempowering.

Mobility rights have a curious status. They are invoked far less frequently than other rights – such as equality and liberty – yet have a lengthy and resonant pedigree. Indeed, the meaning of concepts such as liberty within the writings of early modern liberal thinkers such as Hobbes (1988 [1651]) and Blackstone (1803 [1765]) appears actually to presuppose mobility (Blomley, 1994a). This liberal provenance carries over to contemporary constitutional discourse. In the United States, the Supreme Court has long identified a right to mobility (Baker, 1975: 1140–1141; cf. Houseman, 1979). In Canada, mobility rights have been formally entrenched, as we shall see, and courts have had little difficulty in identifying a direct linkage between liberty, mobility and history.[2]

Mobility, then, is a fundamental right, deeply rooted in the liberal revolution. As such, is it empowering? Could it be used for progressive ends, or is it compromised? I shall argue that the answer to this question rests upon the location within which it is deployed. As the rights-optimists might argue, mobility has an 'expansionary' logic. Progressive groups could extend the meaning of mobility rights whilst remaining within the institutional logic of the liberal tradition. Once in the court-room, however, its meaning is more carefully policed. That judicial policing, moreover, is frequently counter-progressive.

RELOCATING MOBILITY

Mobility rights offer an interesting test case of the claims of the rights-optimists by virtue of their expansionary reach. Without overstepping reasonable bounds, there are several ways in which this extension could occur:

1 Moving and staying: a right to mobility is frequently cast as 'the right to move'. However, it may also imply the right to stay in opposition to certain forms of forced mobility, such as exile. This seems reasonable if it is assumed that the moral grounding for a mobility right is that of the right to choose a place to live, under conditions of one's choosing. Whether that choice necessitates movement seems arguable (Blomley, 1992).
2 Negative/positive: mobility rights can be cast as purely negative; that is, as a protective shield, defending the rights-bearer against those who would deny him or her the right. However, it also seems reasonable to cast mobility rights positively, as providing the means for the advancement of individual or collective goals. As Bowles and Gintis (1987) and others argue, the latter has been the trajectory of recent decades. A concept such as welfare rights, for example, is meaningless in negative terms alone (Fine, 1992).
3 Mobility and other rights: the right to mobility is not necessarily a right in and of itself, but because it makes many other rights possible – such as rights of association or liberty. As Binavince (1982: 341) notes, 'any discussion of mobility rights cuts across the grain of other rights and freedoms'.
4 Moving, travelling, migrating: what forms of movement are protected? Whilst long-distance migration is frequently taken as a pre-requisite, it seems hard to exclude other forms, however unstructured, or routines such as movement within an urban area.

5 Public and private rights: rights are frequently cast as guaranteeing protection from the legitimate actions of public agencies – notably the state – rather than the oppressions of the private sector. Actions confined within the spaces of the home, for example, or the workplace, are removed from the rights-domain. The rationale for distinguishing rights in this manner, however, seems as difficult to sustain as that between migration and movement, or negative and positive rights. Many critics have argued that the public/private distinction itself can only be sustained with reference to conditional claims concerning society and the state (for example liberal characterizations of the sovereign individual and the predatory collective) that are highly contestable (see Blomley and Bakan, 1992).

Such a creative extension on these lines makes it possible to frame many forms of activities and relationships in the language of rights. If the rights-optimists are correct, assigning rights-status to social activity changes – profoundly – the normative and political stakes (Freeden, 1991). We can be more specific: framing social relations in terms of *mobility* rights (as opposed, say, to the rights of liberty or of equality) directs us, in a way that no other rights can, to the vital and poorly documented nexus of space and power. If space is a critical site for the exercise of social power, and if the assumed objectivity of space renders it unusually opaque to critical enquiry, then the linkage of mobility and rights perhaps offers access to a vital and under-explored arena. Let me be clear: I do not wish to suggest that this occurs in isolation from other forms of struggle and the invocation of other rights-claims. Indeed, such linkages would be crucial. However, mobility rights, perhaps, can offer one yardstick by which to measure and contest the colonization and ordering of space for repressive ends. The following examples are meant to hint at these possibilities.

Power, space and mobility rights

In a discussion of United States Supreme Court decisions, Stewart Baker (1975) expresses disquiet at the 'expansionary' potential of the 'right to travel'. The Court, under Chief Justice Warren, is criticized for its 'over-broad' reading. To Baker, the danger is that the right seems to escape its prescribed bounds. Its precise meaning becomes unclear, its geographic reach uncertain. The more restrictive reading of the Burger court – in which, for example, migration is distinguished from 'mere movement' (p. 1150) is welcomed, promising, as it does, containment.

This anxiety is suggestive, hinting at the progressive possibilities of mobility rights when pushed beyond these constraints. I shall claim that, as Baker fears, mobility rights can indeed be deepened and extended and made to reveal a wide range of progressive possibilities. This is important, given the close link between representational and material spaces and the politics of social life, and the imbrication of mobility with other rights (Blomley, 1994b). That extension, moreover, reveals the significance of mobility rights to diverse political struggles. I am not able to do much except alight briefly upon a few such 'spaces'; reflection will reveal many more. The threat of sexual violence in urban areas, for example, could be cast as an affront or denial of women's mobility rights to the extent that their movement within the city is constrained (Pain, 1991; Valentine, 1989). Given Hobbes's injunction to think of mobility in terms, first and foremost, of the liberty of the human body, we might also consider an extended right to mobility as applying to the handicapped (Stewart, 1990). For the moment, however, I wish to focus on mobility rights in relation to homelessness and exile.

The denial of place

Homelessness raises several issues which can, perhaps, be powerfully enframed and thus politicized in relation to mobility rights. Mobility here needs to be cast in both negative and positive terms. The defensive aspect of the right seems especially timely: faced with the increased visibility of the homeless, many American municipalities have reacted by harassment or forms of expulsion (Davis, 1991; Dear and Wolch, 1987). In Los Angeles, for example, where the homelessness problem is acute, the authorities have destroyed semi-permanent homeless settlements and forcibly dispersed the inhabitants throughout the city. Dispersal can work the other way: in Vancouver, for example, the authorities have been charged with 'exiling' the deinstitutionalized mentally ill by funnelling them into inner city neighbourhoods (DERA, 1990). In the United States, many municipalities – such as Phoenix, Dallas, Santa Barbara and St Petersburg – have adopted or begun enforcing ordinances that prohibit outdoor sleeping in public areas. In an intriguing discussion, Paul Ades (1989) suggests that such ordinances deny the right to travel within the state because they seek to discourage the homeless from moving into or remaining in a jurisdiction. The effect is to 'evict the homeless' (p. 619).

Whilst Ades's account is novel, it touches only on the 'negative' mobility rights of the homeless – that is, their right to resist 'eviction'. If, as he implies, mobility implies the right to remain within a place, it can also be extended to include the 'positive' right of an individual not only to shelter, but to a permanent location within a specific community. Homelessness, perhaps, is not simply 'houselessness' but, by its impermanence, constitutes a denial to place – to 'home'. Casting mobility rights as a positive right that sustains the right to stay, redefines those institutions and structures (public *and*, perhaps, private) that deny a home as illegitimate and unjust. This is important: all too frequently, such systematic constraints are naturalized and reified, such that homelessness is deemed an inescapable consequence of the logic of the market or the failings of the individual. Any political anger generated by homelessness which might otherwise fuel mobilization is thus diverted.

Perhaps an appeal to mobility rights might have potential in related instances. For example, critics of a private housing market which can contain people of color within certain spatial enclaves, and deny them access to other neighbourhoods, often cast such practices as a denial of equality rights. However useful, this fails to capture the significance of spatial strategies and representations to racist practice (Anderson, 1991), as well as denying the aspirations of would-be migrants, who may wish not only the right to equal treatment, but also the right to move freely and settle within the city. Adding mobility rights, moreover, draws morally valued questions of choice into the equation which the concept of equality may fail to evoke, especially when defined in narrow terms.

Exile

Casting 'exile' – that is, unwanted mobility – as a denial of mobility rights begs some intriguing questions. Refugees – whether 'economic' or 'political' – can be understood as international exiles, driven from their homes by privation and persecution. To Binavince (1982: 340), the right to leave is 'the last refuge of the oppressed'. However, this does not guarantee a new home: denied a place in the world, the plight of the refugee is especially poignant. The legitimacy of their 'claim-right' on the international community, however, is often belittled. The insistence that refugees have a legitimate right not only to leave, but to a place to settle seems a necessary, if limited claim. This, in turn, folds into a critique of international immigration policy which is very much defined in relation to movement – both in its institutionally desirable and threatening forms. It is noteworthy that recent criticisms of the US embargo of Cuba have invoked the Constitutional freedom to travel (Libertad de Viaje/Freedom to Travel Challenge, *Z Magazine*, September 1993: 63).

Exile within the state, however, can also be identified. Many national governments are guilty of the relocation or confinement of oppressed populations. Japanese–Canadian internment, the Berlin Wall, the South African pass system, the 'ethnic cleansing' of parts of Yugoslavia, and the expulsion of Palestinians by the Israeli state are all object lessons in the importance of the link between space, mobility and state power. An appeal to rights – including those relating to mobility – directs attention to this linkage, as well as providing a 'claim-right' for the populations affected.

It is not only governments, of course, that structure mobility. 'Exile' also operates within the 'private' sphere. The spatial switching of capital within and between urban areas can, directly or indirectly, create forms of spatial expulsion. The effects of gentrification on the displacement of the old and the poor has been well documented (Barry and Derevlany, 1987; Ley, 1981). The power of Allan Sekula's (1991: 146) description of redevelopment in Long Beach, California, as 'one big eviction notice for the elderly, the disabled, for people on fixed incomes, for minorities, for underpaid and unemployed working people of all races', for example, rests precisely on its recognition (and condemnation) of this point.

The denial of a 'right to stay' as a necessary corollary of the right to move can also be extended to another political geography from which it has previously been excluded – that of the economic displacement associated with plant closures, especially in 'single industry' towns. 'Dislocation' – the phrase is apposite – can create a form of 'economic exile'. Despite the sanguine calculations of many regional economists, people are not as mobile as capital, often obliged to stay (because of the collapse of the local housing market, for example, or the limited skills of a newly unemployed workforce) or expressing an active commitment to remain and rebuild the town's economic base (Blomley, 1992; Clark, 1983).

Overall then, an appeal to mobility rights seems to offer considerable progressive potential, precisely because the right can be logically extended, as the rights-optimists suggest. To the extent that its meaning is derived from the vocabulary of the democratic revolution, yet is 'explosive' and multivalent, mobility rights can offer a powerful point of leverage. The effect of a 'rights enframing' is profound, providing the means by which relations of subordination can be politicized. As Laclau and Mouffe (1985) suggest, redefining the eviction attendant upon gentrification or the denial of free movement to women or the disabled as the abrogation of a right, rather than the result of the operation of an indifferent and neutral 'market', or a lamentable burden to be borne by a 'disadvantaged' (and hence 'naturally' marginalized) individual changes the stakes. An appeal to mobility rights also offers something different that perhaps, now more than ever, is needed, ensuring the means by which the politics of space can be brought under scrutiny and contested. To the extent that the space/power relation works on the dispossessed, mobility rights might also provide a novel and intoxicating 'magic wand' by which the marginalized can voice and channel their anger (cf. Cresswell, 1993).

Mobility rights, moreover, cascade into other rights, yet do so in such a way as to give material meaning and force to such rights. Similarly, mobility rights could be thought of as another thread, or 'suture', for disparate struggles. Given their deep roots, the defence of mobility rights could also be the touchstone for collective mobilization; maybe amongst those communities facing plant closures, or even between different struggles – uniting, perhaps, the plight of the homeless and those displaced from gentrified neighbourhoods.

LIMITS ON MOBILITY AND THE CANADIAN JUDICIARY

This argument only goes so far. If mobility rights have progressive potential, recent experience suggests that this does not extend to the court-room. Indeed an appeal to mobility rights might – as the rights-sceptics warn – quite possibly be counter-progressive. This is borne out by the Canadian experience.

For Canada, the 'rights revolution' arrived on 17 April 1982, the day on which the government of Canada entrenched the Canadian Charter of Rights and Freedoms. For a polity that has traditionally cleaved closely to a British model of parliamentary sovereignty and the 'unwritten constitution', the Charter marks a watershed. At its 10-year anniversary, Chief Justice Antonio Lamar described the Charter as 'nothing less than a revolution on the scale of the introduction of the metric system, the great medical discoveries of Louis Pasteur, and the invention of penicillin and the laser' (*Globe and Mail*, 6 April 1992: 12). Progressive movements quickly seized upon Charter rights, seeing them as a vehicle of change. Feminists across Canada, for example, heralded the entrenchment of equality rights in the Charter as 'a symbol of profound political significance around which disparate feminist organizations and women's groups were able to coalesce' (Fudge, 1989: 447), and quickly sought to use the Charter to advance their goals.

Most progressive attention has been directed at sections 7 and 15 of the Charter, with their guarantee of liberty and equality rights. Relatively little, however, has been focused on section 6 of the Charter, which protects mobility rights. Section 6(2) of the Charter provides that:

> Every citizen of Canada and every person who has the status of a permanent resident of Canada has the right
> (a) to move to and take up residence in any province: and
> (b) to pursue the gaining of a livelihood in any province.

This explicit constitutional protection of mobility rights distinguishes the Canadian Charter from other constitutional texts, such as the US Constitution, especially given some recent decisions which suggest that mobility must be protected both at the intra- and the inter-provincial scale (Blomley, 1992). However, I shall argue that, as the sceptics fear, this legal entrenchment has served to structure mobility rights in ways ill-suited – even opposed – to the radical re-imagining of mobility. If mobility rights do have explosive possibilities, the courts seem to have played an active and largely successful role in bomb disposal.

At first glance, the meaning of mobility rights under the Charter seems unclear. Section 6 does not carry with it a mass of case law left by the Canadian Bill of Rights; thus its exact meaning remains somewhat unclear (Laskin, 1982). Indeed, when it was first entrenched, legal commentators struggled with its semantic possibilities, scrambling to nail it down (Binavince, 1982; Laskin, 1982). Their anxiety was unnecessary: if mobility rights exploded in the courts it was not with a bang, but a whimper. The progressive possibilities outlined earlier are notable by their absence in section 6 jurisprudence. Given its potential, the most striking thing about the history of section 6 is the infrequency with which it has been invoked. The Canadian Supreme Court has only discussed it on one occasion. With around 40 significant Charter appeals to the Canadian Supreme Court every year (*Globe and Mail*, 6 April 1992: Al), this is a significant neglect. Higher courts have only directly addressed section 6 a dozen times. Moreover, none of the issues raised earlier appears to be considered. Rather, mobility rights have been invoked in defence of such things as the right of a physician to obtain a provincial medical insurance billing number; the right of lawyers to establish inter-provincial law firms; or the right of companies to legal representation by law firms not registered in a specific province, to name a few examples.[3] In general terms, such litigation as has occurred has been brought by élite social interests, and in cases that relate to allegations of intrusive public sector action.

To date then, the jurisprudence of mobility rights is curiously foreshortened, as is that of other sections of the Charter. Indeed, the 'undue rush of enthusiastic blood to the collective head of legal practitioners' (Glasbeek, 1989: 391) has subsided, as much of the initial Charter optimism has turned to 'frayed patience, disappointment and anger' (Branswell, 1992: A7; cf. Fudge, 1992; Herman, 1990). Why this is so speaks of the dilemmas associated with social activism, rights and the 'legalization' of political life. Several reasons for this are obvious. We should not be surprised, for example, that people such as doctors and lawyers are over-represented, given the high costs of litigation. A 1985 report estimated that those bringing non-criminal cases under the Charter should be prepared to spend at least $200,000 (Petter, 1989: 155).

However, this still fails to explain the selectivity of cases. Why have the courts not addressed the broader meaning of mobility outlined above? The reason that the courts have not explored the progressive possibilities of section 6 – assuming that they could – lies less in the 'meaning' of the Charter itself, than in the set of political and geographic understandings that animate legal interpretation. As the rights-sceptics would argue, expecting otherwise underestimates the interpretative power of the judiciary. An optimism concerning the progressive potential of constitutional documents such as the Charter tends to focus on the 'prescriptive question: "what *should* courts do given the interpretive possibilities of the Charter's rights and freedoms?" Unfortunately, [this] analysis tends to omit consideration of the "predictive" question: "what are courts *likely* to do given the historical and political context in which they operate?"' (Bakan, 1991: 308). In approaching the Charter, it has been persuasively argued that the Canadian judiciary relies (despite protestations to the contrary: cf. Monahan, 1987) on liberal conceptions of social life (Hutchinson and Petter, 1988).

The lineaments of liberalism are well known (Waldron, 1987). One essential component is the assumption of a radical divide between a private and a public realm. By this account, the main enemy of freedom is not the private corporation, disparities in wealth, discrimination, patriarchy or violence, but the 'public sector', which must be kept in constant check. Binavince (1982: 347) sees section 6 as designed 'to create a relatively secure sphere . . . from which the powers of government are banned'. A concept of 'negative freedom' (freedom *from*, rather than freedom to) is thus central to the judicial reading of mobility. The effect of this distinction is powerful, engendering a selective myopia to those offences to mobility rights which originate from the actions of the 'private sector' – such as gentrification, or plant closure. Moreover, to the extent that the public/private vision constitutes a masculine division of space (in which women, being 'nearer nature', are located in the private realm) we should not be surprised to see crimes such as sexual violence being spatially construed. Margaret Thornton (1991) reveals the geographic proscriptions of Australian sexual harassment law, for example, noting that harassment is rendered unlawful in certain spaces (such as those of education and work); conversely 'harassment in the street, a public place, is not proscribed. Inferentially, this is "private" activity which is of no interest to the state' (p. 451).

Not only does the liberal grounding of section 6 lead to an organized forgetting of forms of private power, it also elevates the rights of the individual *versus* the collective. It is partly in these terms, I think, that we can understand the disproportionate judicial emphasis on the right to *move*, as opposed to the associated right to *stay*. To liberals, the act of movement – especially that of migration – seems to embody a willed, masculine rejection of the worrisome ties of place and community. No wonder the frontier myth, the obsession with migration, frictionless distance and the isotropic plain retain such a powerful hold on the political and intellectual imagination.

This point is made clearer when we discover that the Charter's section 6 has an intriguing pedigree. Rather than being intended to serve the broader rights noted earlier, section 6 was put to work to advance the economic integration of the country in the face of provincial 'constraints' and 'barriers' such as preferential hiring schemes or local licensing arrangements (Binavince, 1982; Laskin, 1982). The individualism and selective condemnation of public sector 'distortions' that surround the discussion of the so-called 'economic union' preclude many broader readings of mobility. Not only is this the case, moreover, but the linkage of mobility and the economic union could be made by many social interests actively hostile to the claims of the marginalized and disenfranchised. Recent proposed constitutional changes, for example, would perhaps have allowed employers of developers to use section 6 to advance their right to move capital from one location to another, against the claims of local communities and workers. The fact that section 6 is open-textured and can be claimed by the Left, in other words, means that reactionary interests can similarly appropriate and redefine constitutional meanings. 'People who seek to reinforce hierarchy and perpetuate domination', Klare (1991: 100) notes, 'can speak the language of rights, often with sincerity'. The problem, moreover, is that these people seem to be granted privileged judicial audience.

I could easily go on. My point, I think, has been made. In place of a potentially liberatory reading of mobility, the judicial interpretation of section 6 is a strikingly impoverished and partial one. Section 6 has been taken to include only a protection from state (public) action that disallows inter- and intra-provincial mobility. That this could be extended to include other readings of mobility (including those relating to private sector action) is not in itself an impossibility. However, it would appear to be unlikely given the interpretative context within which mobility is understood. As Deborah Rhode (1990: 635) has argued, the 'central problem with rights-based frameworks is not that they are inherently limiting but that they have operated within a limited institutional and imaginative universe'. The ultimate irony, perhaps, is that it is the judiciary itself which has placed the most individious restrictions on mobility.

CONCLUSION

Where does this leave us? Most importantly, what position should rights – including mobility rights – play in progressive political action? Is rights-talk hopelessly compromised? Can the Charter be used as a vehicle for social change, or are the writings of the rights-optimists unrealistically idealistic? I have tried to argue that, in fact, both rights-sceptics and -optimists are correct. Rights are indeed explosive. Given their potentially broad meaning and their expansionary logic, they can be extended to apply to many areas from which they were previously excluded. As a result, naturalized power relations can be redefined as oppressive and unjust. This can be seen in the context of mobility rights, which also serves to direct critical attention to the under-explored politics of space. If, as indicated, mobility rights have not received detailed attention, yet have a deep resonance within the liberal tradition, their critical deployment may well be powerful. To an extent such a deployment constitutes a form of internal critique – working within the language of rights and its claims of liberty, autonomy, democracy and even mobility, whilst attempting to reclaim the full potential of that tradition and push it beyond its reified boundaries.

The space within which rights are put to work, however, needs to be carefully chosen. It is this, I have suggested, that divides the optimists and the cynics. Most of the pessimism concerning the potential of rights relates to what happens when they are legalized and judicially defined. An important distinction, then, needs to be made between the progressive use of rights for mobilization and critique, and rights-struggles that centre on the court-room. If the containment of mobility rights under the Canadian Charter is anything to go by, such victories will indeed be hard-won. There is a place, however, for progressive struggle around rights. With their special concern for the under-explored linkages between power and space, geographers are well equipped to contribute – carefully – to such struggles.

NOTES

1 I am grateful to participants at the AAG Political Geography Specialty Group gathering in Boulder, Colorado in spring 1992, as well as for the helpful comments of Lynn Staeheli, Joel Bakan, Michael Hayes and three anonymous referees. Any errors are mine alone.

2 Re Mia and Medical Services Commission of British Columbia, [1985] 17 *D.L.R.* (4th) 385 at 412 and 414. For a discussion of this decision, see Blomley (1992).

3 See, respectively: *Wilson* v. *Med Services Comm. of B.C.* [1988] 30 *B.C.L.R.* (2d) 1; Re Mia and Medical Services Commission of British Columbia [1985] 17 *D.L.R.* (4th) 385; *Black* v. *Law Society of Alberta* [1989] 1 *S.C.R.* 591; *Malartic Hygrade Gold Mines Ltd* v. *The Queen* in right of Quebec *et al.* [1982] 142 *D.L.R.* (3d) 512.

REFERENCES

Ades, P. (1989). 'The unconstitutionality of "anti-homeless" laws; ordinances prohibiting sleeping in outdoor public areas as a violation of the right to travel.' *California Law Review* 77, 595–628.

Anderson, K. (1991). *Vancouver's Chinatown: Racial Discourse in Canada. 1875–1980.* Montreal: McGill Press.

Bakan, J. (1991). 'Constitutional interpretation and social change: You can't always get what you want (nor what you need).' *The Canadian Bar Review* 70, 307–328.

Baker, S. A. (1975). 'A strict scrutiny of the right to travel.' *UCLA Law Review* 22, 1129–1160.

Barry, J. and Derevlany, J. (1987). *Yuppies Invade my House at Dinnertime.* Hoboken, NJ: Big River Publishing.

Bartholomew, A. and Hunt, A. (1990). 'What's wrong with rights?' *Law and Inequality* 9, 1–58.

Binavince, E. S. (1982). 'The impact of mobility rights: the Canadian economic union – a boom or a bust?' *Ottawa Law Review* 14(2), 340–365.

Blackstone, W. (1803 [1765]). *Commentaries on the Laws of England*, Book I. London: A. Strahan.

Blomley, N. K. (1992). 'The business of mobility: geography, liberalism and the *Charter*.' *Canadian Geographer* 36(3), 238–253.

—— (1994a). *Law, Space and the Geographies of Power.* New York: Guilford Press.

—— (1994b). 'Representing the Canadian city: space, power and mobility rights.' In *Urban Lives: Fragmentation and Resistance.* (V. Amit-Talai and H. Lustiger-Thaler eds). Toronto: McLelland & Stewart.

Blomley, N. K. and Bakan, J. (1992). 'Spacing out: towards a critical geography of law'. *Osgoode Hall Law Journal* 30(3), 661–690.

Bowles, S. and Gintis, H. (1987). *Democracy and Capitalismn. Property, Community and the Contradictions of Modern Social Thought.* New York: Basic Books.

Branswell, H. (1992). 'Women left to wonder what now as paper rights fail to fix wrongs'. *Vancouver Sun*, 10 April, A7.

Campbell, T. (1983). *The Left and Rights.* London: Routledge.

Clark, G. L. (1983). *Interregional Migration, National Policy and Social Justice.* Totowa: Rowman and Allanheld.

Clark, G. L. (1990). 'The virtues of location: do property rights "trump" workers' rights to self-organization?' *Environment and Planning D: Society and Space* 8, 53–72.

Crenshaw, K. W. (1988). 'Race, reform and retrenchment: transformation and legitimation in anti-discrimination law.' *Harvard Law Review* 101(7), 1331–1387.

Cresswell, T. (1993). 'Mobility as resistance: a geographical reading of Kerouac's "On the Road".' *Transactions of the Institute of British Geographers* 18, 249–262.

Davis, M. (1991). *City of Quartz: Excavating the Future in Los Angeles.* London: Verso.

Dear, M. and Wolch, J. (1987). *Landscapes of Despair: from Deinstitutionalization to Homelessness.* Princeton: Princeton University Press.

DERA (1990). *Downtown Eastside Housing and Residents Survey.* Vancouver: Downtown Eastside Residents' Association.

Fine, S. (1992). 'Activists challenge shackles of poor.' *Globe and Mail*, 25 November, A3, A7.

Freeden, M. (1991). *Rights.* Minneapolis: University of Minnesota Press.

Freeman, A. (1990). 'Antidiscrimination law: the view from 1989.' In *The Politics of Law: A Progressive Critique* (P. Kairys ed.), pp. 121–150. New York: Pantheon Books.

Fudge, J. (1989). 'The effect of entrenching a Bill of Rights upon political discourse: feminist demands and sexual violence in Canada.' *International Journal of the Sociology of Law* 17, 445–463.

—— (1992). 'Evaluating rights litigation as a form of transformative feminist politics.' *Canadian Journal of Law and Society* 7(1), 153–162.

Fyfe, N. (1993). 'Making space for the citizen? The (in)significance of the UK Citizen's Charter.' *Urban Geography* 14(3), 224–227.

Gabel, P. (1984). 'The phenomenology of rights-consciousness and the pact of the withdrawn selves.' *Texas Law Reform* 62(8), 1563–1599.

Gabel, P. and Kennedy, D. (1984). 'Roll over Beethoven.' *Stanford Law Review* 36, 1–56.

Glasbeek, H. J. (1989). 'Some strategies for an unlikely task: the progressive use of law.' *Ottawa Law Review* 21(2), 388–418.

Hall, S. (1983). 'The Great Moving Right Show.' In *The Politics of Thatcherism* (S. Hall and M. Jacques eds), pp. 19–39. London: Lawrence & Wishart.

Harvey, D. (1973). *Social Justice and the City*. London: Edward Arnold.

—— (1992). 'Social justice, postmodernism and the city.' *International Journal of Urban and Regional Research* 16(4), 388–401.

Herman, D. (1990). 'Are we family? Lesbian rights and women's liberation.' *Osgoode Hall Law Journal* 28(4), 789–815.

Hobbes, T. (1988 [1651]). *Leviathan*. London: Penguin Books.

Houseman, G. L. (1979). *The Right of Mobility*. Port Washington, New York: Kennikat Press.

Hutchinson, A. C. and Petter, A. (1988). 'Private rights/public wrongs: the liberal lie of the Charter.' *University of Toronto Law Journal* 38, 278–297.

Klare, K. E. (1991). 'Legal theory and democratic reconstruction: reflections on 1989.' *U. B. C. Law Review* 25(1), 69–103.

Kennedy, M. J. (1971). 'The civil liberties lie.' In *Law Against the People* (R. Lefcourt ed.), pp. 140–149. New York: Vintage Books.

Kropotkin, P. (1885). 'What geography ought to be.' *The Nineteenth Century* 18, 940–956.

Laclau, E. and Mouffe, C. (1985). *Hegemony and Socialist Strategy: Towards a Radical Democratic Politics*. London: Verso.

Laskin, J. B. (1982). 'Mobility rights under the Charter.' *Supreme Court Law Review* 4, 89–106.

Ley, D. (1981). 'Inner city revitalization in Canada: a Vancouver case study.' *Canadian Geographer* 25, 124–148.

Mandel, M. (1989). *The Charter of Rights and the Legalization of Politics in Canada*. Toronto: Wall and Thompson.

Marx, K. (1978 [1843]). 'On the Jewish Question.' In *The Marx–Engels Reader* (R. C. Tucker ed.), pp. 26–52. New York: W. W. Norton.

Mitchell, D. (1992). 'Iconography and locational conflict from the underside.' *Political Geography* 11(2), 152–169.

Monahan, P. (1987). *Politics and the Constitution: the Charter, Federalism and the Supreme Court of Canada*. Toronto: Carswell.

Monet, D. and Skanu'u (Ardythe Wilson) (1992). *Colonialism on Trial: Indigenous Land Rights and the Gitskan and Wet'suwet'en Sovereignty Case*. Philadelphia: New Society Publishers.

Mouffe, C. (1991). 'Democratic citizenship and the political community.' In *Community at Loose Ends* (Miami Theory Collective ed.), pp. 70–82. Mineapolis: University of Minnesota Press.

Pain, R. (1991). 'Space, sexual violence and social control: integrating geographical and feminist analyses of women's fear of crime.' *Progress in Human Geography* 15(4), 415–431.

Petter, A. (1989). 'Canada's charter flight: soaring backwards into the future.' *Journal of Law and Society* 16(2), 151–165.

Rhode, D. (1990). 'Feminist critical theories.' *Stanford Law Review* 42, 617–638.

Rose, G. (1990). 'The struggle for political democracy: emancipation, gender and geography.' *Environment and Planning, D: Society and Space* 8, 395–408.

Rustin, M. (1992). 'Citizenship and Charter 88.' *New Left Review* 191, 37–40.

Sekula, A. (1991). 'People who can't afford to live here should move someplace else'. In *If You Lived Here: The City in Art, Theory and Social Activism* (B. Wallis ed.), pp. 146–147. Seattle: Bay Press.

Smith, D. M. (1977). *Human Geography: A Welfare Approach*. London: Edward Arnold.

Smith, S. J. (1989). 'Society, space and citizenship: a human geography for the "new times"?' *Transactions, Institute of British Geographers* 14, 144–156.

Stewart, D. (1990). *The Right to Movement: Motor Development in Every School*. London: Palmer Press.

Thornton, M. (1991). 'The public/private dichotomy: gendered and discriminatory.' *Journal of Law and Society* 18(4), 448–463.

Tushmet, M. (1984). 'An essay on rights.' *Texas Law Review* 62(8), 1363–1412.

Unger, R. (1983). 'The critical legal studies movement.' *Harvard Law Review* 96(3), 320–432.

Valentine, G. (1989). 'The geography of women's fear.' *Area* 21(4), 385–390.

Waldron, J. (1987). 'Theoretical foundations of liberalism.' *The Philosophical Quarterly* 37(147), 127–150.

Williams, P. J. (1991). *The Alchemy of Race and Rights.* Cambridge, MA: Harvard University Press.

Williams, R. A. Jr (1987). 'Taking rights aggressively: the perils and promise of critical legal theory for people of color.' *Law and Inequality* 5, 103–134.

13

Human rights and development in Africa

Moral intrusion or empowering opportunity?

Giles Mohan and Jeremy Holland

from *Review of African Political Economy*, 2001, 88: 177–196

INTRODUCTION

> Human rights are not, as has sometimes been argued, a reward of development. Rather, they are critical to achieving it.
>
> (UNDP, 2000: iii)

> Human rights in the 1990s, to a greater extent than ever before, set a norm that regulates the relationship between state and society.
>
> (Sano, 2000: 741)

> For many people living in the South, international human rights are understood increasingly as a set of values that support the expansion of global capital, exploitation and control.
>
> (Evans, 1997: 92)

Throughout the 1990s the debates about human rights and development have increasingly converged. Previously, much of the debate around and practice of human rights was confined to the "first generation" of human rights regarding personal or private rights; sometimes referred to as political rights. These are essentially "negative" rights in that a person's freedom should be protected from the actions of other individuals, groups or the state. The struggle has been to enshrine these principles in law, such as a bill of rights, so that a person has the legal means with which to defend their freedom. Such bourgeois ideals grew out of the American and French revolutions of the 18th century as well as the liberalism of the Enlightenment. Subsequently, demands have been made for more positive human rights regarding broader social justice, such as labor rights, and tangible welfare benefits, such as housing and health care. Such rights, sometimes referred to as economic, social and cultural (ESC) rights, are more socially-defined, in that they carry an obligation for society-as-a-whole to ensure a minimum level of well-being for all. Clearly, the two sets of rights may be incompatible, especially for those who see well-being guaranteed through atomistic self-interest as opposed to communal or humanistic principles which emphasize equality. So, the rights arena has been forged out of competing political struggles in specific social and historical circumstances (Shivji, 1999).

Alongside these issues, the major development agencies have pursued a broad anti-poverty agenda, albeit one that is increasingly driven by a neo-liberal market logic. In the post-war period, the international human rights process, led largely by the United Nations, has sought to promote the indivisibility of political rights and economic, social and cultural rights. Since the early 1980s, the adjustment era has seen most Third World countries disciplined via debt conditionality with an emphasis on market-based development. This reduced the scope of the state which has been actively reformed to support marketization. However, the impacts on poverty have been questionable so that some development agencies began to re-assert the need for welfare protection and a more active role for the state.

It is here that the rights agenda has become more mainstream, because it places obligations on the state, amongst other actors, to ensure a minimum level of well-being for all. This differs from the "Basic Needs" of the 1970s, because the poor are encouraged to participate in defining and securing their welfare needs rather than being passive recipients of aid. As DFID (2000: 1) asserts, human rights "provide a means of empowering all people to make decisions about their own lives rather than being the passive objects of choices made on their own behalf." Two related issues arise here. First, is that multilateral and bilateral donor interventions have often usurped sovereignty through debt leverage, so will an emphasis on universal human rights be used as another means of deepening control over developing countries? Second, do the twin discourses of market hegemony and the universality of human rights involve an implicit erasure of cultural specificity and the denial of non-market alternatives to development?

So, given that the human rights agenda has important implications for democracy and sovereignty, and that the donors are championing "rights-based development" (RBD), it seems appropriate to discuss these in an issue on governance in Africa. In this article we want to explore the question of whether the emerging human rights-based approach to development, honed in the period of revisionist neo-liberalism, can deliver meaningful improvements to the African crisis. This article begins by outlining briefly the evolution of the rights-based development agenda from its Enlightenment roots to the present day in order to understand how the present agenda is defined. This has seen the emphasis expand from a personal and civil focus to an international and "developmental" one. The next section examines the theoretical underpinnings of the current rights-based development agenda and goes on to summarize two recent reports which place such concerns at their center. From there we give an overview of the state of play of implementing rights-based procedures in Africa, as a whole, and in individual African countries. The next section assesses the moral and practical implications of the rights agenda for Africa and assesses both the opportunities and threats it presents. We conclude by suggesting the possible future for rights work in Africa and the research agenda attached to it.

EVOLUTION OF THE RIGHTS-BASED DEVELOPMENT AGENDA

The struggles for rights are rooted in the Enlightenment and the emergence of citizenship. This saw the weakening of the monarch–subject relation and the movement away from particular to universal values. The emphasis was on the individual within society which fed into the ascendant liberal philosophy of western politics (see Table 13.1). The French and American revolutions of the 18th century saw rights enshrined in constitutional and legal terms, the most important being the American Bill of Rights of 1791. These civil and political rights primarily benefited the bourgeoisie and protected them from over-bearing state interference on the one hand and the popular participation that a genuine commitment to social equality requires on the other. Hence, as Ake (1987: 6) notes, "the idea of human rights really came into its own as a tool for opposing democracy." Subsequent rights were more socially and economically oriented and related to the welfare agenda of the 20th century whereby the working class could expect a minimum level of protection. The most recent phases of rights have focused on international solidarity through a social movements-led agenda around such issues as the environment and development. Some have argued

Table 13.1 Milestones in rights-based development

American Bill of Rights (1791)
UN Declaration on Human Rights (1948)
The African Charter on Human and Peoples' Rights (1981)
Declaration on the Right to Development (1986)
The African Human Rights Commission (1987)
Vienna Conference on Human Rights (1993)
Copenhagen Conference on Social Development (1995)
South African Government's Bill of Rights & the Truth and Reconciliation Commission (1995)
UNDP Human Development Report (2000)
DFID's Human Rights for Poor People (2000)

(Bobbio, 1996) that we are entering a further phase of genetic rights concerning the integrity of our basic biological identities. The first two phases in the evolution of the rights agenda were largely confined to western democracies and concerned the relationship between the individual and the state. With the growth of international governance, more Third World perspectives are being recognized (though not always incorporated into it) which is re-shaping the relationship between the individual and the global political order (Sano, 2000). Ake (1987) notes that with this evolution the human rights agenda has become a little more relevant to the needs of Africans.

Early debates around development and rights revolved around the political rights of people in colonized countries pushing for independence. In Africa, as Mamdani (1996) has shown, citizenship originally applied to the urban areas where expatriate whites were free to pursue and receive a relatively wide range of civic rights denied to the colonized. Some of these rights gradually spread to the African urban élites, but the rural areas were purposefully divided along and governed by customary laws. It is this "bifurcated state" structure, with its distinction between "rights" and "customs" and the association of rights with colonially-derived privileges, that has led to skepticism and apathy by many Africans towards the promises of the "rights agenda" (Penna and Campbell, 1998).

The latter phases in the evolution of rights cover the post-war period and begin with the 1948 UN Declaration of Universal Human Rights, which enshrined the principles of universality, inalienability and indivisibility of rights. Subsequent Conventions and regional Commissions extended these principles into more areas of social, political, cultural and economic life. The early goals of the Human Rights movement were drawn out of the horrific experiences of World War Two and sought to counter the particularistic and exclusionary racial hierarchy that underpinned Nazism (Shivji, 1999). Such challenges to racially-based discrimination chimed with the demands of anti-colonial struggles in the Third World so that the human rights agenda was supported by these soon to be independent states.

Despite the rapid waning of American interest in the rights agenda (Evans, 1995), the debates were hijacked during the cold war and used as a means of castigating Communism and justifying political, financial and military support for governments who upheld "proper" rights (Slater, 1993; Evans, 1997). Not surprisingly, this ideologically-charged period saw the continued separation of "political" from "economic" rights with the former taking precedence in this global battle over "ways of life." For newly independent African countries, the priority was development so that abstract debates about rights had little relevance to this cohort of modernizers who used centralized mechanisms to push through grandiose development plans. The result was that the developmental and human rights discourses tended to evolve separately. More important, perhaps, was the conflict that emerged over the question of "self-determination" (Shivji, 1999). Initially, the right to self-determination was to be all-encompassing and was clearly an antidote to imperialism in all its guises. However, during negotiations in the mid-1960s over two key Covenants, the interpretation of political self-determination was reduced to the eradication of formal colonization (or colonial-type rule such as apartheid) while economic self-determination was equated with the demands for fairer trade relations and adequate foreign aid.

The thawing of the cold war and the ending of apartheid saw a renewed interest in human rights with people emerging from repressive political structures and demanding economic development alongside political freedoms. Such an environment was ripe for the rights-based development agenda whose institutional architecture had been developing piecemeal for the previous forty years. Sengupta (2000) believes that a consensus now exists over the value of human rights and even suggests that it represents, somewhat ominously, another element in the "end of history."

The rights-based development agenda has risen to prominence in parallel with the emergence of social development notions of participation and entitlements that challenged the "technical fix" development paradigm of the 1950s and 1960s and the delivery of basic needs in the 1970s. The adoption of the 1986 UN Right to Development signaled a unification of the civil and political rights with economic, social and cultural rights and a growing political consensus that was strengthened through subsequent declarations on Environment and Development (Rio), Population and development (Cairo), Social Development (Copenhagen) and the Platform for Action of the World Conference on Women at Beijing (Sengupta, 2000). The perceived indivisibility of rights became increasingly clearly articulated in the development discourse, as illustrated by the language adopted during the fiftieth anniversary of the Universal Declaration in December 1998 and captured in the phrase "All Human Rights for All" (Maxwell, 1999).

THEORETICAL UNDERPINNINGS OF RIGHTS-BASED DEVELOPMENT

> the developmentalists are seeking to reformulate their concerns in the language of rights, while the human rights advocates are taking on board developmental issues without which, they recognize rights-talk can have little meaning to, and legitimacy with the vast majority of the people in the poor countries of the South.
>
> (Shivji, 1999: 262)

The current approach to RBD is iterative and evolutionary in that it learns from and builds upon previous approaches. The move towards RBD has evolved out of the coming together of two strands of development theory and practice which had previously been treated as discrete. On the one hand are human development approaches based around dynamic understandings of poverty and, on the other, human rights approaches based around questions of governance, participation and citizenship.

Dynamic approaches to human development

In recent years, income or commodity-centered conceptions of well-being have been challenged by multifaceted measurements of poverty, such as that underlying the basic needs approach of the early 1980s. Most importantly, from the 1980s, Amartya Sen (1997) conceptualized poverty in terms of human capabilities—(resources that give people the capability to be and to act)—and entitlements (the set of alternative commodity bundles that a person can acquire in any given societal context). By so doing he posed fundamental questions about the quality of life beyond the possession of commodities; the latter having only "derivative and varying relevance." Speaking to these debates have been discussions of vulnerability. This is a more dynamic concept concerning the changing experiences in socioeconomic status relating to survival, exposure to risk, defenselessness and self-respect. Accordingly, vulnerability captures some of the multidimensional, dynamic and structural aspects of poverty: "Vulnerability denotes not simple lack or want, but defenselessness, insecurity and exposure to shock or stress" (Chambers, 1989: 1). Whilst vulnerability is not a concept that has been rigorously theorized, or for which generally accepted indicators exist, there are a growing number of conceptual frameworks for analyzing vulnerability, including Moser (1998) and Bebbington (1999) and the livelihoods analysis frameworks of DFID (Scoones, 1998). These frameworks link entitlements to resources and emphasize the structures, institutions and processes that mediate individual, household and community-level access to a range of assets. The result has been a politicization of the vulnerability discourse, with analytical space created for tackling the policy environment and policy-making institutions.

Institutions, governance and participation

This emphasis on institutional processes links directly with debates around governance and participation (Mohan and Stokke, 2000). The key to RBD is that it attaches political rights and responsibilities to fundamental aspects of human needs and well-being. Gaventa and Valderrama (1999) usefully draw out this entwining of different interpretations of governance and participation. They see two traditions; one driven by community participation and the other by political participation.

The community focused participation is well documented (Chambers, 1983, 1997) and grew out of the realization that formal state-based development programs had yielded limited benefits. Since the 1970s, the acceptance of participation has become widespread, but at its base is a belief that development energies lie outside of the state and are built from local knowledge. It relies on relatively closed and homogenizing notions of community where participation in decision-making is direct and unmediated by representatives. Political participation, on the other hand, has focused on more formal engagement with the state by individuals or organized groups and parties. These political processes tend to be less direct than community participation and involve elections, lobbying, advocacy and the day-to-day interaction with the local state. The good governance agenda of the 1990s (Leftwich, 1994; Rhodes, 1997) focused centrally on this level of participation.

Gaventa and Valderrama (1999) argue that local governance can benefit from the coming together of these two traditions through "citizenship participation" which involves "the direct intervention of social agents in public activities" (Cunill, 1997: 77) and has seen renewed interest in decentralization and political culture. Democratic decentralization has been a perennial tool in development planning since independence, but it holds an important place in RBD, because for the majority of the poor the state is the local state (Mamdani, 1996; Migdal, 1994) and it is where most citizenship claims will be contested. On the other hand questions of political culture have been reawakened through debates around social capital. As the World Bank (1997: 114) notes, local institutions "are valuable not only for their ability to meet basic needs, but also for the role they play in building trust and a sense of public connectedness among those excluded or alienated from the formal political process." So, "there is thought to be a synergistic relationship between the emergence of strong civil society and social capital formation" (McIlwaine, 1998: 418).

THE POLICY CONTENT OF RIGHTS-BASED DEVELOPMENT

So far we have traced the emergence of RBD and discussed the theoretical ideas underpinning it and the political arrangements believed to be necessary to achieve it. The emergence of RBD discourse from its intellectual origins in poverty analysis and participation has created an operational space for an absorption of the rights agenda within the neo-liberal policy frameworks. This is most clearly demonstrated in the transition from the policy analysis in the World Bank's *World Development Reports* of 1990 and the 2000/01. Both WDRs took poverty as their theme, but while the 1990 WDR emphasized labor intensive growth combined with investment in human capital, the WDR 2000/01 signaled a shift in policy analysis towards a concern with "empowerment" through enhanced political participation of poor people in tackling institutional "dysfunctionalities." Even as the ideological climate continues to frame policy imperatives of market provision of goods and services and the attendant erosion of the state's redistributive function, the neo-liberal establishment has successfully repositioned itself with respect to the rights-based agenda by championing accountability, transparency and the role of citizen participation in demanding their rights.

In this section, we look in more detail at the actual policy agenda attached to RBD. We have structured this around different scales and roles for convenience, which reflects a logical division of labor between institutional levels although there are clearly interlinkages between these scales and levels. At the root of RBD is a liberal belief that development is a matter of personal choice and effort, but that this is tempered by the prevailing social and political conditions. It also adds a strong action-orientation, in that people now have a claim or entitlement on other people and institutions which, if it is socially-accepted or legally-defined, gives people a minimum level of expected well-being. The DFID Report (2000) stresses this "obligation" as a key feature of the new framework which takes us well beyond basic needs approaches which were passive and treated the poor as helpless victims. It also provides limits on the damage that individuals should be allowed to bear as a result of externalities generated by other activities, no matter how valuable theses activities appear to be.

As we would expect, the key documents are replete with lessons and action items, or in the UNDP's vocabulary "bold new approaches." Underpinning DFID's policy agenda is a triumvarate of core principles—participation, inclusion and obligation. These involve (DFID, 2000: 3):

- *participation*: enabling people to claim their human rights through the promotion of the rights of all citizens to participation in, and information relating to, the decision-making processes which affect their lives. They acknowledge that action needs to go "beyond and above local-level processes of consultation ... [and] ... linking poor people's perspectives with national and international policy processes" (DFID, 2000: 19);
- *inclusion*: building socially inclusive societies through development which promotes all human rights for all people and encourages everyone to fulfill their duty to the community;
- *obligation*: strengthening state policies and institutions to ensure that obligations to protect and promote all human rights are fulfilled.

Overlying these principles are policy items which map onto different political institutions and scales. These can be summarized thus:

International: The RBD approach takes into account the globalization of the world economy whereby the actions of states beyond their borders are factored into any consideration of rights. At the international level there is a need for commitment, co-operation and co-ordination. The international organizations must be committed across the board to enshrining an RBD approach in their operations. At present, some institutions such as the ILO, UNICEF and UNDP have a strong record of incorporating human rights into project design, but others are less stringent on this, so the challenge remains to bring all institutions in line and up to speed. Practical measures for assisting policy-makers to mainstream RBD revolve around the understanding and measurement of the current state of human rights so that much is made of benchmarking and data collection. Another recommendation regards global governance more generally. In response to an ongoing critique of the internal democracy of the major institutions the UNDP report argues that "all countries—small and weak—have a voice in decisions" (2000: 85). International civil society also has a key role in advocacy, monitoring, and consumer pressure. Civil society has generally been the motive force behind human rights legislation and its role must continue, although this is to be in collaboration with states, international organizations, and corporations which may further erode the "independence" of civil society.

After building RBD into project design, the next challenge remains monitoring and enforcement of rights abuses. The UNDP are careful to stress that strengthening the rights-based approach in development co-operation must be "without conditionality." The emphasis now is on "transparent and open" economic policy formulation which confers "ownership" on the implementing country and where the final decision rests with elected officials. There is a recognition that economic actors have a role to play in promoting RBD. The UNDP (2000: ii) begins its report by arguing that "Rights makes human beings better economic actors" so that a vibrant economy, while not guaranteeing human rights, is a requisite. The report goes on to suggest ways in which economic growth can be balanced with respect for rights. Corporations should not use their wealth for unfair lobbying and should apply codes of conduct in all their operations while states should promote an enabling economic environment which is pro-rights. The DFID makes similar points, but acknowledges that "it has proved equally difficult to hold transnational corporations themselves legally accountable for alleged human rights violations" (2000: 14) although voluntary codes might be a solution.

National: The key role in RBD is given over to the state. The UNDP is at pains to stress that such work must go beyond legislation and actively embed the importance of rights in all social norms. The DFID adds that states are not homogenous entities so that different branches of the state must also show commitment, co-operation and collaboration. The branch which has received most attention is the judiciary since it is an impartial and accessible judiciary which can enforce human rights. Such judicial reforms sit alongside those other elements of good governance which have become accepted elements of policy reform such as increasing bureaucratic accountability and transparency and the holding of competitive elections. However, the practice of democracy must be "inclusive" and go "beyond elections" and include minorities and permit an active civil society and free press.

Locally: At the local level the emphasis is on participation, decentralization and the strengthening of civil society to be more rights-oriented. The onus for this falls on the state to provide

> a legal framework that protects the right to participation . . . the need for continual reform to adapt to changing circumstances . . . put in place decision-making processes that are transparent and open to dialogue, especially with poor people and poor communities.
>
> (UNDP, 2000: 65, 67, 78)

On the other hand civil society must remain vigilant of rights abuses and act as the independent monitor. So, for civil society and NGOs, the emphasis has changed somewhat. The gradual move away from output-based approaches to more process-based ones saw the emphasis shift to capacity building of local NGOs and civil society organizations. With RBD, this has continued, but altered somewhat to enable people to use their rights to ensure their well-being. So NGOs became involved in legal and political literacy, and civic leadership (Fowler, 2000).

IMPLEMENTING RIGHTS-BASED DEVELOPMENT IN AFRICA

In this section we outline the implementation of the rights-based development agenda in Africa as it currently stands. This is important in order to understand the problems facing African countries in realizing a "universal" project and as a baseline against which to assess the desirability and prospects for achieving these goals. A key tension in the rights debates in Africa has been over the timing, balance and importance of political rights on the one hand and ESC rights on the other. Some see democracy as a pre-requisite for any meaningful development, which raises questions about whether there are particularly *African* human rights and consequently a specific *African* democracy (Maluwa, 1997). Others stress that under extreme poverty and marginalization, it is economic and social rights which are more important, and that the operation of imperialism has contributed to this underdevelopment. As Shivji notes:

> This is a dilemma which expresses itself in the dichotomy between the so-called social/economic rights and political/civil rights on the one hand, and various attempts to reconcile the tension by reconceptualizing the jurisprudence of rights, on the other.
> (Shivji, 1999: 260)

The rights-based development approach contends that such a dichotomy is not useful, because only if people are empowered to determine their genuine needs will development occur. This, they contend, simultaneously promotes sustainable democracy and well-being.

The formal rights framework in Africa centers on the 1981 African Charter on Human and Peoples' Rights which came into force in 1987 alongside the establishment of the African Commission of the same name, which is itself a product of the Organization of African Unity (OAU) (Murray, 2000). As with all regional Charters, it is derived from the 1948 Universal Declaration, but takes into account the African experience which saw a greater emphasis on economic, cultural and social rights; that is, those which pertain directly to material well-being or "development" (Maxwell, 1999; Sano, 2000). In keeping with the OAU's beliefs and, more importantly, the tenets of international law, the Charter and Commission have to recognize the primacy of individual states. As we shall see, debates over the limitations of state-centric legal discourses have been paramount given the weak record of accountability of African states and other diverse socio-political entities that co-exist within and alongside formally recognized states:

One of the key elements of the African approach to human rights has been to recognize the particularity of Africa's experiences within a discourse which stresses the universality of human values. African debates stress the role of "tradition," colonialism and imperialism in shaping the constitution and realization of human rights (An-Na'im, 1999a). For example, the African Charter on Human and Peoples' Rights seeks to "eradicate all forms of colonialism from Africa . . . [while taking into consideration] . . . the values of African civilization" (Murray, 2000: 203). Independence marked, in theory at least, the most important conferral of rights in that people became genuine citizens and that their countries were accorded international sovereignty. Clearly, the post-independence record of state decay and neo-imperialism have shown the limitations of these visions as the mass of Africans have been denied, through no fault of their own, some basic aspects of human dignity and social welfare.

A key problem with human rights legislation is that under international law only states are recognized as having "personality"; that is they are the only formally recognized legal bodies. So, any human rights legislation must be embedded within national political and judicial structures. At present the formal policy frameworks for realizing RBD are uneven. It needs re-emphasizing that much of the RBD agenda is iterative and evolutionary and builds upon the good governance and participatory approaches that have become widespread over the past decade. Hence, many of the policy discourses and the institutional architecture already exist. For example, bureaucratic accountability and responsiveness are key elements in RBD, but have clearly been on the agenda for a while. Similarly, gender equality as a central tenet of RBD has been contested since the 1970s. However, current policy stresses certain political and institutional innovations.

Some are part of the general human rights processes led by the United Nations. Since the mid-1960s there have been various Conventions covering discrimination against key groups (race, gender and children) and protection from torture. Countries sign up to these, which indicates a willingness to enshrine these principles in law. The UNDP (2000) reviewed the coverage of these signatories, which shows that some African countries, such as Cameroon and Zambia, have signed and ratified all of the Conventions, whereas Ghana has only signed up to the Conventions covering racial and gender discrimination while Nigeria has signed up to all except that covering torture and degrading treatment. Other policy channels are largely constitutional and involve statements in the constitution or in a separate Bills of Rights. Maluwa (1997) notes that in Southern Africa Botswana, Namibia, South Africa and Malawi have fully-fledged Bills of Rights. Most countries' constitutions include some recognition of fundamental human rights based, to a large extent, in European and American constitutional practice. These see a separation of legislative and judicial branches of the state and include various mechanisms for protecting political freedoms such as freedom of speech and *habeas corpus*. However, as An-Na'im (1999b: 43) notes: "None of the countries surveyed provide full-fledged constitutional protection for economic, social and cultural (ESC) rights," although, crucially, there is no country in the world which does so. A more recent addition, in the wake of South Africa's Truth and Reconciliation Commission, has been the establishment of national Human Rights Commissions to monitor the implementation of human rights legislation and disseminate information on abuses and best practice. Clearly, recognition at the Convention, Commission and Treaty level only stipulates what a state should do and is not a good indication of what it actually does.

The record of human rights protection in Africa, in general, and the work of the Commission, in particular, has been mixed. We do not have the space for a detailed account of these experiences (see Murray, 2000), but some pertinent points, drawn largely from An Na'im (1999b), are worth making. The first problem is the degree to which African countries respect constitutionalism. Many have made important steps in this regard, but often fail to promote economic, social and cultural rights and, more importantly, have a range of means to suspend the constitution. Some of these are legal, such as during States of Emergency or through "claw back clauses," while others are less obvious and range from the selection of judges through to outright intimidation. A second problem relates to the recognition of customary law within the formal legal system. Despite claiming to recognize Africa's uniqueness and diversity, and hence the legitimacy of its customary legal practices, these can conflict with universal principles or are simply not taken seriously by constitutional lawyers (in part because they are not codified). In practice, customary law usually gives way to statutory or common law. A third set of problems relate to the judiciary and legal profession. Training is often poor, selection can be politically motivated, and régimes tend to circumscribe the independence of the legal profession. A fourth set of problems relate to the international organizations which support human rights promotion on the ground. The African Commission is rather élitist, lacks clear reporting structures, and has unclear authority to enforce decisions or condemn violations of human rights (Murray, 2000). Similarly, many of the international NGOs which have been major supporters of human rights causes operate in élitist ways (for example, organizing urban-based workshops for lawyers) and tend to impose, through funding conditionality, their own agendas on local NGOs.

THE PARADOXES OF RIGHTS-BASED DEVELOPMENT IN AFRICA

The implementation of the rights-based development agenda within the context of existing structures of African political economy raises a number of important questions for the future of this project. In this section we highlight some major tensions and contradictions arising out of the articulation of a universal political ideal and the realities of territorial states and embedded cultural practices.

Sovereignty, conditionality and modernity

An overarching set of criticisms relates to the broader agenda of RBD and its relationships to modernity and western imperialism. Turner (1993) argues that modernization involves a progressive move from particularism ("tribe," community, ethnicity, etc.) to universalism and secularism. In this sense, citizenship represents a significant dimension of modernity whereby it initially related to membership of the city-state and later to membership of a nation-state. A possible danger of this reading is that citizenship, as with modernity, becomes equated with the suppression of difference and, hence, open to totalizing or exclusionary practices.

Furedi takes development back to the colonial period and the intertwining discourses of stewardship and civilization which legitimized a "moral intrusion" such that "The right of the West to intervene has become a moral imperative" (1997: 87). Since then all manner of interventions have ensued leading to the present situation of "western proprietorship of human rights" (Penna and Campbell, 1998: 7). As with any discourse, the human rights discourse is based on symbols which confer meaning, but as Penna and Campbell (1998: 9) note: 'In human rights discourse, the majority of positive symbolism used is western." A key element of the evolution of rights thinking is to treat African (and other "non-western") experiences as lacking any relevance for "universal" values, thus effecting a form of Eurocentrism which, as we discuss below, can become an ideological hammer in the face of cultural difference.

Evans (1997) focuses centrally on the paradoxes of universal human rights discourses and sovereignty. He argues that rights are usually discussed in legal or philosophical terms which can mask political and economic interests. He argues that universal human rights are "imposed" because they offer a "coherent claim to authority over the sovereign state" (1997: 91) and "represent a further attempt to forge new structures of colonial dominance" (1997: 92). In this sense human rights might become a new form of conditionality in dealings between the multilateral institutions and recipient countries. Tensions along these lines were clearly visible at the Rio Earth Summit, where Southern delegates felt constrained by western governments bent on protecting their own environment and economic growth while disciplining Southern countries into restrictive environmental codes. Similarly, at the Vienna Human Rights Conference a year later, "some Asian countries questioned external criticism of their human rights records; in particular, they showed their resentment at having imposed on them a set of values based on western traditions" (Potter *et al.*, 1999: 129).

The recent UNDP report is clearly aware of this issue, but is adamant about such responses:

> There is a tension . . . between national sovereignty and the international community's monitoring of human rights within countries . . . Many people still see the promotion of human rights for some groups . . . as a threat to their own values or interests. This divisiveness in values breeds opposition to human rights for all.
>
> (UNDP, 2000: 30)

The implication seems to be that any country that is skeptical about the application of universal human rights may well be using this as a defense mechanism for the pursuance of human rights abuses. However, the blanket refusal to countenance detractors from a universal HR approach does smack of "moral intrusion," and, more importantly, makes it difficult to differentiate between a legitimate and illegitimate rejection of universalistic rights. The emphasis on developing regional human rights codes, such as the African Charter, seems one "best fit" solution which balances universality with politico-cultural specificity.

Universalism, cultural relativity and community

Hence, a key problem is the tension between a universal set of values and a multiplicity of embedded local practices (Nagengast and Turner, 1997; Penna and Campbell, 1998). Debates have arisen around the use and abuse of both "universalism" and "tradition" since both are tied to distinct social and political visions. As we have seen, the universalist argument can conceal western hegemonic aspirations. On the other hand, the discourse of traditionalism has been used by unscrupulous régimes and/or local people to resist external scrutiny and persist with inhumane behavior. In Africa, this tension has been brought to the fore over such matters as female genital mutilation (Penna and Campbell, 1998) and the relationship between customary law and common law whereby the latter usually prevails (Murray, 2000).

One of the sources of these problems is in the philosophical inheritance of the rights discourse. We saw that the dominant conception of human rights and development is based on liberal individualism arising out of the Enlightenment. However, the alternative to liberalism is some form of communitarianism which sees rights shaped by and accountable to a collective (von Lieres, 1999). Talking of Africa, Ake notes "our people still think largely in terms of collective rights and express their commitment to it constantly in their behavior" (1987: 5). It is, in part, this recognition that the African Charter contains the notion of "peoples" although Murray (2000) has discussed the problems of defining and delimiting "peoples" and with it the whole notion of cultures as bounded and identifiable entities.

The African Charter, like the OAU, upholds the sanctity and integrity of colonially-created states, comprised of a multiplicity of "nations." However, in recognizing "peoples," the African Charter creates tensions over self-determination. The first difficulty is that there is no clear definition of what constitutes a "people." It is used to refer to the population of a state, although a people is not the state itself, but it can be something other than the entire population of the state (Murray, 2000). For example, the African Commission has recognized the Katangese of Congo and the Casamance of Senegal as peoples. The second problem is in interpreting the legitimacy of peoples' rights. If a people feel oppressed by the state, how can they press for self-determination within the fixity of a territorial nation-state? Again, this tension has not been resolved and the rights discourse in Africa does not countenance full-blown secession, preferring instead solutions such as participation, decentralization, federalism, and proportional ethnic representation. Indeed, most RBD champions such conflict-reducing solutions which retain a state-based logic, such as the UNDP's idea of "inclusive democracy" (UNDP, 2000).

The state, democracy and accountability

The discussion of freedom and justice at the universal, national, community and individual levels raises further questions about the state. As we have seen, despite the international proclamation of universality, the institutions which oversee international law are relatively weak. This means that the onus for defending human rights claims falls largely on states (An-Na'im, 1999a). So, while various multi-leveled mechanisms exist, or have been proposed, the quality of rights depends upon the nature of the state in which the rights' claimant exists.

In the African context this clearly creates major problems in using a state-centered rights framework for securing development and justice. As An-Na'im (1999b: 22) observes "African societies appear to regard the post-colonial state with profound mistrust and have no sense of ownership of it nor expectation of protection or service from it." Such a problem is compounded when the state is simultaneously the perpetrator of rights abuses and the institution through which grievances should be aired and addressed. In most cases, then, the state in Africa remains a significant generator of human rights abuses as well as holding the key to their protection. For example, the recent report from Human Rights Watch showed that while many régimes have established Human Rights Commissions in order to secure donor support they are largely ineffective and turn a blind eye to rights abuses (*Guardian*, 23 February 2001). However, states can initiate more positive action, as An-Na'im's (1999b) contrasting discussions of Nigeria and South Africa highlight.

The Nigerian state has been a flagrant abuser of human rights. The Constitution contains wide-ranging derogation (the ability to suspend or repeal) clauses which cover most human rights, while much of the Constitution was suspended during the Abacha régime by declaring a state of emergency. Additionally, the state has suppressed *Shari'a* law, which can only be tolerated where it is compatible with the Constitution. There has also been highly flexible and wide-scale abuse of military tribunals which are outside the common law. The poor pay of judges has resulted in them extorting money from litigants in order to get a case to court while human rights lawyers have been stigmatized and threatened by the state authorities. While far from perfect, South Africa has been cited as having an innovative and purposeful approach to human rights. Not only has South Africa passed a Bill of Rights and set up the Truth and Reconciliation Commission, but it has gone furthest in providing constitutional protection for economic, social and cultural rights. For example, customary law has been made expressly subject to the non-discrimination provisions of the constitution although the implementation of this remains to be contested. South Africa has also attempted to increase access to the legal system through a re-structuring of the courts to handle special jurisdictions such as labor and juvenile cases. They have also been at the forefront of experimenting with low-cost delivery of legal services through such mechanisms as para-legal extension.

A further, and more general, problem associated with an overly legalistic and state-centered view of rights is that certain human rights abuses fall outside of the state's purview and authority. Again, this relates to the origin of rights whereby they relate to the "civil" or "public" domain, which, implicitly, meant the political space of men (Assiter, 1999). The private realm fell outside of this discourse yet it has consistently been the site of some of the worst human rights abuses. Clearly, domestic violence against women and the abuse of children are the most significant, yet they hold an ambiguous place in the human rights legislation. Since 1970, the existence of CEDAW (the Convention on the Elimination of All Forms of Discrimination Against Women) provides a commitment to breaching the public–private divide, yet its forceful application in legal systems has been hampered by the feeling that

Western feminists have hijacked gender and development issues on behalf of "universal sisterhood," which only serves to silence and marginalize the voices of Third World women (Mohanty, 1997). Additionally, as expected, national legal processes, dominated by élite men, have tended to stifle gender legislation in the name of its "un-Africaness."

The over-reliance on the legal system in securing rights has seen the opening up of supplementary practices. Maxwell (1999) highlights four of these which are echoed in the major documents from DFID and UNDP. First, monitoring at international, national and local levels can help create a culture of compliance. Second, publicity and advocacy help create political structures and policy changes in support of rights. Third, accountability can be created administratively by specifying delivery standards through such things as Citizen's Charters. Finally, as the DFID were at pains to stress, rights-based approaches can be encouraged by broadening participation and giving more people a "stake" in social decision-making.

Globalization, liberalization and structural underdevelopment

A key debate revolves around the distinction between political and economic rights. These issues have been pre-figured in such debates as those concerning the "developmental state" (Leftwich, 1994) which stresses that economic growth can only be achieved through the suppression of rights, the quashing of civil society and the denial of democracy. On the other hand, the "good governance" agenda, much like the rights agenda, argues that democratic participation is not a reward for a harsh economic transition, but is central to any definition and process of economic development. Either way, there is a clear separation between the "economic" and the "political" which allows states and agencies to focus on one or the other, despite the supposed "indivisibility" of rights. In general, the human rights discourse has privileged the political over the economic, with some going further to suggest that this is because the recognition of political freedoms is relatively costless compared to economic rights which promise tangible material inputs such as housing and health care (Sengupta, 2000). We shall return to this issue below in examining Shivji's (1999) discussion of the right to life.

This in turn means that the discourse of universal rights is relatively mute regarding global capitalism as a generator of inequality. Turner (1993: 2–3) argues that citizenship "is inevitably and necessarily bound up with the problem of the unequal distribution of resources," which is in keeping with the thrust of RBD. However, Evans (1997) is more vitriolic about the relations between global capitalism and rights. He argues that by stressing political and civil rights, the human rights discourse led by "the forces of globalization" has sidelined critical discussions about economic rights. So, while international law stresses sovereignty and self-determination, the actual operation of dependency denies the realization of these rights. He goes on to state that "unless political and economic interests are threatened, the economic imperative of globalization suggests that victims of rights abuses will be ignored" (Evans, 1997: 98). There is a danger, as with recent discourses of democracy and good governance, that by stressing the political realm as distinct from the economic, RBD not only downplays the constraints arising from structural inequalities, but does little to address them.

Nowhere is this more clear than in Africa's experience of colonialism and neo-imperialism (Maluwa, 1997). For many, the adjustment era plunged Africa into deeper dependency and more polarized poverty. As Wanyeki (1999: 104) observes, the lack of respect for the rule of law is evidenced "by the adoption and implementation of structural adjustment programs in disregard of their impact on human rights." This paradox could, charitably, be seen as a failure of "joined-up thinking" in global governance, but more realistically it reflects the neo-liberal urge to impose marketization without consideration of its social and political impacts. The pragmatic question which opens up for the development community is whether such anti-imperialist rights are achievable in the present climate of neo-liberal globalization and geopolitical governance.

Some believe a progressive agenda can and should be realized through rights-based social provision. For example, the exercise of human rights by organized labor and the insistence of "due process" has enabled unions to make political gains against global capital (Bjorn Beckman, personal communication). This opens up wider debates regarding social policy and globalization. In the past, social policy has been an important means for redistributing resources and ensuring social welfare at the level of the nation-state.

Deacon (1997) argues that globalizing forces have forced a "supranational concern" with social policy upon us even though such thinking is in its "primitive stages." In promoting the case for a global social policy, Norton (2000) argues that the freer movement of capital between nations, with capital "régime shopping" for the best conditions, encourages governments to lower standards of labor rights and labor protection. At the same time, liberalization restricts many sources of revenue previously available to fund social expenditures (trade tariffs, labor taxes), producing a "fiscal squeeze," while volatility in capital flows has been shown to lead, under some conditions, to rapidly developing crises of welfare at the regional and national level.

Given these arguments, Norton considers the policy environment to be ripe for change, pointing to the multilateral lending institutions increasing engagement in the 1990s with "classic" social policy areas of concern, such as poverty reduction and social protection. Redistribution between countries already operates at the sub-global level through EU mechanisms and Deacon argues that such programs could be expanded into international development, citing the UNDP's (1992: 78) argument for a global system of progressive income tax from rich to poor nations. However, such initiatives are still open to Furedi's accusations of "moral intrusion." Ferguson (1999) tackles this head on in acknowledging critiques of benign or progressive globalization as ethnocentric or neo-colonialist. The rejection of a raft of global social policy principles, argue its protagonists, serves the interests of class and gender élites in southern contexts as much as it irritates northern governments looking to appease their own constituents. What is stressed by Norton and others is that those with a normative position on social policy principles need to create a broad constituency for those principles in the north and south. That means dialogue and partnership rather than trying to impose measures through policy conditionality. The most encouraging signs of an emerging social agenda are those that are springing up as truly global responses to the challenges of globalization such as the north–south links underpinning the fair trade movement and the movement for debt relief.

Citizenship and social welfare

This emerging debate on global social policy raises further questions regarding what might be termed "thin" citizenship. Marshall (1964) argued that in Britain rights proceeded from civil (legal) in the 17th century, to political (parliamentary democracy) in the 18th and 19th centuries to social (welfare state) in the 20th century. While we can criticize this teleology for presenting a too simplistic view of the evolution of rights and for not specifying whether all rights are of equal importance, its greatest weakness is in failing to specify the linkages between citizenship and capitalism (Turner, 1993). In particular, citizenship rights might be seen as a radical principle of equality or, by providing checks and safety nets, simply a means of promoting solidarity and the stability required for further accumulation. This latter interpretation of rights emphasizes important concrete entitlements such as housing and clean water, but generates a citizen whose political agency is only exercised in pressing for basic needs. The focus on legally-defined welfare provision might preclude alternative trajectories, with RBD becoming another form of neo-liberal market-led development. Such a process will be exacerbated by such initiatives as GATS (Global Agreement on Trade and Services), which moves control away from local people towards global corporations in a mass privatization of welfare (World Development Movement, 2001).

A final paradox of globalization and rights relates to the relative mobility of capital and the control of people (Pettman, 1999). The twin discursive pillars of globalization and universal human rights suggest that the mobility of finance, goods and ideas is greatly enhanced and necessary for continued prosperity while all humans should have the same opportunities and be subject to the same rights. In practice, while certain forms of capital, including some types of labor, have become more mobile, states police their borders like never before. Despite pretensions to global citizenship, in beggar-thy-neighbor global capitalism, clear differentiations are made between citizens and non-citizens. So, despite a discourse which laments the inevitable waning of state power, states still retain authority to territorially define legitimate citizens with valid rights. Pettman (1999) goes on to suggest that we should press states harder to protect rights and expose the myth of powerlessness in the face of globalization.

CONCLUSION

As a creature of liberal individualism, the rights agenda tends to serve the interests of the propertied and the powerful. However, the recent emphasis on economic and developmental rights should be welcomed, because it raises the possibility of cementing the right to a decent standard of living, even if such commitments remain tentative and uneven. Clearly, as with any ideological venture led by the major international development agencies, the potential exists for the rights-based agenda to be used as a new form of conditionality which usurps national sovereignty and thereby further denies the autonomy and freedom which are a *sine qua non* for democratic development. Additionally, by handing the primary responsibility for defending rights to unaccountable and authoritarian states the process does little to challenge the power structures which may have precipitated rights abuses in the first place. Finally, the emphasis on universal rights, as defined through largely Western experiences, limits the relevance of rights to local circumstances and thereby effects yet another form of Eurocentric epistemological violence which seeks to normalize a particular and self-serving social vision. Hence, the balance sheet in favor of rights-based development, as it is currently conceived, is relatively empty. So, do we simply ignore the RBD agenda or can it be used to effect more meaningful solutions to the African crisis of development?

Both Ake (1987) and Shivji (1989, 1999) argue that any discussion of abstract rights pertaining to abstract "humans" is meaningless and unhelpful for Africa and serves only the interests of those protagonists who stand to benefit from the status quo. Shivji argues that:

> human rights talk should be historically situated and socially specific. For the African perspective this ought to be done frankly without being apologetic. Any debate conducted on the level of moral absolutes or universal humanity is not only fruitless but ideologically subversive of the interests of the African masses.
>
> (Shivji, 1989: 69)

He goes on to assert that any concept of rights in Africa must be anti-imperialist, which forces the issue of self-determination back on to the agenda, and it is, in part, for this reason that he welcomes the "new rights agenda" (Shivji, 1999). He states, "imperialism is the negation of all freedom" so that human rights in Africa "must be thoroughly anti-imperialist, thoroughly democratic and unreservedly in the interest of the people" (Shivji, 1989: 70).

Ake (1987) has been equally adamant that any notion of human rights must be grounded in the realities of the African crisis. For him, writing 15 years ago, the specter of fascism in Africa was the paramount political problem so that any articulation of human rights must be "to combat social forces which threaten to send us back to barbarism" (1987: 7). Both authors criticize the individualism of bourgeois liberal rights and assert that African societies are far more socially-oriented so that an African version of human rights must go "beyond the dominant western liberal conception as an individual bearer of rights and include a wide range of more substantive contents" (von Lieres, 1999: 140).

The question remains as to how such a process might be engendered. Mamdani (1996) concludes that more genuine citizenship must acknowledge the bifurcated state and negotiate both rights-based and ethnic identities through a "balance between decentralization and centralization, participation and representation, autonomy and alliance" (Mamdani, 1996: 298). However, beyond that he remains vague. Von Lieres (1999: 146) also sees multiple democratic "spaces" opening up following the retreat of the state, leading to

> a new dynamic model of interaction between multiple, often interdependent socio-political and cultural spaces and groups . . . [moving us away] . . . from the idea of the citizen as a bearer of rights towards the idea of the citizen as participant and claimant, embedded in a series of networks guaranteeing inclusion and preventing marginalization from wider social and political processes.
>
> (von Lieres, 1999: 146)

While she may be right to describe the African political imagination as "survivalist," such a view of political inclusion remains rather voluntaristic and denies the possibility that, in an environment of poverty and political turmoil, the "new spaces" might be filled by warlords, gangsters and other anti-democratic factions.

Both Ake and Shivji also posit alliances between diverse political communities as one means of pressing for meaningful rights. Indeed, Ake's strategic *Realpolitik* leads him to argue, despite his general skepticism towards liberal rights, that we need

> a coalition of all those who value democracy not in the procedural liberal sense but in the concrete socialist sense. This is where the idea of human rights comes in. It is easily the best ideological framework for such a coalition.
>
> (Ake, 1987: 8)

Ake rightly acknowledges that any realization of such rights will involve struggle such that empowerment cannot be handed down by development agencies or the state, but must be taken by the marginalized. Instead, we will see "a protracted and bitter struggle because those who are favored by the existing distribution of power will resist heartily" (1987: 11). As a result, he is somewhat dismissive of any institutional or procedural reforms of the type imagined in the RBD agenda (Maluwa, 1997).

Shivji (1999) is more positive about the potential role of legislation in securing a rights-based approach to development. In keeping with the RBD agenda he believes productive gains can be made by bringing together the developmental and human rights traditions. To support his case he examines the successes of "social action litigation" (SAL) in India, which has pushed for social justice as opposed to the individualistic "natural" justice enshrined in liberal rights discourses. SAL has questioned the issue of what it means to live and broadened it to include a range of rights not normally considered justiciable. In particular, a person not directly affected by an abuse of rights can still bring a petition to court if they feel that basic rights to live have been violated. And by expanding the notion of what it means "to live" to include a right to work, the dichotomy between political and ESC rights is breached. Shivji uses this to press for a new rights régime which asserts that a right to life is the most fundamental human right. However, unlike liberal conceptions, it accepts that "living" involves being part of a wider collective which may be anything from a family to a community organization or trade union.

Allied to this is this right to self-determination which applies in two senses: first, to nations within the uneven global political economy. This marks his approach out as more transformatory and radical than the current RBD approaches which see all nation-states as equals and thus covertly denies the unequal use of power by some states and TNCs. Second, self-determination applies to minorities or "nations" existing within the borders of accepted nation-states. This is not simply about recognition of these groups, but about allowing them "to determine their 'self' politically in terms of participating in major decision-making processes that affect their lives" (Shivji, 1999: 269). In this regard, Shivji is welcoming of the African Charter's inclusion of "people's" rights, so long as these are taken seriously.

Such efforts can justifiably be dismissed as drops in the ocean which will not effect major social transformation. While individual victories in favor of the oppressed and marginalized may have limited impact, the broader process keeps alive the debate about the limitations of existing political structures to deliver development while simultaneously giving credibility to those organized activities aimed at transforming livelihoods and discrediting those who oppose such activities. So, we do not believe that the rights-based development agenda, as currently constructed, will challenge the structures which create underdevelopment. Only by embedding discussions in the locally meaningful struggles that confront impoverished Africans and by promoting broader and direct participation which, crucially, promotes self-determination can a rights agenda more thoroughly promote African development.

REFERENCES

Ake, C (1987), "The African Context of Human Rights," *Africa Today* 34, 1–2, 5–12.

An-Na'im, A. (1999a) "Introduction," in A. An-Na'im (ed.), *Universal Rights, Local Remedies: Implementing Human Rights in the Legal Systems of Africa*, Interights: London, 1–21.

—— (1999b), "Possibilities and Constraints of the Legal Protection of Human Rights under the Constitutions of African Countries," in A. An-Na'im (ed.), *Universal Rights, Local Remedies: Implementing Human Rights in the Legal Systems of Africa*, London: Interights, 22–24.

Assiter, A. (1999), "Citizenship Revisited," in J. Pettman (ed.), *Women, Citizenship and Difference*, London: Zed Books.

Bebbington, A. (1999), "Capitals and Capabilities: A Framework for Analyzing Peasant Viability, Rural Livelihoods and Poverty," in *World Development* 27(12), 2021–2044.

Bobbio, N. (1996), *The Age of Rights*, Cambridge: Polity Press.

Chambers, R. (1983), *Rural Development: Putting the Last First*, Harlow: Longman.

—— (1997), *Whose Reality Counts? Putting the First Last*, London: Intermediate Technology Publications.

—— (1989), "Editorial Introduction: Vulnerability, Coping and Policy," *IDS Bulletin* 2Q(2), 1–7.

Cunill, N. (1997), *Repensando ro Publico a traves de la Sociedad*, Caracas: CLAD.

Deacon, B. (1997), *Global Social Policy*, London: Sage.

Department for International Development (DFID) (2000), *Human Rights for Poor People*, DFID Plans, London: DFID.

Dunne, T. and N. Wheeler (1999), "Introduction: Human Rights and the Fifty Years' Crisis," in T. Dunne and N. Wheeler (eds), *Human Rights in Global Politics*, Cambridge: Cambridge University Press, 1–28.

Evans, T. (1995), "Hegemony, Domestic Politics and the Project of Universal Human Rights," *Diplomacy and Statecraft* 6(3), 616–644.

—— (1997), "Universal Human Rights: Imposing Values," in C. Thomas and P. Wilkin (eds), *Globalization and the South*, Basingstoke: Macmillan Press, 90–105.

Ferguson, C. (1999), *Global Social Policy Principles: Human Rights and Social Justice*, London: DFID.

Fowler, A. (2000), "NGDOs as a Moment in History: Beyond Aid to Social Entrepreneurship or Civic Innovation?," *Third World Quarterly* 21(4), 637–654.

Furedi, F. (1997), "The Moral Condemnation of the South," in C. Thomas and P. Wilkin (eds), *Globalization and the South*, Basingstoke: Macmillan Press, 76–89.

Gaventa, J. and C. Valderrama (1999), "Participation, Citizenship and Local Governance," unpublished paper for workshop "Strengthening participation in local governance," Institute of Development Studies, University of Sussex, June 21–24.

Leftwich, A. (1994), "Governance, the State and the Politics of Development," *Development and Change* 25, 363–386.

McIlwaine, C. (1998), "Civil Society and Development Geography," *Progress in Human Geography* 22(3), 415–424.

Maluwa, T. (1997), "Discourses on Democracy and Human Rights in Africa: Contextualizing the Relevance of Human Rights to Developing Countries," *African Journal of International and Comparative Law* 9(1), 55–71.

Mamdani, M. (1996), *Citizen and Subject: Contemporary Africa and the Legacy of Late Colonialism*, Oxford: James Currey.

Marshall, T. (1964), *Class, Citizenship and Social Development*, Chicago: University of Chicago Press.

Maxwell, S. (1999), "What Can We Do with a Rights-based Approach to Development?," *ODI Briefing Paper*, London: ODI, September.

Migdal, J. (1994), "The State in Society: An Approach to Struggles for Domination," in J. Migdal, A. Kohli and V. Shue (eds), *State Power and Social Forces: Domination and Transformation in the Third World*, Cambridge: Cambridge University Press.

Mohan, G. and K. Stokke (2000), "Participatory Development and Empowerment: The Dangers of Localism," *Third World Quarterly* 21(2), 247–268.

Mohanty, C. (1997), "Under Western Eyes: Feminist Scholarship and Colonial Discourses," in N. Visvanathan (ed.), *The Women, Gender and Development Reader*, London: Zed Books.

Moser, C. (1998), "The Asset Vulnerability Framework: Reassessing Urban Poverty Reduction Strategies," in *World Development* 26(1), 1–19.

Murray, R. (2000), *The African Commission on Human and Peoples' Rights and International Law*, Oxford: Hart Publishing.

Nagengast, C. and T. Turner (1997), "Introduction: Universal Human Rights Versus Cultural Relativity," *Journal of Anthropological Research* 53, 269–272.

Norton, A. (2000), "Can There Be a Global Standard for Social Policy? The 'Social Policy Principles' as a Test Case," *ODI Briefing Paper*, London: ODI.

Penna, D. and P. Campbell (1998), "Human Rights and Culture: Beyond Universality and Relativism," *Third World Quarterly* 19(1), 7–27.

Pettman, J. (1999), "Globalization and the Gendered Politics of Citizenship," in N. Yuval-Davis and P. Werbner (eds), *Women, Citizenship and Difference*, London: Zed Books, 207–220.

Potter, R., T. Binns, J. Elliott and D. Smith (1999), *Geographies of Development*, Harlow: Longman.

Rhodes, R. (1997), *Understanding Governance: Policy Networks, Governance, Reflexivity and Accountability*, Buckingham: Open University Press.

Sano, H.-O. (2000), "Development and Human Rights: The Necessary, but Partial Integration of Human

Rights and Development," *Human Rights Quarterly* 22, 734–752.

Scoones, I. (1998), "Sustainable Rural Livelihoods: A Framework for Analysis," Working Paper 72, Brighton: Institute of Development Studies.

Sen, A. (1997), "Editorial: Human Capital and Human Capability," in *World Development* 25(12), 1959–1961.

Sengupta, A. (2000), "Realizing the Right to Development," *Development and Change* 31, 553–78.

Shivji, I. (1989), *The Concept of Human Rights in Africa*, Dakar: Codesria.

—— (1999), "Constructing a New Rights Regime: Promises, Problems and Prospects," *Social and Legal Studies* 8(2), 253–276.

Slater, D. (1993), "The Geopolitical Imagination and the Enframing of Development Theory," *Transaction of the Institute of British Geographers*, New Series, 18, 419–437.

Turner, B. (1993), "Contemporary Problems in the Theory of Citizenship," in B. Turner (ed.), *Citizenship and Social Theory*, London: Sage, 1–18.

UNDP (2000), *The Human Development Report: Human Rights and Human Development*, New York: UNDP and Oxford: Oxford University Press.

von Lieres, B. (1999), "New Perspectives on Citizenship in Africa," *Journal of Modern African Studies* 25(1), 139–48.

Wanyeki, L. (1999), "Strategic and Thematic Considerations: Proceedings of the Dakar Conference," in A. An-Na'im (ed.), *Universal Rights, Local Remedies: Implementing Human Rights in the Legal Systems of Africa*, Interights: London, 103–126.

World Bank (1990), *World Development Report: Poverty*, Oxford: OUP.

—— (1997), *World Development Report: The State in a Changing World*, Oxford: Oxford University Press.

—— (2001), *World Development Report 2000/2001: Attacking Poverty*, Washington, DC: World Bank.

World Development Movement (2001), Stop the GATSastrophe, www.wdm.org.uk/campaigns/GATS.htm.

14

New world warriors

'Nation' and 'state' in the politics of the Zapatista and US Patriot Movements

Carolyn Gallaher and Oliver Froehling

from *Social & Cultural Geography*, 2002, 3(1): 81–102

INTRODUCTION

This paper is a comparative analysis of two social movements, the Mexican Zapatista and the US Patriot Movements. While these two movements emerged in different countries and under distinct circumstances, they share a common resistance to "the new world order" (Castells 1997). Studies have identified similar movements in South Africa and Japan (Castells 1997; Fredrickson 1997).

In the emerging literature on such movements, scholars have labeled them, variously, as "anti-statist" (Kirby 1997; Luke 1997; Steinberg 1997), as "social movements against the new global order" (Castells 1997), and as "ethnonationalists" (Connor 1993; Smith 1993). Despite the diversity within and between these movements, certain characteristics hold constant across them. First, while these movements oppose globalization, they tend to focus the lion's share of their criticism against their own governments for ushering in its "reforms." This does not imply that such movements have benign views of multilateral organizations and treaties (they do not), but in their rhetoric the state is positioned as the clear enemy. Secondly, while these movements target the state as the enemy, they frame their actions against it in nationalistic terms, invoking traditional symbols of nation to frame their respective causes. In short, these movements have discursively delinked the "nation," a category of meaning, from the state, a governing apparatus, in order to attach the signifier of nation to smaller scales of political control (Steinberg 1997).

Such movements have also shown a willingness to take up arms to further their efforts. On 1 January 1994, for example, while Presidents Salinas and Clinton were celebrating the inauguration of the North American Free Trade Agreement (NAFTA), the Zapatista Army of National Liberation (EZLN) emerged from the jungles of the Lacandon Forest to take over several towns in the southern state of Chiapas. Their battle cry, *Ya Basta!*, or enough already, signaled their discontent not only with NAFTA, but with the previous decade of neoliberal reforms that have undermined the collectivist structure of indigenous village life. These ski-masked rebels were not, however, a leftist guerilla movement intent to overthrow and then replace the existing state, as Castro's guerillas did in Cuba and the *Sendero Luminoso* attempted to do in Peru. Their demands to the government for basics such as housing, electricity and water called for the reconstruction of the Mexican "nation" in opposition to a corrupted Mexican state (Harvey 1998). In the recent 2001 *Zapatour*, in which the rebels marched to Mexico City to present their demands to the Mexican Congress, their eloquent spokesman Subcommandate Marcos told the crowds assembled in the capital's *zócalo* that *autonomía es integracíon* (autonomy is integration). As he explained in a subsequent interview in the political weekly *Proceso* (2001), the only way Mexico's indigenous people may become a functioning part of the nation is for the state to allow them to exist autonomously *within* it.

A comparable movement would appear north of the Rio Grande on 15 April 1995 when Timothy McVeigh, a decorated Gulf War veteran, parked his rented Ryder truck in front of the Murrah federal building in Oklahoma City, Oklahoma. It exploded a few minutes later, destroying the building and killing 169 people. While the government holds that McVeigh developed his plot in relative isolation, his actions were clearly fostered within the burgeoning Patriot Movement (Dees 1996). In the patriot circles McVeigh frequented, from militia meetings in Michigan to patriot expos in Arizona, he was taught that the US government had been co-opted by the new world order and that patriots must prepare to defend the nation in the upcoming war with the government and its international forces (Michel and Herbeck 2001). In the years since the bombing patriots have developed sophisticated networks to further their cause. Through underground operations, such as bombings and assassinations, and above-ground operations, such as rallying for local control ordinances, patriots fight to erase the power and influence of the government in their daily lives. Yet like the Zapatistas, they claim their actions are on behalf of the nation. They call themselves constitutionalists and patriots, and they dig deeply into the writings of the "founding fathers" to defend their position that all power is local.

The purpose of this paper is to analyze and evaluate the projects of "national" resistance to the state these movements offer. In doing so we follow the lead of Kirby (1997), who argues that analyzing "anti-statist" movements can tell us much about the place of the globalized state in our everyday lives, and Steinberg (1997), who has called on scholars to deepen our understanding of such groups through comparative work. For our part, we hope to contribute to this emerging debate in two ways. First, we heed Steinberg's advice by undertaking a comparative analysis of two anti-statist groups, the Mexican Zapatistas and the US Patriot Movement. We choose these particular movements because several scholars have indicated that their comparison would be compelling (Castells 1997; Murray 1998), and because we have conducted prior research on them (Froehling 1997, 1999; Gallaher 2000). Second, we hope to contribute to this debate by proffering a framework for evaluating the political content of such "national" projects in terms of their ability to offer progressive political change.

In this paper we argue, as Kirby does, that the claims of these movements should be evaluated "at face value." As such, while both movements may be termed nationalistic, and thus dismissed as anachronistic, dangerous and/or insupportable, we argue that the category of nation is constructed (rather than essential) and may take any number of forms, from the liberatory to the repressive. We hold that the best way to assess nationalistic projects is to assess whether antagonism infuses the construction of the category of nation and/or its categorical and spatial policing. Following Laclau and Mouffe (1985), we argue that radical democracy is found not in the elimination of "difference" but in rooting out and domesticating antagonistic identity formation—where difference is regarded as threatening and worthy of destruction. Through our empirical analysis we illustrate that *Zapatismo*, though not without its problems, is "agonistically" constructed—that is, difference is acknowledged but not viewed as threatening—creating a plural space for nation (re)building. In contrast, we demonstrate that the *Patriotism* espoused by the militias is antagonistically defined, fostering an exclusive view of nation and a rigid policing of its categorical and material borders.

The remainder of this paper is organized in the following manner: The paper begins by establishing the comparability of these two movements and the criteria we use to evaluate them. The second section outlines a brief history of each movement so that readers not familiar with them have the necessary context for understanding our argument. The third and fourth sections of this paper examine, respectively, the Zapatista and Patriot "national" projects. The final section proffers tentative conclusions about what these movements tell us about the nation and the state under neoliberalism, and suggests areas for further research.

ESTABLISHING COMPARABILITY, EVALUATING DIFFERENCES

In this section we lay out the theoretical framework we use to compare these two movements and to evaluate their respective projects. While establishing careful theoretical grounds for comparison is always necessary, it is especially important in the case of the two movements we have chosen to analyze. These movements both tend to invoke passionate, even visceral responses that make both easy "targets" for careless analysis. In the case of the Zapatista Movement, commentators have tended to romanticize its ski-masked rebels. Indeed, an almost cult-like following has developed around the Zapatistas. This is especially the case for their charismatic spokesman, Subcomandante Marcos, whose image adorns pins, curios and T-shirts, demonstrating that surely this revolution is "for sale." Such hero worship is not only incommensurate with the rebels' message, but tends to gloss over the movement's complexities and contradictions (Subcomandante Marcos 2001). In the case of the US Patriot Movement, observers routinely castigate its adherents for being "cracked and loaded," "malcontents" and "angry white men with guns" (Buchwald 1996; Junas 1995). While such labels may crudely capture the profile of some patriots, they clearly fail to explain the widespread resonance the patriot message has for various segments of the population. By establishing a clear framework for their comparison, and their evaluation, we hope to avoid these pitfalls, focusing instead on the place of such movements between the nation and the state.

There is a wide body of scholarship on the Zapatista Movement (Collier 1994; Harvey 1998; Katzenberger 1995; Pena 1995; Rajchenberg and Heau-Lambert 1998; Womack 1999) and a growing body of literature on the Patriot Movement (Aho 1990; Dyer 1997; Gallaher 2000; Stern 1996; Stock 1996). These studies come from a wide variety of disciplines, including sociology, history and geography, and interdisciplinary studies such as new social movements theory. To date, however, most of this literature has examined these movements in separation from one another, and has thus failed to articulate the larger common context out of which they arise. There is, however, a small but growing body of literature emerging to study the common roots of our case studies and movements like them.

Of particular importance is the work of Manuel Castells (1997), who argues that groups like the Zapatistas and the Patriots represent a new breed of "new social movements." He labels these groups "social movements against the new global order" and argues that what these otherwise diverse groups have in common is their opposition to global flows of capital and information, and the vertigo they create. As Castells notes, however, while these movements are reacting to globalization, they have focused their protest against their own state governments for allowing it to happen. In the discourse of such movements, the *state* (the government) is generally regarded as betraying the *nation* (the people) (Castells 1997).

Political geographers have also recently focused their attention on the emergence of such movements, although to date most of their work has focused on movements in the USA. In 1997, for example, a special edition of *Political Geography* was organized to analyze the question of organized violence against the state (Kirby 1997; Luke 1997; Steinberg 1997; Tabor 1997). Its lead author, Andrew Kirby, queried, "is the state our enemy?" and concluded with a tentative "yes." As Kirby notes, while we may be uncomfortable with such violence, the emergence of "anti-statist" groups is predicated on a continuation of state control over the means of violence, and on its growing use in the domestic arena (1997).

While the exchange between Kirby and his responders is not without disagreement, the dialogue resulted in several theoretical points of convergence. First, although political violence against the state has existed for as long as the state has, its current resurgence is rooted in the pressures attendant with the globalizing political economy. Indeed, while the global reach of capitalism is nothing new, and is in fact associated with the initial rise of the nation-state in the first place, its current form has significant differences from previous manifestations, especially as it concerns the state. The initial consolidation of the European nation-state was marked by the ability to structurally control and regulate the circulation of capital at larger scales of accumulation (Giddens 1985; Tilly 1992). The capacity of the state to accumulate capital led to its ability to reinvest it, and eventually to redistribute some of it to the citizenry. By the 1930s the redistributive powers of the state were well consolidated, with the modern welfare state guaranteeing social privileges to the citizenry, such as land for peasants in Mexico and price supports to consumers and producers in the USA. In contrast, today's global extension of capital is marked by the accumulation of capital by corporations, which operate within, but increasingly across, state apparatuses. These changes, brought on in part by states themselves, involved a significant paring down of the modern welfare state in both countries (see Cockcroft 1998; Harvey 1996).

The unpopularity of restructuring, especially among certain segments of the population, was clear in the rhetorical justification governments used to "sell" it to the citizenry. Margaret Thatcher's now famous quip "there is no alternative" set the rigid tone governments would use when negotiating (or refusing to negotiate) with labor and producer constituencies (Harvey 1996, 2000). The harsh rhetoric, when coupled with the loss of important social guarantees, made the state an understandable target of frustration. In Mexico these reforms would anger peasant communities, especially those in the South, whose access to government largess has always been limited, and their effects would prove fertile ground for the Zapatista Movement to emerge (Collier 1994). Likewise, in the USA, factory workers and farmers alike would see their social and economic gains lost to the harsh logic of reforms, and they would later prove to be the movement's earliest and most loyal advocates (Abanes 1996; Dyer 1997).

Secondly, while neoliberal reforms weakened the ability of the state to provide social guarantees, they would not diminish its traditional control over the means of violence. Indeed, during the 1990s governments poured increasing amounts of money into militarization efforts that were focused domestically. In the case of Mexico this entailed an escalation of the US-backed drug war internally. In Chiapas, for example, many peasants saw their coca and marijuana crops (bound for US markets) destroyed by spraying operations encouraged and partially funded by the USA. The government also militarized the southern border with Guatemala and key transportation routes across Southern Mexico (Castro Apreza 1999). A similar fortification occurred in the USA when the government militarized domestic agencies such as the Bureau of Alcohol, Tobacco, and Firearms (BATF) and the Drug Enforcement Agency (DEA) in the early 1990s, and coordinated efforts between them (Kirby 1997). Patriots (and others) saw the death of civilians at both Ruby Ridge and Waco (where the BATF, the FBI and the US Marshals Service worked in tandem) as the outcome of such co-ordination.

Finally, while such movements emerged in quite different national contexts, a common discursive framework structures their projects. While these groups protest their respective governments for permitting and in some cases expanding neoliberal reforms, each frames its protest as on behalf of the nation, and invokes traditional symbols of "nation" to legitimize its actions. The Zapatistas have elevated the use of such symbolism to an art form. Pena (1995), for example, has called Marcos the "Subcomandante of performance" for his sophisticated use of revolutionary history not only to ground the rebels' concerns but also to sell them to the wider public. Patriots also have borrowed liberally from American revolutionary history to frame their cause and justify their existence. They sell the Bill of Rights and the Constitution at their meetings, they decorate their meeting halls with portraits of the founding fathers, and most can defend any patriot position with a quote from George Washington, Patrick Henry or Thomas Jefferson (Gallaher 2000).

The liberal use of such symbolism is, however, more than tactical. Indeed, Steinberg argues that what is extreme about such groups is not that they enact violence but that they have divorced "nationalism from the state" (1997: 16). In so doing they have chosen to reconstitute national identity, and to ground it spatially at a "smaller, more controllable scale" (1997: 16). It is here that the political projects of the Zapatistas and the Patriots become comparable. Both groups have taken a national identity and argued that it can be understood at and through smaller spaces of governance and culture. In the case of the Zapatistas, its rebels have argued that the only way for Mexico's indigenous population to truly become a part of the nation is to grant its various groups political autonomy within it. Only then can they exist rather than merely survive. For the Zapatistas, Mexico is not one nation, but many nations within one (Subcomandante Marcos 2001). In the US, Patriots have also argued that nation can be deployed at scales smaller than the federal state. Most Patriots, for example, believe that the county is the locus of power and that its laws trump state and federal laws (Abanes 1996; Dyer 1997).

While the situation of such movements' projects between the nation and the state makes them comparable, there has to date been no agreement on how, or even whether, to evaluate the political content of their projects. For his part, Castells resolutely refuses evaluation. As he argues, "there are no 'good' and 'bad,' progressive and regressive social movements" (1997: 3). To give them such labels, he concludes, is imprudent because we can neither know nor predict "where" their projects will lead.

Most scholars, however, agree that evaluation is important. The leaders of anti-statist movements have constructed an alternative trajectory for consideration and have mobilized thousands behind their respective causes. To refuse to consider the trajectory posited by their alternatives, even when acknowledging its potential for change, is a bit like an ostrich sticking its head in the sand. Among those willing to evaluate their projects, however, no consensus has emerged on how best to do so. Most of the literature on anti-statist movements has tended to evaluate these movements in a broader context, examining what they can tell us about the role of the state in a globalizing political economy. Focusing on the Patriot Movement, for example, anti-statist scholars have argued that while patriots frequently espouse a "politics of hate," they are also responding to a militarization of domestic agencies that undermine basic civil liberties, such as freedoms of speech, religion and association (Kirby 1997). Sparke argues that while McVeigh's actions were horrific, they were "ideologically underpinned by the very same inside/outside analytics" long espoused by the US government in its actions abroad (1998: 2). And, Luke (1997) has argued that such movements reflect a weakened state no longer willing or able to grant social guarantees. Other scholars have argued that we should evaluate these movements by analyzing the social positionality of their constituents. Steinberg, for example, argues that the form such movements take is due in large part to the social positionality of their key members. As he notes,

> To fully appreciate both the progressive and regressive tendencies in these groups (and many groups contain elements of both progressive and regressive politics), one must treat their politics not as incidental but as lying deep within their members' social positions.
>
> (Steinberg 1997: 15)

We agree that the social positionality of anti-statists is crucial to their form, and when such groups are considered in isolation from one another, and within their national contexts, is a necessary ingredient for analysis. In a comparative context, however, we hold that focusing on social positionality confounds rather than clarifies analytic comparison. At a practical level, the radical differences in the social positionality of Zapatistas and Patriots makes them difficult to compare. Zapatistas are indigenous peasants who live on the brink of survival. Patriots are white farmers and workers with a well-developed social security system to rely on in times of need.

At a more fundamental level, however, we also argue that social positionality does not guarantee a certain social trajectory. In the case of Chiapas, we might reasonably assume that being a peasant predisposes one to support the Zapatista cause. However, many peasants in Chiapas oppose the Zapatista uprising. Some of this opposition may be tracked to fear of reprisals from the government, but studies suggest that extant divisions within indigenous communities also play a role (Collier 1994). In the case of the USA, commentators routinely assume Patriots have taken up arms because they feel their dominance as white men is under question (Dees 1996). Yet a closer look reveals that most Patriots occupy conflicting social positions. As white males in American society Patriots are the beneficiaries of social privilege, yet as members of the working class, they suffer the harsh logic of neoliberal economic policies. Their mobilization, therefore, may take progressive or regressive form, and depends on how activists tap into their contradictory social positions. That the right has effectively mobilized them does not mean that the left could not do the same, and to better cause (Gallaher 2000).

Given the indeterminacy of positionality to social action, we propose that these groups are better compared and evaluated by what they hold in common—their attempts to delink notions of nation, as a category of identity, from the state, a governing apparatus, and their desire to reattach the nation to smaller scales of place. As such, we argue that their evaluation should focus on *how* each group (re)defines the category of nation in their politics and how they discursively suture it to smaller spaces of governance.

To frame our analysis of anti-statist projects we use the work of Ernesto Laclau and Chantal Mouffe (1985) and geographers who have spatialized their work (Massey 1995; Natter 1995; Natter and Jones 1997). Their work is relevant to our comparison because it asks us to consider how categories of identity, and the politics that derive from them, may create radical democracy (as opposed to democracy defined merely by the ability to vote).

Several key themes arise out of this work that are directly relevant to the case at hand. First and foremost, these scholars recognize that categories of identity are constructed rather than essential in nature. Social identities are given meaning by their opposition to an "other." Secondly, because all identities require an "other" for their existence, we must ask how social identification may be constructed in such a way that "others" may co-exist peacefully and fairly under extant forms of governance. Such a proposition may be considered radical because it represents a break with standard liberal politics that seek to erase differences, as illustrated, for example, in liberal calls to create "a color blind society." Rather, these scholars argue that the goal for radical democracy should be to root out antagonistic forms of identification, where the "other" is considered dangerous and in need of extermination, and actively work to "agonize" them. Indeed, as Laclau and Mouffe note, "agonizing" or domesticating politics entails creating a context where articulating difference(s) is seen as *crucial to* rather than *dangerous for* the political whole (1985). This recognition is especially important for the analysis of nationalism. Antagonistic forms of nationalism are generally buttressed spatially as proponents define spaces as belonging to a given group and actively root out "others" for extermination or deportation (Natter and Jones 1997).

Given this history, it is not surprising that traditional critical theory has viewed nationalism as dangerous, and has called for vigilance against all its forms. Post-structural identity scholars argue, however, that the nation is, like any category, constructed, and as such may take antagonistic or agonistic form. As Mouffe notes,

> It is very dangerous to ignore the strong libidinal investment that can be mobilized by the signifier "nation"; and it is futile to hope that all national identities could be replaced by so-called "post conventional" identities. The struggle against the exclusive type of ethnic nationalism can only be carried out by articulating another type of nationalism, a "civic" nationalism expressing allegiance to the values specific to the democratic tradition and the forms that are constitutive of it.
>
> (Mouffe 1995: 264)

As such, rather than dismissing our case studies as unsupportable by virtue of their invocations to nation, we propose, instead, to evaluate both the way each movement constructs the idea of nation and the way each polices its categorical and spatial boundaries, searching for potential antagonisms therein. The presence of antagonism serves, therefore, as our evaluative guide, allowing us to label these groups indicative of, or contrary to, radical democracy.

We now turn to a brief historical overview of each movement so that readers unfamiliar with them have a context in which to place the coming analysis. In the interest of space, however, we only lay out the broad political and economic context in which these movements emerged in order to illustrate their situation within globalization, and their reason for targeting the state as "enemy." As such, we do not discuss the mobilization of these two movements. We note, however, that the initial mobilizers of each group (radical university students in Chiapas, white supremacists in the USA) as well as the social positionality of each group's constituent base clearly influenced the form each movement's identity politics takes today (Dyer 1997; Harvey 1998).

ZAPATA RESURRECTED

While the Zapatistas timed their initial uprising to coincide with the signing of NAFTA, the processes that led up to the Zapatista uprising may be traced to two intertwined processes: long-standing racism and discrimination against indigenous populations, and macro-economic shifts in capital accumulation beginning in the 1970s and their negative effects on indigenous groups in Chiapas (Cockcroft 1998; Collier 1994; Esteva 1994; Harvey 1998). On the whole, these changes served to undermine collective forms of social and economic organization characteristic of indigenous villages in Southern Mexico. They were particularly devastating, however, to the integrity of the *ejido*, the communal land tenure system practiced by indigenous villages for centuries, and given constitutional protection in the 1917 Constitution as an appeasement to the original Zapatistas of the Mexican Revolution.

In 1972 OPEC launched its first oil embargo. Within a ten-year period the price for crude oil would skyrocket almost twenty dollars a barrel (Collier 1994). With established wells, and known reserves yet untapped, the Mexican government decided to develop its oil for export industry. At the time, Mexico's prospects for expansion were good: the USA was anxious to find oil outside of OPEC's control, and financial markets, awash in petro dollars, were happy to provide the capital to finance it (Cockcroft 1998). Most of the new wells were tapped in Chiapas and neighboring Tabasco, and young, indigenous, mostly male peasants soon flocked to the oil fields in search of work.

While jobs in the oil fields were a boon to the peasants who obtained them, they tended to create divisions within indigenous communities. As Collier (1994) notes, those left behind saw their productive capacities decline. They also saw their relative wealth diminish because peasants who entered the cash economy were able to build concrete houses for their families, to purchase consumer goods, and even to buy the family plots of other villagers and hire them to work them. Such sharecropping arrangements on *ejidos* had been virtually unheard of before the oil boom.

When the oil industry went bust in 1982 many peasants returned to their lands. They did so, however, with relative advantage (Collier 1994). Returning peasants had cash, trade skills and the ability to speak Spanish. Some peasants returned to agriculture and purchased green revolution technology, which increased their relative wealth. After 1982 illegal share-cropping-style relationships between peasants also increased. Other returning peasants used their skills to start trucking enterprises, allowing them to remain in the cash economy and further differentiating them from those tied to the land (Benjamin 1989).

Government policies also contributed to exacerbating tensions, although they tended to divide community against community rather than separating communities from within. As Womack (1999) notes, after the debt crisis of the 1980s government development aid to Chiapas steadily increased, yet its distribution tended to favor peasant communities aligned with the ruling party, the PRI. As such, while aid was on the rise, its unequal distribution had the effect of creating an even stronger peasant underclass in Chiapas as non-PRI aligned communities saw their wealth decline in relative terms.

Perhaps most important to the rebellion's emergence was President Carlos Salinas' decision in 1992 to rewrite Article 27 of the Mexican Constitution. When it was initially written, Article 27 granted constitutional protection to *ejidos*, forbidding their sale, or the sale of family plots within them. The "reforms" of Article 27, however, legalized the sale of *ejidos*. For the poorest peasants, this "liberalization" meant not only that foreign interests could now purchase large chunks of indigenous lands for speculation of natural resources, but more importantly that on a smaller scale peasants could also do the same. The illegal share-cropping arrangements that were already occurring on *ejidos* were now legal and set to increase. Likewise, PRI-aligned communities, who had benefited most from government largess, were in a better position to buy out neighboring *ejidos*. Landless communities, on the other hand, were left at a comparative disadvantage with no legal title to the land they occupied and no chance for it in the future.

The divisions within the indigenous population of Chiapas have a long history. Neoliberal reforms, beginning with the oil for export expansion, however, increased the intensity of the divide, giving its existing political character a distinctly economic edge (Harvey 1998). The decision by the Zapatista Committee of Clandestine Indigenous Revolution to order its military branch, the EZLN, to revolt, then, was based on the trajectory that neoliberal reforms would take already divided indigenous communities (Subcomandante Marcos 1994a). It was read as the proverbial nail in the Revolution's coffin: its goals of land reform and the inclusion of indigenous peasants into the economic body of the nation were over in real *and* symbolic terms.

OUT OF THE RUBBLE

When the Murrah Federal Building was bombed in 1995, most Americans had never heard of the Patriot Movement. It seemed, literally, to emerge out of the rubble. As a variety of scholars note, however, the Patriot Movement had been brewing for almost twenty years (Dyer 1997). The most comprehensive accounts of the Patriot Movement situate its emergence during the farm crisis of the 1980s. The story begins, however, a decade earlier, when the US government enacted policies designed to modernize US agriculture by encouraging farmers to expand their acreage and use of technology. And, lest the message go unheeded, Nixon's Secretary of Agriculture, Earl Butz, issued the blunt warning to farmers to "get big or get out" (Davidson 1996: 15). These policies, though domestic in nature, represent the beginning of neoliberal reforms in the US agricultural sector. They were designed to restructure American agriculture in order to benefit from economies of scale in the global market, while simultaneously leveling its growing trade deficit, due in part to the OPEC oil embargo (Davidson 1996; Dyer 1997).

To facilitate expansion, the Farm Home Administration (FHA), a division of the Department of Agriculture, began offering floating interest loans to farmers. To ensure the program's success the government set minimal qualification standards for borrowers. Interest rates were also low at the time, which further encouraged borrowing. As farmers tried to outbid one another to acquire more land, however, the value of farmland became artificially inflated. To keep expansion going, lenders simply encouraged farmers to borrow more, and they made it easy to do so (Davidson 1996).

The crisis hit in 1979 when Paul Volcker, then chairman of the Federal Reserve, raised interest rates sharply to curb inflation. The result for farmers was devastating. Because FHA loans had floating interest rates, farmers witnessed sharp increases in their monthly payments (Dyer 1997). Some farmers saw their interest rates climb as much as 6 percent in four years (Davidson 1996). Rising interest rates also had the effect of bursting the property bubble created during the "land grabs" of the 1970s. After the bubble burst, many farmers held mortgages worth more than the current value of their land. In Iowa alone the value of the state's farmland dropped 63 percent (Davidson 1996). The combination of sky-rocketing interest rates and collapsing property values set the stage for widespread farm failure. By 1987 over one million family farms were lost to foreclosure (Dyer 1997).

Reactions were varied. Many turned their anger and grief inward. Suicide surpassed farm accidents as the number one cause of rural death (Dyer 1997). Others turned their anger outwards, and towards the government (Abanes 1996). They formed citizens militias, designed initially to protect farms on foreclosure day. They also conducted "seminars" to teach farmers how to delay foreclosure by filing illegal liens (Dyer 1997). And they constructed elaborate conspiracy theories to explain why the government would abandon its farmers. Borrowing from well-established right-wing discourses, they explained that the government had been overtaken by Jews who intended to seize control of US land and labor, and they nicknamed the government the Zionist Occupation Government, or ZOG for short (Diamond 1995). While such theories were racist, and such measures rarely worked, it is not difficult to see why farmers would target the government. The government had encouraged them to acquire irresponsible debt loads, but did little to protect them when the crisis hit (Davidson 1996).

The bad will engendered by the government during the farm crisis provided a ready context for patriot interpretations of its actions several years later at Ruby Ridge and Waco.[1] Militias saw government actions as purposeful assaults against the citizenry rather than mistakes borne of poor judgment and planning (Dees 1996; Mozzochi 1995). Such interpretations, when coupled with white working-class anger over job loss and their status anxiety in the wake of globalization, proved fertile ground for the formation of hundreds of new militias (Junas 1995). Given its focus on republicanism, however, the movement lacks vertical integration. There is, therefore, no one key organization such as the EZLN that represents the movement. Nonetheless, Patriots recognize one another across these divides by their commonly held belief that the federal government has been hijacked by "outside" forces who will take away their land, livelihood and traditions. And while those in the movement often disagree on the methods necessary to solve the "problem" of government, most believe it will ultimately come to violence, and are arming themselves in preparation for it (Abanes 1996; Dees 1996).

ZAPATISMO

To name their movement the leaders of the EZLN invoked Emiliano Zapata, a popular hero of the Mexican Revolution. The choice of Zapata to name their cause is significant. Like those who invoke his name today, Zapata was a poor peasant born to *campesino* parents. Yet like the indigenous communities who claim his name today, Zapata saw himself as a part of the Mexican nation. *Zapata* fought for Mexico and a place within it for peasants and Indians. By invoking his name, the movement places itself within Zapata's legacy, while highlighting the unfinished nature of his revolutionary project. As such, while the Zapatistas denounce their dire material circumstances, they place not only the responsibility for them, but also the solutions to them within a national context. As Subcomandante Marcos noted early on in the struggle,

> *Mexican* brothers and sisters, we are a product of 500 years of struggle: first against slavery, then during the War of Independence against Spain . . . then to avoid being absorbed by North American imperialism, then to promulgate our constitution and expel the French empire from our soil . . . and the people rebelled and leaders like Villa and Zapata emerged, poor men just like us.
> (Subcomandante Marcos 1994a; emphasis added)

Indeed, the Zapatistas are adamant that their struggle is a national one, and they decry detractors and sympathizers alike who define them as separatist. In an open letter written a year after the rebellion began, for example, Subcomandante Marcos belittled the Mexican government for seeking to dismiss the

Zapatistas as the latest in a long line of insurgencies tied to "anachronistic" Marxist ideology. They have, he argued,

> denied us our fundamental essence: The national struggle. For them we continue to be provincial citizens, capable of a consciousness of our own origins and everything relative to it, but incapable without "external" help of understanding and making ours concepts like "nation," "homeland," and "Mexico."
>
> (Subcomandante Marcos 1994b)

As the insurgency has matured, the rebels have continued to hammer home the assertion that their cause is national rather than separatist. In March 2001, for example, Comandanta Susana, a member of the Clandestine Indigenous Revolutionary Committee, used a portion of her speech at the celebration for the International Day of the Rebel Woman to reiterate the point:

> It's not true that we want to separate from Mexico. What we want is for them to recognize us as Mexicans, as the indigenous we are, but also as Mexicans, since we were born here, we live here.
>
> (Irish Mexico Group 2001)

While the Zapatista identity is given form through the boundaries of the Mexican nation, and the promises of its revolution, these boundaries are not antagonistically constructed. Indeed, the ability of the Zapatistas to survive is due in large part to their ability to cultivate links with other oppressed peoples across the world. They have struggled to create an identity politic that lives within the national but flows easily across its categorical bounds. Relying on *communiques* issued by Subcomandante Marcos, who is at turns poetic, witty and acerbic, and the internet to spread them, the Zapatistas have developed a global network among those resisting the social and economic policies of neoliberalism (Castells 1997; Froehling 1997, 1999). The Zapatistas have, for example, hosted delegations from the Kensington Welfare Rights Union, an organization of inner-city women in Philadelphia, Pennsylvania, demanding economic rights for those hurt first by globalization and later by welfare reform. They have also hosted indigenous delegations from other Latin American nations. In 1999 Subcomandante Marcos even wrote a letter to US prisoner Mumia Abu Jamal expressing solidarity with his fight against the prison industrial complex in the USA. Such discursive extensions have rightly led observers to remark that it is the Zapatistas' war of words, fought on the global stage, that has kept these otherwise under-armed rebels and their cause afloat.

The balaclava, or mask, worn by Zapatistas has come to embody the porousness of their categorical struggle. In his film documenting the Zapatista struggle, Saul Landau (1996) asks Subcomandante Marcos to discuss the man behind the mask. Landau is curious to know who Marcos is in familiar terms—where he is from and who his family is. Shrugging in response, Marcos replies that who he is, is not important. Marcos, he informs Landau, is not an individual. Rather, he is many people. Looking intently into the camera, he says,

> Marcos is gay in San Francisco, black in South Africa, Palestinian in Israel, Jew in Germany, pacifist in Bosnia, woman alone in the metro at 10 p.m., campesino without land, unemployed worker—Marcos is all the minorities saying enough!
>
> (Landau 1996)

While the Zapatista identity is global in reach, it has clear spatial boundaries. As detailed above, the Zapatistas place their particular struggle, and solutions to them, within the bounds of the Mexican nation. Their choice of nation as the categorical boundary for their struggle, however, is mediated by their recognition that traditional calls to national unity by the Mexican state have long allowed Chiapas to serve as an internal colony for the rest of the country (Subcomandante Marcos 1992). Subcomandante Marcos (1995) is clear that Zapatista invocations to nationalism do not represent a return to this past. Indeed, in an early letter to the Mexican newspaper *La Jornada* Subcomandante Marcos' *alter ego* Durito, a cartoon beetle, even makes light of the Zapatistas' use of the national to frame their cause:

> Vale, health, and know that for love, a bed is only a pretext, for dance a tune is only adornment. And, for struggle, nationalism is merely a circumstantial accident.
>
> (Subcomandante Marcos 1995)

While such a pithy statement is a common feature of Marcos' (and Durito's) *communiques*, it represents more than mere wordplay. Rather, it indicates a strategic moment in the Zapatista identity politic. The nation is a social and spatial fact, the result of a historical process beyond Zapatista control, yet like any category its deployment is open to anyone and for any cause. By invoking nation the Zapatistas remake it as their own. And, by invoking it, rather than signifiers such as "indigenous," the Zapatistas impel the state to consider the negative effects of neoliberal economic policy *across* social categories.

Yet, because of the historical place of the indigenous within traditional discourses of nationalism (i.e. Indians must abandon their culture for the national one), the Zapatistas have made indigenous autonomy within the Mexican state a central part of their demands as they have re-opened peace negotiations with the new government of Vicente Fox. In order to reopen dialogue with the new president, the Zapatistas issued several conditions. They called for the release of Zapatista prisoners from jail; they insisted that the government close down military bases opened in Chiapas after 1994; and they called on the government to honor the 1996 San Andreas Peace Accords, which outlined a plan for indigenous autonomy. To date, President Fox has made significant concessions to the Zapatistas' demands, but the real decision on the implementation of the San Andreas Accords lies with the Mexican Congress, and until they are approved the Zapatistas have refused to disarm.

While the Zapatistas are unyielding in their demands for spatial autonomy organized around indigenous difference, their insistence is not based on antagonism. Indeed, the Zapatistas do not wish to banish the federal government from indigenous territory completely (something, as we note momentarily, the militias hope to do). Rather, they argue that the only way that indigenous communities can interact with the federal government is to first establish an equal footing between them. And, this equal footing may only be developed if indigenous communities are given the space and the freedom to rebuild collective forms of decision making, governance and economic subsistence integral to indigenous ways of life. Yet this process does not involve the elimination of the nation-state. Rather, as Subcomandante Marcos noted in a recent interview with *Proceso*, "we believe that Mexico should reconstruct the concept of nation." The new nation, he continued, must be "rebuilt on different foundations, and these foundations are based in the recognition of difference" (2001: 12). Indeed, Subcomandante Marcos sees the reconfiguration of nation not only as central to its survival, but also key to the domestication of political antagonism across the globe. As he argued in the same interview:

> When we suggest that the new century and the new millennium are the millennium and the century of difference, we are noting a fundamental rupture with respect to the twentieth century: The grand battle between hegemonies. The ultimate thing we remember between the socialist camp and the capitalist one is that it occasioned two world wars. If this is not recognized, the world will terminate into being an *archipelago* in continuous war, from outside and inside territories.
>
> (Subcomandante Marcos 2001: 12–13; authors' translation)

For the Zapatistas, then, the harbinger for peace, whether in Mexico or outside of it, lies in the recognition that the nation-state is not the space of a homogenous citizenry but instead a space built on and through the recognition of difference.

While we argue that the Zapatista struggle represents a domesticated form of nationalism, it is not without potential problems. As a variety of scholars have noted, indigenous decision making at the village level (known as *usos y custombres*) often excludes women, rendering them second-class citizens within village life (see, for example, Nader 1991). Support for indigenous autonomy could, therefore, support antagonisms inherent to traditional indigenous political forms. Based on recent comments by Zapatista women, however, we believe that their embrace of indigenous autonomy is constructed on the premise that autonomy will provide a sphere for transformation rather than attempt to "fix," in time and space, the meaning and form of indigenous identity and politics. Comandanta Esther's words to the Mexican Congress on 28 March 2001 urging support of the San Andreas Accords illustrate this point well:

> We know which are good and which are bad *usos y custombres.* The bad ones are hitting and beating a woman, buying and selling, marrying her by force against her will, not being allowed to participate in assembly, not being able to leave the house . . . [yet] [o]ur rights as women are also included in this law [San Andreas Accords], so that no one will any longer be able to prevent our participation, our

> dignity and safety in any kind of work, the same as men.
>
> (Comandanta Esther 2001)

As such, while the Zapatistas harken to a traditional form of decision making with its own set of problems, they are engaged concurrently in changing problems associated with it.

PATRIOTISM

In the aftermath of the Oklahoma City bombing, the press widely condemned the Patriot Movement. They blamed the movement for creating a monster and they queried what should be done to eradicate its politics of hate. In response, the Patriot Movement moved quickly to defend itself. Leading the effort were the most organized militias at the time, the Michigan Militia and the Militia of Montana. Just days after the bombing, Norman Olsen, then commander of the Michigan Militia, issued a public statement repudiating reports that McVeigh had been a member of his militia, although he added that he could understand McVeigh's frustration and anger. John Trochmann, co-founder of the Militia of Montana, also issued a public statement denouncing the bombing. That same year, both men also testified at a hearing of the Senate Subcommittee on Terrorism where they again condemned the violence in Oklahoma.

While the militia leaders assembled for the congressional hearing were intent to distance themselves from McVeigh, they were also determined to affirm their role as a national one. In the wake of government strong-arming at Ruby Ridge and Waco, they argued that militias were citizens' last defense against a government all too willing to overstep its constitutional bounds. And, like the Zapatistas, they were quick to decry those who tried to write them off as "fringe elements" or as separatists. One particularly testy exchange between Arlen Specter (Republican, Pennsylvania) and Norman Olson during the 1995 hearings illustrates the seriousness with which the movement sees itself as working on behalf of the nation:

> *Specter*: . . . I cannot understand how anybody could understand why someone would bomb the Oklahoma City federal building as a matter of retribution.
>
> *Olson*: . . . well, then you don't understand the problem . . .
>
> *Specter*: . . . well, Mr. Olson. I may not understand, and that's why we've had these hearings, so that you could have a full opportunity to express yourself.
>
> *Olson*: May I make a correction for the record, too? Senator Kohl [Democrat, Wisconsin] raised a poster a moment ago showing Hitler with his hand raised in the air—so that's a copyrighted poster produced by Jews for the Preservation of Firearm Ownership. It is not the work of some militia organization. So just to make that comment for the record.
>
> *Specter*: . . . well, we'll pick up your comment about copyrights and about Jews in a few minutes—
>
> *Olson*: No sir. I believe you're trying to lay at the feet of the militia some culpability as a responsibility. You're trying to make us out to be something that we are not, much like the press has tried to do over this last year. . . . We are people who are opposed to racism and hatred. We are people who love our government and love the Constitution . . . We're proud of the United States of America. But the thing that we stand against is corruption. We stand against oppression and tyranny in government. And we, many of us, are coming to the conclusion that you best represent that corruption and tyranny.
>
> (US Senate 1995)

Since the bombings, militia members have continued their efforts to combat popular media images of themselves as racist and exclusionary. Many militias' internet homepages now contain "educational" components designed to counter such unflattering depictions. A popular essay by Patriot Kenneth Maue entitled "What is the Militia?," for example, can be found on militia homepages across the country. In it Maue argues that the movement is constitutionally sanctioned and he is careful to disavow racism and exclusion (see Militia of Montana 2001). Similar essays appear on dozens of other militia homepages (see California Militia 2001; Michigan Militia 2001).

While the movement's rhetoric preaches the openness of the patriot category as inclusive of all citizens, its spatial bounding of the category of nation belies an antagonistic form of identity construction. These antagonisms become apparent when one examines not only how Patriots define "federal space" and the citizenship attached to it, but also how they view the relationships between federal space and smaller scales of place within it.

In militia circles it is widely believed that the federal government only has territorial jurisdiction over the District of Columbia and US territories such as Guam and Puerto Rico (Abanes 1996; Dyer 1997; Stern 1996). While patriots consider states, counties and cities to be a part of the USA, they hold that the only power the federal government has over these entities is the power to regulate commerce between them. According to patriots, the federal government has no constitutional mandate to regulate intra-state and county activity or to dictate activities therein. That the federal government regularly does both is regarded as pernicious in Patriot circles.

To explain how the federal government could overstep its constitutional bounds and dupe an "unsuspecting" populace, the movement has developed a complicated theory involving the notion of citizenship and the fourteenth amendment [of the US Constitution] (Dyer 1997). The fourteenth amendment, ratified shortly after the Civil War, was designed to guarantee equal protection under the law to all citizens—rights denied to African-Americans via the institution of slavery. In militia circles many people believe there are two types of citizens in the USA, natural citizens and fourteenth amendment citizens. Natural citizens are individuals born in any of the fifty [US] states. They are considered citizens of the state in which they were born, and bear no responsibility to the US government and federal laws issued from it. Fourteenth amendment citizens, on the other hand, are people living within federal territory and/or those granted "citizenship" by the fourteenth amendment. Unlike natural citizens fourteenth amendment citizens are bound to the laws of the federal government. Not surprisingly, people in the movement generally regard African-Americans as fourteenth amendment citizens. Indeed, the distinction between natural and fourteenth amendment citizenship was initially a way for racist leaders to "rationalize" (and nationalize) their supremacist ideas (Abanes 1996). Militia leaders argued, however, that whites could become fourteenth amendment citizens by engaging in contractual relationships with the state, such as using social security cards or obtaining driver's permits. To reclaim natural citizenship members are instructed to revoke all contractual relations with the state (Dyer 1997).

Once Patriots are "naturalized" they are expected to adhere to Common Law. There is no universally accepted notion of Common Law in the movement, but Common Law courts are meant to operate at the county level, so it is expected that local law will vary from place to place. During the farm crisis farmers convened Common Law Courts to try those held responsible for taking away their farms (Abanes 1996; Davidson 1996). More recently a Common Law Court, convened in Wichita, Kansas, ordered President Clinton, Congress and the Supreme Court to explain why the constitution had been "suspended" (Dyer 1997; Stern 1996).

It is important to note that not all militia members believe that only whites can be natural citizens. This is especially the case now that many in the movement have attempted to mainstream the cause (Gallaher 2000). Many militias now hold that any person born in the USA is a natural citizen. As such, African-Americans, Jews and other minorities can revoke their fourteenth amendment citizenship just as whites can. Nonetheless, "non-racist" militias often refuse to sanction their more extremist brethren, arguing that it is up to locals to decide their norms and the job of the militia to defend free expression of them. At an obvious level, the tacit approval of white supremacists within the movement represents the level of power still held by its original organizers. By disavowing racism while refusing to sanction it, these leaders reap the rhetorical benefits of refusing racism while bearing none of its responsibilities. At a more fundamental level, such rhetoric allows Patriots to collapse antagonisms against women, minorities and gays (and the government for granting protections to them) within "safe" and "palatable" national coding.

The antagonism apparent in the movement's identity politic is also illustrated in how it views the relationships between the nation (understood as a loose federation of sovereign counties) and the federal government. In short, the Patriot Movement espouses eliminating all vestiges of federal presence from within county boundaries. Such efforts have taken varying forms, from the legal to the illegal. In western ranching communities, for example, US Forest Service rangers have been shot at, "deported" and in some cases detained by militia units "protecting" county lands from federal agents (Stern 1996). In several ranching communities, militia leaders, with the help of sympathetic lawmakers, have also enacted local control ordinances giving county sheriffs the legal right to "deport" federal employees working in the county without permission (Kenworthy 1995). The federal government maintains these ordinances are illegal and has brought suit against several offending counties. While the issue is tied up in the courts, however, operations at agencies such as the Forest Service and the Bureau of Land Management find their efforts severely curtailed (Stern 1996). Perhaps most importantly, Patriots now run for state and local office on explicitly "Patriot platforms." In Kentucky, for example, a Patriot leader ran for Governor in 1999, garnering 12 percent of the vote (Gallaher 1998). His campaign was organized around calls to eliminate a federal presence in the state.

In agricultural communities in the Midwest, Patriots have also articulated positions that seek to "oust" the federal government from Patriot "territory." The antagonism of these positions is further highlighted by the fact that in many cases the government actions they seek to eliminate are economically beneficial to them. Several scholars have noted, for example, that Patriots in farming communities are frequently opposed to subsidies and price supports, even though these government-driven programs are often what allow Patriots in farm communities to survive (Dyer 1997; Gallaher 2000; Stock 1996). When considered in the context of the farm crisis and the distrust it created for the government, such views are based in lived experience. Moreover, farm communities have long resented the position of dependency into which subsidies put them (Stock 1996). However, Patriots' antagonistic depictions of the government as without mandate, co-opted and otherwise "occupied" keep them from demanding aid packages on their own terms and drive them to reject all such packages that emanate from the government (Gallaher 2000).

While the Patriot identity politic is fraught with antagonisms, we see potential for meaningful intervention within it. In particular, we note that many of the issues around which Patriots galvanize are rooted in class-based anxieties. To date, however, they have been articulated through right-wing discourses, which obscure class as a category of action and foster instead an inward-looking nationalism. There is, however, room for the left to address these concerns, and to better effect. Such a politic requires not only re-embracing class as a category of action, but also resisting the temptation to "write off" such groups because of the current form their politics takes (Gallaher 2000; Harvey 2000). It also requires addressing at the grassroots level how whiteness as a social position has been, and continues to be, used to thwart class politics in the USA (see Ignatiev 1995; Kincheloe and Steinberg 1998; Lipsitz 1998; Roediger 1994).

CONCLUSIONS

The struggles embodied in both the Zapatista and Patriot Movements have a rich history. As our discussion indicates, while both social movements are relatively new, the concerns each grapples with are rooted in long-standing debates. For their part, the Zapatista's struggle is an attempt to finally exorcise the demons of colonialism, from the racism instilled against indigenous people to the poverty it confined them to. Patriots are also grappling with a long-standing debate, in their case about what the balance of power should be between federal and state governments, particularly as it relates to rural producers.

In both instances, however, the effects of neoliberal policy have led each movement to address these respective debates in new ways. Unlike earlier indigenous struggles in Mexico, for example, the Zapatistas no longer view the state as able to solve indigenous problems. Indeed, the Mexican state is seen as the handmaiden of neoliberalism, and thus no longer willing to act on behalf of its most needy citizens. As such, rather than call for the reform of state programs (such as the National Solidarity Program) or a simple reinstatement of the original Article 27, the Zapatistas ask for formal autonomy from (albeit within) the state. In the USA, patriots also proffer new interpretations of extant debates, specifically the question of federalism. Indeed, the battle cry for state's rights, long popular among rural conservatives, has been replaced in Patriot circles by calls for local sovereignty. Today's Patriots no longer view federal *or* state governments as willing to keep global flows at bay, and thus they invoke local sovereignty in an effort to divorce themselves structurally from the government entities that facilitate it in their everyday lives.

In sum, both movements are creating new discursive spaces that require civil society to fundamentally rethink the role of the modern nation-state. It is here that evaluation becomes crucially important because, while the Zapatistas and Patriots may disappear, the discursive space they have created will not do so as easily, and may be "occupied" by other social movements in the future. Rather than attempting to destroy, eliminate or co-opt such groups (because they are nationalist, or because they are armed), therefore, we must actively grapple with the alternatives they present. And, most importantly, we must be prepared to evaluate these alternatives for potential antagonisms. Only by identifying antagonisms and working to domesticate them may we eliminate political violence.

This concern is, obviously, most pressing for the US Patriot Movement. Commentators and activists alike have long sought to destroy the movement by exposing its crimes, discrediting its leaders and embarrassing its adherents. When we recognize that such actions are predicated on the attempt to destroy difference (in this case from a cosmopolitan norm), however, we may discern the futility in such measures. As post-structuralism reminds us, all categories of meaning are defined through difference. We cannot therefore simply embark on an effort to destroy the difference invoked by the Patriot identity. Rather, it behooves us to consider alternatives for those attracted to the movement, and to domesticate the antagonisms inherent to its current form.

While we argue that the Zapatista Movement is constructed agonistically, the only ever temporarily fixed character of any identity category requires us to be vigilant of their politics as well. Indeed, the on-going hero worship of Subcomandante Marcos obscures the ever-evolving nature of the movement. And, while the movement's current form may be regarded as progressive, changes to it are not guaranteed to follow the same trajectory. This is particularly the case concerning the role of women in the indigenous autonomous spaces the Zapatistas call for. While Comandanta Esther's comments to Congress support gender equity within autonomous zones, long-standing biases against female participation are not guaranteed to disappear overnight, and must be actively negotiated within indigenous communities. As such, while Zapatista women are optimistic, they are acutely aware, as Comandanta Yolanda noted at the International Day of the Rebel Woman, that "the men are struggling to totally understand what we are asking for as women" (Irish Mexico Group 2001). It is for this reason that the women associated with the Zapatista struggle continue to seek outside attention for their struggle within a struggle—so that gender equality becomes part and parcel of the struggle, and the gains, of the Zapatista Movement.

We conclude by noting the limitations to this study. First, we recognize the limited nature of this comparison. Both of these movements are complex, and our analysis has, at best, only scratched the surface. While an overview paper like this one can only go so far, it is worth reiterating this point in closing because, as scholars of both contexts indicate, the insurgencies discussed here have deep, sometimes convoluted, but always constitutive roots (see Stock [1996] on the Patriot Movement and Castaneda [1993] on Latin American movements). We are also aware that further comparison of these groups might be fruitfully undertaken in several areas. Both movements have been successful at invoking their national revolutions to situate their cause. A thorough analysis and comparison of such invocations could tell us much about their discursive strategies and about the unfinished nature of national revolutions in general (see Negri 1999). We also suggest that further comparison should be conducted on the role of land tenure to each movement's identity politic. In particular, it would be useful to compare and contrast the rural policies instituted during the 1930s by Lazaro Cardenas and Franklin Roosevelt—the two men responsible for building "welfare states" in their respective countries. These are the policies which neoliberalism has undone, yet in neither case do our movements ask for a return to them. Lastly, we would suggest comparisons with and between other anti-statist movements as well. Some movements that come to mind include, but are not limited to, the Corsican nationalist movement and the more extreme factions of both the Republican and Unionist movements in Northern Ireland.

ACKNOWLEDGEMENTS

We would like to thank the three anonymous reviewers of this manuscript for their knowledgeable and helpful comments. Their insights helped us sharpen our theoretical and empirical analysis, and we are very grateful to them.

NOTE

1 There is little documentation on Patriot activity between the farm crisis of the mid-1980s and Ruby Ridge and Waco in the early 1990s. The existing scholarship suggests, however, that many of the same right-wing activists who exploited the farm crisis took similar advantage of the mishaps at Ruby Ridge and Waco (Dees 1996). Given that these events were not concentrated in one geographical region (as the farm crisis was), activists were also able to use these events to launch the movement on to the national scale.

REFERENCES

Abanes, R. (1996) *Rebellion, Racism and Religion: American Militias.* Downers Grove, IL: Inter Varsity Press.

Aho, J. (1990) *The Politics of Righteousness: Idaho Christian Patriotism.* Seattle: University of Washington Press.

Benjamin, T. (1989) *A Rich Land, a Poor People: Politics and Society in Modern Chiapas.* Albuquerque: University of New Mexico Press.

Buchwald, A. (1996) "Cracked and loaded," *Washington Post*, 9 April, p. B l.

California Militia (2001) "Who we are," http://www.geocities.com/CapitolHill/Congress/2608/whoarewe.htm.

Castaneda, J. (1993) *Utopia Unarmed: The Latin American Left After the Cold War.* New York: Knopf.

Castells, M. (1997) *The Power of Identity.* Oxford: Blackwell Publishers.

Castro Apreza, I. (1999) *Quitarle el agua al pez: la guerra de baja intensidad en Chiapas (1994–1998).* Mexico City: Chiapas 8 Institute de Investigaciones Economicas—ERA.

Cockcroft, J. (1998) *Mexico's Hope.* New York: Monthly Review Press.

Collier, G. (1994) *Basta! Land and the Zapatista Rebellion in Chiapas.* Oakland: Food First Books.

Comandanta Esther (2001) "Speech to the Congress of the Union," http://www.ezln.org/marcha/20010328a.en.htm.

Connor, W. (1993) *Ethnonationalism.* Princeton, NJ: Princeton University Press.

Davidson, O. (1996) *Broken Heartland: The Rise of America's Rural Ghetto.* Iowa City: University of Iowa Press.

Dees, M. (1996) *Gathering Storm: America's Militia Threat.* New York: HarperCollins.

Diamond, S. (1995) *Roads to Domination: Right Wing Movements and Political Power in the United States.* Guilford Press: New York.

Dyer, J. (1997) *Harvest of Rage: Why Oklahoma Is Only the Beginning.* Boulder, CO: Westview Press.

Esteva, G. (1994) *Cronica del Fin de una Era.* Mexico City: Editorial Posadas.

Fredrickson, G. (1997) *The Comparative Imagination: On the History of Racism, Nationalism, and Social Movements.* Berkeley: University of California Press.

Froehling, O. (1997) "A war of ink and internet: cyberspace and the uprising in Chiapas." *Geographical Review* 87: 129–138.

—— (1999) "Internauts and guerilleros: the Zapatista rebellion in Chiapas, Mexico and its extension into cyberspace," in Crang, P. and May, J. (eds) *Virtual Geographies: Bodies, Space and Relations.* New York: Routledge, pp. 164–177.

Gallaher, C. (1998) "America's New Patriots: livelihood and the politics of identity," unpublished PhD thesis, Department of Geography, University of Kentucky.

—— (2000) "Global change, local angst: class and the American Patriot Movement," *Environment and Planning D: Society and Space* 18: 667–691.

Giddens, A. (1985) *The Nation-state and Violence: Volume Two of a Contemporary Critique of Historical Materialism.* Cambridge: Polity Press.

Harvey, D. (1996) *Justice, Nature and the Geography of Difference.* London: Blackwell.

—— (2000) *Spaces of Hope.* Berkeley: University of California Press.

Harvey, N. (1998) *Chiapas Rebellion: The Struggle for Land and Democracy.* Durham, Duke University Press.

Ignatiev, N. (1995) *How the Irish Became White.* New York: Routledge.

Irish Mexico Group (2001) "March 8—International Day of the Rebel Woman," http://flag.blackened.net/revolt/ mexico/ezln/2001/march/women_mar8.html.

Junas, D. (1995) "The rise of citizen militias: angry white guys with guns," in Berlet, C. (ed.) *Eyes Right! Challenging the Right Wing Backlash.* Boston: South End Press, pp. 226–235.

Katzenberger, E. (ed.) (1995) *First World Ha Ha Ha! The Zapatista Challenge.* San Francisco: City Light Books.

Kenworthy, T. (1995) "Angry ranchers across the West see grounds for an insurrection," *Washington Post*, 21 February, p. A3.

Kincheloe, J. and Steinberg, S. (1998) "Addressing the crisis of whiteness: reconfiguring white identity in a pedagogy of whiteness," in Kincheloe, J., Steinberg, S., Rodriquez, N., and Chennault, R. (eds) *White Reign: Deploying Whiteness in America*. New York: St. Martin's Press, pp. 3–29.

Kirby, A. (1997) "Is the state our enemy?", *Political Geography* 16: 1–12.

Laclau, E. and Mouffe, C. (1985) *Hegemony and Socialist Strategy: Towards a Radical Democratic Politics*. London: Verso.

Landau, S. (1996) *Sixth Sun: Mayan Uprising in Chiapas*. New York: Cinema Guild.

Lipsitz, G. (1998) *The Possessive Investment in Whiteness: How White People Profit from Identity Politics*. Philadelphia, PA: Temple University Press.

Luke, T. (1997) "Is the state our enemy? Or is our state the enemy?", *Political Geography* 16: 21–26.

Massey, D. (1995) "Thinking radical democracy spatially," *Environment and Planning D: Society and Space* 13: 283–288.

Michel, L. and Herbeck, D. (2001) *American Terrorist: Timothy McVeigh and the Oklahoma City Bombing*. New York: Regan Books.

Michigan Militia Corps 9th Division (2001) "Statement of purpose and mission," http://www.michiganmilitia.org/html/Mission2.html.

Militia of Montana (2001) "What is the Militia?," http://www.militiaofmontana.com/whomom.htrn.

Mouffe, C. (1995) "Post-Marxism: democracy and identity," *Environment and Planning D: Society and Space* 13: 259–265.

Mozzochi, J. (1995) "America under the gun: the militia movement and hate groups in America," in Berlet, C. (ed.) *Eyes Right! Challenging the Right Wing Backlash*. Boston: South End Press, pp. 236–240.

Murray, J. (1998) "Chiapas and Montana: tierra y libertad," *Race Traitor* 8: 39–50.

Nader, L. (1991) *Harmony Ideology: Justice and Control in a Zapotec Mountain Village*. Palo Alto, CA: Standford University Press.

Natter, W. (1995) "Radical democracy: hegemony, reason, time and space," *Environment and Planning D: Society and Space* 13: 267–274.

Natter, W. and Jones, J. P. (1997) "Identity, space, and other uncertainties," in Benko, G. and Strohmayer, U. (eds) *Space and Social Theory: Geographical Interpretations of Postmodernism*. Oxford: Blackwell, pp. 141–161.

Negri, A. (1999) *Insurgencies: Constituent Power and the Modern State*. Minneapolis: University of Minnesota Press.

Pena, G. (1995) "The Subcomandante of performance," in Katzenberger, E. (ed.) *First World Ha Ha Ha! The Zapatista Challenge*. San Francisco: City Light Books, pp. 89–96.

Rajchenberg, E. and Heau-Lambert, C. (1998) "History and symbolism in the Zapatista Movement," in Holloway, J. and Plaez, E. (eds) *Reinventing the Revolution in Mexico*. London: Pluto Press, pp. 19–38.

Roediger, D. (1994) *Towards the Abolition of Whiteness*. New York: Verso.

Smith, A. (1993) *National Identity: Ethnonationalism in Comparative Perspective*. Las Vegas: University of Nevada Press.

Sparke, M. (1998) "Outsides inside patriotism: the Oklahoma Bombing and the displacement of heartland geopolitics," in Dalby, S. and O'Tauthail, G. (eds) *Critical Geopolitics: A Reader*. London: Routledge, pp. 198–223.

Steinberg, P. (1997) "And are the anti-statist movements our friends?," *Political Geography* 17: 13–19.

Stern, K. (1996) *A Force Upon the Plain: The American Militia Movement and the Politics of Hate*. New York: Simon and Schuster.

Stock, C. (1996) *Rural Radicals: Righteous Rage in the American Grain*. Ithaca, NY: Cornell University Press.

Subcomandante Marcos (1992) "Chiapas: the southeast in two winds, a storm and a prophecy," http//www.ezln.org/documentos/1994/199208xx.en.htm.

—— (1994a) "First declaration of the Lacandon jungle: today we say enough is enough!," http://www.ezln.org/fzln/lst-decl.html.

—— (1994b) "Carta de Marcos a remitentes que aún no obtienen respuesta," http//www. ezln.org/marcos.12-13.html.

—— (1995) "Letter from Durito, September 4th 1995," http://www.ma.utexas.edu/users/guilfoyl/Chiapas/let.html.

—— (2001) "La entrevista insolita," *Proceso*, 11 March.

Tabor, J. (1997) "A failed Utopian vision," *Political Geography* 16: 27–32.

Tilly, C. (1992) *Coercion, Capital, and European States.* New York: Routledge.

US Senate. Senate Subcommittee on Terrorism (1995) *Hearings on Militias, June 15, 1995 (Part 2).* Washington, DC: Government Printing Office.

Womack, J. (1999) *Rebellion in Chiapas: An Historical Reader.* New York: The New Press.

15

Social theory and the de/reconstruction of agricultural science

Local knowledge for an alternative agriculture[1]

Jack Kloppenburg, Jr.

from *Rural Sociology*, 1991, 56(4) (Winter): 519–548

> [We] have a dual task before—a deconstructive project and reconstructive project that are intimately linked. Our deconstructive task requires close attention to, and the dismantling of, technostrategic discourse. The dominant voice of militarized masculinity and decontextualized rationality speaks so loudly in our culture, it will remain difficult for any other voices to be heard until that voice loses some of its power to define what we hear and how we name the world—until that voice is delegitimated.
>
> Our reconstructive task is a task of creating compelling alternative visions of possible futures, a task of recognizing and developing alternative conceptions of reality, a task of creating rich and imaginative alternative voices—diverse voices whose conversations with each other will invent those futures.
>
> (Carol Cohn, "Sex and Death in the Rational World of Defense Intellectuals," 1987: 717–18)

INTRODUCTION

No less than those who would challenge the way in which the defense intellectuals have defined our world, we who believe that contemporary agricultural production is neither socially just nor ecologically benign also face dual tasks. In part, the deconstructive task entails the demonstration that agricultural science as currently constituted provides neither a complete, nor an adequate, nor even a best possible account of the sphere of agricultural production.[2] Indeed, it is in large measure an historical overreliance on this partial knowledge—and a failure to recognize how specifically situated that knowledge is—that has brought our agriculture into its present straits.

The reconstructive task will be the more difficult, for it will entail the identification and legitimation of alternative sources of knowledge production for agriculture—sources which now have no voice, or speak without authority, or simply are not heard in contemporary agroscientific discourse. It is out of conversations among this fuller range of knowledge sources—conversations that should include, but must not be limited to, what is now known as agricultural science—that an alternative and a truly sustainable agriculture may emerge.

The deconstructive project has enjoyed considerable success since what the Agricultural Research Institute dubbed "Hurricane Rachel" Carson appeared on the horizon in 1962. In no other sector of science has as much space been opened for reconstructive possibility as in agriculture. And this space has been created at a time when new resources are available both for extending the deconstructive project and for initiating the reconstructive task. In contemporary sociological interpretations of science and in feminist analyses we have new theoretical resources for challenging that voice of decontextualized rationality which agricultural science has used to such dominating effect. And in the diverse literatures on what I will provisionally call "local knowledge" and in the knowledge contained in the heads of farmers and agricultural workers, we have the material resources for a plausible reconstruction of what Sandra Harding (1986) has termed "successor science."

THE GREENING OF THE NATIONAL RESEARCH COUNCIL?

The deconstructive project—actually, it is more a diffuse historical tendency than a coherent project—has been gathering momentum for nearly three decades now. Rachel Carson (1962) was midwife to the birthing of a wide variety of initiatives directed to forestalling the kind of ecological apocalypse described in *Silent Spring*. Subsequent critiques have focused not only on the social and environmental externalities associated with modern agricultural technologies (Berry 1977; Commoner 1972; Strange 1988), but also on the manner in which particular social interests gain differential influence over the institutional structure of knowledge production. There is concern that corporations and agribusinesses have managed to shape to their own advantage the choice of the problems that public agricultural science has undertaken and the way solutions to those problems are expressed in technologies (Busch and Lacy 1983; Buttel 1986; Friedland 1978; Hightower 1973; Kenney 1986; Kloppenburg 1988).

More recently, criticism has been directed not simply at the priorities to which agricultural science has been directed, but at the validity and utility of the methodologies employed in research and the epistemic constitution of knowledge production itself. Suppe (1988) argues that agricultural research of the sort performed by experiment stations can have only limited applicability to actual farming operations because of limitations intrinsic to the probabilistic extrapolation of experimental results to highly variable biological and social systems. A growing number of biological scientists are concerned that the reductionistic and positivistic approaches characteristic of modern science constrain pursuit of unorthodox but potentially productive research initiatives, obscure important connections between organisms and phenomena, and actively inhibit achievement of holistic understanding of ecological systems (Allen and Starr 1982; Levins and Lewontin 1985; MacRae *et al.* 1989; Odum 1989; Prigogine and Stengers 1984).

This discursive opposition by academics has helped to inform and complement activists who seek to transform the scientific and technical bases of agricultural production and who have found a great deal of support in growing popular disaffection with the continuing deterioration of the environment. This activist movement has been given institutional expression not only in national level environmental groups (e.g. Greenpeace, Natural Resources Defense Council), but also in organizations with a specifically agricultural focus (e.g. Pesticide Action Network, Rodale Institute, the Land Institute) and, most importantly, in countless local groups organized around a wide variety of issues of local concern. In whatever terms these organizations may frame their particular vision of a transformed agriculture—reduced-input, biological, sustainable, organic, permanent, ecological, regenerative—it is clear that they are seeking an *alternative* to conventional agricultural practice.

To the list of organizations calling for an alternative to conventional agricultural practice we may now append the National Academy of Sciences. In a book length analysis titled *Alternative Agriculture*, the National Research Council's (NRG) Board on Agriculture has affirmed the benefits of "alternative systems." The chair of the committee that conducted the study goes so far as to say in the preface that "the committee believes that farmers, researchers, and policymakers will perceive the benefits of the alternative systems described in this report and will work to make them *tomorrow's conventions*" (NRC 1989: vi, emphasis added). The nation's premier scientific body has placed itself in support of an approach to agricultural production the designation of which is "alternative." Such a designation reflects a long sojourn in the wilderness of scientific marginality.

The NRC's report is a clear indicator of just how successful the deconstructive project has been. Activists and academic critics have struggled long and hard to illuminate and focus attention on the link between contemporary modes of agricultural production and contaminated waters, eroded lands, human cancers, pesticide residues, foreclosed farms, and declining rural communities. The NRC has been led to see the connection between the hard tomatoes and the hard times. And with this partial delegitimation of conventional production practices has come a concomitant questioning of the scientific and technical bases of those practices. The NRC's recognition of the need for an alternative agriculture is evidence that the deconstructive project has succeeded in opening up a space in which the hegemonic forms of science, though still powerful, are no longer completely secure. Given the existence of this space, it may be possible to initiate the reconstructive task of building an alternative science as part of the process of building an alternative agriculture.

But agroindustry and elements of the public agricultural science community have already begun their counterattack. A central theme of this counterattack is that if an alternative agriculture is necessary, it will be up to *scientists* to determine what that alternative will be. The objective is to control the shape that alternative agriculture will take by insisting upon the hegemony of existing science and thereby limiting the type and range of knowledges that can be brought to bear upon the construction of an alternative agriculture. For example, in its critique of the NRC's report, the agribusiness-oriented Council for Agricultural Science and Technology (CAST) asserts that "The extensive coverage and dependence on [farm] case studies reflects the paucity of solid factual information . . . this renders certain findings and related recommendations more philosophic than scientific" (CAST 1990: 2). CAST will allow knowledge produced by farmers no credibility independent of validation by scientists. The deconstructive project may have opened space, but there is no question that the use that is made of that space will be contested. And the object of that contest is not simply what should constitute alternative agriculture but—even more fundamentally—who is even to have the power to speak authoritatively in that debate, who is to have a voice at all.

The NRC's report itself reflects the same class of sin, though it is less one of commission than of omission; the NRC does not take farmers' knowledge seriously enough. Fully half of *Alternative Agriculture* is indeed given over to eleven farm case studies. When the NRC study committee wanted to see how alternative farming worked, it had little choice but to seek out farmers who had themselves developed alternative practices since the agricultural science establishment had virtually nothing to offer. Far from disparaging this farmer-generated knowledge as CAST does, the NRC staff praises its richness and creativity. But the conclusions reached in the report relate almost entirely to the need for the application of more *scientific* effort to the development of alternative agriculture, and the report's recommendations focus on how this scientistic strategy might best be accomplished. Farmers are regarded as recipients of technology, advice, and information. The authors of the NRC report simply do not conceive of any potential for farmer-generated knowledge except in connection with or translation through agricultural "science."

There is broad agreement that American agriculture should move toward some "alternative" form. But the extent to which this alternative future will be a change in kind rather than degree—that is, the extent to which it approaches reconstruction rather than reproduction—will depend in significant measure upon whether agricultural science itself is reconstructed or simply reproduced. Despite their differences, both CAST and the NRC propose to achieve an alternative, sustainable, regenerative, low-input, diversified agriculture through the application of the same methods and institutions of knowledge production that have given us a conventional, non-sustainable, non-regenerative, high-input, homogeneous agriculture in the first place. If we are to achieve a truly alternative agriculture, might we not also require an alternative science? And should not that alternative science encompass—at a minimum—the knowledge production capabilities of farmers who by their very survival outside conventional agriculture have already demonstrated their capacity for the generation of useful and workable alternatives?

Now is the time for bold hypotheses and innovative research. Rural sociologists can and should play a central role in the struggle to create a truly alternative science, as well as an alternative agriculture. What follows is an effort to survey the resources available for both the deconstructive and the reconstructive projects, to suggest ways in which these resources might be used, and to outline productive areas for research.

FROM DECONSTRUCTION TO RECONSTRUCTION

In contemporary society, what we call science enjoys a privileged status among the possible ways of establishing knowledge about the world (Aronowitz 1988; Marcuse 1964; Mulkay 1979). And for a long time, social theorists joined the public, the business community, and policymakers in treating scientists as virtual truthsayers. By virtue of its methodological foundation and normative characteristics, the community of scientists was held to be capable of generating knowledge that—unlike the products of any other way of knowing—bears no traces of its birthing in a particular social context (Merton 1973; Polanyi 1962). But, just as the assertion that Adam had no navel because he was created and not born was challenged, so now has the absence of the scars of social contingency on the bodies of scientific "facts" been brought into question.

Over the past fifteen years there has emerged a wide variety of provocative new sociological interpretations of science which constitutes a rich and diverse body of theoretical and empirical resources to draw upon in challenging positivist and realist epistemologies of scientific knowledge (see, e.g., Barnes and Bloor 1982; Callon and Law 1989; Cozzens and Gieryn 1990; Knorr-Cetina 1981; Latour 1987; Latour and Woolgar 1979; Longino 1990). The analytic frameworks associated with the new sociology of science are theoretically and methodologically diverse. Still, all of these programs share a distinctive point of departure: the central insight that the mental productions we call scientific knowledge are no less subject to social influences than are the products of any other way of knowing and are, therefore, the fruits of a scientific enquiry that must be envisioned as, in Knorr-Cetina's (1981: 3) succinct phrasing, "constructive rather than descriptive." A number of important points follow from this characterization.

First, the recognition that the "facticity" of science is not comprised of objective descriptions of a determinate natural world but of socially contingent constructions provides a foundation for a powerful new critique of science. Socially contingent objectives can be recognized not just in the uses to which science is put, but in scientific facts themselves. Second, the inadequacy of criteria for the epistemic demarcation of science as a uniquely legitimate way of knowing means that what we call modern science is itself an historical product of continuous social struggle not only to define science in a particular way, but also to exclude other ways of producing knowledge from that definition (Gieryn 1983). Third, if scientists do not have a uniquely privileged capacity to speak authoritatively on nature's behalf, then knowledge claims arising outside the institutions of science can no longer be summarily dismissed because they are "nonscientific." And fourth, if science is socially constructed and is therefore subject to social deconstruction, then certainly it must also be amenable to social *reconstruction*. The boundaries of what we might call "actually existing" science are in fact negotiable and might be redrawn to include other ways of producing knowledge, to effect new articulations and combinations between modes of knowledge production whose essential complementarity is now obscured, or even to produce a radically transformed science whose contours we can now only dimly foresee.[3]

But, for the most part, the new sociologists of science have not applied themselves to the transformative task to which their work appears to logically lend itself (Amsterdamska 1990; Kleinman 1991; Restivo 1988). In contrast, feminist analysts bring a commitment to activist social change to their own distinctive yet constructivist approach to the critique of actually existing science. Some feminists came quickly to the realization that the source of the evident hostility of science to women and the earth is located in the very fabric of scientific rationality itself, in the language and the concepts and the methods deployed in the construction of scientific meanings and scientific "facts." As Keller (1987: 37–38) points out, this conclusion follows logically from a central premise of feminist studies generally: just as gender is a socially constructed representation (rather than a precise reflection) of sex, so is science a socially constructed representation (rather than a precise reflection) of nature.

This essentially constructivist premise guides feminists to the sociopolitical implications that practitioners of the new sociology of science have only partially acknowledged and have generally failed to pursue. Feminists regard constructivist insights not simply as a foundation for the interpretation of actually existing science, but as tools for social as well as epistemological criticism. They recognize the role that the exclusion of other knowledges has played in the historical demarcation of science and understand that this creates space for legitimation of other ways of knowing (especially women's knowledges). And, most importantly, motivated by a desire to transform what they perceive as an intrinsically androcentric mode of knowledge production, many feminists are actively engaged in the search for alternatives to the way in which hegemonic science is presently constituted.

This is not to say that there is complete agreement among feminists regarding the shape that a transformed science might best assume. Harding (1986) identifies three tendencies in feminist thinking about the production of scientific knowledge: "feminist empiricism," "feminist postmodernism," and "feminist standpoint" theory. Feminist empiricism recognizes the distortion introduced into the construction of scientific facts by the historical legacy and contemporary reality of sexism, but it holds that this androcentric bias can be, in large measure, mitigated through more rigorous adherence to the existing scientific method. What is sought is "a larger canon rather than a different one; a richer, perhaps even multi-faceted, representation of reality, but not a separate reality" (Keller 1987: 46). If feminist empiricism maintains a faith in the possibility of scientific consensus enriched by multiple voices, feminist postmodernism recognizes the diversity of voices but denies their commensurability. A maximally objective view of the world is to be sought not in essentializing universals, but in alliances between partial knowledges which are capable of generating richer understandings when "federated in solidarity" (Harding 1986: 55; see also Haraway 1988 and Smith 1987). Finally, feminist standpoint theory proposes a transition to a "successor science" which would be superior to actually existing science because it would be founded on a feminist epistemology that is itself intrinsically superior (Bleier 1986; Merchant 1980). This epistemic superiority is derived not from biological differences between men and women, but from the distinctive *experience* associated with women's lives in gendered society. Hartsock (1983) and Rose (1983, 1986) suggest that it is specifically the character of women's *labor*—especially caring labor and manual labor—that structures and shapes a feminist as opposed to a masculinist understanding of the world.

The differences among these approaches to the feminist deconstruction and reconstruction of science are perhaps less important than the characteristics they share. Women's distinctive historical experiences—of their bodies, of oppression, of caring (about and for)—make possible alternative ways of thinking about nature and knowing the natural world. A feminist science would be one in which

> no rigid boundary separates the subject of knowledge (the knower) and the natural object of that knowledge; where the subject/object split is not used to legitimize the domination of nature; where nature itself is conceptualized as active rather than passive, a dynamic and complex totality requiring human cooperation and understanding rather than a dead mechanism, requiring only manipulation and control.
>
> (Fee 1986: 47)

Knowledge production founded on a feminist epistemology would draw—as contemporary patriarchal science allegedly does not—on the "integration of hand, brain, *and* heart" (Rose 1983: 90).

But to what extent is this emphasis on *experience* uniquely feminist? Sandra Harding (1986: 165) has noted the "curious coincidence of African and feminine 'world views'" regarding the relationship between the self and the phenomenal world. Elizabeth Fee (1986) extends this insight and identifies parallels between feminist epistemology and Native American, Chinese, and even working-class perspectives on nature and knowledge. What feminists criticize as masculinist science is also criticized from other standpoints—differently situated knowledges, one might say—as European science, or imperialist science, or bourgeois science. Elizabeth Fee (1986: 53) concludes that "Clearly these different critiques need to be brought together . . . It seems to me that any one of these critiques provides a partial, but incomplete, perspective—and each adds important elements otherwise missing in the analysis."

I agree with Fee that this spirit of eclecticism—of "shared conversations in epistemology" (Haraway 1988: 584)—is the most fruitful analytical approach in a world of multiple identities and hyphenated commitments. One of the central themes in the feminist analysis of science is the importance of legitimating and reaffirming the value of producing knowledge through "sensuous activity" (Rose 1986: 72) and "personal experience" (Harding 1986: 240) that is necessarily and specifically "local" (and therefore neither universalizing nor essentializing) in character (Smith 1987). I suggest that what I will call "local knowledge" is an expression of such production and that it is the global ubiquity of this form of knowledge production that accounts at least in part for the curious coincidences noted by Harding and Fee.

Feminists are neither the only nor the first analysts to mark reliance on sensuous activity and personal experience as a fundamentally different kind of knowledge production than that commonly called scientific. True, in one sense all knowledge is both personal and sensuous inasmuch as it must be obtained by individuals who have no access to the natural world except through their senses. But while the Ojibwa herbalist and the NIH biochemist both rely on sensuous observation to obtain knowledge, they do so from quite different epistemological stances (as well as within quite different social contexts, with quite different objectives, and with quite different tools). A wide variety of analysts from the phenomenologist philosophers to contemporary anthropologists have tried to illuminate this epistemic distinction through elaboration of a range of paired concepts: "tacit knowledge/scientific knowledge" (Polanyi 1966), "science of the concrete/science" (Levi-Strauss 1962), "life-world knowledge/scientific knowledge" (Bohme 1984; Husserl 1970), "craft knowledge/scientific knowledge" (Braverman 1974), "practical labor/science" (Bittner 1983), "folk wisdom/processed knowledge" (Krimsky 1984), "indigenous knowledge/scientific knowledge" (Richards 1985), "working knowledge/scientific knowledge" (Harper 1987).

In providing the foregoing list I do not mean to imply that these analysts are saying precisely the same thing. They are not. However, their thoughts are clustered in such a way as to constitute an identifiable constellation of analysis that provides a rich set of resources for exploring the production of knowledge by obstetrical nurses in Chicago, blacksmiths in Nairobi, Jivaro shamans in the Peruvian Amazon, and hog farmers in Iowa. The practical, sensuous, personal labor of such people "is always controlled by full regard for the timely and local features of the environment within which it takes place" (Bittner 1983: 253). It is the *locality* of such knowledge production which most completely intimates the many dimensions of its character. Such knowledge is *local* in the sense that it is derived from the direct experience of a labor process which is itself shaped and delimited by the distinctive characteristics of a particular place with a unique social and physical environment.[4]

One dimension of locality is an intimacy between the worker and the materials and objects of labor. The "many-sided gestalt of theoretical, tactile, and auditory input" which constitutes the craft skill of Harper's mechanic/bricoleur, Willie, enables him to "reduce the gap between the subject—the worker —and the object—the work" (Harper 1987: 133). Thus, like the tribal bricoleurs of Levi-Strauss (1962), Willie "speaks not only *with* things . . . but through the medium of things." This (feminist) elimination of the boundary between subject and object and the intimacy of the conversation between the knower and the known permits the craftsperson to "see beyond the elements of a technique to its overall purpose and coherence" (Harper 1987: 21). This holistic sense of the substance and context of the labor process produces a unified field of knowledge that is finely tuned to the concrete exigencies, needs, and requirements of local conditions.

It is the central importance of *local* knowledge to women and to African peoples that accounts for the curious coincidence of African and feminist world views. And it is *local* knowledge that informs the birthing skills of the *sages-femmes* studied by Bohme (1984). It is *local* knowledge that enables the competent farmer to master the "intricate formal patterns in ordering his work within the overlapping cycles—human and natural, controllable and uncontrollable —of the life of a farm" (Berry 1977: 44). It is *local* knowledge that allows Robert Pirsig to keep his bike running through *Zen and the Art of Motorcycle Maintenance* (Pirsig 1974). It is *local* knowledge that enables machinists to "make out" on the shop floor (Burawoy 1979). And it is the *local* knowledge produced by workers that is the object of appropriation and control in both Taylorist and "postindustrial" strategies of industrial management (Braverman 1974; Hirschhorn 1984).[5]

But today it is not the herbalist but the biochemist, not the midwife but the obstetrician, not the craftsperson but the engineer, not the campesino but the agronomist who dominates knowledge production and deployment. What we all know as scientific knowledge has attained virtually undisputed intellectual hegemony, while local knowledge has been pushed to the epistemic peripheries, its utility so poorly recognized that we have difficulty even labeling it. Until recently, the scientific method was held to be not just a different, and not just a better, but the best and the only consistent way of producing reliable knowledge of the world. It is precisely this epistemic uniqueness that has now been so powerfully challenged. But to say that scientific knowledge is not epistemically unique is *not* to say it is not *different* from that produced by other ways of knowing and, in particular, different from local knowledge.

That such a difference between local and scientific knowledge should now exist is rather ironic since science, in fact, grew out of local ways of knowing (Braverman 1974; Gieryn 1983). Indeed, prominent progenitors of the scientific method such as Bacon, Descartes, and Boyle explicitly saw their task as explaining why craftworkers could do what they could (Merchant 1980: 179–89). But the emergent scientists sought their explanations not in order to understand a particular phenomenon or labor process in all its idiosyncratic complexity, but in order to understand singular identities recognizable across phenomena and labor processes. They were interested not in locality but in translocality. They were interested not in complete understanding of a specifically situated phenomenon, but in partial understandings of widely dispersed but similar phenomena. They were interested in the production not of local knowledge, but of what Latour (1986: 7–14) calls "immutable mobiles," information which is invariant through any change in spatial or social location.

The methodological approach which has historically characterized the production of immutable mobiles—or scientific facts—is Cartesian reductionism. This is the practice of breaking a problem down into discrete components, analyzing these separate parts in isolation from each other, and then reconstructing the system from the interpretations of the parts (Levins and Lewontin 1985: 2; Merchant 1980: 182). There can be no doubt that this approach has been exceedingly powerful, but it also appears to be flawed in a number of ways. For many feminists, the detachment from nature and the objectification of the natural world that are characteristic of the Cartesian method fit all too well with the premium placed on power and control by authoritarian and patriarchal society and have served to reinforce the domination of women and nature (Bleier 1986; Longino 1990; Merchant 1980).

The reductionistic dissection of problems is also seen to involve a loss of context (social and political as well as physical and biological) which encourages a hierarchical and linear rather than an interactive and ecological view of nature (Aronowitz 1988; Odum 1989: 177; Prigogine and Stengers 1984). Inasmuch as it relies on models which are necessarily partial and selective, Cartesian reductionism is biased toward those elements of nature which yield to its method and toward the selection of problems most tractable to solutions with the knowledge thereby produced (Krimsky 1984; Levins and Lewontin 1985). In pursuing the paths along which it realizes successes, Cartesianism neglects those areas where other approaches might prove fruitful. And as its successes and achievements have mounted, Cartesianism has come to appear as the *uniquely* effective mode of knowledge production and is increasingly regarded not just as a methodological tactic, but as "an ontological stance . . . more than simply a method of investigation; it is a commitment to how things really are" (Levins and Lewontin 1985: 2–3).

Corollary to the commitment to Cartesianism is the neglect and delegitimation of local knowledge production and, as Husserl (1970) expressed it, the progressive separation of science from the "life-world." As Cartesian science is elaborated and institutionalized in laboratories, it loses touch with the local knowledge and everyday experience of concrete labor processes which might have informed and shaped its development and application. Science is "no longer guided by a live intelligence, fallibly tuned to actual circumstances; instead it is determined by a detached and externalized intelligence embodied in a formula" (Harper 1987:20). That is, the application of immutable mobiles to particular geographic or social places may fail to respect the exigencies and needs of a specific locality. Because it is reductive, abstracting, and interested in the immutable components of a phenomenon, science loses connection with the variability of local systems. On the other hand, the contextual detail that local knowledge brings to the understanding of a particular place or event has little utility outside that place or event. And because it must be intimately tuned to the totality of continually changing circumstances that define a particular locality, the content of local knowledge is relatively plastic. Indeed, if Cartesian science produces immutable mobiles, local knowledge produces "mutable immobiles."

What we are confronted with, then, is distinctive ways of knowing the world, each with particular strengths and weaknesses. Yet one of these has achieved a hegemonic position from which it dominates epistemic discourse and enjoys a virtual monopoly on the resources that society allocates to the production of knowledge. The new sociology of science has provided a foundation for the deconstruction of actually existing science by demonstrating that Cartesian reductionism has no unique claims to truth. Feminist analysis has produced a similar insight, but uses the deconstructive opening to work toward reconstructive possibility. The recovery of local knowledge should be an important component of such reconstruction.

And what better place to explore the theoretical and practical opportunities for using local knowledge to reconstruct science than the agricultural sector? The deconstructive challenge has brought even the National Research Council to the recognition that the re-doctrinism of actually existing science is not adequate to the task of achieving a sustainable agriculture:

> Most of the new knowledge has been generated through an intradisciplinary approach to research. Scientists in individual disciplines have focused their expertise on one aspect of a particular disease, pest, or other agronomic facet of a particular crop. Solving on-farm problems, however, requires more than an intradisciplinary approach. Broadly trained individuals or interdisciplinary teams must implement the knowledge gained from those individual disciplines with the objective of providing *solutions to problems at the whole-farm level.*
>
> (NRC 1989: 137, emphasis added)

The route to solutions to problems at the whole-farm level—at the local system level—runs not through agricultural scientists, but through those who think in terms of whole farms, those whose experiences are of whole farms, and whose knowledge has been developed by the integration of hand, brain, and heart in caring labor on whole farms—that is, through farmers. We should be exploring how to bring farmers and their local knowledge back into formal knowledge production for agriculture.

RECONSTRUCTION: BRINGING THE FARMER BACK IN

It is profoundly ironic to be suggesting now that farmers be brought *back* into rather than simply into knowledge production for agriculture. For until 1862 farmers not only were in, they were just about the *only* ones who were in since prior to that year neither the USDA nor the land grant universities had been established and only the most embryonic forms of what would come to be known as agribusiness had yet emerged. At least through the turn of the century it was farmers, agricultural laborers, and associated craftspeople who were the chief developers of new practices and technologies for U.S. agricultural production.

While a good deal is known of the tactics deployed by agricultural scientists in their efforts to establish the superiority of their way of knowing (Marcus 1985; Rosenberg 1976), much less is known of the process by which farmer-generated knowledge was simultaneously delegitimated and subsequently hidden from history. The principal contours of a relationship in which "scientists preached and farmers applied what they preached" had been established by the last two decades of the nineteenth century (Marcus 1985: 31). Thereafter, the accelerating "academicization of agriculture . . . led to the feeling that the expert knew more than the farmer, and that therefore the communication flow was from the expert to the practitioner" (Bennett 1986: 367). Natural scientists came early to the treatment of farmers as recipients rather than generators of knowledge and, given the evidence of the proliferation of adoption/diffusion studies, it is apparent that by 1950 social scientists had embraced this viewpoint as well.

Neither the existence nor the persistence of this social scientific myopia in regard to agricultural producers as producers of knowledge as well as commodities should be surprising. Rural sociologists—like other social scientists—do, after all, regard themselves as *scientists* and are no less captive of the epistemological assumptions of hegemonic science than are biochemists. Prodded now and then to treat people as subjects rather than objects, to engage in a pedagogy "*with*, not *for* the oppressed" (Freire 1970: 33), or to try "reverse learning . . . to learn from farmers" (Chambers 1983), even the best and the brightest of us cling instead to our own form of scientific hubris. Painfully cognizant of the problems generated by modern agricultural science, even the most progressive rural social scientists have tended to see solutions to these problems in the leavening effect of social science on the natural sciences rather than in challenging the nature of the scientific enterprise itself.

Thus, Buttel and Gertler (1982: 117) conclude "that cooperation between social and biological scientists" is the key to developing "solutions to pressing problems of agricultural resource management, and for the long-term security of the farm population of North America." Busch and Lacy (1983: 237–38) opt for interdisciplinary enlightenment as well. The logical extension of this confidence in the palliative effect of social scientific expertise is its application in forms of social impact assessment (SIA). Friedland (1978: 11) proposes the prevention of "social sleepwalking" in regard to new technology development through the use of predictive assessment of impacts by a "university public" acting on behalf of the wider public. Hightower (1973: 64) expresses a populist faith in the ability of the wider public to act on its *own* behalf, but he anticipates this action as indirect pressure on the land grant complex through direct pressure on politicians. Only Busch (1984: 310) goes beyond populist initiative and the technocratic review to call for "democratization of the problem formation process."

Now there is no question that the social sciences have interesting things to say, that interdisciplinary cooperation is desirable, that social impact assessment can be useful, that political pressure is an important tool, and that actually existing science needs democratization. These are worthy objectives, worth working for. But while all these proposals imply a critique of what established science has accomplished and how it is organized, they take not just the political and intellectual hegemony but also the epistemological hegemony of that type of knowledge production as a given. They do not see beyond the "democratization" of science at the most radical.

But what ought now to be apparent is that if what we now call science is one socially constructed interpretation among many possible interpretations, then we can ask for more. We can ask not simply for the democratization of actually existing science, but for its transformation. It is perhaps appropriate that among all contemporary critics of agricultural science, it is left to a farmer (an uncommon farmer, it is true, but insight is not evenly distributed among scientists either) to grasp what social scientists have missed. Wendell Berry asserts that since

> what we have now in agriculture . . . is a modern scientific orthodoxy as purblind, self-righteous, cocksure, and ill-humored as Cotton Mather's, our history also forbids us to expect it to change from within itself. Like many another orthodoxy, it would rather die than change, and may change only by dying. . . . If change is to come, then it will have to come from the outside. It will have to come from the margins.
>
> (Berry 1977: 173–74)

For Berry, those margins are the largely unexplored landscapes of knowledge and skill shaped and maintained by the intelligence and labor of farmers themselves.

What appears as radical revisionism to the U.S. agroscientific orthodoxy—recognizing farmers as sophisticated knowledge producers and bringing them back into the process of technology generation—has already achieved a measure of legitimacy in the field of international agricultural development. While using linguistic analysis to study systems of classification and cognition among indigenous peoples, ethnoscientists found that traditional farmers the world over are, in fact, exceedingly keen observers of the natural environment (Brokensha *et al.* 1980). Not only has it become clear that traditional farmers have accumulated large bodies of empirical knowledge which they apply with great skill and imagination in their agricultural operations, it has also been found that they are frequently engaged in trying out changes in their technologies or practices of production, changes that are informed by simple experimental methodologies and which merit being described as forms of research (Richards 1989). The development and deployment of this stock of "local technical knowledge"[6] is thus a dynamic, living tradition which John Hatch (1976: 17) argues constitutes "the single largest resource not yet mobilized in the development enterprise."[7]

The last decade has seen an outpouring of work from researchers sensitive to the need for more effective and equitable international agricultural development policies and committed to exploring the potential of local technical knowledge for achieving an "indigenous agricultural revolution" that is both ecologically sustainable and socially just (Altieri *et al.* 1987; Chambers 1983; McCorkle 1989; Richards 1985; Thrupp 1989). Though intellectually and disciplinarily heterogeneous, this set of work has enough in common that at least some of its practitioners have proclaimed the emergence of a "new paradigm" of knowledge production for agriculture which replaces "transfer of technology" with "farmer first" (Chambers *et al.* 1989: xiii–xiv). This point of view has by no means been universally embraced, but sufficient progress has been made for Chambers *et al.* (1989: xiii) to suggest that taking local technical knowledge seriously might not only provide a new way forward for resource poor farming in the Third World, but might have "lessons also for all agriculture."

Now "lessons also for all agriculture" surely must be read as a tentative suggestion that a "farmer first" approach to knowledge production for agriculture might be as appropriate for the industrialized nations as it is for the Third World. The U.S. agricultural science community will certainly find this difficult to accept. It is one thing to argue that the technical knowledge of resource poor farmers should be taken seriously precisely because they are resource poor and therefore not in a position to take advantage of the technologies that science has to offer. It is quite another thing to argue that farmers who do have the material and intellectual resources to make use of science-based technologies possess—in addition—knowledge that should be used to alter the way science develops and deploys those very technologies.

Still, even among farmers whose operations are most isomorphic to the "best management practices" promulgated by extension cadres, there must exist a substantial reservoir of local knowledge. And at the margins and in the interstices between technological convention and scientific orthodoxy there are all manner of traditionalists and visionaries—Amish, Mennonites, Native Americans, new alchemists, organic farmers, perennial polyculturists, low input producers, seed savers, biodynamicists, horse farmers—who continuously produce and reproduce a landscape of alternative agricultural possibilities. This landscape comprises institutional as well as technical alternatives, for unconventional producers have been supported in their efforts by a set of unconventional institutions, some of which are of their own making (e.g. Practical Farmers of Iowa, Southwest Wisconsin Farmers Research Network) and others of which have been established by apostates who have defected from conventional science to pursue alternative paths of knowledge production (e.g. the Rodale Institute, the Land Institute, the Seed Savers' Exchange).

Through all the lean decades of official neglect and an agricultural policy environment actively hostile to their interests, many alternative farmers and alternative institutions managed not only to survive but even to thrive. Their persistence, coupled with the increasingly conspicuous failings of conventional industrial agriculture and the pressures applied by agro-environmental public interest groups, have created an intellectual and political space in which the potentials of an improved goodness of fit, or substantive interaction, between scientists and farmers appears even to the NRC and the USDA as a means of developing kinder and gentler agricultural technologies and production practices.

An emerging interest in the potentials of on-farm research is clearly apparent in the agroscientific community (Lightfoot 1987; NRC 1989; Francis *et al.* 1990; Lockeretz and Anderson 1990). As part of the 1985 Farm Bill, Congress passed the Agricultural Productivity Research Act (Public Law 99–198), which actually required the systematic initiation of cooperative research with agricultural producers. In fulfilling its congressional mandate, the USDA has established a Low Input/Sustainable Agriculture (LISA) program intended to fund research projects that take "a whole farm or SYSTEMS approach" and involve "FARMER PARTICIPATION" (USDA 1988). Many state level institutions—land grant universities, state departments of agriculture—have also identified farmer participatory research as an important component of their initiatives in the area of sustainable agriculture. Prominent among these are the Center for Integrated Agricultural Systems at the University of Wisconsin and the Aldo Leopold Center at Iowa State University. The efforts taken by such organizations to enhance articulation and cooperation between scientists and producers are encouraging. But they also reveal just how little expertise and experience there is with such ventures and how difficult it is to counter the powerful forces and incentives that hold scientists to established patterns (Thornley 1990; Stevenson *et al.* 1991).

There is indeed growing interest in bringing the farmer back in. But we need to be clear about what it is that we are bringing together before we can decide how that is to be accomplished. There now exists a window of opportunity in which to reverse the historical marginalization of local knowledge and to move the development of agricultural science out of its established trajectory and onto a reconstructive path. But the existence of this window may only be transitory, and its transparency is already contested as agribusiness mobilizes its resources in an attempt to dominate discourse and to make its meaning of "alternative agriculture" the universal meaning (Kleinman and Kloppenburg 1991; Kloppenburg 1991). How can we foster support for and understanding of local knowledge production in agriculture? What kinds of articulations might it be desirable to establish between local knowledge and scientific knowledge? How might such articulations be achieved? Are these the right questions? Where can we look for guidance?

ACCEPTING PARTIALITY: ARTICULATING SITUATED KNOWLEDGES

The purpose here has been to suggest that there are a variety of places in which to find the guidance required. No one of these intellectual locales by itself offers sufficient resources, and all may be necessary in varying degrees. Haraway's (1988: 583) central precept seems appropriate here: "The moral is simple: only partial perspective promises objective vision." What we need to do is to establish conversations among these partial perspectives and ground them in the specific and material context of the agricultural sector. Some of the principal topics that might be the subject of such conversations are now outlined.

Defining "local knowledge"

The central question, of course, is "What is 'local knowledge'; is it different from scientific knowledge and, if so, how?" Much of the preceding discussion has been devoted to establishing that such knowledge exists and that investigating its character and content would reward both theory and practice and possibly even provide a basis for the transformation of actually existing science. So far, the concern has been with presenting the various resources available for an exploration of this sort rather than with the exploration itself. Nevertheless, by expressing a preference for the term "local knowledge," the implication is that "locality"—in the sense of inseparability from a particular *place* in the sense of embeddedness in a particular *labor process*—is the key distinguishing feature of this type of knowledge.[8]

Surely, given the theoretical resources available, many other productive interpretations are possible. Anthropologists have begun to examine American agriculture (Chibnik 1987) and have even begun to touch upon the contours of local knowledge (Bennett 1982; Wells 1991). Wells' conclusion that California farmers' knowledge systems are constructed through the operation of specifiable social networks and her description of the ways in which the character of these knowledge systems varies among ethnic groups seem particularly promising. A diverse set of nonacademic analysts—principally farmers or activists working with farmers—also provides a rich fund of information on local technical knowledge in American agriculture (Berry 1984; Irwin 1990; Logsdon 1984; Strange 1988). The focus of this body of work is on understanding the production and reproduction of local knowledge as a "live tradition" (Berry 1984: 25). What should characterize sociological efforts to explore and define the parameters of local knowledge is careful attention to both theorization and the observed evidence of local knowledge production gained through direct contact with farmers and agricultural workers. Harper's (1987) superb study of a rural mechanic is a model of the sort of work that could be accomplished on the farm in order to define local knowledge and to understand the social context in which it is generated, transmitted, and used.

Recovering the historical farmer as a knowledge producer

The understanding gained through direct analysis of contemporary local knowledge production on the farm should guide the recovery and reintegration of the historical farmer as a knowledge producer. Privileging the written records left by the evangelists of an emergent agricultural science, historians have too often accepted and promulgated the image of the "reluctant farmer" and celebrated the rise of cooperative extension as a "victory of change and progress over traditionalism and apathy" (Scott 1970: 3). Like women's knowledge, the skills, practices, and wisdom developed by farmers have been, in Rowbotham's (1973) words, "hidden from history." But if they are hidden, perhaps they are not completely lost. What sorts of information might we be able to recover from primary and secondary historical materials simply by altering our perspective and purposefully searching out what we have so far neglected? Such investigations could result in the recovery of practices and technologies that might constitute "a resource, a fund of experience, a lexicon of proven possibilities and understood mistakes" (Berry 1977: 180) on which an alternative science can draw in developing an alternative agriculture. The connection between the achievement of agricultural scientific legitimacy and the delegitimation and marginalization of local knowledge should prove to be a rich field for socio-historical research.

The curious coincidence of agroecology and feminism

Of the conversations that it may be possible to foster, perhaps the most intimate will be that between feminism and the emergent field of agroecology. In looking to the agricultural sector, feminists will uncover a variety of standpoints with considerable affinity to their own. Wendell Berry (1977: 123) observes that "no matter how much one may love the world as a whole, one can live fully in it only by living responsibly in some small part of it . . . We thus come again to the paradox that one can become whole only by the responsible acceptance of one's partiality." This seems very close to the point Haraway (1988: 583) makes when she argues for "partial perspective . . . limited location and situated knowledge." Now knowledge produced from a "limited location"—what I have been calling "local knowledge"—provides an alternative to the immutable mobiles of Cartesian science: "All these pictures of the world should not be allegories of infinite mobility and interchangeability but of elaborate specificity and difference and the loving care people might take to learn how to see faithfully from another's point of view" (Haraway 1988: 583). And that is pretty much what Robert Chambers (1983: 201) asks us to do when he argues that "putting the last first" in agricultural development requires epistemic "reversals in learning." Such affinities between feminism and agroecology are multiple and articulable. Feminist concerns for context dependence, diversity, affection, responsibility, accountability, and dialogue in knowledge production find counterparts in the thoughts of agroecologists such as Berry (1977, 1984), Jackson (1980), Altieri *et al.* (1987), Norgaard (1987), and Odum (1989). I would not argue that these multiple points of view are homogeneous. I do believe that they are the kinds of related stances which could be "federated in solidarity" (Harding 1986), and it is precisely solidarity that we need if we are to actually achieve a sustainable agriculture.

Reformed science, successor science, or decentered science?

Feminist theory should prove extremely useful in framing conversations regarding the possible ways in which local knowledge might be involved in transforming actually existing science. In particular, the various feminist interpretations of science can be seen to imply distinct sets of hypotheses about the relationship between local and scientific knowledge, and between farmers and scientists. Feminist empiricism suggests that while the existing canon needs to be enlarged and enriched, modern science is not irremediably flawed. From this perspective, local knowledge and scientific knowledge are fundamentally complementary. The implication is that agricultural scientists need to take what farmers know seriously, but that such knowledge is more or less translatable into existing scientific frameworks (though those frameworks themselves may be partially restructured by such translation). Feminist standpoint theory eschews reform in favor of fundamental epistemological reconstruction. Women's experience does constitute a separate reality and, by extension, local knowledge also constitutes a separate reality. The point is not to establish complementarities or translations (which simply reinforce the hegemony of Cartesianism), but to foster so complete a deconstruction of existing science that the emergence of a successor science on a new epistemological base becomes not only possible but necessary. The practical means for achieving such an epistemic birthing are difficult to imagine, but would surely involve the dissolution of the institutional and intellectual boundaries now separating farmers and agricultural scientists.

Feminist postmodernism suggests that the transition to a successor science is a mistaken project. Multiple and separate realities do exist and to suggest that a universal epistemological stance is possible and desirable—however feminist, holist, or organicist it might be—is simply to replace one hegemony with another (Haraway 1986). While difference must be recognized and valued, productive interactions between ways of knowing can be established through partial connection and "decentered knowledge seeking" (Harding 1986: 55). Farmers know something that agricultural scientists do not know and cannot completely know; and vice versa. Articulations between these different ways of knowing need to be established not in order to combine the knowledges, and not to translate the knowledges, but to permit mutually beneficial dialogue. The problem is not one of choosing between scientific knowledge or local knowledge, but of creating conditions in which these separate realities can inform each other.

Alternative methods for an alternative science

One product of such a struggle has already been criticism of existing methodologies of Cartesian science and the slow emergence of alternative techniques for learning about the world and of articulating differently situated knowledges. Haraway (1988: 584) suggests that "there is a premium on establishing the capacity to see from the peripheries and the depths." In his book *Rural Development: Putting the Last First* Robert Chambers (1983) details a wide range of practical steps—learning reversals—that can be taken by scientists to learn how to "think from below." How might his work be applied to the rural sectors of the advanced industrial nations? How can we foster the engagement of rural peoples' own knowledge in self-development and self-empowerment? The more we learn about local knowledge and the social integument of its generation and transmission, the better we will be able to respond to those questions.

Further, agroecologists have begun to explore the possibilities of research methods that respect the integrity of farming systems as ecological and social unities (MacRae *et al.* 1989). Proponents of "hierarchical theory" have begun to generate methods which "combine holism and reductionism" to address the structure, function, and interrelation of the different levels of organization which they believe characterize complex systems (Allen and Starr 1982: Odum 1989). Sociological attention to the social constitution of research methodology should provide some interesting insights into this process. Just as technology is a product of social choice, the techniques used to produce knowledge are also selections from among a range of possibilities (Knorr-Cetina 1981; Latour 1987). If as social analysts we must be alert to the lost possibilities and foregone alternatives to the technologies that ultimately emerge from the laboratory, we must also recognize that research methods are also being lost or foregone. And if we now know little about how and why scientists select or construct their methods, we know even less about farmers' methods of experimentation and trial.

Women and the transformation of agricultural science

Finally, the role of women scientists as vectors bearing social codes of epistemological transformation should be an interesting topic for conversation. Keller (1983) has described the distinctive vision and practice—a feeling for the organism—that Barbara McClintock brought to genetics. The degree to which enlarging the participation of women in science can itself be a potent catalyst for epistemic transformation is an important strategic issue for the feminist reconstructive project (Harding 1986: 247; Keller 1988: 241). Hrdy (1986) and Haraway (1989) argue that the accumulation of feminist consciousness that accompanied the increasing number of women in the field of primatology resulted in the toppling of long held disciplinary paradigms and traditions of narrative. There is now occurring a rapid growth in the number of women in the agricultural sciences. And, while there is not yet any substantial population of internal critics within the agricultural sciences, what ferment does exist inside the disciplines appears to be substantially female, and the most expansive and creative thinking is, in fact, explicitly feminist (Crouch 1990a, 1990b; Handelsman 1991; Handelsman and Goodman 1991). Could what happened in primatology be recapitulated in, say, plant pathology?

CONCLUSION

The agricultural sciences and the agricultural sector as a whole stand now at a pivotal conjuncture. More space is available now for moving agricultural technoscience onto new trajectories than at any time in American history. A critical rural sociology has played a key role in pushing forward the deconstructive project that has been instrumental in creating this space. Many rural sociologists are interested in participating in the reconstructive project as well. But in this effort we need to enlarge not only the canon of our colleagues in the natural sciences, but our own canon as well. In this article I have suggested what the theoretical resources for such an enlargement might be.

Sociological constructivism provides a set of tools for the deconstruction of actually existing science, but has not developed the political or social conscience that would direct the reconstructive use of those tools. Feminist analysis brings such a conscience to bear and actively imagines alternative régimes of knowledge production, but has so far not addressed the agricultural sector as a concrete terrain for the working out and testing of theory and practice. The literature on what I have labeled "local" knowledge constitutes a rich conceptual and empirical resource, but analyses are widely dispersed across time and discipline and lack explicit points of contact and comparison. Studies of indigenous technical knowledge provide a wealth of information on the actual activities of local knowledge production in agriculture, but the field lacks the theoretical base that would give it a self-conscious epistemic stance and a developed awareness that local knowledge might be more than just a complement to Cartesian science in the Third World.

Articulated as partial realities, these perspectives may accomplish in conversation what none of them can alone. The new sociology of science has opened for us a crucial deconstructive door. It is feminist theory that speaks most clearly as to how to proceed through that door. In turn, the agricultural sector provides a uniquely appropriate concrete terrain for the testing of a whole range of theoretical propositions drawn from both the sociology of science and feminism, and for the necessary work of developing and elaborating the "here-and-now prefigurative forms" (Rose 1986: 73) of what might one day be a transformed science. And that transformed science will need to encompass the distinctive contributions to understanding the world that can be provided by "local knowledge."

Wendell Berry (1977: 160) has written that Cartesian science "accumulates information at a rate that is literally inconceivable, yet its structure and its self-esteem institutionalize the likelihood that not much of this information will ever be taken *home*." That is, it is not sufficiently relevant *locally*; it fails to take home —the distinctiveness of particular cultural, social, and ecological spaces—sufficiently into account. A truly alternative agriculture must be based on a truly alternative science that articulates multiple ways of knowing. Rural sociologists can and should participate in this articulation. We can go home again.

NOTES

1 For support of this research I am grateful to the MacArthur Foundation—Social Science Research Council, Program in International Peace and Security and to the University of Wisconsin, College of Agricultural and Life Sciences. I would also like to thank Jess Gilbert, Daniel Kleinman, and Cynthia Truelove for their critiques of preliminary versions of this article. The comments of Steve Murdock and four anonymous referees helped me make my arguments with increased clarity.

2 The analysis contained in this article is "deconstructive" in the sense that, as Jane Flax (1986: 195) put it, it seeks "to distance us from and make us skeptical about ideas concerning truth, knowledge, power, the self, and language that are often taken for granted within and serve as legitimation for contemporary Western culture."

3 Several of the referees for this article gained the impression that I do not believe that what I call actually existing science is capable of producing valid knowledge. This is a serious misreading of my position and that of the constructivists as well. Let me be as clear as I can. Scientific facts are socially contingent, just as are the conclusions of all other ways of knowing. But "socially contingent" does not mean "false." As Busch (1984: 309) correctly emphasizes, "the problem is not that scientific and technical truths are relative, but that they are partial." Loss of its unique epistemological status does not imply a wholesale invalidation of science. It does imply the creation of space for the consideration of competing modes of knowledge production, which themselves represent partial understandings.

4 In affirming the importance of such locally based experiential knowledge, I do not mean to imply that it is free of social contingency. The scientific method does not produce a reading off of nature unmediated by social relations, and neither does direct experience. The "facticity" of experience is every bit as socially constructed as is the "facticity" of science (see Scott 1991).

5 As one referee correctly noted, scientists produce local knowledge too. It is the existence of such local knowledge—"the largely inaccessible idiosyncrasies of the individual or the laboratory"—that explains, for example, the inability to easily replicate hybridomas across molecular biology labs (Cambrosio and Keating 1988). But, as I hope will shortly become clear, I believe that the knowledge that enables the technician to synthesize the hybridoma is quite different from what I am going to call scientific knowledge.

6 In the anthropological and international development literature "indigenous knowledge" has been the most common term used to refer to what I call "local knowledge." My own analysis of locality and choice of terminology has been influenced by McCorkle's (1989: 4–5) and Thrupp's (1989: 14) assertions that "local knowledge" most fully captures the sense in which this type of knowledge is distinctive. See Chambers (1983: 82–83) for discussions of the relative utility of other terms.

7 Because local technical knowledge is "pre-adapted to its physical and human ecology" (McCorkle 1989: 8), its elaboration and improvement are more likely than exogenous innovation to be environmentally and socially appropriate and therefore more likely to be sustainable in the long term. Moreover, intimate and sustained engagement with their means and conditions of production endow farmers not only with deep knowledge of local particularities, but also with a holistic and systemic understanding of local agriculture that reductionistic science cannot easily approximate. While cautioning that local technical knowledge is not free from error (of course, neither is science), Chambers (1983: 75) concludes that "Rural people's knowledge and scientific knowledge are complementary in their strengths and weaknesses. Combined they may achieve what neither would alone."

8 I will briefly outline the thrust of my own current thinking. A dairy farmer produces new knowledge about milk production in the process of producing milk. But the physical and temporal space available to a farmer for knowledge-producing activity is defined by the nature of commodity production on the farm. For dairy farmers, generally, a necessary condition of new knowledge production is success in milk production. The knowledge production activity of farmers is thus a secondary process which is necessarily simultaneous with, embedded in, inextricable from, and constrained by the primary process of commodity production. The resources available for use by the farmer in knowledge production are limited to those which are also *locally* available for use in commodity production. Such local knowledge production depends on the unaided senses, accumulates in time-bound fashion through aggregative experience, and is holistic.

The scientific labor process is quite different and, in fact, is unique inasmuch as it makes the generation of new knowledge its primary objective rather than a secondary epiphenomenon (Whitley 1977: 25). Scientists are not more rational than farmers, they have no capacity to think more abstractly, they are not necessarily even better experimenters. What dairy scientists do enjoy is release from the constraints of milk production. No longer completely bound by the locality of their labor process, they develop tools which uncouple knowledge production from "situatedness" and "personal" perception (i.e. the microscope which permits access to subperceptual entities or the survey form that permits collection of data at a supra-perceptual dimension). The scientific laboratory can (within certain persistent limits of locality) be everywhere and nowhere and the knowledge generated therein is relatively immutable and mobile, whereas local knowledge, bound to the locality of a particular labor process, is relatively mutable and immobile.

REFERENCES

Allen, Timothy F. H., and T. B. Starr 1982 *Hierarchy: Perspectives for Ecological Complexity*. Chicago, IL: The University of Chicago Press.

Altieri, Miguel A., with contributions by Richard B. Norgaard, Susanna B. Hecht, John G. Farrell, and Matt Liebman 1987 *Agroecology: The Scientific Basis of Alternative Agriculture*. Boulder, CO: Westview Press.

Amsterdamska, Olga 1990 "Surely you are joking, Monsieur Latour!" *Science, Technology, & Human Values* 15 (4) (Autumn): 495–504.

Aronowitz, Stanley 1988 *Science as Power: Discourse and Ideology in Modern Society*. Minneapolis, MN: University of Minnesota Press.

Barnes, Barry, and David Bloor 1982 "Relativism, rationalism, and the sociology of knowledge." Pp. 21–47 in Martin Hollis and Steven Lukes (eds), *Rationality and Relativism*. Cambridge, MA: MIT Press.

Bennett, John W. 1982 *Of Time and the Enterprise: North American Family Farm Management in a Context of Resource Marginality*. Minneapolis, MN: University of Minnesota Press.

—— 1986 "Research on farmer behavior and social organization." Pp. 367–402 in Kenneth A. Dahlberg (ed.), *New Directions in Agriculture and Agricultural Research*. Totowa, NJ: Roman & Allanheld.

Berry, Wendell 1977 *The Unsettling of America: Culture and Agriculture*. New York, NY: Avon Books.

—— 1984 "Whose head is the farmer using? Whose head is using the farmer?" Pp. 19–30 in Wes Jackson *et al.* (eds), *Meeting the Expectations of the Land: Essays in Sustainable Agriculture and Stewardship*. San Francisco, CA: North Point Press.

Bittner, Egon 1983 "Technique and the conduct of life." *Social Problems* 30 (3): 249–61.

Bleier, Ruth 1986 "Introduction." Pp. 1–17 in Ruth Bleier (ed.), *Feminist Approaches to Science*. New York, NY: Pergamon Press.

Bohme, Gernot 1984 "Midwifery as science: an essay on the relation between scientific and everyday knowledge." Pp. 365–85 in N. Stehr and V. Meja (eds), *Society and Knowledge*. New Brunswick, NJ: Transaction Books.

Braverman, Harry 1974 *Labor and Monopoly Capital: The Degradation of Work in the Twentieth Century*. New York, NY: Monthly Review Press.

Brokensha, David W., D. M. Warren, and Oswald Werner (eds) 1980 *Indigenous Knowledge Systems and Development*. Lanham, MD: University Press of America, Inc.

Burawoy, Michael 1979 *Manufacturing Consent: Changes in the Labor Process under Monopoly Capitalism*. Chicago, IL: The University of Chicago Press.

Busch, Lawrence 1984 "Science, technology, agriculture, and everyday life." Pp. 289–314 in H. K.

Schwarzweller (ed.), *Research in Rural Sociology and Development*, vol. 1. Greenwich, CT: JAI Press.

Busch, Lawrence, and William B. Lacy 1983 *Science, Agriculture, and the Politics of Research*. Boulder, CO: Westview Press.

Buttel, Frederick H. 1986 "Biotechnology and agricultural research policy: emergent issues." Pp. 311–47 in Kenneth A. Dahlberg (ed.), *New Directions for Agriculture and Agricultural Research: Neglected Dimensions and Emerging Alternatives*. Totowa, NJ: Rowan & Allanheld.

Buttel, Frederick H., and Michael E. Gertler 1982 "Agricultural structure, agricultural policy, and environmental quality: some observations on the context of agricultural research in North America." *Agriculture and Environment* 7: 101–19.

Callon, Michel, and John Law 1989 "On the construction of sociotechnical networks: content and context revisited." Pp. 57–83 in *Knowledge and Society: Studies in the Sociology of Science Past and Present*, vol. 8. New York, NY: JAI Press.

Cambrosio, Alberto, and Peter Keating 1988 "'Going monoclonal': art, science, and magic in the day-to-day use of hybridoma technology." *Social Problems* 35 (3) (June): 244–60.

Carson, Rachel 1962 *Silent Spring*. Boston, MA: Houghton Mifflin.

Chambers, Robert 1983 *Rural Development: Putting the Last First*. New York, NY: John Wiley & Sons, Inc.

Chambers, Robert, Arnold Pacey, and Lori Ann Thrupp (eds.) 1989 *Farmer First: Farmer Innovation and Agricultural Research*. New York, NY: The Bootstrap Press.

Chibnik, Michael (ed.) 1987 *Farm Work and Fieldwork: American Agriculture in Anthropological Perspective*. Ithaca, NY: Cornell University Press.

Cohn, Carol 1987 "Sex and death in the rational world of defense intellectuals." *Signs: Journal of Women in Culture and Society* 12 (4): 687–718.

Commoner, Barry 1972 *The Closing Circle: Nature, Man, and Technology*. New York, NY: Alfred A. Knopf.

Council for Agricultural Science and Technology (CAST) (ed.) 1990 *Alternative Agriculture: Scientists' Review*. Special Publication No. 16, July. Ames, IA: CAST.

Cozzens, Susan E., and Thomas N. Gieryn (eds) 1990 *Theories of Science in Society*. Bloomington, IN: University of Indiana Press.

Crouch, Martha L. 1990a "Debating the responsibilities of plant scientists in the decade of the environment." *The Plant Cell* 2 (April): 275–77.

—— 1990b "Biotechnology and sustainable agriculture: philosophical musings by an ex-genetic engineer." *Journal of New World Agriculture* 8 (2) (Fall): 5, 13.

Fee, Elizabeth 1986 "Critiques of modern science: the relationship of feminism to other radical epistemologies." Pp. 43–56 in Ruth Bleier (ed.), *Feminist Approaches to Science*. New York, NY: Pergamon Press.

Flax, Jane 1986 "Gender as a problem in and for feminist theory." *Amerikastudien/American Studies* 31 (2): 193–213.

Francis, Charles, James King, Jerry DeWitt, James Bushnell, and Leo Lucas 1990 "Participatory strategies for information exchange." *American Journal of Alternative Agriculture* 5 (4): 153–60.

Freire, Paulo 1970 *Pedagogy of the Oppressed*. New York, NY: The Seabury Press.

Friedland, William H. 1978 *Social Sleepwalkers*. Research Monograph No. 13, Davis, CA: Department of Applied and Behavioral Sciences, University of California—Davis.

Gieryn, Thomas N. 1983 "Boundary-work and the demarcation of science from non-science: strains and interests in professional ideologies of scientists." *American Sociological Review* 48 (December): 781–95.

Handelsman, Jo Emily 1991 "Changing the image of agriculture through curricular innovation." *Proceedings of Investing in the Future: Professional Education for the Undergraduate*. Washington, DC: National Academy Press.

Handelsman, Jo Emily, and Robert Goodman 1991 "Banning biotechnology." In Richard Klemme (ed.), *Proceedings of the Conference on Rural Wisconsin's Economy and Society: The Influence of Policy and Technology*. Madison, WI: Center for Integrated Agricultural Systems, University of Wisconsin.

Haraway, Donna 1986 "Primatology is politics by other means." Pp. 77–118 in Ruth Bleier (ed.), *Feminist Approaches to Science*. New York, NY: Pergamon Press.

—— 1988 "Situated knowledges: the science question in feminism and the privilege of partial perspective." *Feminist Studies* 14 (3) (Fall): 575–99.

Haraway, Donna 1989 *Primate Visions: Gender, Race, and Nature in the World of Modern Science*. New York, NY: Routledge.

Harding, Sandra 1986 *The Science Question in Feminism*. Ithaca, NY: Cornell University Press.

Harper, Douglas 1987 *Working Knowledge: Skill and Community in a Small Shop*. Chicago, IL: The University of Chicago Press.

Hartsock, Nancy 1983 "The feminist standpoint: developing the ground for a specifically feminist historical materialism." Pp. 283–310 in Sandra Harding and Merrill B. Hinikka (eds), *Discovering Reality*. Dordrecht, The Netherlands: D. Reidel.

Hatch, John K. 1976 *The Corn Farmers of Motupe: A Study of Traditional Farming Practices in Northern Coastal Peru*. Land Tenure Center Monograph No. 1. Madison, WI: Land Tenure Center, University of Wisconsin.

Hightower, Jim 1973 *Hard Tomatoes, Hard Times*. Cambridge, MA: Schenkman Publishing Co.

Hirschhorn, Larry 1984 *Beyond Mechanization: Work and Technology in a Postindustrial Age*. Cambridge, MA: The MIT Press.

Hrdy, Sarah Blaffer 1986 "Empathy, polyandry, and the myth of the coy female." Pp. 119–46 in Ruth Bleier (ed.), *Feminist Approaches to Science*. New York, NY: Pergamon Press.

Husserl, Edmund 1970 *The Crisis of European Sciences and Transcendental Phenomenology*. Evanston, IL: Northwestern University Press.

Irwin, Mike 1990 *From the Ground Up: Wisconsin Sustainable Farmers Tell of Their Practice and Vision*. Madison, WI: Madison Area Technical College.

Jackson, Wes 1980 *New Roots for Agriculture*. San Francisco, CA: Friends of the Earth.

Keller, Evelyn Fox 1983 *A Feeling for the Organism: The Life and Work of Barbara McClintock*. New York, NY: W. H. Freeman.

—— 1987 "The gender/science system: or, is sex to gender as nature is to science?" *Hypatia* 2 (3) (Fall): 37–49.

—— 1988 "Feminist perspectives on science studies." *Science, Technology, & Human Values* 13 (3 & 4) (Summer & Autumn): 235–49.

Kenney, Martin 1986 *Biotechnology: The University-Industrial Complex*. New Haven, CT: Yale University Press.

Kleinman, Daniel Lee 1991 "Conceptualizing the politics of science: a comment on Cambrosio *et al.*" *Social Studies of Science* 21(4): 769–74.

Kleinman, Daniel Lee, and Jack Kloppenburg, Jr. 1991 "Taking the discursive high ground: Monsanto and the biotechnology controversy." *Sociological Forum* 6 (September): 3.

Kloppenburg, Jack R., Jr. 1988 *First the Seed: The Political Economy of Plant Biotechnology, 1492–2000*. New York, NY: Cambridge University Press.

—— 1991 "Alternative agriculture and the new biotechnologies." *Science as Culture* 2 (4) (October): 13.

Knorr-Cetina, Karin 1981 *The Manufacture of Knowledge: An Essay on the Constructivist and Contextual Nature of Science*. New York, NY: Pergamon Press.

Krimsky, Sheldon 1984 "Epistemic considerations on the value of folk-wisdom in science and technology." *Policy Studies Review* 3 (2) (February): 246–62.

Latour, Bruno 1986 "Visualization and cognition: thinking with eyes and hands." Pp. 1–40 in *Knowledge and Society: Studies in the Sociology of Culture Past and Present*. Greenwich, CT: JAI Press.

—— 1987 *Science in Action: How to Follow Scientists and Engineers through Society*. Cambridge, MA: Harvard University Press.

Latour, Bruno, and Steve Woolgar 1979 *Laboratory Life: The Construction of Scientific Facts*. Los Angeles, CA: Sage.

Levi-Strauss, Claude 1962 *The Savage Mind*. Chicago, IL: The University of Chicago Press.

Levins, Richard, and Richard Lewontin 1985 *The Dialectical Biologist*. Cambridge, MA: Harvard University Press.

Lightfoot, Clive 1987 "Indigenous research and on-farm trials." *Agricultural Administration and Extension* 24: 79–89.

Lockeretz, William, and Molly D. Anderson 1990 "Farmers' role in sustainable agriculture research." *American Journal of Alternative Agriculture* 5 (4): 178–83.

Logsdon, Gene 1984 "The importance of traditional farming practices for sustainable modern agriculture." Pp. 3–18 in Wes Jackson *et al.* (eds), *Meeting the Expectations of the Land: Essays in Sustainable Agriculture and Stewardship*. San Francisco, CA: North Point Press.

Longino, Helen C. 1990 *Science as Social Knowledge: Values and Objectivity in Scientific Inquiry*. Princeton, NJ: Princeton University Press.

McCorkle, Constance M. 1989 "Toward a knowledge of local knowledge and its importance for agricultural RD&E." *Agriculture and Human Values* 6 (3) (Summer): 4–12.

MacRae, Rod J., Stuart B. Hill, John Henning, and Guy R. Mehuys 1989 "Agricultural science and sustainable agriculture: a review of the existing scientific barriers to sustainable food production and potential solutions." *Biological Agriculture and Horticulture* 6 (3): 173–218.

Marcus, Alan I. 1985 *Agricultural Science and the Quest for Legitimacy*. Ames, IA: Iowa State University Press.

Marcuse, Herbert 1964 *One Dimensional Man*. Boston, MA: Beacon Press.

Merchant, Carolyn 1980 *The Death of Nature: Women, Ecology, and the Scientific Revolution*. New York, NY: Harper & Row.

Merton, Robert K. 1973 *The Sociology of Science: Theoretical and Empirical Investigations*. New York, NY: Harper and Row.

Mulkay, Michael 1979 *Science and the Sociology of Knowledge*. London, UK: George Allen & Unwin.

National Research Council (NRC) 1989 *Alternative Agriculture*. Washington, DC: National Academy Press.

Norgaard, Richard B. 1987 "The epistemological basis of agroecology." Pp. 21–28 in Miguel Altieri *et al.* (eds), *Agroecology: The Scientific Basis of Alternative Agriculture*. Boulder, CO: Westview Press.

Odum, Eugene P. 1989 "Input management of production systems." *Science* 243 (13 January): 177–82.

Pirsig, Robert M. 1974 *Zen and the Art of Motorcycle Maintenance*. New York, NY: William Morrow and Company, Inc.

Polanyi, Michael 1962 "The republic of science." *Minerva* 1 (1): 54–73.

—— 1966 *The Tacit Dimension*. Garden City, NY: Doubleday & Company, Inc.

Prigogine, Ilya, and Isabelle Stengers 1984 *Order Out of Chaos: Man's New Dialogue with Nature*. New York, NY: Bantam Books.

Restivo, Sal 1988 "Modern science as a social problem." *Social Problems* 35 (3) (June): 206–25.

Richards, Paul 1985 *Indigenous Agricultural Revolution: Ecology and Food Production in West Africa*. Boulder, CO: Westview Press.

—— 1989 "Farmers also experiment: a neglected intellectual resource in African science." *Discovery and Innovation* 1 (1) (March): 19–25.

Rose, Hilary 1983 "Hand, brain, and heart: a feminist epistemology for the natural sciences." *Signs* 9 (1): 73–90.

—— 1986 "Beyond masculinist realities: a feminist epistemology for the sciences." Pp.57–76 in Ruth Bleier (ed.), *Feminist Approaches to Science*. New York, NY: Pergamon Press.

Rosenberg, Charles E. 1976 *No Other Gods: On Science and American Social Thought*. Baltimore, MD: Johns Hopkins University Press.

Rowbotham, Sheila 1973 *Hidden from History: 300 Years of Women's Oppression and the Fight Against It*. London, UK: Pluto Press.

Scott, Joan 1991 "The evidence of experience." *Critical Inquiry* 17 (Summer): 773–97.

Scott, Roy V. 1970 *The Reluctant Farmer: The Rise of Agricultural Extension to 1914*. Urbana, IL: University of Illinois Press.

Smith, Dorothy E. 1987 *The Everyday World as Problematic: A Feminist Sociology*. Boston, MA: Northeastern University Press.

Stevenson, G. W., Kathleen Duffy, and Richard Klemme 1991 *Interdisciplinary, Farmer-involved Research Teams: Principles and a Case Study of a Rotational Grazing Project in the Upper Midwest*. Center for Integrated Agricultural Systems, Madison: University of Wisconsin Press.

Strange, Marty 1988 *Family Farming: A New Economic Vision*. Lincoln, NE: University of Nebraska Press.

Suppe, Frederick 1988 "The limited applicability of agricultural research." *Agriculture and Human Values* 5 (4) (Fall): 4–14.

Thornley, Kay 1990 "Involving farmers in agricultural research: farmer's perspective." *Journal of Alternative Agriculture* 5 (4): 174–77.

Thrupp, Lori Ann 1989 "Legitimizing local knowledge: from displacement to empowerment for Third World people." *Agriculture and Human Values* 6 (3) (Summer): 13–24.

United States Department of Agriculture (USDA) 1988 *Information Bulletin: On-Farm Research Opportunity*. Washington, DC: U.S. Department of Agriculture.

Wells, Miriam J. 1991 "Ethnic groups and knowledge systems in agriculture." *Economic Development and Cultural Change* 39 (4): 739–71.

Whitley, R. D. 1977 "The sociology of scientific work and the history of scientific developments." Pp. 21–50 in Stuart Blume (ed.), *Perspectives in the Sociology of Science*. New York, NY: John Wiley & Sons.

PART 2

Justice-based goals

Introduction

In singling out "justice" as a goal of social struggle our intention is to indicate certain endemic *aspects* of social struggle. We key in on that word "aspects" because our intent is not necessarily to define justice, as in "Justice will have been served when *x* and *y* criteria have been met." As will be seen, there are multiple definitions (concepts) of justice, all serviceable under different sorts of conditions. By "justice" we mean to ask instead: What is the situation of justice? Or better, what sorts of situations call for something called "justice"? This seems to us to be an appropriate way to frame the issue, because justice in our view needs to be seen as struggle of a certain kind, worth isolating for purposes of analysis.

Setting our sights then on situations that call for justice, the first of these is the struggle to bring a claim to some sort of resource (e.g. a material good, a protection, participation in decision-making), when the right to make the claim is not necessarily in question or has been positively secured. That is to say, because the right to make a claim does not guarantee that a claim has been won, justice requires that there be access to the apparatuses of distribution and decision-making (a process that may alter what is judged right—remember Box 1 on pp. 198–200). A corollary to this is that situations arise in which there are multiple, simultaneously competing claims and a struggle for a just resolution ensues. Second, the question sometimes arises of how collaborations and alliances may be formed. This is a question about how different struggles may become joined together, and what happens when they do. Sometimes we see situations where competing struggles find a way to articulate with each other in some way; sometimes we see situations where common claims can be identified and social agents then embark upon collaborative efforts. Either way, within these very cases a question of justice seems to be involved—that is, inside of collaboration emerges the issue of who is empowered to make decisions or to distribute scarce resources. But let's bring these abstractions down to earth and think them out through the struggles that bring them to life. As before, the role of scholarship will be a primary concern—we want to know how scholarship does its work and how it might ally itself to progressive struggle. The various roles of scholarship we have already identified sometimes remain operative here, but we will leave readers to that discovery for themselves. So as not to be redundant other roles will be identified now.

The aim therefore is twofold: to identify some justice-based struggles and see how scholarly work can be joined to them. A note before continuing: It is not easy to demarcate a boundary between the sorts of struggles described in this group of essays and those recounted in the previous group on rights. Certainly we do not suggest that struggles for rights come first and if won enter the field of justice. It was mentioned earlier that the two are locked in an ontological embrace.

ASSESSING THE ADEQUACY OF DIFFERENT CONCEPTS OF JUSTICE TO PARTICULAR INSTANCES OF STRUGGLE; ASSESSING CHANGING SOCIAL-SPATIAL SITUATIONS AND NAMING NEW POLITICAL STRATEGIES

For a politics of difference, distributive and procedural justice are insufficient to the cause of just social arrangements but this is *not* to say they are wholly inadequate or have not served certain purposes. (Box 1 on pp. 198–200 reveals this point. Readers may also wish to refer back to the "exchange" between Iris Marion Young, in Chapter 4, and Nancy Fraser, in Chapter 5.) Over time and across space the situation is one whereby the adequacy and practice of different justice concepts fluctuate. Another role that scholarship can play, therefore, is to take a step back and assess the adequacy of different concepts of justice to particular and dynamically changing instances of struggle, noting how and why certain such concepts serve their purpose only to become less useful once conditions on the ground change. As with Gathorne-Hardy's research in Box 1 on pp. 198–200 this demands attention to the iterative nature of theory and practice, but now with a difference: it can be illuminating to understand just how the iterative process happens, how and why a particular concept becomes less useful and certain others become more salient. This theme is played out very strongly in Laura Pulido's research on "Restructuring and the contraction and expansion of environmental rights in the United States" (Chapter 16).

Along with struggles for national liberation, civil rights, sexual freedom, or emancipation from labor exploitation, the struggle for "environmental rights" and for their just enforcement has a prominent position on the political landscape. In Pulido's eyes (she is a geographer at the University of Southern California) the struggle for environmental rights and justice is conceptually and geographically dynamic: she examines the difference it makes when environmental struggles are carried out in the U.S.–Mexico border region, whose particularities play a role in the applicability of different social justice strategies, concepts, and goals. (Readers will note that the attention Pulido devotes to the limits of particular justice concepts, most especially procedural justice, makes Iris Young a powerful figure for her.)

As Pulido recounts, the successes of environmental activists have been many, if limited. Through the work of mainstream environmentalism significant rights have been extended across species, involving humans and non-humans alike, and over space and time, running the gamut from urban to wilderness areas. The 1960 Clean Water Act, the 1963 Clean Air Act, and the 1973 Endangered Species Act stand as examples of what was achieved, even if these pieces of legislation are regularly embattled. Then, in the 1980s and 1990s, a new round of activism was spurred on by the discovery of "toxics" at Love Canal, New York, and of a racially biased distribution of hazardous waste in Warren County, North Carolina. The latter was a signal event in galvanizing minority communities in the pursuit of "environmental justice," as that movement has come to be called. Environmental justice activists have worked especially hard to gain access to the environmental decision-making process, particularly regarding the distribution of environmental hazards. Several important victories have been gained as a result.

Pulido nonetheless argues that "[t]he environmental justice movement has focused largely on procedure and has not significantly tackled underlying structural inequality, regional capital investment patterns, or pollution reduction, and as such can only achieve marginal gains" (see p. 275). She therefore calls for a stronger linkage between environmental justice activism and other types of struggles. Three in particular stand out. First is the battle against capital flight and "uneven development." Capital's search for ever lower-cost places to do business involves avoiding (or disinvesting from) places with strict pollution regulations, and gravitating toward places with fewer or weakly enforced regulations. Second is the fight against social and political inequality, be this poverty, racism, and/or other kinds of bigotry that leave people in highly vulnerable circumstances, subject to various sorts of abuses and power-plays. Third is the struggle to gain access to decision-making regarding production. Decisions regarding what to produce, with what inputs and outputs, and at what cost to human and non-human environments are still very much a private affair. If a primary goal of the environmental justice movement is pollution reduction, Pulido argues that the movement must take on the sphere of production itself. Environmental quality stands at the intersection of these struggles, Pulido argues (cf. Iris Young on the multiple "faces of oppression", in Chapter 4). For this reason, procedural justice has become a limited concept and goal for the movement.

An object lesson is the U.S. Southwest–Mexico cross-border region. Pulido documents what she calls a "contraction" of environmental rights in California, the result of an attack on that state's environmental regulations, and an "expansion" of rights south of the border in Mexico, where activists have awakened to the polluting of their own sphere. This contraction and expansion have awakened activisits on both sides of the border to the need for transborder environmental justice activism that engages in the multiple struggles noted above. Procedural environmental justice remains part of the political imaginary and part of political practice, but it is articulated closely to those other goals.

RECONCILING (ARTICULATING) THE DIVERSE POLITICS OF DIFFERENCE?

A primary aim of Pulido's essay is to identify a "new political landscape in which activists must renegotiate the terms of struggle" (see p. 274). As we saw, these terms involved allying procedural justice as a strategy to other struggles that engage, both critically and constructively, the forces of structural power. Pulido cites several instances of this collaborative work. But the necessity for alliance, collaboration, or some other strategic joining of forces, is such a constant refrain within social struggles that it must be wrestled with as a distinct dynamic which produces problems of its own. The essence of the problem seems to be that the very identities of "us" and "them" that structure the relationship between a political movement and what it moves against also structure politics *inside* of struggle. This, at least, is one way of stating what the problem is. Another and quite different way of putting it is to say that social struggle produces new sorts of identities. That is, for those who see political action in this way it is important to note that articulations of movements and political agents actively produce identities, discourses, practices, and aims that were not there prior to their articulation, at least not overtly. (For theorists such as Ernesto Laclau and Chantal Mouffe this is one of the constructive promises of an agonistic politics.) Justice struggles may, in short, be seen as ongoing, reflexive struggles to make a "we"; they are about the struggle to constitute struggle. We will examine this kind of justice situation in the last two articles of this cluster. The first returns us to this business of environmental justice in the U.S. West but notes how complicated it becomes when conjoined with the issue of American Indian sovereignty.

It is no news that the U.S. West is extraordinarily complicated politically. As Noriko Ishiyama describes in Chapter 17, based on her dissertation research in the Geography Department at Rutgers University, during the post-World War II period the question of what to do about radioactive wastes produced by nuclear-based commercial energy production and military-industrial buildup has intensified. These wastes have an excessively long life as an ecological hazard. And the question of where to put them only intensifies with the continued production of nuclear-based energy and armaments. One temporary solution sought by the Federal Government has been (and continues to be) to offer payment in return for storage. But, as Ishiyama notes, Congressional funding for these payments has been insufficient and commercial producers of nuclear energy have had to contract for waste storage on their own. In the early 1990s this situation led the Skull Valley Band of Goshutes, a small American Indian tribe with a reservation in Tooele County, Utah, an hour's drive from Salt Lake City, to contract for the storage of high-level radioactive waste on the reservation. This was against the desires of environmental activists, environmental justice activists inclusive, and against the wishes of the state government. (In an ironic twist pointed out by Ishiyama, the state government, which had long played a role in environmental despoliation without regard for tribal concerns, was against storage partly for reasons of environmental injustice.) Contrary to the assertions by environmental activists that the Skull Valley Band was a victim of environmental injustice, tribal leaders claimed to be acting in the name of tribal sovereignty, that is, claimed to be acting not as victims but as knowledgeable, self-determining actors. Their struggle, as they saw it, was not for environmental justice; it was the continuation of a struggle for sovereignty (see Bebbington 2004 for a comparable case in Ecuador). As Ishiyama indicates, no one was claiming there was no such thing as environmental injustice. Tribal leaders were well aware of this problem as a defining aspect of their history in Utah. Instead, tribal leaders were framing the incident in a quite different, more encompassing manner, i.e. the struggle for sovereignty. At stake, they argued, was economic development, notoriously difficult to achieve in Tooele County's desolate stretch of desert. Furthermore, it was through such development that an appropriate environmental management infrastructure could be built. As Ishiyama explains, the development of such an infrastructure has lagged, not least because of the Federal Government's approach to American Indian affairs. The situation prompts Ishiyama to ask, "What is environmental justice in the context of questions of tribal sovereignty?" (see p. 293) The question becomes more complicated still when one considers that the Skull Valley Band was itself not unified on the storage issue: the struggle over sovereignty and environment was staged as much within the tribe as between tribe and non-tribe. (Indeed, this is only a short-list of the struggles. See pp. 300–2 of the article.)

As is true in Laura Pulido's analysis, Ishiyama concludes that the case in Utah resolves neither to distributive justice nor procedural justice, although both could potentially have their uses. As conventionally pursued by environmental justice advocates, distributive justice relies too simply on matching the location of hazards to the location of tribes in such a way as to maximally reduce the risk of exposure. This is too simplistic for the reasons already given: it ignores the issue of tribal sovereignty. And procedural justice is not a simple matter of opening tribal access to environmental decision-making in Utah by letting the tribal leadership manage such decision-making. What Ishiyama gets us to see is that a constellation of different identities (different permutations of "we," if you will) have emerged within the Skull Valley Band of Goshute Indians. Some of the identities link members of the Skull Valley Band to people outside the tribe, some of whom are Native American and others not: there are therefore environmental justice activists within the Skull Valley Band who regard themselves as people of color, an identity inclusive of non-American Indians. They work with other environmental justice activists of color, and contest the platform of the tribal leadership. But there are also members of the Skull Valley Band who emphasize "the significance of sovereignty, which makes American Indian tribes distinctive from other ethnic minorities fighting against environmental injustice" (see p. 300). While some of these members then agree with the tribal leadership that the issue of sovereignty is indeed at stake, they see the tribal leadership as having forsaken a cultural truth: "Indigenous Peoples . . . are only caretakers of this great sacred land" (see p. 301). Sovereignty for them is not simply asserted to an antagonistic outside; it is also part of an internal, agonistic struggle, set in motion in the first place by an already unjust choice set. For its part, then, the tribal leadership assumes the mantle of "we," in so far as it represents the Skull Valley Band of Goshute Indians at large.

There are then multiple collective identities, each salient for different but related struggles and each capable of composing political networks that are inclusive of some people but not everyone. It is a crucial point for Ishiyama that the identities seized upon actively direct political strategies, as when Goshute opponents of the tribal leadership seize upon the idea of indigeneity, claim greater ecological legitimacy, and thereby make cause with environmental justice networks more broadly. What does this particular struggle illustrate? Individuals do not have one and only one identity; there is no single identity capable of embracing the numerous affinities that persons may have. Therefore, the various things that happen in the real world often tug single individuals in different directions. (Note that this is a notion of identity that some Skull Valley Band members would likely have real trouble with!) Interestingly, though, alliances that are formed to broaden the cause of progressive social change cannot even be possible without these internal differences, but nor is the purpose of alliance to dissolve difference (Bystydzienski and Schacht, 2001). But this returns us straight to the situation of justice: We are returned to the struggle over how to structure differences—different struggles and different struggles for difference—as agonistically as possible. (And as vexing as this is, it is not a call for the flattening of difference as a goal of radical politics. These politics rely upon difference and aim as a goal to constructively produce it [see Chapter 19.]) On agonistic and procedural grounds, Ishiyama argues that justice will not have been served if the tribe opts to exercise sovereignty and contracts for storage of the nuclear waste. The historical geographic context is that such a decision will have been reached under highly constrained, unjust conditions. Ishiyama in fact refuses to try to settle the controversy, opting instead for a note of caution: "environmental-justice scholars are encouraged to reframe their research questions to articulate the truly complex practices of political economy and historical colonialism over communities' struggles to self-determination" (see p. 303). Our sense, though, is that she underplays her hand, that she raises questions of importance for scholars and also for how movements may or may not be linked, and as accompanied by what sorts of social and geographical knowings: What sort of claims can different activists and activisms make on each other? How might progressive social actors waken to the possibility of shifting their terms of struggle and their conceptual armament? In what ways might one sort of struggle be continuable by or through another? What the Pulido and Ishiyama cases tell us is that justice is not simply achieved by the articulation of movements; justice, as discussed in Box 2, concerns additionally that struggle to articulate.

BOX 2: JUSTICE AND THE STRUGGLE TO FORM STRUGGLES: THE INTER-CONTINENTAL CARAVAN

As we will see in the contributions of Noriko Ishiyama (Chapter 17) and Laura Pulido (Chapter 16), situations that call for justice may call for coordination by different activists and activisms. How can this coordination happen when differences may be deeply entrenched? What tools and resources, concepts and practices, might be necessary? This is the question posed by David Featherstone, a geographer at University of Liverpool, in his research on the Inter-Continental Caravan (ICC), a transnational movement against capitalist globalization (Featherstone, 2003). Like Gathorne-Hardy's analysis of the housing project in Box 1 on pp. 198–200, Featherstone sees in the ICC a skepticism toward an assimilationist model of fairness—that is, a skepticism toward the idea that in order to attain justice one must assimilate one's demands and one's identity either to the prevailing social-spatial order or to the dominant actors within social movements. An alternative is to be open to the *generative* nature of coordinated political practice. What does this mean? As he puts it, "Geographies of solidarities need to be seen as . . . actively shaping political identities, rather than merely bringing together different movements around 'common interests'" (Featherstone, 2003: 405).

The Inter-Continental Caravan for Solidarity and Resistance evolved out of a transnational support network that grew in response to the Zapatista uprising in Chiapas, Mexico. The particular ICC project that Featherstone writes about took place in 1999 and involved a series of traveling protests through various European cities. Participants included the Karnataka State Farmer's Union (KRRS), from India, and a variety of other grassroots groups largely from the global South and East. The protests were especially aimed against biotechnology-induced changes to agriculture, targeting European centers of power from which such changes have emanated. The movements that gathered into the ICC were committed to both transnational social activism and to counter-globalization. (Featherstone distinguishes between counter- and anti-globalization.) There was also a commitment to a horizontal form of organization. That is, power within the ICC was not to be defined vertically, from the top down. Rather, it was to consist of the linkages across the social actors and movements involved in the ICC.

From the onset a problem—and opportunity—was differences of opinion over whom and where to protest: biotechnology corporations? Bankers and financiers? The seat of national governments? At stake in deciding protest sites was the ability to develop an account of structural power, the conditions not of our making that structure our positionality in the world, yet also pose opportunity for social action (see Chapter 18). While these accounts differed among Caravan participants, Featherstone traces the capacity to develop new political analyses to the geopolitical arrangement of the ICC itself. Participant encounters produced surprising reevaluations of the most basic concepts through which activists understood their worlds: people found themselves questioning their hostility to "development," or found themselves reassessing their notions of what counts as "traditional farming" or "environmentalism." Featherstone also found instances of unmovable opinion: The ICC was not utopian, it was experimental. He concludes:

> The forms of commonality mobilized by the ICC . . . were more diverse, multiple and productive than is suggested by a fixed notion of a common good or interest, pre-existing in the formation of these political alliances . . . The bringing together of different activist cultures was a process that was generative of debate, negotiation and contestation rather than a simple coming together of homogeneous action or pre-existing political wills.
>
> (Featherstone, 2003: 416)

In short, these transnational activists did not know who they were with respect to structural power before entering into encounters with each other.

At the same time, emerging political awareness became indivisible from spatial practice. It mattered that the ICC traveled to the very centers of financial power into which are networked the many localities of the global South and East that experience agricultural changes. It mattered too that activists met and confronted each other at the ICC itself, thus making visible the relations among the participating activists. (The ICC was a geographical place/site, as much as the centers of power to which the Caravan traveled.) Geographical encounters entered into the political knowledge and into the identities that *became* (emerged) during the ICC: "The practices through which geographies of power are contested have effects on the identities formed through political struggles" (Featherstone, 2003: 409). Featherstone concludes that "solidarities are not just the amalgamation of fixed interests, but are productive practices that form equivalences between different struggles." That is to say, when I understand your struggle I understand my own struggle and myself in a new way. Equivalences, then, are not pre-formed, nor automatic: "Equivalences are here understood as practices of solidarity which unsettle fixed and particularistic identities to produce new, open and relational political identities" (Featherstone, 2003: 409). And as just noted, the geography of such practices is constitutive. The task for critical geography, therefore, "is to find ways of experimenting with geographies of power and practices of solidarity that make alliances between different struggles against neo-liberal globalization more rather than less possible and productive" (Featherstone, 2003: 409).

USING SOCIAL THEORY, AND THE DEBATE BETWEEN "STRUCTURE" AND "AGENCY," TO EXPLAIN MOVEMENT SUCCESS OR FAILURE IN SPECIFIC TIMES AND SPACES

What makes the time right to engage in a particular struggle? Impassioned feelings may be necessary, but are they a sufficient condition for sustaining and winning a struggle? And if there is the sense that the time is right, what is this sense about exactly? For example, a judgment that the time was right to revivify one aspect of the struggle for an alternative food production system in the U.S. was an explicit, motivating factor for Jack Kloppenburg's research. University- and industry-sanctioned agricultural science had been sufficiently called into question by activists and members of officialdom alike, such that the struggle over agricultural science could be given greater attention. It was *not* Kloppenburg's purpose, however, to develop a *conceptual* understanding of why movements gain success at particular times, and within particular spaces, and not others. Such an understanding is a legitimate concern for social theory and is one of the uses to which it can be put. James F. Glassman, a geographer at the University of British Columbia, undertakes this effort in "Structural power, agency and national liberation: the case of East Timor" (Chapter 18). After an extraordinarily long and violent struggle, East Timor finally won independence from Indonesia in 1999: what ultimately made the struggle for a right to independence from the occupying power successful? (Note, however, that the process of political reconstruction is ongoing.) Glassman argues that even "though the maneuverings of different actors in the Timorese resistance struggle were necessary conditions of liberation, they were not sufficient and required the enabling context created by shifts in structural forces that had sustained the basis for the Indonesian invasion and occupation" (see p. 308). What theoretical resources do we have at our disposal in order to conceptualize these "conditions of liberation" and "shifts in structural forces"? To engage with this question requires no less than engaging with a central and very longstanding problem in the social sciences, the relationship between social structure and agency. Let's begin very simply by taking a page from Karl Marx, who argued, in a pithy study of the return of an emperor to 19th-century France (*after* a period of democratic revolution), that people make history but not under conditions of their choosing (Marx, 1926). On the surface this seems entirely self-evident. Who would disagree? People do indeed have a certain capacity to act but no one can act entirely as they please; there are manifold opportunities and constraints on what people can do. Agency is socially "conditioned," so to speak. Yet how are social conditions themselves produced if not in part by people's actions? Moreover, are these conditions the result of conscious design? The result of unintended consequences? How do these conditions change—through slow evolution?

Sudden rupture? And is agency something that only individuals have, or can it be a property of social collectivities? In a nutshell, this is the problem of "structure" and "agency" and these are some of the questions that swirl around the problem (see Cloke *et al.*, 1991 for a review).

As opposed to a purely abstract debate over the relative power and constitutive features of structure and agency, James F. Glassman asks: What would people need to understand about the social world if changing their circumstances in it, in a durable way, was a goal? Isn't getting a grip on structure and agency essential in trying to account for movement success or failure? In fact, it is an axiom of activist organizations and of pragmatically minded social movement participants that attention needs to be paid to the present and what its possibilities and constraints are (e.g. Barndt, 1989, 1996). This includes developing a conceptual understanding of the actors and forces that shape the particular oppressions and wrongs that are being struggled against. The possibilities are rife for geography to place its own interest in structure and agency within this activist context.

In developing his argument Glassman assesses the relative strengths and weaknesses of two of the reigning models in Geography that have been used to explain the complex interactions between "agency" and "structure." Both of these models, the structuration approach developed by Anthony Giddens and the Marxist notion of structural power that Glassman finds more convincing, eschew a strict dichotomy between structure and agency. That is, neither sees the world as composed of purely autonomously acting individuals or as composed purely of structures that determine individual behavior. On the contrary, both models attempt to discern and analyze the middle ground. A key difference for Glassman though is how much these two approaches address the capacity of social agents to engage in actively *transforming social structures as opposed to only reproducing them* (cf. Brown, 2002). The emphasis on transformation is what draws Glassman to the idea of structural power. An understanding of structural power, he argues, is an effective way to account for how organized acts of resistance may (or may not) escalate into social change. As readers will see, Glassman extends the concept of structural power by introducing a spatial account of it. It is crucial to understand that social agents and the conditions that account for their capacity to act are territorially and temporally delimited. The implications are interesting indeed. Glassman writes: "what I am suggesting here is that the complex territoriality of global capitalism makes the actions of specific groups of people in particular locations the structural conditions constraining and enabling agency by other groups elsewhere" (see p. 313). (And this is regardless of whether these constraints or opportunities are intended.) For this reason, the account of the East Timorese struggle provided by Glassman involves reviewing a very diverse and spatially scattered array of actions: foreign investors, IMF officials, and U.S. functionaries pressuring the Indonesian government to open up to global capital; actors within Indonesia (e.g. local capitalists, professionals, and students) seeking various sorts of political and economic reforms of their own; and the various events leading to an economic crisis in Indonesia, which played its own role in making for a successful referendum for independence for East Timor. In no way, however, was that independence guaranteed, no matter how tenacious the national liberation movement. But the point, as Glassman puts it,

> is not to encourage skepticism about the prospects of national liberation struggle or other forms of resistance . . . The point here, rather, is to note that since the necessary conditions of successful struggle include structural transformations not under the control of resistance groups, awareness of structural constraints and potential openings is crucial to resistance strategy. It is for this reason that actors in class struggles and national liberation struggles . . . have paid careful attention to the opportunities created by economic crises and changing configurations of geo-political power. Resistance struggles cannot control such developments, but by being alert to their evolution they can construct strategies and time actions in ways that maximize impact.
>
> (see p. 320)

We note that geography journals are replete with analyses and debates about structure and agency. To his credit Glassman gives readers a good sense of what these have entailed. (His review is far from comprehensive though.) What we find especially important is that the article gives credence to the idea that structure and agency are not simply about academic theory; they are forms of social knowledge developed in practice. If you will, Glassman seeks to understand what political agents seeking justice must themselves understand about the world.

16

Restructuring and the contraction and expansion of environmental rights in the United States

Laura Pulido

from *Environment and Planning A*, 1994, 26: 915–936

INTRODUCTION

One of the most important recent developments in environmentalism in the United States is the rise of the environmental justice movement. This loose grouping of activists and organizations emerged from the antitoxics movement of the early 1980s and has become wedded to the language, actions, and rhetoric of the civil rights and social justice movements. The merging of environmentalism and social justice has resulted in a new sense of environmental rights, with important implications for changing notions of citizenship.

Although the environmental justice movement is national in scope, activism in California and the Southwestern United States must contend with special issues. Virtually all environmental justice activists deal directly with questions of racism, participatory democracy, and justice. Southwestern US activists, however, must also confront the international border and its many implications.[1] Because of the realities of immigration, uneven development, capital flight, cultural differences, economic restructuring, and severe pollution, environmental justice activists are engaged in a continual effort to redefine environmental rights.

Many of the struggles of the environmental justice movement have been framed in terms of procedural justice, which means making the process of environmental decisionmaking more open and accessible to all people, especially marginalized communities. Achievement in this area has been critical to the success of the environmental justice movement, but it is also limited, particularly in the face of powerful global and economic forces which have created a series of political and economic changes. These changes have created a new political landscape in which activists must renegotiate the terms of struggle.

One change can be seen in Los Angeles, where activists are fighting for clean air but are finding that some companies are threatening to leave rather than comply with environmental regulations. These activists are encountering the limits to procedural justice. A second shift, related to the first, is the erosion of hard-won environmental rights through deregulation and political retrenchment. As companies find themselves facing greater competition, they are pressuring for a reduction in regulation and citizen participation, arguing that these are costly and affecting their competitive advantage. Last, there are efforts by US activists to extend environmental rights and considerations to citizens of another country. Southwestern environmental justice activists are responding to internationalization by making contacts across the Mexican border and organizing binational campaigns. Both the contraction and expansion of environmental rights represent qualitative changes in the types of rights being asserted and in the strategic and philosophical basis of the environmental justice movement.

In the next part of this paper I examine the conception of rights guiding the environmental justice movement. Using Iris Young's framework given in her book *Justice and the Politics of Difference* (1990), I argue that, although the environmental justice movement has fought for several kinds of justice, the emphasis has been on procedural justice which, although essential, is insufficient to extend environmental rights and quality to everybody, a stated goal of the movement. In the third section I examine political and economic restructuring and how it has produced both a contraction and an expansion of environmental rights. These opposing developments underscore the fact that, although procedural justice is crucial in the fight for environmental quality for oppressed people, it is insufficient in the face of global realities. I conclude by discussing some of the implications of these developments in terms of community rights, citizenship, and empowerment.

ENVIRONMENTAL RIGHTS AND PROCEDURAL JUSTICE

Procedural justice

In her recent book, Young (1990) challenges traditional conceptualizations of citizenship, participation, and difference. She emphasizes that both the policy and the scholarship of inequality have centered on distribution and have ignored "the social structure and institutional context that often help determine distributive patterns" (1990: 15). As a corrective, she urges greater attention to social and cultural structures and processes which create and maintain "otherness," and thus serve to keep certain groups outside of decisionmaking circles. She is especially critical of the emphasis placed on jobs by the traditional distributive agenda. Indeed, jobs are at the heart of many civil rights and affirmative action programs. Thus, rather than concentrating on programs to increase one's allotment of "good jobs," she urges us to consider other factors that cause, for example, the poor representation of Mexican Americans in recent White House appointments. In particular, she explores how decisionmaking, the division of labor, and culture all work to create inequality, *regardless* of material inequality. Young calls for a renewed attention to procedural justice, whereby "different" will no longer be considered inferior and whereby a deeper understanding of citizenship will be reached through attention to equality and participation, particularly in the nonmaterial realm.

Young's argument is important for understanding the current oppression and exclusion of racial-ethnic[2] groups. Because of the Black civil rights movement of the 1960s and 1970s, formal legal rights were extended to all. No longer could one discriminate or deprive any community of political and civil rights. But this guarantee did not necessarily translate into equality, full participation, or even appropriate representation, precisely because of social difference. Therefore, regardless of legal advancements and improvements in political rights, inequality still exists because of social differences which are manifest in often subtle prejudice or exclusionary practices. Young's argument clearly applies to the status of African Americans today. Even though African Americans have been granted all the formal rights of equality, few would deny that they still encounter discrimination and as a consequence are therefore not treated as equal members of society.

The environmental justice movement *has* focused largely on procedure and has not significantly tackled underlying structural inequality, regional capital investment patterns, or pollution reduction, and as such can only achieve marginal gains. As long as severe material inequality remains, other inequities, such as environmental inequities, will continue, despite the recent and laudable efforts of the environmental justice movement. The limits of this framework can be seen more clearly in light of economic globalization which has accelerated the trend towards both greater deregulation and increased capital mobility. One consequence of these developments is environmental justice activists' realization that their success in repelling a local environmental hazard may only push it across a national border. Faced with such an obvious contradiction, environmental justice activists are confronting the limits to our modern construction of citizenship and allegiance to the nation-state. Economic internationalization, shifting demographics, and the spatial characteristics of pollution are beginning to create a new notion of rights and citizenship which is more international and potentially more radical in its nature.

Environmental rights

Over the past several decades environmental rights in the USA have been identified and expanded to include legal protection as well as philosophical and policy considerations (Nash, 1989). Rights have been expanded across species (Callicott, 1989; Leopold, 1988; Stone, 1974), space, and time (Berkovitz, 1992;

Partridge, 1981). Early conservation history centered on protecting natural resources and wilderness areas and had a strong rural emphasis (Fox, 1981; Hays, 1987). Not until the 1960s and 1970s, inspired by Rachel Carson's influential book, *Silent Spring* (1962), and the Earth Day in 1970, did the focus shift to the urban environment (Borrelli, 1988).[3] By focusing on pollution, environmental activists attracted a wide following and used the courts and legislative system to achieve change, producing a watershed of environmental regulation. Examples of such legislation include the 1963 Clean Air Act, the 1960 Clean Water Act, the significant revision of the 1972 Federal Insecticide Fungicide Rodenticide Act, and the 1970 National Environmental Policy Act.

Legislation protected the rights of "natural objects" by requiring such things as environmental impact reports, mitigation efforts in the case of damage, and, in some instances, such as under the 1973 Endangered Species Act, the abandonment of disruptive plans. But by far the most significant extension of rights was to humans. This occurred in two ways: First, the formal right to a clean environment was solidified through the establishment of exposure levels for criteria pollutants. Second, the rights of activists were greatly expanded through the development of detailed procedures requiring citizen participation in the case of projects and expansions which would have a "significant" impact on the environment. Thus the public was granted far more rights to intervene both in private production processes and in state regulatory procedure.

Although concern for the environment became an established part of social practice, participation by racial-ethnic minorities in the mainstream movement, as characterized by the Sierra Club, was minimal. Certainly it was not because racial-ethnic minorities were not heavily affected by pollution (Berry *et al.*, 1977; Burke, 1993; CBE, 1989; Gelobter, 1992; Hurely, 1988; McCaull, 1976; UCC, 1987; USGAO, 1983; Wernette and Nieves, 1992; Zupan, 1973) nor because they had no stake in the outcome of environmental policies and decisions (Asch and Seneca, 1978; Freeman, 1972; Gianessi *et al.*, 1979; Lazarus, 1993). Rather, racial-ethnic activists involved in environmental issues did not always articulate them as such (Pulido, 1994) and others were simply opposed to the environmental movement itself, seeing it as a challenge to civil rights activism (see Ruffins, 1991: 56; Scheffer, 1991: 19). Besides these factors, social difference has also contributed to in the limited participation of low-income and minority communities in mainstream environmentalism. Despite the existence of universal *formal* rights, not all communities have enjoyed either equal environmental protection (*EPA Journal* 1992) or equal access to the regulatory process (Young, 1983). In the case of environmental enforcement and protection, it has been shown that the US Environmental Protection Agency (EPA) imposes lower fines against polluters located in minority communities in comparison with those in Anglo communities (Lavelle, 1992). With regard to the regulatory process, English-language-only public hearings in a community which is monolingual Spanish clearly limit the residents' ability to comprehend and participate in decisions affecting their daily lives.

This situation changed dramatically in the 1980s. Although the environmental legislation of the 1970s created structures for increased participation and rights, it has been the environmental justice movement of the 1980s and 1990s which has addressed procedural justice and which has thus effectively extended those rights to the poor and to racial-ethnic groups (Bullard, 1990: chapter 1; Taylor, 1993). Two events were catalysts in the development of the environmental justice movement: Love Canal and Warren County. Love Canal refers to the New York State community which was built on a hazardous waste site and which suffered severe contamination (Gibbs, 1982). Love Canal was the first and best publicized of such incidents, as communities across the country increasingly came to realize that they were directly affected by hazardous wastes (Freudenberg, 1984). The term "Love Canal" is often synonymous with the antitoxic movement, not only because it was the location of the first recognized incident involving toxic waste, but also because activists from Love Canal later founded Citizens' Clearing House for Hazardous Waste, which has been instrumental in building the environmental justice movement.

Although the environmental hazards in question are generically called "toxics," people may in fact be fighting a diversity of land-use and environmental threats, such as abandoned hazardous waste dumps, treatment and storage facilities, or polluting industries. These environmental hazards may vary considerably in their origins and locational processes, but what the battles over them do share, from the activists' point of view, is an emphasis on equity, justice, and the right to participate (Capek, 1993). The resulting movement is highly popular and grass-roots-oriented (Gottlieb and Ingram, 1988), with activists sharing information, creating networks, and assisting other communities in their struggles. Initially, many activists were concerned solely with their own communities. As they began to understand the process by which they became vulnerable in the first place, as well as the larger economic and social assumptions underlying the production of hazardous wastes, the level of consciousness of some activists has expanded beyond the boundaries of their own neighborhood (Heiman, 1990).

The second key event in shaping the environmental justice movement was an effort to place contaminated soil in Warren County, NC, in a predominantly Black community. This prompted residents to realize that racial-ethnic minorities were disproportionately bearing the brunt of toxic pollution (Lee, 1992). Local Black leaders organized around the issue, giving it a civil rights framework. Walter Fauntroy (Democrat, DC) commissioned a study to examine the relationship between racial-ethnic minorities and hazardous waste sites in the Southeastern USA (USGAO, 1983). Fauntroy's study triggered a series of other investigations which explored patterns of environmental equity. The most influential of these was the United Church of Christ's (UCC's) "Toxic Wastes and Race in the United States" (1987). Ultimately, these studies cast the issue firmly within the realm of social justice and civil rights, helping to solidify minority participation and further energize the environmental justice movement (Lee, 1992). These activists have focused on environmental racism, a concept which refers to the fact that people of color are disproportionately impacted by pollution (Bullard, 1993), have historically been excluded from the mainstream environmental movement (SWOP, 1990) and have not received equal consideration from regulatory agencies (Bullard, 1992).

These events collectively gave rise to the environmental justice movement. The movement has largely sought to change the environmental decisionmaking process and culture in order to make it more accessible to low-income and minority communities. This new activism has led to the continued expansion of formal environmental rights and at the same time has addressed social difference through procedural justice. The resulting discourse of the environmental justice movement, particularly among minority populations, embodies two identifiable sets of rights: the right to participate in the regulatory process and the right to live free of pollution. Both of these rights were made explicit in the First National People of Color Environmental Leadership Summit, held in October 1991, in which a set of "Principles of Environmental Justice" was adopted:

> *Environmental Justice* demands the right to participate as equal partners at every level of decision-making including needs assessment, planning, implementation, enforcement and evaluation . . . *Environmental Justice* calls for *universal* [my emphasis] protection from nuclear testing, extraction, production and disposal of toxic/hazardous wastes and poisons . . . that threaten the fundamental right to clean air, land, water, and food.
>
> (UCC, 1991, unpaginated)

Although substantial progress has been made in terms of participation, via procedural justice, the right to live in a clean environment will remain elusive until material inequality, uneven development, and greater democracy in production are also addressed.

The right to participate "as equal partners" has been operationalized through an increase in formal community environmental rights, particularly in the realm of citizen access and participation. Much of the push for greater access stemmed from the secretive way in which toxins were (and continue to be) handled (Greenberg and Anderson, 1984). Communities, outraged by the backdoor approach, organized and successfully demanded greater accountability to and increased participation by the public in policy decisions about toxins. One result is the 1986 Emergency Planning and Community Right to Know Act, which requires that certain manufacturers publicly report toxic chemical release and transfer information.[4] The act was intended to embarrass polluters into reducing their emissions. A state-level example is the 1987 California Toxic Hot Spots Act (Assembly Bill 2588), which requires manufacturers emitting over a specified hazardous level of air toxins to notify local residents and in some cases to hold community meetings with appropriate translation. Besides laws which provide greater access, environmental justice lawyers have also used the courts to ensure equal access through procedural justice. For example, California Rural Legal Assistance recently sued Kings County in California for attempting to place a hazardous waste incinerator in Kettleman City, a rural community in which 95 percent of the population is Latino. The suit argued that the civil rights of Kettleman City residents were being violated by denying them equal rights and protection (Cole, 1992).[5] In addition to these specific examples, there has also been a general trend toward mandatory community meetings, appropriate translation services, and the need to entertain public input.

The second set of rights claimed by environmental justice activists and articulated by Jesse Jackson is the "Right to Breathe Clean Air" (Stammer, 1990: A28). The "Principles of Environmental Justice" adopted at the People of Color Environmental Leadership Summit clearly call for *universal* protection from pollution. This is a sweeping demand which will be far harder to achieve than equal participation. As minorities and poor people have become aware of their disproportionate exposure to pollution, they have increasingly seen a clean, nonhazardous environment as a right to which they are entitled (Bryant and Mohai, 1992: 6). As a community is forced to fight the siting of a dump or incinerator, or a polluter, participants undergo a process of politicization. Whereas before they were not, perhaps, fully aware of the specific impacts of pollution, or did not realize that they could, in fact, resist an undesirable project, low-income groups have seen their expectations and their notion of rights enhanced. Many groups have always regarded clean air as a right, because of their sense of entitlement and privilege, but other communities have not exhibited this belief. One activist from the environmental justice group the Mothers of East Los Angeles (MELA) recounted how she did not believe she had the power to challenge the series of freeways which destroyed her community. But after successfully resisting an incinerator and prison, the women of her group automatically challenged all perceived environmental threats to their community. "We hope to . . . be treated like first-class citizens. Ya basta! Enough is enough of being treated like second-class citizens . . . If we . . . leave a legacy it is that the Mothers of East Los Angeles struggled for [their children] so they would be treated as first-class citizens . . . all of our successes help our reputation and inspire others to get involved" (interview with A. Castillo, 12 July 1993).

The environmental justice movement does not have a clear strategy to achieve universal protection, particularly on a global scale. There have been a few attempts, however, to assure that marginal and oppressed groups in society have access to a clean environment, which is a step in that direction. Most of these endeavors fall under the rubric of both procedural justice (Colquette and Robertson, 1991) and traditional civil rights strategies (Bullard, 1992; Godsil, 1991; Lazarus, 1993). For example, several environmental equity bills have been introduced (HR 1924 and HR 2105, for instance), and an Office of Environmental Equity (OEE) has been created within the EPA (S 171) (Gaylord, 1993). Both the legislation and the OEE require the government to study patterns of environmental equity and to redress the most glaring injustices. Sociologist and activist Robert Bullard has explained the need for a comprehensive, proactive bill modeled on the civil rights framework:

> Current government practices reinforce a system where environmental protection is a "privilege" and not a "right." Some communities receive "special" benefits and privileges by virtue of the skin color of their residents. The many facets of discrimination *persist despite laws banning such practices.* It should not be a surprise to anyone that discrimination exists in environmental protection.
>
> (Bullard, 1992: 5, my emphasis)

The italicized line in Bullard's quote speaks to Young's theory, the actual practice of the environmental justice movement, and the limits to both. Bullard recognizes that inequality will not be erased by legislation alone, as Young has pointed out; nor will it be erased by continually fighting on the grounds of social difference and procedural justice. Although the environmental justice movement has made great strides in democratizing the decisionmaking process, the fact remains that procedure alone will not translate into environmental equity for all, nor will the right to a clean environment become a reality without addressing social and spatial inequities and the absence of public accountability in private production decisions.

Obstacles to universal environmental rights and quality

Exactly what is necessary to achieve universal environmental quality is a crucial question, but one that has not been sufficiently analyzed by the movement. As one activist explained, "A lot of people know what they're for, but they are not as consolidated on what the problem is that prohibits them from getting it" (interview with C. Mathis of LCSC, 18 August 1993). One effort to conceptualize and achieve universal rights has come from the EPA, which has been under intense pressure from environmental justice activists to respond to their concerns (see Bryant and Mohai, 1992). Unfortunately, the EPA defines environmental equality as everyone sharing the same degree of environmental burdens and amenities (EPA, 1993). This is not the objective of the environmental justice movement, which is committed to pollution reduction (Roque and Tau Lee, 1993).

Aside from procedural justice, there are at least three other types of inequalities which must be addressed in order to reduce overall pollution and to ensure that particular groups and places are not disproportionately impacted. They are: a lack of democracy over private production decisions, uneven development, and material and social inequality. The environmental justice movement has considered these issues in only a cursory way. Although some groups include these concerns in their rhetoric, to build a struggle around them is far more difficult because it entails challenging fundamental notions of private property and capital mobility, and not all activists agree beyond the lowest common denominator of procedural justice.

The first major obstacle to addressing effectively the production of pollution is a lack of public accountability in private production activities. Environmental justice activists have made only a few gains in this area. When a manufacturer or industrial facility poses a hazard to a community a struggle must be waged to alter that production process. This could mean the use of substitute materials, employing different procedures and processes, or installing pollution-control devices. It is entirely possible that a facility be in full compliance with the law and yet still pose a threat to a community, often making legal remedies inadequate. Additionally, such a struggle may be particularly difficult because community members employed at the facility may feel their livelihood could be threatened should they mount a protest (Kazis and Grossman, 1982).

One successful challenge to private production decisions is the Southwest Organizing Project's (SWOP's) battle against a particle board company, Ponderosa Products of Albuquerque. Local residents had long complained of the smells and fumes emanating from the site, contaminated groundwater, incessant noise, and health problems. SWOP waged a campaign against Ponderosa and through community involvement, pressure tactics, and negative publicity, including offering the media "toxic tours," they were able to extract concessions (interview with R. Moore, SWOP, 1990). In 1987, as part of a larger package, Ponderosa signed a "Ground-water Reclamation Plan" to pump out the contaminated water and reduce noise (Martinez, 1991: 64). This is one example where environmental justice activists successfully demanded a socialization of private decisions. Such victories, however, are still too rare.

Uneven development is the second inequality which must be addressed in order to reduce both pollution and its inequitable distribution. Uneven development is a term which captures the dynamic nature of capitalism and which acknowledges that its unfolding is spatially expressed. Because investment patterns are such powerful forces, they have tremendous consequences for different places, whether at the local, regional, or international level. Capital is attracted to certain locales because of real or perceived attributes, which can include natural resources, cheap labor, a lax regulatory environment, or proximity to markets and/or transportation systems. Investment of capital in one place is often accompanied by disinvestment in another, thus creating a landscape of uneven economic development and contributing to concentrations of poverty.

Because impoverished areas often lack political, in addition to economic, power, the concentration of environmental hazards in poor and less powerful neighborhoods, cities, regions, and countries will continue, despite progress in procedural justice. This is partly because inequality is built into the physical and social history of a community and cannot be erased overnight. In an effort to promote environmental equity, for example, one could declare that all Los Angeles communities will host the same number of environmental hazards. Thus, the affluent Westside and the Eastside barrio would be equally considered for the next hazardous facility.[6] However, some Westsiders would inevitably point out that the zoning and land use of the Eastside, which is largely industrial, is better equipped to handle such a facility. In response, a truly progressive city council might change the zoning citywide to make it more equitable. But it would be unreasonable to relocate all existing land uses in order to achieve geographic balance.[7] In accordance with standard planning practice, there would be a grandfather clause, whereby only new facilities would have to comply. Thus, it would take decades for any semblance of environmental equity to be achieved.

The final inequality which must be addressed is social. This broadly refers to cultural, political, racial-ethnic, and economic differences which exist and cause some social groups to be less politically powerful than others. Using the previous example of Los Angeles, the significance of social inequality can be seen. In Los Angeles, Eastside Latino residents are heavily involved in low-wage manufacturing (Morales and Ong, 1993). The fact that Latinos serve as a large, low-wage labor pool ensures their exposure to a polluted environment (Freed, 1993a, 1993b, 1993c; Ong and Blumenberg, 1993). This particular inequality is a function of Latinos' place in the division of labor, which is a form of social inequality. In this case, division of labor is exacerbated by such things as substandard educational institutions, and, for noncitizens, the inability to vote, which in turn reinforce and perpetuate unequal economic and power relations. Thus, various forms of social inequality cause certain individuals and groups to have only limited power and efficacy in their efforts to build a better life.

Simply put, one cannot avoid the fact that the right to a clean environment is largely unobtainable without an explicit distributive agenda which incorporates more than procedural justice. But conversely, as the phrase "Not in Anybody's Backyard" (Heiman, 1990) suggests, material equality, although essential, is also not enough. A redistribution of wealth and power may ensure that currently oppressed communities are not disproportionately impacted, but it does not address the problem of pollution production itself. Thus, there is need for action on several levels: democratizing private production decisions as well as addressing uneven development and social inequality, which can include procedural justice.

RESTRUCTURING AND ENVIRONMENTAL RIGHTS

The processes of political and economic restructuring are impacting most aspects of public and private life, including environmental rights. Changes in both the US national and the global economy have local political and economic repercussions (Kratke and Schmoll, 1991; Logan and Swanstrom, 1990). The net effect of these changes is to create a climate which is increasingly competitive and transnational, and which operates with little regard for the needs of local communities. Restructuring has posed new political obstacles to environmental justice activists through deregulation and capital flight, which have, in turn, produced changes both in the activism of the movement and in the types of rights being asserted. Such changes have closed certain opportunities to environmental justice activists and have simultaneously presented new possibilities.

California and Mexico

Three fundamental and related processes are currently affecting California and other Southwestern states: economic restructuring, internationalization, and immigration. Although all states are experiencing these changes to varying degrees, they are more pronounced in California. The first, economic restructuring, refers to the transition from a Fordist to a post-Fordist, or flexible, economy (Storper and Scott, 1989). Economic restructuring reflects the decline of old manufacturing industries, the externalization of production, the growth of the service sector, and the subcontracting of work. In terms of people's lives, these changes translate into the loss of steady, high-wage employment, a societal income polarization, and local fiscal crises leading to reduced social services and a political retrenchment.

The second and related process, which has economic and social consequences, is internationalization, in which the economies of the world are becoming increasingly tied together (Sassen, 1991). As part of the Pacific Rim, California and other states are poised to take advantage of the rapid expansion of the Asian and, to a lesser extent, Latin American economies (Szekely, 1993). As part of this trend, the USA is seeking to create a North American trade bloc, which would further integrate the Canadian, US, and Mexican economies. Although barriers remain to the flow of people, there is a concerted effort to reduce global barriers to the flow of capital.

Regardless of whether the agreement is ratified, the USA is becoming economically and socially integrated with surrounding nations. One result of these developments is greater competition among communities to attract capital, allowing firms to locate in the most "friendly" business climate.

The third significant process is immigration. During the 1980s the USA experienced tremendous immigration which was especially profound in southern California (Turner and Allen, 1990). Immigration, particularly from Latin America, has two specific implications in terms of environmental rights. First, it is an overt expression of increasingly interdependent economies (Rubio and Trejo, 1993). Second, it represents social integration in a profound way (Pastor and

Ayon, 1992). In many communities along the 2000-mile US–Mexican *frontera* the border is simply a marker. Organic communities and families straddle the region, creating a dense network of social relations. This reality is important in that it facilitates the development of a more international identity and politics, based not only on a common heritage but also on a common vulnerability to the global economy and its pollution.

Air pollution in southern California and the contraction of environmental rights

In response to Los Angeles' ranking as the "smog capital" of the USA, a concerted effort has been underway at the federal, state, and local level to clean up the air. The South Coast Air Quality Management District (SCAQMD), which encompasses parts of five counties in southern California, was created by the California Air Resources Board to bring the region into compliance with federal regulations. The SCAQMD developed a three-tier plan known as the Air Quality Management Plan (AQMP) (SCAQMD, 1989). It is based on technological availability and is intended to bring the region into compliance by 2010.

SCAQMD rules have affected southern California in various ways, from minor changes, such as a ban on certain barbecue fluids, to more significant inconveniences, such as employer-developed ride-sharing plans and a reduction in local control over land use and development. Although many municipalities and individuals resent the intrusion into their lives, the SCAQMD has until recently enjoyed widespread public support, primarily because most realize that local air pollution must be more carefully controlled. Despite the public health risk posed by the polluted air, there has been opposition by segments of the business community who argue that environmental regulation is responsible not only for the current economic stagnation but also for the loss of industry to other less expensive areas, such as Utah, New Mexico, Arizona, and Mexico.

The Los Angeles environmental justice movement arose out of struggles over air pollution when the MELA and Concerned Citizens of South Central Los Angeles (CCOSCLA) fought proposed incinerator sitings (Blumberg and Gottlieb, 1989; Russell, 1989). Because both organizations consist of working-class women of color, they attracted great attention, became celebrities in the environmental community, and carried tremendous moral authority. However, after the initial victories of MELA and CCOSCLA, environmental justice activism temporarily subsided in Los Angeles. In the interim, in 1989, a more regionally oriented organization, the Labor/Community Strategy Center (LCSC), joined the scene (LCSC, 1989).[8] Since 1990 the LCSC has become a leader in the fight against pollution and deregulatory efforts, working with key members of MELA, CCOSCLA and, more recently, more traditional environmental organizations, such as Citizens for a Better Environment and Greenpeace. Whereas MELA is a Latino organization, and CCOSCLA African American, the LCSC is multiracial and multiethnic, with a far more explicit class analysis:

> While some frequently talk about "workers" and "people of color" as completely separate categories, in Los Angeles far more than half of the working class is composed of Latino, Black, and Asian workers—many of whom are immigrants—who suffer because of their class position in society and because of their race . . . There is a need for a new social movement, one that demands democratic control over basic corporate production decisions to stop the pollution from these industries, and that demands the production of non-polluting alternatives.
>
> (Mann, 1991: 28 and 35)

The ultimate goal of the LCSC is to build a multiracial progressive movement to challenge corporate capital in southern California. One organizer from the LCSC shared with me his analysis of corporate power:

> I looked around at my own situation, at other organizations, and it became very frightening to me when I looked at how overwhelming corporate power is in this society. How little the chance is that we can really impact anything. But if you look back at history, the only way that working people and people of color have gotten any justice is by organizing.
>
> (interview with C. Mathis, 18 August 1993)

The last four years in the LCSC have been spent organizing, developing political consciousness, and waging procedural justice fights. These are all seen as precursors to mounting a countywide corporate campaign against a major polluter.

Environmentalism, and public health in particular, provides an entry to questions of corporate responsibility and community rights. It is fair to say that over the past few years, because of the heavy involvement of the LCSC, the entire environmental discourse of the SCAQMD has shifted towards community rights and public health. Together, the local groups have persistently made the claim that clean air is a right, that public health cannot be compromised, that rules and regulations should not come at the expense of the oppressed and poor of southern California, and that residents of the region have a right to participate in the formulation of environmental regulation. As members of the environmental justice movement, they demand greater citizen participation and access, as well as reduced levels of pollution.

The coalition has enjoyed several successes largely based on procedural justice. In 1991 they pressured the SCAQMD board to adopt a resolution barring employers from developing trip-reduction plans which were discriminatory or posed undue hardship on minority, female, low-income, or disabled workers. The resolution was known as the Social Justice Amendment, and the coalition spent over six months organizing around the issue, developing allies, and lobbying members of the board. The victory was an important extension of local community environmental rights, as the amendment explicitly acknowledged that, though people all enjoy the same formal rights, inequality remains and colors the implementation of policy, underscoring Young's point of social difference (1990). Thus, by creating this explicit social justice amendment, activists sought special accommodations to ensure that the less powerful will not be regressively impacted. This was the first time in the USA that such a policy had been instituted.

Despite such impressive gains, the structural forces impacting California have created a strong deregulatory drive. Because of capital flight, local recessionary conditions, national policy, immigration, and greater competition (Levy and Arnold, 1992), the political discourse of the region has shifted to the political right (Davis, 1992).[9] According to a poll by the *Los Angeles Times*, in 1991 38 percent of the population felt business was overregulated, but in 1993 that figure had jumped to 58 percent (Stall, 1993: A28). Local governments, which face declining revenues and a greater demand for services (partly because of immigration), and local industry, which faces greater competition, are both seeking to make California more conducive to capital accumulation (CCC, 1992; SRC, 1992). Although few would dispute the need to streamline and improve the state's environmental regulations, the recent efforts of the "regulated community" have been more akin to a complete roll-back of any commitment to public health (see SCAQMD, 1992a, 1992b). Efforts have centered on reducing both environmental regulation and citizen participation. This can be seen as both a contraction of established environmental rights and a loss of procedural justice.

One example of the erosion of environmental rights can be seen in the arena of translation. Over the years, the environmental justice coalition, led by the LCSC made significant gains in terms of demanding and receiving Spanish translation services. This was one of the few areas in which the SCAQMD staff and board, business, and environmental justice activists agreed (SCAQMD, 1992b, appendix 2: 3). Given the demographics of Los Angeles, which is 40% Latino, the demand for Spanish language translation was considered reasonable. The preferred method of translation from the activists' point of view was simultaneous translation, in which Spanish-speaking individuals would wear headphones while an SCAQMD staff member translated. For Spanish-speaking individuals wishing to testify, a bilingual activist would translate their comments. This method was felt to offer the greatest inclusion and access to Spanish-speaking persons.

However, even this seemingly mild issue soon fell victim to the larger political retrenchment. At a 1993 hearing, numerous monolingual Spanish speakers were in attendance, but no translation services were provided. The activists confronted the board, who assured the activists that simultaneous translation would be provided next time. At the following board meeting simultaneous translation was again not provided. Instead, the SCAQMD placed all persons in need of translation in a separate room, where a staff person translated out loud, because simultaneous translation disturbed the board members. The activists were outraged and interpreted these actions as racism against Latinos. It was not felt that monolingual Spanish speakers could participate as equals if they were located in another room. A heated debate ensued and one board member made a motion to *study* the possibility of adjusting the room and identifying the necessary equipment in order to allow for translation that would not disturb the board. The motion failed.

The implication of this decision was that the board no longer felt compelled to provide full participation. What had once been a common arena of agreement had become contested terrain in which previously won rights by activists were rolled back. To conduct without proper translation major public hearings in a region with millions of immigrants can be seen as a gross erosion of environmental and procedural rights. This is but one way in which environmental rights are being impacted by restructuring, as business interests are increasingly demanding that the state curtail special services in this time of fiscal austerity. However, these same forces are also creating an expansion of environmental rights.

Internationalization and the expansion of environmental rights in the US Southwest

Although procedural justice is critical in the struggle for a clean environment, its significance is diminished in light of internationalization. One consequence of this trend is industry's greater mobility in its search for lower wages and regulatory relief. Because of the imperatives of global capitalism, not only has the rhetoric of capital flight been used to gain concessions from both the state and workers, but also within certain industrial sectors flight has become a reality.

The economic integration of the USA and Mexico has a long history (Heyman, 1991; Lowenthal and Burgess, 1993; Morales and Tamayo-Sanchez, 1992). This includes US firms developing Mexican resources and infrastructure, the prominence of the USA in the global economy, and worker immigration from Mexico to the USA. This historical relationship took on added significance with the *maquiladora* program, a strategy to promote Mexican economic development along the US–Mexican border (Sklair, 1992). Since the *maquiladora* program began in the 1960s, capital flight of labor-intensive operations, such as textiles, electronics assembly, and furniture making, has increased (Herzog, 1991). What makes this latest round of activity unique is that, although seeking lower wages, some firms are also attracted to Mexico because of its more lax environmental regulatory system, resulting in heavy pollution. Thus the border region has assumed a new significance. In addition to traditional concerns of poverty and uneven development, pollution has become a major issue (*TEF* 1993).

As US activists demand greater environmental protection and access to decisionmaking, they often find themselves confronting industries which threaten to move to Mexico. Most firms leave to take advantage of lower wages, but other factors can also be important. Environmental regulations are increasingly cited as a cause of flight; they are said to contribute to an "unfriendly" business climate, although they are often used for strategic political reasons. Nevertheless, some industries have been heavily impacted by environmental regulations, and this has contributed to their decision to relocate. The Los Angeles furniture industry is an example of this. The furniture industry, which is characterized by undercapitalized firms (Hise, 1992), was hard hit by SCAQMD regulations in 1988. Several studies have sought to document the reason for the exodus of the furniture industry and, although lower wages and workers' compensation were primary reasons for leaving, environmental regulations also played a role (Bloch and Keil, 1991; Hise, 1992; USGAO, 1991).

Although Mexico has strong environmental protection laws, enforcement is weak, especially in light of Mexico's fiscal and development crisis (Barkin, 1990: chapter 3; Mumme *et al.*, 1988). Because of a lack of enforcement and industry's disregard for the Mexican environment, it is difficult to exaggerate the pollution of the border. In both Matamoros and Tijuana there are anecdotal stories of children being born with encephalitis (Pasternak, 1991), but researchers are only just beginning to document the extent of the problem. Given the extreme hazard posed by the *maquiladoras*, environmental activism has been growing among border residents since the 1980s.[10]

As US environmental justice activists began to realize the contradictions between capital flight and the goal of environmental quality for all, they were forced to reconsider their strategy and its implications. Clearly, they were facing the limits of procedural justice, particularly given a nation-state framework. "When we look at the fact that there's an internationalization of trade and commerce and communication, then also we, at the grassroot level need to internationalize our struggle in order to put a halt to those things that are killing us right now" (Solis, 1993). One example of activists' response to internationalization is the Southwest Network for Economic and Environmental Justice (SNEEJ), a regional coalition consisting of environmental justice organizations throughout eight Southwestern states and several

Indian nations. California groups, including MELA, CCOSCLA, and the LCSC, all belong to the SNEEJ. The SNEEJ was created after a regional dialogue among activists in 1989, and at the time of its formation decided to focus on five particular issues: the accountability of the EPA, campaigning for sustainable communities, the sovereignty of Native lands and the prevention of dumping on them, youth development and leadership, and campaigning for border justice. Membership in the SNEEJ offers organizations active assistance and support in local struggles, as well as access to resources, solidarity, and the opportunity to be part of a network which is gaining recognition.

The SNEEJ is structured so that each State has a representative on the Coordinating Council. In 1991 it was decided that Mexico should be invited to participate and be allotted a representative to the Council. This was an important decision for the activists, because a central theme of the environmental justice movement is "We speak for Ourselves" (Alston, 1990). One US activist explained the thrust of bilateral organizing: "we put not only the U.S. government on notice, but also the Mexican government, that the *colonia* that we are visiting and that our people from Matamoros, Nogales, and other communities in Mexico will not be by themselves anymore. We will not be standing in front of them, or behind them, but in fact we'll be standing side-by-side them" (Moore, 1993). This is in contrast to more traditional environmental groups which are often charged with speaking *for* grassroots activists rather than working *with* them.

A graphic example of solidarity came in 1992 when the US firm Chemical Waste Management sought to locate a toxic waste incinerator in Tijuana (Leal, 1993; Rotella, 1992). The SNEEJ was active in working with the Mexican opposition who eventually prevented the siting of the incinerator. The significance of the fact that the firm was Chemical Waste Management was not lost on US activists. This same firm was responsible for many of the battles US environmental justice activists had found themselves in. This pattern of internationalization extends beyond waste companies to many other industries, illustrating the permeable nature of the border and the futility of organizing on one side only. "Many times the companies that are polluting us are on both sides of the border. Therefore, we have a common cause. Here, just above us in the industrial park there is a lead smelter that is polluting. This foundry is the same company that is polluting people in Dallas Texas, in an African-American and Latino community. In this unity we have created power" (Solis, 1993, my translation).[11]

The Border Justice Campaign also held an *Encuentro sin Fronteras* (Meeting without Borders) in Tucson. At this meeting, a common agenda was developed between Mexican and US environmental justice activists. They decided to focus on the environmental and health crisis facing the border region and began developing a "border network" to address such problems. Again, it is important to note the formation of a common agenda. Mexican activists complained that US organizations and activists often consulted with them only when they wanted a "toxic tour" or the like (Sanchez, 1993). They were not operating as equals. "Many times the agenda in terms of border work is brought already from the United States and basically implemented on the Mexican organizations without a common agenda being built which is representative of both sides" (Solis, 1993). As part of the organizing plan, the SNEEJ funded a Mexican activist to work full-time in the Mexican border region on such issues as wages, working conditions, and community pollution (Leal, 1993).

Last, it was decided that the 1993 annual gathering should be binational, spending one day in Tijuana and one in San Diego. This accentuated the coming together of US and Mexican activists as equals in the fight for environmental and economic justice. As part of the conference activists marched and protested against the firm Metales y sus Deritivos, a Tijuana *maquiladora* that recycles batteries from the USA. Residents of the *colonia* Canon del Padre considered Metales to be the worst polluter in the area and had been fighting it for ten years. The joint nature of border environmental problems was highlighted because the batteries come from the USA.

Even among individual groups, there is a growing emphasis on internationalization and the expansion of environmental rights. For instance, the LCSC has sent an organizer to Mexico City. Using the Spanish translation of their book *L.A.'s Lethal Air* (Mann, 1991) as a tool, the organizer hopes to make contacts with Mexican activists and to identify areas for future collaboration. This development fits in with the LCSC's larger vision of building a broad-based oppositional movement, one which does not see Mexico as a threat to US standards but rather as a potential ally, one which when *united* with those in the USA could pose a formidable challenge to mobile polluting sources (interview with L. Duran, LCSC organizer, 6 October 1992).

The North American Free Trade Agreement (NAFTA), or El Tratado Libre Comercio (TLC), has also facilitated bilateral organizing. US environmental justice activists oppose the NAFTA not only because they believe it will cause job loss, but also because they know Mexican workers will receive miserly wages, ensuring their continued poverty. Worse, they believe it will exacerbate already severe pollution. Mexican environmental justice activists have also opposed the TLC. They are convinced it will lead to greater pollution and death. The SNEEJ and its member organizations oppose NAFTA not only for its potential consequences, but also on philosophical grounds. It represents another instance of corporate control coming before the needs of local communities. Together, US and Mexican grass-roots activists have written letters to government officials to oppose the NAFTA (TLC), have protested against it, and have even testified in Washington, DC (Rhee, 1993; Sanchez, 1993).

These instances of binational activism are based both in procedural justice and in attempts to democratize production decisions and address uneven development. US activists wish to extend the right to participate to Mexicans. Hence they call for public hearings and a modification of Mexican and US environmental laws to give Mexicans more access, information, and rights. The struggle over procedural justice will continue, but will also be complemented by more radical efforts, including direct campaigns against specific polluters, as well as efforts to address the wages and working conditions of Mexican workers.

THE GLOBAL CITIZEN? SOME CONCLUSIONS

Southern California and the US–Mexican border offer a unique opportunity to examine recent developments in environmental rights. Southern California demonstrates the limits of procedural justice, as such rights are being eroded by economic restructuring and internationalization. The border illustrates the limits to procedural justice because of the extreme inequality of the two nations. This contradictory situation inevitably raises difficult questions of capital flight, the environmental rights of Mexicans, and more international forms of citizenship and rights, questions which are fundamental to recreating the political left (Smith, 1989).

There are two distinct lessons to be drawn from these cases in terms of citizenship. One is the expansion of rights to citizens of other countries, and the second is a more meaningful form of citizenship, that of "active engagement" (Dietz, 1987). In order to understand citizenship and its attendant bundle of rights, one must recognize that these are socially constructed terms which vary over time and space. Throughout the history of the USA there has been a continual expansion of citizenship and rights to marginalized groups, such as women (Sapiro, 1984), slaves, persons without property, Native Americans, and Blacks (Marston, 1990). All of these developments, however, took place within the confines of the nation-state, a unit which was instrumental in the rise of capitalism (Kearney, 1991). Indeed, the concept of the nation-state is essentially modern, as it is a unifying structure. Accordingly, political boundaries coincide with the geographic limits of a particular set of laws, customs, language, economy, and culture (Gupta and Ferguson, 1992). As the cases examined in this paper illustrate, borders are becoming increasingly fuzzy in an age of transnational economies, migration flows, and a general "disjuncture of place and culture."

Because of transnationalism, it is not surprising that traditional notions of citizenship and rights will also undergo change. In fact, the expansion of humanitarian rights and concerns to others across the globe is an increasingly common phenomenon, what Carol Gould (1988) calls "cosmopolitan democracy." In this paper, it is shown that US environmental justice activists are growing increasingly concerned about the environmental rights of people from another country, Mexico. There is a recognition that humans, regardless of their nationality, should be granted certain basic rights, including the right to be free from a poisoned environment. Because the USA has more established environmental rights and is facing capital flight, US activists are making overtures towards Mexican activists. Simultaneously, Mexicans have become increasingly active and outspoken in their efforts to reduce border pollution. There is a clear understanding that Mexico needs jobs and meaningful forms of economic development, but this does not mean it is acceptable for firms to relocate to Mexico to produce more pollution. Both the problem of "pollution flight" and the consciousness surrounding it at the grass-roots level are relatively new. Some favor legislative reform and procedural justice in order to address the problem, but others recognize it is fairly meaningless to speak of laws, voting, and procedure in light of extreme regional inequality and capital mobility.

This brings us to the second point of citizenship: the emphasis on active engagement. Traditional analyses of citizenship center on formal legal structures and institutions, such as laws, rights, and the franchise. Although these are all critical to the practice of citizenship and democracy, they are limited in both their political and economic implications. Many scholars have recognized that any meaningful notion of democracy must include economic well-being. Judith Shklar (1991), for example, has argued that there are two components to citizenship, the conventional right to vote, as well as the right to make a living. Gould (1988), picking up a different thread, suggests that democracy must be spread to the economic arena, via workers' rights. However, neither of these issues, including John Rawls' 1971 emphasis on distributive justice, offers an appropriate framework in which to understand material inequality and the rights of people living *across* a national border and subject to the vagaries of global capitalism.

These conditions require a more useful form of citizenship and democracy, what Mary Dietz has called the active engagement of citizens, that is, "the collective and participatory engagement of citizens in the determination of the affairs of their community" (1987: 14). In the case of the border, the terms "citizens" and "community" are being redefined and expanded in new and creative ways. Voting, for example, has not been a major force in the recent expansion of environmental rights. Instead, most advances have been won via social movements in efforts to extend procedural justice to marginalized groups, such as the Los Angeles struggle for a social justice amendment. Along the border there is a void of state-sanctioned vehicles or institutions. Thus, in a very real sense, concerned "citizens" of a nonstate place must practice alternative, but perhaps more meaningful, forms of democracy.

To date, noncitizen residents in the USA do not have the right to vote, and there exists no means for people to vote on binational issues. Although I believe that all persons residing in a given area should be able to vote and participate in affairs of community (see Pincetl, 1994), there is a need to reconsider the power of formal rights and procedural justice, as these have limited power against the forces of capital flight and transnational pollution. Given the binational nature of the problem, there is little choice but to employ new categories of political participation. These are changes which could only have come about in a period of fundamental social and economic change.

Environmental justice activists in the USA have articulated at least two sets of claims: the right to participate as equals, and the right to a clean environment. Most efforts to attain these objectives have been based on procedural justice, efforts to include those who are marginalized and excluded from decisionmaking. Procedural justice has been an important tool in achieving greater participation but has been less successful in achieving universal rights to a clean environment.

The limits to procedural justice have become apparent as a result of restructuring. The processes of restructuring, internationalization, and immigration have closed off some opportunities but have opened up new avenues. In southern California, environmental justice activists are finding that previously established environmental rights are being eroded. At the same time, internationalization has forced US environmental justice activists to consider the international dimensions of their actions, and this reconsideration has led to greater collaboration with Mexican environmental justice activists and to efforts to expand environmental rights to citizens of another country.

Although procedural justice will remain essential—a strategy which does not address material inequality, uneven development, and greater democracy in private production decisions—the goal of environmental quality for the oppressed and exploited will remain elusive. The relocation of polluting industries to Mexico poses important challenges to questions of citizenship and rights. Traditional models of citizenship and democracy are less applicable in a new world characterized by internationalization. Structural changes require greater emphasis on active engagement as well as a reconsideration of the alignment between rights and the nation-state. These developments require a new conceptualization and practice of rights which may be the first step not only in eliminating environmental inequity and in reducing pollution but also in encouraging activists to address uneven development and other forms of inequality.

ACKNOWLEDGEMENTS

Many thanks to Sallie Marston and anonymous reviewers for their useful comments. The author is responsible for all interpretations and errors. The empirical basis for this paper is based on my own participation in the environmental justice movement over the past four years.

NOTES

1 The US–Mexican border is quite different from the Canadian–US border: it divides a rich from a poor country, it also separates people by language, culture, "race," and history.
2 I borrow the term racial-ethnic from Nakano Glenn (1992) because it signifies that, in current politics, in the USA, "racial" groups are also ethnic groups.
3 Hays (1987) identifies the turning point from conservation to environmentalism as World War II. Regardless of the date, the defining characteristics are an urban, affluent, middle-class movement concerned with public health, safety, aesthetics, and the preservation of wildlife and wilderness.
4 Under the Emergency Planning and Community Right to Know Act of 1986, manufacturers who employ at least ten people and use any of 300 chemicals in amounts of 25,000 pounds or more annually in production, or 10,000 or more annually in nonproduction, must report their releases and transfers (EPA, 1991).
5 Chemical Waste abandoned their plans to build an incinerator in Kettleman City in September 1993 (Grossi, 1993).
6 The Westside of the city has traditionally been home to an (affluent) Anglo population, and the Greater Eastside has the nation's largest concentration of Latinos (Soja, 1989).
7 After an explosion of a plating shop in the residential community of Boyle Heights in Los Angeles in 1988, an effort was made by then Los Angeles City Councillor Gloria Molina to relocate all metal-platers to inner-M3 zoning (Topping, 1988). The ordinance was never passed because of the opposition of the metal-platers and the logistical problems it would have posed (MFASC, 1989).
8 LCSC was actually founded in Los Angeles in the early 1980s in an effort to keep the Van Nuys General Motors plant open.
9 One manifestation of a more conservative trend can be seen in recent legislation. Presently, there are thirty bills in the California legislature aimed at curtailing environmental regulations, plus an additional fifty directed at the 1970 California Environmental Quality Act (CEQA) (White *et al.*, 1993). These bills range from abolishing local air districts (AB 716) to exempting all developments in riot-affected areas from CEQA and SCAQMD regulations (SB 1007). There were also over twenty anti-immigration bills introduced in 1993 (Bailey and Morain, 1993; NCC, 1993).
10 There is a history of binational cooperation between the Mexican and US government, as well as activism on the part of mainstream environmental groups and other organizations (Bath, 1982). But much of this activism, such as the Border Ecology Project, is largely composed of professionals, not grass-roots working-class people (see IHERC, 1993).
11 "[P]orque muchas veces las companias . . . que estan contaminando solas mismas en aquel lado de frontera y en este lado de la frontera. Entonces tenemos una causa comun. Aqui arribe en el parque industrial hay una fundadora de plomo que esta contaminando. Esta compania que es fundicion es la misma compania que esta contaminando la gente en Dallas Texas. Una comunidad Afro Americana y Latina. En esa unidad hemos creado poder."

REFERENCES

Alston D. (ed.) 1990, *We Speak for Ourselves: Social Justice, Race, and Environment* (Panos Institute, Washington, DC).

Asch, P. and Seneca, J. 1978, "Some evidence on the distribution of air quality," *Land Economics* 54: 278–297.

Bailey, E. and Morain, D. 1993, "Anti-immigration bills flood legislature," *Los Angeles Times*, 3 May, pp. A3, A22, A23.

Barkin, D. 1990, *Distorted Development: Mexico in the World Economy* (Westview Press, Boulder, CO).

Bath, R. 1982, "Health and environmental problems: the role of the border in El-Paso Ciudad Juarez Coordination," *Journal of Interamerican Studies and World Affairs* 24: 375–392.

Berkovitz, D. 1992, "Pariahs and prophets: nuclear energy, global warming, and inter-generational justice," *Columbia Journal of Environmental Law* 17: 245–325.

Berry, B., Caris, L., Gaskill, D., Kaplan, C., Piccinini, J., Planert, N., Rendall, J. III, de Sainte Phalle, A. (eds)

1977, *The Social Burdens of Environmental Pollution: A Comparative Metropolitan Data Source* (Ballinger, Cambridge, MA).

Bloch, R. and Keil, R. 1991, "Planning for a fragrant future: air pollution control, restructuring, and popular alternatives in Los Angeles," *Capitalism, Nature and Socialism* 2: 44–65.

Blumberg, L. and Gottlieb, R. 1989, "Saying no to mass burn," *Environmental Action* (January/February): 28–30.

Borrelli, P. 1988, "Environmentalism at a crossroads," in *Crossroads: Environmental Priorities for the Future*, ed. P. Borrelli (Island Press, Washington, DC), pp. 3–25.

Bryant, B. and Mohai, P. 1992, "Introduction," in *Race and the Incidence of Environmental Hazards*, eds B. Bryant and P. Mohai (Westview Press, Boulder, CO), pp. 1–9.

Bullard, R. 1990, *Dumping in Dixie* (Westview Press, Boulder, CO).

—— 1992, "Unequal environmental protection: incorporating environmental justice in decision-making," paper presented at the Resources for the Future Conference on Setting National Environmental Priorities: The EPA Risk-based Paradigm and Its Alternatives, Annapolis, MD, November; copy available from R. Bullard, Sociology Department, University of California-Riverside, Riverside, CA 92521.

—— 1993, "Anatomy of environmental racism and the Environmental Justice Movement," in *Confronting Environmental Racism: Voices from the Grassroots*, ed. R. Bullard (South End Press, Boston, MA), pp. 15–39.

Burke, L. 1993, "Environmental equity in Los Angeles," MA thesis, Department of Geography, University of California, Santa Barbara, CA.

Callicott, B. (ed.) 1989, *In Defense of the Land Ethic* (State University of New York Press, Albany, NY).

Capek, S. 1993, "The "Environment Justice" frame: a conceptual discussion and an application," *Social Problems* 40: 5–24.

Carson, R. 1962, *Silent Spring* (Riverside Press, Cambridge, MA).

CBE 1989, "Richmond at risk: community demographics and toxic hazards from industrial polluters," Citizens for a Better Environment, 122 Lincoln Boulevard, Suite 201, Venice, CA 90291.

CCC 1992, *California's Jobs and Future*, Council on California Competitiveness; copy available from California State Senator Leonard, 1020 N Street, Suite 534, Sacramento, CA95814.

Cole, L. 1992, "Empowerment as the key to environmental protection: the need for environmental poverty law," *Ecology Law Quarterly* 19: 619–683.

Colquette, K. and Robertson, E. H. 1991, "Environmental racism: the causes, consequences, and commendations," *Tulane Environmental Law Journal* 5: 153–207.

Davis, M. 1992, *City of Quartz: Excavating the Future of Los Angeles* (Verso, London).

Dietz, M. 1987, "Context is all: feminism and theories of citizenship," *Daedalus* 4: 1–24.

EPA 1991, *1991 Toxic Release Inventory, Public Data Release* (US Environmental Protection Agency, Office of Pollution, Prevention and Toxics, Washington, DC).

—— 1993, *Environmental Equity: Reducing Risk for All Communities* (US Environmental Protection Agency, Office of Policy, Planning and Education, Washington, DC).

EPA Journal 1992, "Environmental protection—has it been fair?," special issue, ed. J. Heritage, 18: 1–64.

FitzSimmons, M. and Gottlieb, R. 1988, "A new environmental politics," in *Reshaping the U.S. Left: Popular Struggles in the 1980s*, eds M. Davis and M. Sprinkler (Verso, London), pp. 114–130. (Reference of particular interest to this paper which is not cited in the text.)

Fox, S. 1981, *John Muir and His Legacy: The American Conservation Movement* (Little, Brown, Boston, MA).

Freed, D. 1993a, "The dangers of life on the line," *Los Angeles Times*, 5 September, pp. Al, A26–A28.

—— 1993b, "Few safeguards protect workers from poisons," *Los Angeles Times*, 6 September, pp. Al, A24, A26.

—— 1993c, "Revived Cal fails to inspect most factories," *Los Angeles Times*, 7 September, pp. Al, A14, A15.

Freeman, A. 1972, "Distribution of environmental quality," in *Environmental Quality Analysis*, eds A. Kneese and B. Bower (Johns Hopkins University Press, Baltimore, MD), pp. 243–278.

Freudenberg, N. 1984, *NOT in Our Backyards!* (Monthly Review Press, New York).

Gaylord, C. 1993, "Statement of Dr. Clarice Gaylord before the Subcommittee on Civil Rights and Constitutional Rights Committee on the Judiciary,

US House of Representatives. 3 March" (US Government Printing Office, Washington, DC).

Gelobter, M. 1992, "Toward a model of environmental discrimination," in *Race and the Incidence of Environmental Hazards*, eds B. Bryant and P. Mohai (Westview Press, Boulder, CO), pp. 64–81.

Gianessi, L., Peskin, H., and Wolff, E. 1979, "The distributional effects of uniform air pollution policy in the United States," *The Quarterly Journal of Economics* XCII: 281–301.

Gibbs, L. 1982 *Love Canal: My Story* (State University of New York Press, Albany, NY).

Godsil, R. 1991, "Remedying environmental racism," *Michigan Law Review* 90: 394–427.

Gottlieb, R. and Ingram, H. 1988, "The new environmentalists," *The Progressive* 52: 14–16.

Gould, C. 1988, *Rethinking Democracy* (Cambridge University Press, Cambridge).

Greenberg, M. and Anderson, R. 1984 *Hazardous Waste Sites: The Credibility Gap* (Center for Urban Policy Research, Rutgers University, New Brunswick, NJ).

Grossi, M. 1993, "Chemwaste won't build Kettleman incinerator," *The Fresno Bee*, 8 September, pp. Al, A8.

Gupta, A. and Ferguson, J. 1992, "Beyond "culture": space, identity, and the politics of difference," *Cultural Anthropology* 7: 6–23.

Hays, S. 1987, *Beauty, Health and Permanence* (Cambridge University Press, Cambridge).

Heiman, M. 1990, "From "Not in My Backyard!" to "Not in Anybody's Backyard!" Grassroots challenge to hazardous waste facility siting," *Journal of the American Planning Association* 56: 359–362.

Herzog, L. 1991, "Cross-national urban structure in the era of global cities: the US–Mexico transfrontier metropolis," *Urban Studies* 28: 519–533.

Heyman, J. 1991, *Life and Labor on the Border* (University of Arizona Press, Tucson, AZ).

Hise, L. 1992, "The role of environmental regulations in industrial location: furniture manufacturing in Southern California," client project, Graduate School of Architecture and Urban Planning, University of California, Los Angeles, Los Angeles, CA.

Hurely, A. 1988, "The social biases of environmental change in Gary Indiana, 1945–1980," *Environmental Review* 12: 1–19.

IHERC 1993, *BorderLines* 3, Inter-Hemispheric Education Resource Center, PO Box 4506, Albuquerque, NM 87102.

Kazis, R. and Grossman, R. 1982, *Fear at Work* (Pilgrim Press, New York).

Kearney, M. 1991, "Borders and boundaries of state and self at the end of empire," *Journal of Historical Sociology* 4: 52–74.

Kratke, S. and Schmoll, F. 1991, "The local state and social restructuring," *International Journal of Urban and Regional Research* 15: 542–552.

Lavelle, M. 1992, "The minorities question," *National Law Journal*, 21 September, p. S2.

Lazarus, R. 1993, "Pursuing 'Environmental Justice': the distributional effects of environmental protection," *Northwestern University Law Review* 87: 787–857.

LCSC 1989, "Labor/community watchdog: a three year organizing strategy to impact the Los Angeles Clean Air Plan," Labor/Community Strategy Center, 3780 Wilshire Boulevard, Suite 1200, Los Angeles, CA90010.

Leal, T. 1993, Presentation at the Gathering of the Southwest Network for Economic and Environmental Justice, Tijuana, Mexico, 20 August; copy available from Southwest Organizing Project, 211 10th Street SW, Albuquerque, NM 87102.

Lee, C. 1992, "Toxic waste and race in the United States," in *Race and the Incidence of Environmental Hazards*, eds B. Bryant and P. Mohai (Westview Press, Boulder, CO), pp. 10–27.

Leopold, A. 1988, *A Sand County Almanac*, 25th printing (Ballantine Books, New York).

Levy, S. and Arnold, R. 1992, *The Outlook for the California Economy*, Center for Continuing Study of the California Economy, 610 University Avenue, Palo Alto, CA94301.

Logan, J. and Swanstrom, T. (eds.) 1990, *Beyond the City Limits* (Temple University Press, Philadelphia, PA).

Lowenthal, A. and Burgess, K. (eds.) 1993, *The California–Mexico Connection* (Stanford University Press, Stanford, CA).

McCaull, J. 1976, "Discriminatory air pollution," *Environment* 18: 26–31.

Mann, E. 1991, "L.A.'s Lethal Air," Labor/Community Strategy Center, 3780 Wilshire Boulevard, Suite 1200, Los Angeles, CA90010.

Marston, S. A. 1990, "Who are 'the people'?: gender, citizenship, and the making of the American nation," *Environment and Planning D: Society and Space* 8: 449–458.

Martinez, E. 1991, "When people of color are an endangered species," *Z-magazine* 4: 61–65.

MFASC 1989, "Memorandum on City Council proposed ordinance requiring relocation of metal finishing facilities, council file 88–0983," Metal Finishing Association of Southern California, 20 January; copy available from author.

Moore, R. 1993, Presentation at the Gathering of the Southwest Network for Economic and Environmental Justice, San Diego, CA, 21 August; copy available from Southwest Organizing Project, 211 10th Street SW, Albuquerque, NM 87102.

Morales, R. and Ong, P. 1993, "The illusion of progress: Latinos in Los Angeles," in *Latinos in a Changing U.S. Economy*, eds R. Morales and F. Bonilla (Sage, Beverly Hills, CA), pp. 55–84.

Morales, R. and Tamayo-Sanchez, J. 1992, "Urbanization and development of the United States–Mexico border," in *Changing Boundaries in the Americas*, ed. L. Herzog (Center for US–Mexico Studies, University of California, San Diego, San Diego, CA), pp. 49–68.

Mumme, S., Bath, C. R., and Assetto, V. 1988, "Political development and environmental policy in Mexico," *Latin American Research Review* 23: 7–34.

Nakano Glenn, E. 1992, "From servitude to service work: historical continuities in the racial division of paid reproductive labor," *Signs* 18: 1–43.

Nash, R. 1989, *The Rights of Nature* (University of Wisconsin Press, Madison, WI).

NCC 1993, "Anti-immigrant legislation: 1993 California legislative session," New California Coalition; copy available from the Committee for Human and Immigrant Rights, Coalition for Humane Immigrant Rights of Los Angeles, 621 S Virgil Avenue, Los Angeles, CA 90005.

Ong, P. and Blumenberg, E. 1993, "An unnatural trade-off: Latinos and environmental justice," in *Latinos in a Changing U.S. Economy*, eds R. Morales and F. Bonilla (Sage, Beverly Hills, CA), pp. 207–225.

Partridge, E. (ed.) 1981, *Responsibilities to Future Generations* (Prometheus, Buffalo, NY).

Pasternak, J. 1991, "Firms find a haven from U.S. environmental rules," *Los Angeles Times*, 19 November, pp. Al, A24.

Pastor, M. and Ayon, D. (eds.) 1992, *Encuentro: Mexico in Los Angeles* (International and Public Affairs, Occidental College, Los Angeles).

Pincetl, S. 1994, "Challenges to citizenship: Latino immigrants and political organizing in the Los Angeles area," *Environment and Planning A* 26: 895–914.

Pulido, L. 1994, "Reconstructing Chicano environmental history: the early pesticide campaign of the UFW, 1965–71," *New Scholar*, forthcoming.

Rawls, J. 1971, *A Theory of Justice* (Belknap Press, Cambridge, MA).

Rhee, S. 1993, "Southwest Network campaign update," *Voces Unidas* 3: 16–17.

Roque, J. and Tau Lee, P. 1993, "Environmental Justice and comparative risk: a report of the Environmental Justice Workgroup of the California Comparative Risk Project," September; copy available from the Department of Health Services, State of California, 2151 Berkeley Way, Berkeley, CA 94704–1011.

Rotella, S. 1992, "Activists unite to oppose new incinerator," *Los Angeles Times*, 30 March, pp. A3, A21.

Rubio, L. and Trejo, G. 1993, "Reform, globalization, and structural dependence: new economic ties between Mexico and California," in *The California–Mexico Connection*, eds A. Lowenthal and K. Burgess (Stanford University Press, Stanford, CA), pp. 51–65.

Ruffins, P. 1991, "Environmental commitment as if people didn't matter," in *Environmental Racism: Issues and Dilemmas*, eds B. Bryant and P. Mohai (Office of Minority Affairs, University of Michigan, Ann Arbor, MI), pp. 51–57.

Russell, D. 1989, "Environmental racism," *The Amicus Journal* (Spring): 22–32.

Sanchez, M. 1993, Presentation at the Gathering of the Southwest Network for Economic and Environmental Justice, Tijuana, Mexico, 20 August; copy available from Southwest Organizing Project, 211 10th Street SW, Albuquerque, NM 87102.

Sapiro, V. 1984, "Women, citizenship, and nationality: immigration and naturalization policies in the U.S.," *Politics and Society* 11: 1–26.

Sassen, S. 1991, *The Global City* (Princeton University Press, Princeton, NJ).

SCAQMD 1989, "The path to clean air: attainment strategies, summary of the Air Quality Management Plan," copy available from South Coast Air Quality Management District, 21865 Copley Drive, Diamond Bar, CA 91765.

—— 1992a, "South Coast Air Quality Management District's report to the governing board regarding the recommendations of the Special Commission on Air Quality and the Economy and recommendations to adopt certain recommendations," agenda 35, 7 August; copy available from South Coast Air Quality Management District, 21865 Copley Drive, Diamond Bar, CA 91765.

—— 1992b, "Reconsideration of the nine rejected recommendations of the Special Commission on Air Quality and the Economy," agenda 18, 18 September; copy available from South Coast Air Quality Management District, 21865 Copley Drive, Diamond Bar, CA 91765.

Scheffer, V. 1991, *The Shaping of Environmentalism in America* (University of Washington Press, Seattle, WA).

Shklar, J. 1991, *American Citizenship: The Quest for Inclusion* (Harvard University Press, Cambridge, MA).

Sklair, L. 1992, "The maquila industry and the creation of a transnational capitalist class in the United States–Mexico border region," in *Changing Boundaries in the Americas*, ed. L. Herzog (Center for US–Mexico Studies, University of California, San Diego, San Diego, CA), pp. 69–88.

Smith, S. 1989, "Society, space and citizenship: a human geography for the 'new times'?," *Transactions of the Institute of British Geographers: New Series* 14: 144–156.

SNEEJ 1992, "Statement of solidarity," Southwest Network for Economic and Environmental Justice; copy available from the Southwest Organizing Project, 211 10th Street SW, Albuquerque, NM 87102.

Soja, E. 1989, *Postmodern Geographies* (Verso, New York).

Solis, R. 1993, Presentation at the Gathering of the Southwest Network for Economic and Environmental Justice, Tijuana, Mexico, 20 August; copy available from Southwest Organizing Project, 211 10th Street SW, Albuquerque, NM 87102.

SRC 1992, "Are we losing our competitive edge?," Senate Republican Caucus; copy available from California State Senator Leonard, 1020 N Street, Suite 534, Sacramento, CA 95814.

Stall, B. 1993, "Regulations, taxes blamed for state's woes," *Los Angeles Times*, 25 March, pp. Al, A28.

Stammer, L. 1990, "Jackson calls on minorities to take up environmental battle," *Los Angeles Times*, 3 April, pp. A3, A28.

Stone, C. 1974, *Should Trees Have Standing? Toward Legal Rights for Natural Objects* (William Kaufmann, Los Altos, CA).

Storper, M. and Scott, A. J. 1989, "The geographical foundations and social regulation of flexible production complexes," in *The Power of Geography*, eds J. Wolch and M. Dear (Unwin Hyman, Winchester, MA), pp. 21–40.

SWOP 1990, "Major national environmental organizations and problems of the 'environmental movement,'" briefing paper; copy available from Southwest Organizing Project, 211 10th Street SW, Albuquerque, NM 87102.

Szekely, G. 1993, "California and Mexico: facing the Pacific Rim," in *The California–Mexico Connection*, eds A. Lowenthal and K. Burgess (Stanford University Press, Stanford, CA), pp. 113–128.

Taylor, D. 1993, "Environmentalism and the politics of inclusion," in *Confronting Environmental Racism: Voices from the Grassroots*, ed. R. Bullard (South End Press, Boston, MA), pp. 53–61.

TEF 1993, "Will NAFTA protect the environment?," *The Environmental Forum* (March/April): 28–35.

Topping, K. 1988, "Transmittal of report and ordinance restricting metal plating to the inner M3 zone to City Council," 14 July; copy available from author.

Turner, E. and Allen, J. P. 1990, *An Atlas of Population Patterns in Metropolitan Los Angeles and Orange Counties* (Department of Geography, California State University, Northridge, Northridge, CA).

UCC 1987, "Toxic wastes and race in the United States," United Church of Christ, 105 Madison Avenue, New York, NY 10016.

—— 1991, "Principles of environmental justice," Proceedings of the First National People of Color Environmental Leadership Summit, Washington, DC, October; copy available from United Church of Christ, Commission for Racial Justice, 105 Madison Avenue, New York, NY 10016.

USGAO 1983, "Siting of hazardous waste landfills and their correlation with racial and economic status of surrounding communities," US General Accounting Office, Washington, DC.

—— 1991, "US–Mexico trade: some US wood firms relocated from Los Angeles area to Mexico," report to the Chairman, Committee on Energy and Commerce, House of Representatives GAO/NSIAD 91–191, US General Accounting Office, Washington, DC.

Wernette, D. and Nieves, L. 1992, "Breathing polluted air," *EPA Journal* 18: 16–17.

White, J., Holmes, B., and Goodwin, E. 1993, "Regulating reform issues," memorandum to Environmental Recovery Group; copy available from Labor/Community Strategy Center, 3780 Wilshire Boulevard, Suite 1200, Los Angeles, CA 90010.

Young, I. M. 1983, "Justice and hazardous waste," in *The Applied Turn in Contemporary Philosophy. Bowling Green Studies in Applied Philosophy, Volume 1*, eds M. Bradie, T. Attig, and N. Reseller (Bowling Green State University Press, Bowling Green, OH), pp. 171–183.

—— 1990, *Justice and the Politics of Difference* (Princeton University Press, Princeton, NJ).

Zupan, J. 1973, *The Distribution of Air Quality in the New York Region* (Johns Hopkins University Press, Baltimore, MD).

17

Environmental justice and American Indian tribal sovereignty

Case study of a land-use conflict in Skull Valley, Utah

Noriko Ishiyama

from *Antipode*, 2003, 35(1): 119–139

INTRODUCTION

Since the early 1990s, the Skull Valley Band of Goshute Indians, a small tribe in a desolate desert in Tooele County, Utah, has antagonized the governor, politicians, and citizens of Utah, environmentalists, and environmental-justice advocates with a proposal to host temporary storage of high-level radioactive waste on the reservation. Having been historically neglected and isolated in what has come to be a toxic desert in northwestern Utah, the Goshute leaders argue that this is the only choice left for the tribe to survive. They emphasize the notion of tribal sovereignty and reject assertions that they are simply the powerless victims of environmental injustice. The current land-use controversy reveals unresolved dilemmas regarding self-determination and environmental justice as they are structured in the capitalist political economy and the history of colonialism.

This paper examines the development of the Skull Valley land-use debate and ultimately poses a challenging question: What is environmental justice in the context of questions of tribal sovereignty? Based on archival research and interviews with various players in the conflict, the case study challenges the predominant tradition of environmental-justice scholarship, which emphasizes the inequitable distribution of hazards in low-income minority communities (Bryant 1995; Bryant and Mohai 1992; Bullard 1990, 1993, 1994). The complicated politics of environmental justice in relation to tribal sovereignty as situated within the history and geography of Skull Valley contradicts oversimplified tales of distributive injustice. Rather, political-ecological dynamics intersect closely with issues of self-determination and identity formation, at and across different geographic scales, to shape the geography of environmental injustice.

The Skull Valley case study complicates widely accepted understanding of environmental justice. When the Department of Energy (DOE) began looking for a place to site interim storage of high-level radioactive waste, some communities showed interest in accepting the facility in return for economic compensation. The majority of such interested communities have been American Indian tribes. Calling attention to the historical pattern of colonialism, some academics and activists have emphasized the notion of environmental racism in explaining the correlation between the locations of ecological contamination and tribal nations (Churchill 1997; Grinde and Johansen 1995; Laduke 1999; Laduke and Churchill 1992). These writers' contribution to illuminating the environmental and social problems threatening Native America is significant. However, the discourse of environmental racism has a pitfall, in that it develops a theory of environmental justice based primarily on the dichotomy of racist white society on the one hand and victimized tribes on the other. Struggles for environmental justice, as they are intertwined with politics of tribal sovereignty and identities developed in a political-geographic context of colonialism, raise a much more complex set of issues.

Early environmental-justice literature concerning environmental destruction on tribal lands did not properly address the issue of sovereignty, in spite of this issue's significance for achieving social justice for tribes. Some academics even questioned the legitimacy of tribal governments on the grounds that they made harmful environmental decisions for economic benefits (Hall 1994; Shrader-Frechette 1996). In contrast, an American Indian legal scholar, Dean Suagee (1994, 1999), harshly criticized some environmental-justice advocates for not understanding the unique sovereign status of tribes. Other legal scholars have explored the potential conflict between environmental-justice advocacy and the tribal governments' protection of their sovereign rights (Foster 1998; Gover and Walker 1992; Louis 1997; Sachs 1996). If environmental-justice scholars do not acknowledge the intersection of tribal sovereignty and environmental justice in the context of historical colonialism, they fail to address the issue of community self-determination, potentially leading to an uneasy relationship between a struggling tribe and environmental-justice advocates.

This paper first presents a critical review of the existing scholarly literature of environmental justice, to clarify the potential contribution of the study. After a brief discussion of US nuclear-waste policies in relation to American Indian tribes, which sets up an analytical basis for the study, the paper examines the locational conflict in Skull Valley, Utah. The analysis starts with an introduction of the history of the Skull Valley Goshute tribe and the historical geography of Skull Valley as a toxic haven. This overview establishes a significant context in which to illustrate the struggles for self-determination at different geopolitical scales in relation to the politics of environmental justice, tribal sovereignty, and American Indian identities. The section focused on the pursuit of self-determination examines the politics of tribal sovereignty broadly in the environmental-justice movement and then specifically in the Skull Valley land-use debate. The final part of the paper explores the central and yet extremely difficult question: What is environmental justice? The conclusion does not aim to provide a simple answer, but encourages environmental-justice scholars to redirect their focus to address historical and structural contexts.

ENVIRONMENTAL-JUSTICE RESEARCH

The environmental-justice movement pursues a wide range of agendas that have significant implications for politics, legislation, and social activism. The majority of the existing academic literature, however, does not address the political and historical complexity of environmental justice, simplifying activism within a dichotomous framework of environmental racism and/or its focus on a superficial distribution of hazards. The following section critically examines these issues in order to elucidate the significance of (1) developing studies on ideological and structural racism, (2) going beyond the notion of distributive justice, and (3) refining the concept of procedural justice in relation to communities' access to and capacity for self-determination.

The notion of environmental racism requires a critical clarification. As employed in some of the literature, this notion implies the existence of a simple, clear dichotomy between racist white society and communities of color, neglecting the difficult dilemmas regarding serious issues of internal power structure, identity politics, and ideological disparities that confront communities of color. Laura Pulido's (1996a) critique of the earlier literature is helpful for the reconsideration of environmental racism. She points out the theoretical flaws in simplifying racism as overt actions, neglecting racism as an ideology, and portraying racism as fixed without mobility or change. By contrast, her own recent study of environmental racism in the context of urban development in Los Angeles (Pulido 2000) illustrates the hegemonic nature of racism, explaining the distinction between racism expressed through direct words and actions and subtle racism expressed through ideas and structural forces. Studies of environmental justice carry the potential to elaborate the analysis of institutional, ideological, and structural dynamics and practices of racism.

Furthermore, environmental-justice studies need to go beyond the concept of distributive justice. The existing scholarship has been dominated by this theory, which problematizes the unequal allocation of hazards based on the racial and economic characteristics of communities. This theory avoids issues of social relations and historical, cultural, and ideological contexts that are inherent within capitalist geographies. Moreover, as observed by Dobson (1998: 20), environmental-justice scholarship within the framework of distributive justice has reduced the concept of environment to "no more—and certainly no less—than a particular form of the goods and bads that society must divide among its members." This theoretical reduction of the notion of environment makes the environmental-justice scholarship vulnerable to easy criticism. For example, some academics have challenged the environmental-justice movement with chicken-and-egg logic, according to which the key question becomes: Which came first, the minorities or the facilities? (Been 1994; Been and Gupta 1997; Huebner 1998). Others purport to show that hazardous facilities are not disproportionately located in low-income minority neighborhoods (Anderton *et al.* 1994; Anderton *et al.* 1997). Recent studies, rejecting such limited scope of analysis focused on superficial outcome of distribution and with a critical eye toward political economy, articulate the social contexts that underlie and produce environmental-justice problems (Faber 1998; Heiman 1996; Hunold and Young 1998; Lake 1996; Low and Gleeson 1998).

In order to clarify the contexts of environmental justice, the concept of procedural justice should be elaborated. Lake (1996: 169) notes that "redistributing outcomes will not achieve environmental justice unless it is accompanied and, indeed, preceded by a procedural redistribution of power in decision-making." Examination of the social processes and the political-economic structure through which communities participate in or are excluded from decision-making (for example, in regards to the production, siting, and management of radioactive waste) illustrates the determinants of environmental justice. Moreover, as emphasized by Schroeder (2000), the conception of justice raises a variety of questions with regards to, for instance, rights to livelihood, enhancement of infrastructure for democratic political processes, and acknowledgement of cultural diversity. Accordingly, the notion of procedural justice should convey a broad range of social processes that develop the scope of communities' rights to self-determination at different geographic scales. The following case study may contribute to the theoretical understanding of environmental justice by elucidating complicated socio-historical and political-economic contexts for the Skull Valley land-use debate.

US NUCLEAR-WASTE POLICY AND AMERICAN INDIAN TRIBES

Post-World War II industrial prosperity and the rise of nuclear technology have left American society an unwanted and lethal legacy—radioactive waste, accumulating every day in both military and commercial nuclear sites. Most high-level radioactive waste is produced in commercial power plants, which generate 20 percent of the electricity for the American public.

In the 1970s, a growing sense of urgency finally compelled Congress to address the nuclear-waste issue, since it was evident that numerous electric power plants would otherwise have to be closed down. In 1982, Congress passed the Nuclear Waste Policy Act (NWPA), mandating that the DOE find a permanent repository site for spent nuclear fuel. The NWPA authorized the federal government to take responsibility for radioactive-waste disposal (Davenport 1993; Raeber 1989). A 1987 amendment to the NWPA terminated the further investigation of potential sites for a permanent high-level radioactive waste repository, except for Yucca Mountain in Nevada. This congressional decision has been controversial, and Nevadans charge that their state was selected for political rather than scientific reasons. Not surprisingly, the state of Nevada, the Western Shoshone tribe (which has traditionally treasured this region as a sacred place), and environmental and antinuclear organizations have vigorously opposed the federal government's political maneuver.

In addition to proposing a permanent disposal site in Nevada, the 1987 act enabled the DOE to build monitored retrievable storage (MRS) facilities for the temporary storage of radioactive waste and spent nuclear fuel. At the same time, the Office of the Nuclear Waste Negotiator (ONWN) was established to find communities that would accept MRS facilities in return for monetary grants. On 3 May 1991, the ONWN sent requests for proposals, including three stages of study grants plus compensation packages, to all states and counties and to 535 federally recognized American Indian tribes.

The majority of the communities that showed interest in the MRS project were American Indian tribes (Erickson *et al.* 1995; Sachs 1996). For the Phase I study grant, which provided US$100,000 to each community, 16 tribes and four nontribal communities applied.[1] Two of the nontribal communities were interested in further study,[2] but state governors issued vetoes to prevent them from doing so. Five months after the issue of the Phase I study grant, nine tribes applied for Phase IIA study grants of $200,000.[3] In August 1993, the Mescalero Apache Tribe of New Mexico and the Fort McDermitt Paiute-Shoshone Tribe of Nevada and Oregon applied for the Phase IIB grant, which was supposed to offer $2,800,000. However, this grant was never awarded, because Congress pulled funding to the ONWN for the 1994 fiscal year.

The federal government's failure to site an MRS facility under the ONWN project did not prevent the electric utilities from seeking a community to host the radioactive waste facility. Shortly after the DOE's failure to receive adequate funding from Congress, nuclear-energy corporations and several tribes began pursuing direct negotiations, unmediated by the federal government. The nuclear-power utilities failed in their negotiation with the Mescalero Apache tribe in April 1996. As a result, Private Fuel Storage (PFS), a limited-liability company composed of eight electric utilities,[4] started negotiating with the Skull Valley Band of Goshute Indians for a leasing contract to locate an interim storage facility on their reservation in Tooele County, Utah. Despite the keen competition for study grants earlier on, by 1996 the Skull Valley Goshute Indians were the only entity still seeking to accept a temporary storage site for commercial high-level radioactive waste.

THE SKULL VALLEY BAND OF GOSHUTE INDIANS

In the contemporary cultural landscape cultivated predominantly by the Mormon population in Utah, Indian tribes have been the most invisible population. The executive director at the State of Utah Division of Indian Affairs described the invisibility of Utah Indian tribes as follows:

> Indians are invisible in Utah. Indians do not even exist here. If you go to Arizona or New Mexico, for instance, you look at the landscape and you can tell that Indians have existed in the past and the present. You can see that Indians live there. There are Indian art crafts and Indian shops. Here, you can see only one monolithic cultural landscape dominated by one religion. People have only superficial and paternalistic understanding of Indian tribes.
>
> (interview, Salt Lake City, 4 November 1999)

The sanitized landscape of today's Salt Lake City hardly reminds us of Utah's colonial history. In the 19th century, Indian tribes were first militarily subjugated and then made the target of Mormon conversion efforts. Since then, the state's Indian tribes have largely been neglected, if not forgotten, by Utah political leaders and the general public.

On the other hand, American Indian tribes occupy a significant role in Mormon doctrine. As Mormon historian Juanita Brooks (1944: 1), pointed out, "the Mormon philosophy regarding the Indians is unique: the Mormon treatment of their dark-skinned neighbors was determined largely by that ideology." The following quote from *The Book of Mormon* represents the fundamental core of Mormon ideology with regard to the history and anthropology of American Indian tribes:

> And he [Lord God] had caused the cursing to come upon them [Lamanites (later American Indians)], yea, even a sore cursing, because of their iniquity. For behold, they had hardened their hearts against him, that they had become like unto a flint; wherefore, as they were white, and exceedingly fair and delightsome, that they might now be enticing unto my people the Lord God did cause a skin of blackness to come upon them. And thus saith the Lord God: I will cause that they shall be loathsome unto thy people, save they shall repent of their iniquities . . . And because of their cursing which was upon them they did become an idle people, full of mischief and subtlety, and did seek in the wilderness for beasts of prey.
>
> (2 Nephi 5: 15–24, pp. 66)

Lamanites represent the Euro-American Mormons' lost brothers, who became a "fallen people awaiting the arrival of their white brothers who would once again redeem them and make them a great people" (Gottlieb and Wiley 1986: 158). Accordingly, the traditionally indifferent Euro-American attitude towards Indians is, in Mormon Utah, intertwined with a paternalistic and racist philosophy supported in detail by *The Book of Mormon*.

The Skull Valley Band of Goshute Indians is a small tribe of 124 members, maintaining an 18,000-acre reservation located approximately 45 miles southwest of Salt Lake City, Utah. Goshute Indians have lived in today's Tooele County since AD 1200 (Blanthorn 1998). The harsh environment of the region shielded the tribe from outside influence, encouraging maintenance of a traditional lifestyle and protecting them from the encroachment of Euro-Americans until large-scale Mormon settlement started in the 1840s. Later, in spite of an 1863 treaty which acknowledged the existence of the Skull Valley Band, the federal government tried to relocate the tribe from its territory during the late 19th century to consolidate them with the Confederated Tribes of the Goshute Reservation in Ibapah, which is located at the border of Utah and Nevada, and with the Ute tribe (Crum 1987). As a result of the tribe's resistance to relocation, a 1917 executive order finally approved the Skull Valley Band of Goshute Reservation as an independent tribal nation. The nomadic tribe, who used to roam a wide range of Salt Lake, Tooele, and Skull Valleys, ended up being provided with only a small portion of land useless to agricultural settlers.

The Goshutes live modest lives, isolated in the arid desert and forgotten by mainstream society after the initial period of colonization. They used to hunt and gather the limited resources available in their homeland, efficiently adjusting to the ecology of the desert and relocating themselves as seasons rotated (Defa 1979). The Euro-American settlement transformed the ecological system of the desert, introducing horses and mules, which overgrazed the grasses and thus lessened the prevalence of seeds that the tribe gathered. Finding hungry Goshutes digging roots from the ground, settlers called them "diggers." Other Indian tribes did not acknowledge the Goshutes as equals, either—Utes used to call them poor people (Papanikolas 1995: 12). Despised, feared, and neglected, Goshutes have demanded little from the outside society, living quietly for a long time.

The reservation community has been struggling to survive, suffering the legacies of colonialism. Only 24 of the 124 members currently live on the reservation; many Goshutes have left, seeking opportunities and jobs elsewhere. The reservation has not attracted major business or industry, except for a small Pony Express convenience store and the tribe's leasing contract with a rocket-engine testing facility on the site. The tribal leaders argue that the PFS project may enable the tribe to pursue sustainable development as a united community.

In the 1990s, therefore, the Skull Valley Band of Goshutes became significantly more visible, due largely to the leaders' declaration to welcome a temporary storage facility for high-level radioactive waste into their reservation. Suddenly, the invisible tribe isolated in a desolate desert presented a political and ecological threat to Utah politicians and the public. The decision to sign a contract with PFS appeared out of the blue for Utah politicians, but for the Skull Valley Goshute tribe, the project was completely consistent with the environmental history of Tooele County, Utah.

TOXIC DESERT: HISTORICAL GEOGRAPHY OF COLONIALISM

The landscape of Skull Valley symbolizes the historical subjugation of people and the environment pursued within a capitalist political economy by federal and state governments as well as by commercial industry. Skull Valley, and Tooele County as a whole, first became home to ecologically undesirable facilities in the post-World War II period (Figure 17.1). A rancher living next to the Goshute Indian Reservation called his homeland "a toxic box" (interview, Skull Valley, Tooele, Utah, 29 October 1999). Mike Davis (1998: 35) described Tooele County as the "nation's greatest concentration of hyperhazardous and ultradeadly materials." Sociologist Valerie L. Kuletz (1998) illustrated the federal government's creation of "national sacrifice zones" in the American West for the purpose of fulfilling the military and industrial interests over those of local communities. In her analysis, Tooele County occupies a significant part of this sacrificed geography.

Several federal military territories surround the Skull Valley Goshute Reservation. Open-air nerve agent tests, chemical and biological weapon tests, and incineration have been conducted on these military reserves. The Deseret Chemical Depot in Tooele County stores 768,400 artillery shells filled with sarin gas, 29,600 artillery shells of mustard gas, and 22,700 land mines filled with VX gas (Center of the American West 1997: 136). As indicated by Kevin Fedarko (2000: 117), the Pentagon estimates that a serious accident could kill as many as 89,000 people in the surrounding area. In 1968, more than six thousand sheep died after a nerve-gas leak from an airplane conducting open-air experiments with hazardous chemical and biological agents; their dead bodies were buried in the reservation territory.

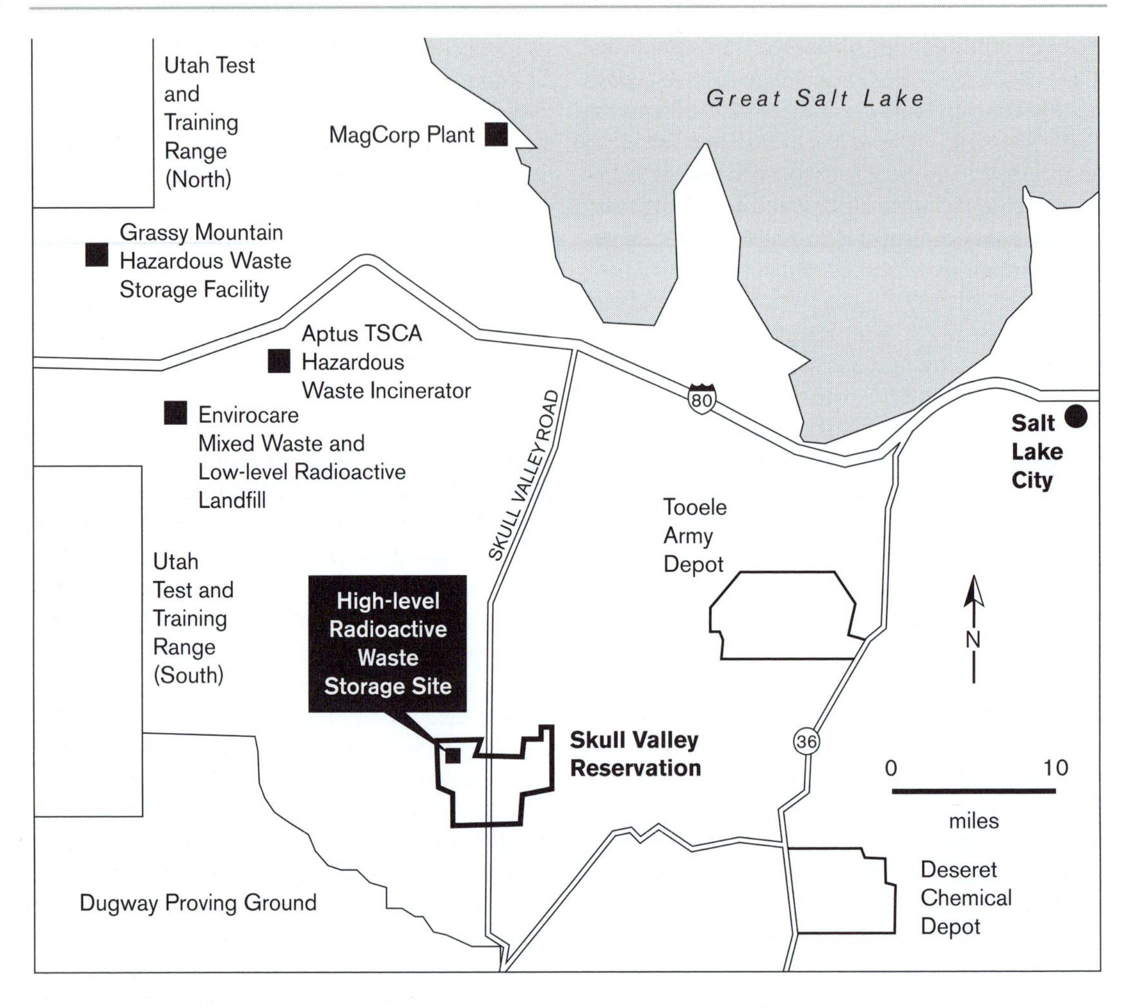

Figure 17.1 Hazardous waste and military facilities in Tooele County

Seeking to promote job growth and increase revenues, both Tooele County and the state of Utah host numerous environmental hazards in the region. Commercial facilities, including hazardous-waste incinerators and low-level radioactive and mixed waste-disposal facilities, are located in the vicinity of the reservation. According to environmental law scholar Michael Gerrard (1995), Tooele County commissioners allowed chemical weapons to be incinerated in the Tooele Army Depot in exchange for $20 million to build a hospital. They also established the West Desert Hazardous Industry Area, which "created more than nine hundred new jobs and brings in $2 million in annual 'mitigation' fees, which have allowed the county to freeze its property taxes" (Gerrard 1995: 119). The state of Utah supported this zoning project in the early 1980s in order to relocate uranium tailings from densely populated Salt Lake County to the barren desert in Tooele County.

Tooele County politicians have acted in a fashion consistent with the county's previous environmental policy, encouraging the PFS project to move forward. In the beginning, county commissioners did not openly make public statements concerning the Goshute project, apparently because they were trying to get a contract of their own with PFS. In May 2000, the commissioners signed an agreement with PFS that would provide the county up to $300 million. According to an article published in *The Salt Lake Tribune* (Fahys 2000), one of the commissioners explained that the commission would not be able to prevent the facility from coming and, therefore, they agreed to a deal which would provide the county with some legal and financial rights. The county became entitled to receive lucrative financial benefits, all the while asserting that public safety would be its highest priority.

The existing waste and military facilities have caused serious pollution problems, while providing some short-term economic benefits. According to Chip Ward (US Nuclear Regulatory Commission 1998), spokesperson for a local environmental advocacy group, West Desert HEAL, MagCorp's magnesium refinery in Tooele County emits 85 percent of the point-source chlorine gas emitted in the nation. "More than 33 pounds of toxic pollution per capita is emitted each year in Utah," he points out, "compared to a national average of just under 6 pounds per capita a year" (US Nuclear Regulatory Commission 1998). However, since the facilities provide tremendous economic benefits, these facilities have been tolerated and even sought after by local municipalities.

Neither the federal nor the state governments invited the participation of the Goshute Indian tribe in the decision-making processes that have resulted in a contaminated Tooele environment. The desert was seen as desolate, and its residents were invisible to policy-makers at larger geographic and political scales. The Skull Valley Goshute chairman pointed out the exclusion of the tribe in the spatial construction of environmental injustice: "They've never asked *us* for our permission when they built all these facilities around our reservation" (interview, Salt Lake City, UT, 2 November 1998). The tribe's distrust of outside communities has therefore grown severe, which has, in turn, caused the Skull Valley Band to be even more politically isolated.

In keeping with this political tactic of the federal and state governments, Goshute leaders did not consult with any neighboring communities in the process of developing their plan to host a temporary storage facility for high-level radioactive waste. As a result, the tribe has encountered harsh objections from the state of Utah, environmentalists, environmental-justice advocates, other American Indian tribes and organizations, and even some of its own members. Despite the state government's notorious environmental policies, some of the strongest opposition to the Goshute tribal project has come from the state. Governor Mike Leavitt issued a state executive order in April 1997 creating a task force opposed to the PFS facility. At the same time, he established the Office of High-Level Nuclear Waste Storage Opposition within the Utah Department of Environmental Quality (DEQ). Since then, Leavitt has been pursuing an "over-my-dead-body" policy to prevent high-level radioactive-waste storage.

The governor and other Utah policy-makers resent the fact that the state of Utah has been excluded from the environmental decision-making processes developed at both tribal and federal levels. Since federally recognized Indian tribes have environmental regulatory authority over their own land, the state does not have jurisdiction within the Skull Valley Goshute Reservation. Not having any legal authority over the tribe's political and environmental decision-making processes has troubled Utah political leaders. In addition, the history of Utah's exclusion from nuclear policies goes back to the 1950s, when the US military's nuclear-bomb tests in Nevada were conducted only when the wind was blowing in the direction of Utah. As a result, southern Utah citizens have been victimized by the Cold War defense policies of the US government and suffer high rates of cancer and reproductive problems (Ball 1986; Fuller 1984). This history of nuclear-weapons tests and the downwinders, described by regional writer Terry Tempest Williams (1992) as the painful experiences of the "Clan of One-Breasted Women," has, quite reasonably, made Utah policy-makers extremely sensitive toward nuclear facilities. Ward (1999) witnessed Utah policy-makers sharing their personal stories of losing relatives from cancer at the Utah Legislature meeting in 1998. The governor himself has experienced such pain and has developed strong suspicion toward the federal government. He stated: "I am from there [Southern Utah]. They [people from the federal government] told us that it was safe. It was clearly not safe. I saw my schoolmates dying from cancer and leukemia. Herds of sheep died in one day" (interview, Salt Lake City, UT, 5 July 2000). Behind the strong testimonies of the governor as he protested the PFS nuclear waste facility lies Utah's historical antipathy towards federal nuclear policies that have sacrificed Mormon downwinders to the national interest.

The historical geography of Tooele County and the state of Utah as a whole, reflecting years of mass destruction and exploitation of the environment—including peoples and societies—provides the context for the contested political ecology and the pursuit of environmental justice. Having witnessed Tooele County's environmental history of colonialism, the Skull Valley Goshute tribal leaders realize that Utah politicians have adamantly fought against their project for political rather than ecological concerns. The tribe's decision to welcome nuclear waste has raised difficult issues involving environmental justice, tribal sovereignty and retention of Goshute community and identity, the state government's fear of not having control over tribal land and its resentment against federal nuclear policies, the federal government's legal responsibility to find a dumping place for nuclear waste, friction among tribal members, and the political ecology of the production of nuclear waste.

STRUGGLES FOR SELF-DETERMINATION AND THE POLITICS OF TRIBAL SOVEREIGNTY

In addition to the paradoxical dynamics cultivated in the historical geography of colonialism, the Skull Valley locational conflict conveys challenging questions for the environmental-justice movement, with its complex politics of tribal sovereignty. In order to explicate tensions over rights to tribal sovereignty and their social as well as ethical implications, this section illustrates (1) the politics of tribal sovereignty in the environmental-justice movement and (2) struggles over tribal sovereignty and environmental justice specifically in the Skull Valley land-use dispute.

Politics of tribal sovereignty in the environmental-justice movement

The concept of tribal sovereignty has essentially defined the politics regarding struggles for self-determination among actors involved with the Skull Valley environmental management as well as the siting of environmental hazards in tribal nations in the United States. Sovereignty recognized by treaties and the US constitution does not represent something given to the tribe; rather, it is what tribes have retained throughout the tragic history of colonialism. As Vine Deloria and Clifford Lytle (1984: 15) state, self-government of tribes has been "a product of the historical process" required for tribal political survival. Retention of sovereignty, therefore, means the survival of tribes, which hold unique legal and political status as independent nations within the United States.

American Indian activists engaged in the environmental-justice movement have explicitly addressed the importance of sovereignty (Pulido 1996b). They participated in the process of drafting the "Principles of Environmental Justice" during the first National People of Color Environmental Leadership Summit in 1991. Principle 11 (Newton 1996: 156–158) makes a clear statement: "Environmental justice must recognize a special legal and natural relationship of Native People to the US government through treaties, agreements, compacts, and covenants affirming sovereignty and self-determination." This principle emphasizes the significance of sovereignty, which makes American Indian tribes distinctive from other ethnic minorities fighting against environmental injustice in terms of their political strategy for activism.

Environmental-justice activists have demanded that tribes get appropriate federal assistance in order to establish infrastructure equivalent to that of state governments so that they can develop sound environmental programs to protest outside industry's attempt to damage the ecology of tribal land. The director of the Indigenous Environmental Network, Tom Goldtooth (1995), points out that the Environmental Protection Agency failed to acknowledge the sovereign status of tribal governments entitled to receive funds to develop environmental management of their own until 1984. Consequently, he clarifies (1995: 147), most tribes were severely behind state governments in environmental infrastructure development. Grassroots activists hold a strong mandate to protect tribal sovereignty, struggling to develop the political-economic and legislative infrastructure for participatory democracy.

At the same time, American Indian activism for social and environmental justice faces a challenging question over who has the legitimate right to realize tribal sovereignty. Grassroots environmental activists tend to question the legitimacy of tribal representatives when they do not share the same ecological philosophies. Goldtooth reveals an ideological and political disagreement over tribal sovereignty among indigenous communities:

> We as indigenous grassroots are the most protective of our sovereignty and do not hide behind it or use it as a cloak or shield like some of our Tribal governmental leaders. Some of our Tribal leaders use sovereignty to protect them from criticism or legal attack on tribal developments that are environmentally unsound.
>
> (personal communication, 27 September 1996)

Grace Thorpe (1996: 720), another key American Indian figure in the environmental-justice movement, criticizes tribal leaders who support nuclear waste sites in Indian nations for "selling our sovereignty." Thus, while recognizing the significance of protecting tribal sovereignty from various outside threats, tribal and grassroots leaders hardly share a consensus on the strategic process to fortify it.

Some indigenous grassroots activists have adopted the romanticized ethnic identity of American Indians as stewards of the environment for the purpose of justifying their right to self-determination in environmental management. They have utilized what Pulido (1998) calls "ecological legitimacy" rooted in cultural essentialism to empower and establish solidarity in the movement. Goldtooth asserts:

> There are ideological differences with the mentality of our Indian "relatives" who have decided to follow the "American dream." The American dream is about money and power. It is about owning the land . . . Indigenous Peoples don't think this way. We are only caretakers of this great sacred land.
>
> (Personal communication, 27 September 1996)

Cultural essentialism in the formation of American-Indian identities has played a significant role in developing the environmental-justice movement. Nevertheless, it has promoted the problematic generalization of a culturally diverse population according to the stereotype and has dismissed the voices of those who do not share the same ecological views. This tension concerning tribal sovereignty intertwined with the definition of American-Indian ethnic identities establishes the context of intensified political battles concerning the Skull Valley land-use debate.

Struggles over tribal sovereignty and environmental justice in the Skull Valley conflict

In the Skull Valley land-use debate, struggles for self-determination enmeshed with the politics of tribal sovereignty and Goshute ethnic identity contain the potential to clash with the agenda of the environmental-justice movement. The Goshute tribal leaders have explicitly expressed their apprehension regarding the paternalistic implications suggested by environmental-justice advocates that Goshutes have been the victims of environmental injustice. Instead, they have emphasized the tribe's capacity for environmental management and its right to self-determination based on tribal sovereignty. The Skull Valley Band of Goshute Tribe Executive Office (1995: 66) published a forceful statement arguing that "the charges of 'environmental racism' and the need to 'protect' and 'save' us smack of patronism. This attitude implies we are not intelligent enough to make our own business and environmental decisions." Their position contradicts the stereotype of American Indians as helpless victims and upholds the claim that the tribe is in charge of the use of its land.

The combination of tribal identity politics and the process of defining sovereignty has clearly influenced the environmental decision-making of the Skull Valley Goshutes. The leaders' demand for political acknowledgment of sovereignty has been a principal step in tribal identity formation. The *New York Times* reported a sensational remark made by the tribal chairman: "I don't belong to two nations. I belong to one—the Skull Valley Goshute Nation" (Egan 1998: A1). The strength of his identity as a Goshute citizen has influenced his use of tribal sovereignty: "We are alive and well and a sovereign nation. And we're using that sovereignty to attract the only business we can get to come here" (Egan 1998: A1). The tribe's utilization of sovereignty for a business deal contradicts the socially constructed image of American Indians as perfect preservationists.

Not everyone in the tribe, however, shares the ideology of the tribal leaders. Opponents within the tribe have utilized the perspective of indigenous ecology and environmental justice to justify their position as traditionalists opposed to tribal leaders. Identifying herself as a tribal traditionalist, Margene Bullcreek organized an opposing group in 1997. With financial support from the state of Utah, another antagonist, Sammy Blackbear, has initiated litigation against the Bureau of Indian Affairs, which has approved the governance of the present leadership. Like Bullcreek, Blackbear stresses his spiritual tie with the Goshute land, which has held sacred meaning to him and his family.

Identity politics and the social process of defining the meaning of tribal sovereignty have played significant parts in the development of political actions pursued by tribal members with distinct beliefs. The opponents have different visions of tribal sovereignty than do their leaders. Bullcreek explained her perception of sovereignty as follows:

> Sovereignty means who we are. We need to protect who we are. Our tribal leaders are taking traditional cultures away from us, using the corporation language. They are taking away some spirit, which has always been in the tribe.
>
> (interview, Tooele, UT, 17 November 1999)

Although she considers the maintenance of sovereignty important, she is fundamentally against the PFS–tribal project to host nuclear waste.

In contrast to the tribal leaders, the Goshute opponents have been networking broadly with grassroots environmental-justice groups and individual activists, including the Indigenous Environmental Network and Thorpe, to confront tribal leaders' policies. The opponents' claims to represent the traditional Goshutes, entitled to the right to defend their ancestral land, parallel the position of grassroots activists in the national environmental-justice movement. The philosophical differences within the tribe and the contradictory relationship developed between various tribal members and the environmental-justice advocates have made the Skull Valley land-use conflict even more difficult to resolve.

The issue of tribal sovereignty in Skull Valley has played a crucial role in determining the relationship between the tribe and the state of Utah. The tribe has been frustrated that the state has tried to intervene in the PFS project on the reservation, even though the state has no legal authority. As the tribal chairman argues, "The reservation isn't part of Utah. Utah doesn't tax it, and has no business on it unless we invite them. Utah has to understand our position as a sovereign nation" (Fedarko 2000: 122). Political representatives from the state of Utah, however, have little understanding of the complex meanings and practices of tribal sovereignty. For example, a Republican Congressman from Utah, Merril Cook, has opposed the tribal project, arguing that "something is dead wrong when a small group of people can ignore the will of 90% of our state . . . I don't think this is what the Founding Fathers had in mind. It's just not right, this use of sovereignty. The implications are frightening for us as a nation" (Egan 1998: A1). Cook's attack on tribal sovereignty illustrates the classic conflict between state governments and tribes over the control of policy-making on tribal land.

Ironically enough, Utah state officials have used the notion of environmental justice to justify their opposition against the PFS–tribal project. For example, the executive director of the Utah Department of Environmental Quality, attending the public hearing organized by the NRC on 2 June 1998, asserted that the Skull Valley project is ethically unjust. This political use of environmental-justice language by state policy-makers disregards Skull Valley Goshute tribal sovereignty, striking at the most fundamental principle of tribal justice. The debate over tribal sovereignty and its implications for the land-use issue illustrates the state government's hypocrisy and its veiled agenda to erode tribal sovereignty to get tribal land under its control.

CONCLUSION: WHAT IS ENVIRONMENTAL JUSTICE?

The Skull Valley case study entails a far more intricate story than that presented in the majority of existing literature dominated by analytical frameworks of environmental racism and distributive environmental justice. The historical colonialism grounded in the Skull Valley landscape has structurally limited the capability of the tribe to achieve economic and environmental self-determination. Conflict over the definition and practice of tribal sovereignty at different geographic scales reveals the social, historical, and political-economic complexity of environmental justice while implicating structural influences in both the production and distribution of nuclear waste and the economic survival strategies available to the Skull Valley Band of Goshute Indians.

Within the contexts of colonialism and the politics of tribal sovereignty, it is not possible to conclude that actually siting the PFS facility on the Skull Valley reservation means that justice will have been achieved for the tribe, simply because the tribe made the decision in its sovereignty capacity. Instead, the following question needs to be answered: In what context has the tribe made the decision to accept high-level radioactive waste?

The inevitable answer is that a prolonged process of historical colonialism over people and land has produced a landscape of injustice in which the tribe's choices have been severely structurally limited. Even if the tribe makes an informed decision to host the PFS facility, working from consensus among tribal leaders reached through a democratic process, they never participated in the decision-making process leading to production of nuclear waste or to the absence of alternate means of economic survival in the desert landscape. Lack of economic autonomy, due largely to the protracted environmental degradation of Skull Valley, has prevented the tribe from pursuing robust political-economic sovereignty as an indigenous nation. Whether the tribe ends up hosting high-level radioactive waste on the reservation or not, therefore, this land-use conflict represents procedural environmental injustice conditioned by Skull Valley's historical geography.

The Skull Valley land-use debate reveals theoretical defects in the predominant discourses of environmental racism and distributive environmental justice, raising significant conceptual questions. As indicated in this case, it is not necessarily useful to prove racist intention on the part of electric utilities and the federal government to site nuclear waste on the tribal reservation. The disenfranchisement of the tribe through institutional exclusion from and isolation in the environmental decision-making processes indicates the structural aspects of racism embedded in hegemonic ideologies. Rather than seeking equity in the distribution of hazards or eliminating intentional racist actions, therefore, environmental justice requires the participation of communities in various decision-making processes that are conditioned by and intertwined with the political-economic processes and social relations at different geographical scales. In the context of Indian country, environmental justice depends on tribes' sovereign capacity to pursue politically, economically, and ecologically sound options for sustainable development. Accordingly, reinforcement of both political and economic sovereignty of tribes will lead to the long-term accomplishment of environmental justice.

No easy answer exists to resolve the Skull Valley conflict concerning the siting of high-level radioactive waste. Making a simple judgment regarding environmental justice solely in the context of the present siting of the PFS facility leads us nowhere. This paper does not provide specific suggestions to resolve the immediate conflict, which is complicated by a variety of difficult historical, social, and political-economic questions. Instead, environmental-justice scholars are encouraged to reframe their research questions to articulate the truly complex practices of political economy and historical colonialism over communities' struggles for self-determination. The landscape and the peoples who play active roles in the Skull Valley conflict would not then be subject to the influence of the simplistic analyses of environmental justice that have restricted the terms of this debate thus far.

ACKNOWLEDGMENTS

I would like to acknowledge the following sources for funding this project: Fulbright All Grant Scholarship, Association of American University Women International Fellowship, Matsushita International Foundation Research Grant, National Science Foundation Doctoral Dissertation Improvement Award, and Rutgers University Dissertation Fellowship. I would like to thank Robert Lake for his guidance in developing this paper and project. Many thanks to Sandra Baptista, Matthew Gandy, Kazuto Oshio, Laura Pulido, Rick Schroeder, Phil Steinberg, Kimberly Tallbear, Mervyn Tano, and Elvin Wyly for their helpful suggestions on earlier drafts. I would also like to thank Mike Davis for providing background information and materials and Mike Siegel for technical assistance. Acknowledging their generous assistance, all shortcomings are the sole responsibility of the author.

NOTES

1 The tribes that applied included Mescalero Apache (NM), Chickasaw Indian Nation (OK), Prairie Island Indians (MM), The Sac and Fox Nation (OK), Yakima Indian Nation (WA), Skull Valley Band of Goshute Indians (LIT), Alabama/Quassarte Tribe (OK), Eastern Shawnee Tribe (OK), Lower Brule Sioux Tribe (SD), Ponca Tribe (OK), Fort McDermitt Paiute-Shoshone Tribe (NV and OR), Tetlin Village Council (AK), Akihiok-Kaguyak Inc./ Akhiok Traditional Council (AK), Apache Development Authority (OK), Absentee Shawnee (OK), and Caddo Tribe (OK). The counties that applied included Grant County (ND), Fremont County (WY), San Juan County (UT), and Apache County (AZ).

2 San Juan County (UT) and Fremont County (WY).

3 These tribes included Mescalero Apache (NM), Skull Valley Band of Goshute Indians (UT), Fort McDermitt Paiute-Shoshone Tribe (NV and OR), Ponca Tribe (OK), Eastern Shawnee Tribe (OK), Prairie Island Indians (MN), Ute Mtn. Ute Tribe (CO), Miami Tribe (OK), and Northern Apache Economic Development Community (WY).

4 These companies included American Electric Power, Consolidated Edison Company of New York, Inc.,

Dairyland Power Cooperative, Southern California Edison, GPU Nuclear Corporation, Northern States Power, Illinois Power Company, and Southern Company.

REFERENCES

Anderton, D. L., Anderson, A., Oakes, J., and Fraser, M. (1994) "Environmental equity: Demographics of dumping." *Demography* 31: 229–248.

Anderton, D. L., Oakes, J. M., and Egan, K. L. (1997) "Environmental equity in Superfund: Demographics of the discovery and prioritization of abandoned toxic sites." *Evaluation Review* 21: 3–26.

Ball, H. (1986) *Justice Downwind: America's Atomic Testing Program in the 1950s.* New York: Oxford University Press.

Been, V. (1994) "Locally undesirable land uses in minority neighborhoods: Disproportionate siting or market dynamics?" *The Yale Law Journal* 103: 1383–1407.

Been, V. and Gupta, F. (1997) "Coming to the nuisance or going to the Barrios? A longitudinal analysis of environmental justice claims." *Ecology Law Quarterly* 24: 1–56.

Blanthorn, O. (ed.) (1998) *A History of Tooele County.* Salt Lake City: Utah Historical Society.

Book of Mormon: Another Testament of Jesus Christ (1830) "An account written by the hand of Mormon upon plates taken from the plates of Nephi." Translated by Joseph Smith. Salt Lake City: The Church of Jesus Christ of Latter-day Saints.

Brooks, J. (1944) "Indian relations on the Mormon frontier." *Utah State Historical Society* 7: 1–48.

Bryant, B. (ed.) (1995) *Environmental Justice: Issues, Policies, and Solutions.* Washington, DC: Island Press.

Bryant, B. and Mohai, P. (eds) (1992) *Race and the Incidence of Environmental Hazards: A Time for Discourse.* Boulder: Westview Press.

Bullard, R. D. (1990) *Dumping in Dixie: Race, Class, and Environmental Quality.* Boulder: Westview Press.

—— (ed.) (1993) *Confronting Environmental Racism: Voices from the Grassroots.* Boston: South End Press.

—— (ed.) (1994) *Unequal Protection: Environmental Justice and Communities of Color.* San Francisco: Sierra Club Books.

Center of the American West (1997) *Atlas of the New West: Portrait of a Changing Region.* New York: W. W. Norton Company.

Churchill, W. (1997) *A Little Matter of Genocide: Holocaust and Denial in the Americas, 1492 to the Present.* San Francisco: City Lights Books.

Crum, S. J. (1987) "The Skull Valley Band of the Goshute tribe—Deeply attached to their native homeland." *Utah Historical Quarterly* 55: 251–267.

Davenport, J. (1993) "The federal structure: Can Congress commandeer Nevada to participate in its federal high-level waste disposal program?" *Virginia Environmental Law Journal* 12: 540–571.

Davis, M. (1998) "Utah's toxic heaven." *Capitalism, Nature, Socialism: A Journal of Socialist Ecology* 9: 35–39.

Defa, D. (1979) "A History of the Goshute Indians to 1900." Masters thesis, Department of History, University of Utah.

Deloria, Jr. V. and Lytle, C. (1984) *The Nations Within: The Past and Future of American Indian Sovereignty.* New York: Pantheon Books.

Dobson, A. (1998) *Justice and the Environment: Conceptions of Environmental Sustainability and Dimensions of Social Justice.* New York: Oxford University Press.

Egan, T. (1998) "New prosperity brings new conflict to Indian country." *The New York Times* 8 March: Al, 24.

Erickson, J., Chapman, D., and Johnny, R. (1995) "Monitored retrievable storage of spent nuclear fuel in Indian Country: Liability, sovereignty, and socioeconomics." *American Indian Law Review* 19: 73–103.

Faber, D. (ed.) (1998) *The Struggle for Ecological Democracy: Environmental Justice Movements in the United States.* New York: The Guilford Press.

Fahys, J. (2000) "Tooele signs deal for N-waste." *The Salt Lake Tribune*, 25 May. Available online by subscription at www.sltrib.com (last accessed 28 September 2002).

Fedarko, K. (2000) "In the valley of the shadow." *Outside*, May. Available online at http://web.outsideonline.com/magazine/200005/200005skullvalleyl.html (last accessed 28 September 2002).

Foster, S. (1998) "Justice from the ground up: Distributive inequities, grassroots resistance, and the transformative politics of the environmental justice movement." *California Law Review* 86: 775–841.

Fuller, J. G. (1984) *The Day We Bombed Utah: America's Most Lethal Secret.* New York: New American Library.

Gerrard, M. (1995) *Whose Backyard, Whose Risk: Fear and Fairness in Toxic and Nuclear Waste Siting*. Cambridge, MA: The MIT Press.

Goldtooth, T. B. K. (1995) "Indigenous nations: Summary of sovereignty and its implications for environmental protection." In B. Bryant (ed.) *Environmental Justice: Issues, Policies, and Solutions* (pp. 138–148). Washington, DC: Island Press.

Gottlieb, R. and Wiley, P. (1986) *America's Saints: The Rise of Mormon Power*. New York: Harvest/HBJ.

Gover, K. and Walker, J. (1992) "Escaping environmental paternalism: One tribe's approach to developing a commercial waste disposal project in Indian country." *University of Colorado Law Review* 63: 933–943.

Grinde, D. and Johansen, B. (1995) *Ecocide of Native America: Environmental Destruction of Indian Lands and Peoples*. Santa Fe, NM: Clear Light Publishers.

Hall, K. (1994) "Impacts of the energy industry on the Navajo and Hopi." In R. Bullard (ed.) *Unequal Protection: Environmental Justice and Community of Color* (pp. 130–154). San Francisco: Sierra Club Books.

Heiman, M. K. (1996) "Race, waste, and class: New perspectives on environmental justice." *Antipode* 28: 111–121.

Huebner, S. B. (1998) "Are storm clouds brewing on the environmental justice horizon?" *Policy Study* 145: 1–29.

Hunold, C. and Young, I. M. (1998) "Justice, democracy, and hazardous siting." *Political Studies* 46: 82–95.

Kuletz, V. (1998) *The Tainted Desert: Environmental and Social Ruin in the American West*. New York: Routledge.

Laduke, W. (1999) *All Our Relations: Native Struggles for Land and Life*. Boston: South End Press.

Laduke, W. and Churchill, W. (1992) "Native North America: The political economy of radioactive colonialism." In M. A. Jainies (ed.) *The State of Native America: Genocide, Colonization, and Resistance* (pp. 241–266). Boston: South End Press.

Lake, R. W. (1996) "Volunteers, NIMBYs, and environmental justice: Dilemmas of democratic practice." *Antipode* 28: 160–174.

Louis, L. (1997) "Sovereignty, self-determination, and environmental justice in the Mescalero Apache's decision to store nuclear waste." *Boston College Environmental Affairs Law Review* 24: 651–688.

Low, N. and Gleeson, B. (1998) *Justice, Society, and Nature: An Exploration of Political Ecology*. New York: Routledge.

Newton, D. E. (1996) *Environmental Justice: A Reference Handbook*. Denver: ABC-CLIO.

Papanikolas, Z. (1995) *Trickster in the Land of Dreams*. Lincoln: University of Nebraska Press.

Pulido, L. (1996a) "A critical review of the methodology of environmental racism research." *Antipode* 28: 142–159.

——(1996b) "Development of a 'people of color' identity in the environmental justice movement of the southwestern United States." *Socialist Review* 96(3&4): 145–180.

—— (1998) "Ecological legitimacy and cultural essentialism: Hispano grazing in the Southwest." In D. Faber (ed.) *The Struggles for Ecological Democracy: Environmental Justice Movements in the United States* (pp. 293–311). New York: The Guilford Press.

—— (2000) "Rethinking environmental racism: White privilege and urban development in Southern California." *Annals of the Association of American Geographers* 90: 12–40.

Raeber, J. D. (1989) "Comment: Federal nuclear waste policy as defined by the Nuclear Waste Policy Amendments Act of 1987." *St. Louis University Law Journal* 34: 111–116.

Sachs, N. (1996) "The Mescalero Apache Indians and monitored retrievable storage of spent nuclear fuel: A study in environmental ethics." *Natural Resources Journal* 35: 661–672.

Schroeder, R. (2000) "Beyond distributive justice: Resource extraction and environmental justice in the tropics." In K. Zimmerer (ed.) *People, Plants and Justice: The Politics of Nature Conservation* (pp. 52–66). New York: Columbia University Press.

Shrader-Frechette, K. (1996) "Environmental justice and Native Americans: The Mescalero Apache and monitored retrievable storage." *Natural Resources Journal* 35: 703–714.

Skull Valley Band of Goshute Tribe Executive Office (1995) "Native Americans have the rights to make their own land-use decisions." In J. Petrikin (ed.) *Environmental Justice* (pp 65–69). San Diego: Greenhaven Press.

Suagee, D. (1994) "Turtle's war party: An Indian allegory on environmental justice." *Journal of Environmental Law and Litigation* 9: 461–497.

Suagee, D. (1999) "The Indian country environmental justice clinic: From vision to reality." *Vermont Law Review* 23: 567–604.

Thorpe, G. (1996) "Our homes are not dumps: Creating nuclear-free zones." *Natural Resources Journal* 36: 715–723.

US Nuclear Regulatory Commission (NRC). 1998. *Scoping Meeting for Preparation of an EIS for the Private Fuel Storage Facility License Application.* Tuesday, 2 June. Salt Lake City: USNRC. Formerly available online at www.nrc.gov/OPA/reports/pfseostr.htm.

Ward, C. (1999) *Canaries on the Rim: Living Downwind in the West.* New York: Verso.

Williams, T. T. (1992) *Refuge: An Unnatural History of Family and Place.* New York: Vintage Books.

Structural power, agency, and national liberation

The case of East Timor

James F. Glassman

from *Transactions of the Institute of British Geographers*, NS, 2003, 28: 264–280

INTRODUCTION

On 30 August 1999, more than 400,000 East Timorese, nearly all of the adult population of East Timor, voted in a UN-monitored referendum on the Indonesian government's autonomy proposal. Nearly 80 percent of those who voted rejected special autonomy within the Indonesian state, opting instead for independence (Robinson 2001, 58). In doing so, they articulated electorally the position they had previously expressed through nearly 25 years of resistance to Indonesian occupation. The referendum set in motion a process of transition, under United Nations auspices, that led to a formal consummation of this resistance with the inaugural independence day celebration on 20 May 2002, and the swearing in as President of Xanana Gusmão, long-standing leader of the independence struggle.

The cost of Timorese resistance was high to the bitter end, as it had been throughout the occupation. From the time of Indonesia's invasion in late 1975 until the beginning of the 1980s, it is estimated that warfare and militarily imposed starvation and disease cost at least 150,000–200,000 Timorese lives, and possibly over 300,000, making the war in East Timor the most relatively genocidal of any in modern history (Budiardjo and Liong 1984, 51; Tanter *et al.* 2001, 260). The entire period of resistance was marked by massacres—the most noted of these in the outside world being the 1991 Santa Cruz massacre of hundreds of Timorese at a Dili cemetery (Carey 1995b, 48–55)—along with routine repression and torture (Carey and Bentley 1995; Taylor 1999; Tanter *et al.* 2001).

When the possibility of East Timor gaining independence through a referendum finally emerged in 1999 the response of the Indonesian military (TNI) was to immediately mobilize local militias that had been used throughout the occupation in order to intimidate Timorese into voting for autonomy (Kingsbury 2000, 70–1; Kammen 2001, 179–82). The death toll inflicted by these militias prior to the election has been estimated by the Timorese Catholic church to be at least 5000–6000 people (Taylor 1999, xiii). When the intimidation tactics failed and the referendum yielded a decisive victory for the independence struggle, these same militias—under direct supervision by TNI members—rampaged through the territory, killing untold thousands, burning virtually every structure in the major cities and towns, forcing over 200,000 people into hiding in the mountains, and forcibly evacuating more than 200,000 across the border and into refugee camps in Indonesian West Timor (Taylor 1999, xvii–xix; Kammen 2001, 156–57; Chomsky 2001, 144; Tanter 2001, 193). When the destruction was finally ended weeks later, by the intervention of a UN peacekeeping force, East Timor's independence seemed humbling and bitter. Moreover, the painfully slow and sometimes politically otiose process of reconstruction leaves significant doubts about the meaning and consequences of independence in the new state of Timor Leste (Tanter *et al.* 2001).

All of this having been said, however, it would be cynical and callous to assert that nothing has been accomplished through the achievement of independence, even if it takes decades of continued effort for many people in Timor Leste to benefit from the possibilities that have been created. Moreover, the liberation struggle in East Timor and the final shedding of Indonesian occupation stands as a remarkable—if seemingly untimely—concluding chapter to a century of revolutionary anti-colonial struggle, coming long after the era of national liberation movements has presumably ended. It is thus a process worthy of not only respect but careful analysis.

In this paper, I argue that a viable analysis of how, why and when East Timor gained independence requires an account of the workings of structural power. More specifically, I argue that though the maneuverings of different actors in the Timorese resistance struggle were necessary conditions of liberation, they were not sufficient and required the enabling context created by shifts in structural forces that had sustained the basis for the Indonesian invasion and occupation.

As a prelude to this analysis, I present a basic outline of Marxist conceptions of structural power, contrasting these with the conceptions of structure put forward in structuration approaches, showing that a Marxist conception can both account for agency and avoid economic reductionism. Indeed, I use the term "structural power" to refer not only to economic power narrowly conceived, but to what, following John Agnew and Stuart Corbridge, we might refer to as "geo-political economic power" (Agnew and Corbridge 1995, 6)—a conception I explain in the first section of the paper. The notion that the outcomes of independence struggles may hinge on such power has been suggested by various authors, including world systems theorists (e.g. Taylor and Flint 2000, 129–44, 225–6). But whereas world systems theorists have identified the structural context largely with long waves of economic activity (Taylor and Flint 2000, 116–18), I want to suggest that the structural context can be taken to include somewhat more specific political economic developments. Those developments may be less patterned or cyclical than economic long waves, thus making prediction of when and where independence struggles will be enabled hazardous at best, but this specificity is precisely what is required to account for a case like that of East Timor.

After outlining this basically Marxist theoretical framework in the first section of the paper, I turn in the second section to an analysis of the Timorese liberation struggle. I focus in some detail on two crucial turning points in this struggle: (1) the period from 1974 to the beginning of the 1980s, when the liberation struggle developed, but was very nearly annihilated by Indonesian aggression; and (2) the period from the mid-1990s to the present, when the protracted resistance struggle drew increasing international support and finally achieved its immediate goal of ending Indonesian occupation and creating the conditions for formal political independence. In the conclusion, I will suggest some theoretical as well as political implications for the reading of national liberation struggles as enabled and constrained by structural power.

RECUPERATING STRUCTURE FROM THE CRITIQUE OF "STRUCTURAL MARXISM"

Endorsing a notion of structural power is not equivalent to endorsing structuralism. The latter, in its many forms, has come in for a variety of criticisms over the years, and most Marxists who have been labeled in these terms—notably Louis Althusser—have denied that they are structuralists. I am in agreement with this rejection of the equation between "structural Marxism" of the sort that Althusser championed and the structuralism critiqued by Giddens and various post-structuralists.[1] What is important for the theoretical argument here, however, is merely to note that whether or not one considers "structural Marxism" structuralist, there are no grounds for considering analyses of structural power to be either (1) in tension with notions of human agency or (2) economically reductionist. Moreover, precisely because Marxist analyses of structural power are not economically reductionist, they require theorization of (3) the political forms and territoriality of structural power.

Structural power and human agency

The first of these two points has been somewhat obscured by the debate about structure and agency inaugurated in the work of Anthony Giddens and followed up in wide-ranging discussions in the social sciences throughout the 1980s (Giddens 1979, 1984). I need not rehearse those debates here, but instead I want to assert one basic claim that sidelines much of the supposed problem of structure and agency—namely, that from a Marxist perspective structure is merely the agency of large collectivities of people.[2] This means, among other things, that the exercise of agency referred to under the heading of structure connects groups of people across time and space, in ways often neglected by the conceptions of agency that have prevailed in the debates around structuration.

The perception involved in this claim is deeply sociological: humans may act individually, but they always do so as social beings, and the actions they undertake that shape the sorts of phenomena under study by social scientists—as opposed to, perhaps, biographers—are always collective to a greater or lesser degree. The structure–agency debate, in which Marxist notions of structure have sometimes been improperly identified with economic forces beyond conscious human control (e.g. Thrift 1983), thus mistakes the accurate claim that individuals and sometimes even collectivities cannot change the behavior of larger or more powerful collectivities of humans for an assertion of human impotence in the face of "impersonal" forces.

If the view that structural power represents something like "the impersonal forces of history" is asserted, it is easy enough to disprove. But few if any advocates of "structural Marxism" have ever seriously endorsed such a view. On the other hand, if structure is simply seen as the agency of large collectivities of people, then there is no special structure–agency problem to be resolved. The only issues are whether, where and how some subset of a larger collectivity can gain enough support in its actions to substantially alter relatively long-standing features of the social relations that constitute "the structure."

While it is beyond the scope of the argument here to critique structurationist perspectives in any detail, it is worth noting briefly the crucial differences between Marxist understandings of structure and the concept of structure elaborated in Giddens' work. First, Giddens' reading of Marxist historical materialism as a form of functionalism (Giddens 1979, 111–15)—an inaccurate reading in the view of most Marxists—leads him to focus on the problem of how social actors can be understood as consciously and intentionally producing and reproducing the world around them. Within Marxist conceptions of structural power, this is not a fundamental problem. While Althusser's loaded reference to humans as "supports" and "bearers" ("*Träger*") of class positions (Althusser and Balibar 1970, 180, 252) has led to the claim that this conception evacuates any notion of conscious agency, such a claim would only be true if it were the case that agency can only be conceived in individualist terms rather than in class and other collectivist terms.

Marxist notions of structural power, instead, center on the shared conditions of various collectivities of humans within the processes of material production and social reproduction. Althusser's rather hyperbolic assertion notwithstanding, most Marxist theory has insisted that individual humans can and do—as members of classes and class fractions—act to both reproduce and change social structures. Moreover, whether or not they do so intentionally is somewhat beside the point. Marx clearly appealed to workers to understand their shared interests in a particular way (i.e. to develop class consciousness and revolutionary commitments), but historical materialist analyses have recognized that while people act intentionally their actions also frequently produce unintended consequences because of the ways humans relate to one another as members of collectivities with differing projects and interests. Thus, the important analytic task is not to show what people intended—as Marx put it, "our opinion of an individual is not based on what he thinks of himself" (Marx 1977c, 390)—but rather to show how in the working out of the projects of different collectivities with different interests societies are either reproduced or transformed.

Giddens' reading of the issue of structure also differs from Marxist readings in a second basic sense, one already implied in the discussion of this first difference. From a Marxist perspective, Giddens' reading is not only implicitly too individualist but also too idealist. Giddens identifies structures as "rules and resources, recursively implicated in the reproduction of social systems" (Giddens 1979, 64). The conception of rules as crucial to structure is worked out in such a way as to support Giddens' contention that humans are conscious (i.e. rule-comprehending) actors in the reproduction of society.

Marxist conceptions of structure, while not denying that individuals and collectivities act consciously and intentionally, do not construe structures in terms of idealist conceptions of rules consciously comprehended but rather in terms of webs of social relations centered on those productive activities without which life and livelihood would be impossible. As Marx puts it,

> Men do not in any way begin by "finding themselves in a theoretical relationship to things in the external world." Like every animal, they begin by *eating, drinking*, etc. that is, not by "finding themselves" in a relationship but by behaving actively, gaining possession of certain things in the external world by their actions, thus satisfying their needs. (They thus begin by production.)
>
> (Marx 1977a, 581)

Moreover, humans do not simply construct the social relations of production *ex nihilo*. Rather,

> In the social production of their life, men enter into definite relations that are indispensable and independent of their will, relations of production which correspond to a definite stage of development of their material productive forces.
>
> (Marx 1977c, 389)

None of this means that humans act unconsciously or without choice but, rather, that those choices—and consciousness of the possible choices—are both enabled and constrained by what can be done in the way of producing and reproducing material social life.

A third and final sense in which Marxist conceptions of structure differ from that of structurationists is in the degree to which the discussion of structure focuses on either reproduction or social change. Giddens certainly addresses both in his writings, but much of the theory of structuration is devoted to explaining how it is that social structures are reproduced through the conscious and intentional behavior of individuals. Giddens' own intentions notwithstanding, this has the effect of imparting a somewhat conservative cast to the discussion in that it leads to weak theorization of the conditions that lead to social transformation.

Marxist conceptions of structure have, by contrast, been focused typically on the issue of what would lead to structural transformation. While a variety of Marxist approaches to this issue can be identified, most such approaches take various kinds of crises—political, economic and ecological, for example—as crucial. This is not because crises are taken by Marxists to lead automatically to system change (cf. Gramsci 1971, 184), but rather because they manifest disruptions of the "normal" processes that make social reproduction occur more smoothly. For this reason, a significant amount of research by historical materialists has focused on the ways in which crisis tendencies develop and/or are countered within capitalist societies, as well as the ways different collectivities both act to produce and act to capitalize on crises, rather than on the ways in which workers knowingly reproduce the social structures around them. For most Marxists, the routine and knowing reproduction of social structures is not surprising. As Marx put it, "mankind always sets itself only such tasks as it can solve" (Marx 1977c, 390), and the task of radical social change is insoluble in contexts where most of the population can, exploitation notwithstanding, successfully reproduce itself—and where it would be faced with severe repression in attempts to promote social change. Nonetheless, for Marxists, it is not these "normal" conditions that are of the most interest but rather the comparatively more rare circumstances in which larger collectivities than usual may act to transform social structures. It is in relation to this concern with the forces of change—not reproduction—that historical materialists study structure, and such a concern animates my analysis of independence struggle in East Timor, below.

It is worth noting that critical realist analyses such as those put forward by Roy Bhaskar (1989, 1993), which are broadly compatible with a Marxist account of structural power, have made more of the potential of social agents to produce change. Bhaskar develops a transformational model of social activity (TMSA) that indicates the variety of ways in which social actors can intentionally or unintentionally transform the social structures that emerge from these actors' relations with each other. While Bhaskar's TMSA provides a useful way around the stasis threatened by the structure–agency duality of Giddens' approach, it is also exceedingly general and is established through the continued elaboration of theoretical principles rather than through recursive analysis of empirical cases and development of abstract theory. In contrast to Bhaskar's approach, I rely here on the kind of approach taken by Althusser in his discussions of contradiction and overdetermination (Althusser 1977), who himself follows the approach taken by Lenin and Mao in analyzing the Russian and Chinese revolutions. This approach emphasizes concrete analysis of concrete situations, with the theoretical generalizations that are then framed in order to explain the relationship between agency, structural power and social transformation being built by abstracting from the concrete situation. Put another way—one that is theoretically congenial to Bhaskar's own favored methods—the approach here is "transcendental." That is, my starting point is not the question of whether or not dramatic forms of social change can occur, given a particular theorization of social forces; rather, starting from a concrete case of such social change, I try to theorize the kinds of conditions that made such change possible. It is this approach that I will take in analyzing the case of East Timor, though some of the results might be consistent with what would be argued by advocates of the TMSA.

Structural power and economic reductionism

A conventional view of structural power within Marxist theory identifies it with position in the class structure (e.g. Isaac 1987; Winters 1996). Capitalists are seen as having structural power in this context because of their control over investable surplus, which allows them to substantially shape society merely through exercising their politically institutionalized right to invest or withhold investment (Block 1987, 58–9). While these accounts sometimes seem to one-sidedly identify structural power with capitalist control over investment, the same basic conception of structural power could be used to identify the power of workers, whose power is constituted by their ability to collectively supply or withhold their labor power. The historical conditions that gave birth to capitalism have made the latter option particularly challenging for workers, which helps account for capitalist dominance within the structural relations of capitalist society. But the structural power of workers is nonetheless a basis for the struggles they are able to launch in challenging exploitation.

This conception of structural power may be read as economistic—but only insofar as class is equated with "economic," an equation I will challenge.[3] Economics refers simply to the social processes by which the material (including social) requirements of existence are produced, and thus the processes by which society is reproduced over time. Class refers to the specific position of collectivities within these processes of production and reproduction, indicating the degree to which one or another group is able to appropriate the surplus labor time of other groups. Class thus mediates (and for Marxists drives) economic processes—but it is not reducible to them. As E. P. Thompson notes, classes are always simultaneously economic, political and cultural entities (Thompson 1978, 287–9). Classes are thus defined by their role in economic processes, but are not themselves merely economic.

This point speaks to the refusal of most Marxist analysis to concede the existence of an economic "realm" that is separate from politics or other moments of society (e.g. Rupert 2000, 2–4). The analytical categories of liberal social science—which construct society as divided into discrete realms of politics, economics, culture and society—are impossible to completely avoid within capitalist society, and Marxist approaches can do so no more than any others. Yet the fundamental commitment of Marxist theory has been to an approach that refuses the idea of an economy that exists distinct from politics or that operates according to immutable economic "laws." Rather, even if economics and politics are admitted to be construed as separate realms within capitalist society, this is a phenomenon that has to be explained by the historically specific development of capitalism, and which is not a function of economics and politics having separate "laws of motion" or dynamics but rather a function of the dynamics of capitalist class relations.

Such a contention has important implications for the concept of structural power within Marxism. If structural power is a function of position within class relations, and if classes are always simultaneously socio-cultural and political-economic, then structural power is itself a socio-cultural, political-economic phenomenon, not a narrowly economic one. Thus, in discussing structural power we cannot settle merely on the investment behavior of capitalists or the strike activity of workers, even if these retain crucial importance. Instead, the whole panoply of interpenetrating social processes through which class structures are maintained, reproduced or challenged need to be seen as implicated in the exercise of structural power. In this sense, the activities of states, of families, of community organizations and other collectivities are all part of the exercise of structural power insofar as they bear on class issues.[4]

The position for which I am arguing here is consistent with John Agnew and Stuart Corbridge's assertion of the need for not just a political economy but a "geo-political economy"—one that analyses relations such as the power struggles within and between states as integral to the development of political economy, and thus of class processes (Agnew and Corbridge 1995, 6). It is from this sort of position that one can best begin to analyze imperialism and national liberation struggles in relation to structural power. For the political economy of imperialism is always intimately bound up with geo-politics (including cultural politics), and thus an adequate account of structural power cannot end with the structural power of capitalist investors and wage laborers but must extend to the ways that class projects are embedded in and carried out by actors within states and resistance movements.

Such a claim also implies that the analysis of structural power must deal with the spatiality of class processes in a global capitalist system that is mediated (and fragmented) by states and resistance movements. I will thus turn briefly, in the next section, to Marxist theories of state power and territoriality, including the way these have been developed in light of the changing geography of global capitalism.

Structural power, the power of the state, and social form in the era of "globalization"

Marxist theories of the state have by now generated an enormous amount of literature that cannot be reviewed here. Instead, I focus on just a few basic issues that are central to fleshing out a conception of geo-political economic power in an era when nation-states have putatively been outstripped by the "economic" forces of "globalization."

A basic point made by Marxist theorists—and especially forcefully by those of a Gramscian persuasion—is that the notion of strictly economic forces circumventing the state is a *non sequitur.* Class power always both involves and extends beyond the state (Gramsci 1971; Poulantzas 1978). Since there are no purely economic phenomena, for Marxists, even the most instrumentalist or structuralist conceptions of the state (neither of which are endorsed here) do not in fact reduce the power of the state to economic power. Rather, Marxist theories ground state power in—and connect it to—processes of class struggle, while versions of Marxist theory appropriately sensitized to issues such as gender and race (certainly not all Marxist theory, here) have also regarded struggles central to state power as always already gendered and racialized in specific ways.

The important questions from a Marxist perspective are thus not about whether or not political (as opposed to economic) power is exercised—power always being political-economic and socio-cultural at the same time—but rather in what form such power is exercised. While the neo-liberal "globalization" thesis asserts declining state power (Ohmae 1995), a more plausible thesis put forward by Marxists is that *forms* (not necessarily *amounts*) of state power are changing because of the changing territoriality of global capitalism. As a consequence, the national, territorial state is arguably giving up some of its power to statist forms at other scales. A now popular version of this thesis has it that state power and economic processes are simultaneously becoming more localized and more globally interconnected, the process referred to as "glocalization" (Swyngedouw 1997).

I will not enter debates about "glocalization" here, but will merely point out that the territorial reach of the nation-state has itself been a longstanding topic of conversation within Marxism, both because of the history of theorizing about imperialism and because of more recent interest in how state power is being transformed by new patterns in the internationalization of capital. Nicos Poulantzas' arguments are important in this regard, both for the general conception of state power that he develops and for his more specific claims about the "internationalization of the state." For Poulantzas, the state is grounded in definite ways in class struggles and thus in processes of production. Yet the state is no mere epiphenomenon within this class struggle. Rather, as Poulantzas sees it, the state has a specific role in the social division of labor, one that he identifies in particular with the division of manual from mental labor (Poulantzas 1978, 54). In this sense, Poulantzas' views echo Gramsci's characterization of certain members of the state as "organic intellectuals" of specific social classes (Poulantzas 1978, 56). States are thus arenas in which struggle takes place over the overall processes of production and social reproduction. They are part of the social division of labor but not reducible to some other presumably privileged part of the production process.

If this broad conception of the relationship between class power and the state suffices to avoid charges of economic reductionism, it still does not address how one should interpret relationships between classes and state power in an era when class power is increasingly transnational yet state power is confined, by definition, to the national territory. Poulantzas addressed this issue early in the evolution of transnational corporate power within Europe by analyzing the "internationalization of the state" (Poulantzas 1975, 80). Put most simply, this refers to the claim that states do not automatically have a privileged relationship with capital of any particular national origin and can exercise the roles they play in support of capital accumulation on behalf of capitalists based within different national territories. What this in effect means is that nation-states are not truly national, and particularly not in the era of "globalization." Rather, like capitalist classes, states have differing territorial ranges—from the sub-national to the supra-national—dependent not upon the official definition of their sovereign territory but rather on their actual political reach and effective power.

A direct consequence of this, that is important for the discussion below, is that the networks of geo-political economic power exercised by classes and fractions of classes extend across varied geographical scales, often beyond the boundaries of given societies. Moreover, they are linked to one another across national boundaries in ways far more complex than might be implied by images of territorial states as containers of "domestic" class processes. Class struggles in one location of the global political economy can have immediate if unintended repercussions for struggles elsewhere in the global system, including through the mediation of state activities in the international arena. Indeed, the evolution of struggles in particular localities may depend crucially upon the outcomes of struggles elsewhere and their effects on the exercise of state power—especially those struggles occurring in locations of greater power and global reach. This is to say that structural power—encompassing not only its economic but its political, cultural and other dimensions—is transnational (if not fully global) in scale. Thus, the structural forces that need analysis in Marxist accounts of geo-political economy and struggles for change are not related to one another externally as so many independent, national class struggles to be summed additively. Instead, they are unevenly (and sometimes unpredictably) internationalized social processes connecting classes and segments of classes both across and within borders.

This point can be usefully elaborated in relation to Margaret Archer's claim that structural power links different generations across time, thus making the constraints and enablements relevant to a particular collectivity's actions a function of earlier choices and actions that present generations cannot control. As one example, Archer notes that the development of literacy (or lack thereof) in a particular generation will enable and constrain what can be done a generation or more later in the way of various social and educational projects (Archer 1995, 66–79). Thus, present collectivities encounter structural conditions for the exercise of agency that are not of their own making. These conditions, however, are not the agentless presence of impersonal forces but rather the contemporary manifestation of past forms of collective agency.[5]

In a parallel fashion, what I am suggesting here is that the complex territoriality of global capitalism makes the actions of specific groups of people in particular locations the structural conditions constraining and enabling agency by other groups elsewhere. The "territorial trapped" tendencies of the social sciences (Agnew and Corbridge 1995)—in particular, the conception of societies as contained territorially by the states that exercise formal sovereignty over the national territory—prevent this point from emerging as clearly as it should. Much debate about agency proceeds as if the major actors relevant to social reproduction or social transformation all exist fundamentally within the spaces of the society in question. Yet in the transnational geo-political economy created by capitalism, such an assumption is generally problematic and often wrong. Transnational political, economic, social and cultural linkages between different collectivities—some of these linkages being consciously constructed, as with trade networks, others being unconsciously evolved, as with the global absorption of "Western" consumer norms—fracture national social spaces and make actors in given locations the producers of conditions for the agency of others.

In the analysis of the East Timorese liberation struggle that follows, I will employ this broad, non-reductionist and transnationalized geo-political economic sense of structural power. I will suggest that the ability of the Timorese struggle to transcend resistance and attain its major objective of independence hinged crucially on changes in the structural context of struggle, in this broad, geographically expansive sense.

OCCUPATION, STRUGGLE AND LIBERATION IN EAST TIMOR[6]

As of only a few years ago, it was common to read lamentations to the effect that East Timor had received little popular or scholarly attention. If that was the case at a crucial point in the Timorese struggle—namely, the 1970s and early 1980s—it can no longer be said, especially in the wake of the significant amount of work that has been done since the Santa Cruz massacre of 1991. In this section, I rely on this substantial body of description and analysis, focusing on providing an interpretation of two crucial turning points in the Timorese independence struggle: the first, the period from 1974 to the early 1980s; the second, the period from the mid-1990s to the present.

The rise of East Timorese nationalism in the 1970s

It was during the first of these periods that the basic contours of East Timorese nationalism—as understood today—began to emerge. Recent scholarship on East Timor has emphasized that though there had been a long history of popular resistance to colonialism, including a major uprising in the early twentieth century (Gunn 2001), the often highly localized precolonial social and political economic structures of Timor remained intact throughout most of the Portuguese colonial period, weakening the prospects for any modern form of anti-colonial nationalism (Anderson 1998; Taylor 1995, 1999). Structural, geopolitical economic changes that rippled through Portugal and the Portuguese overseas empire in the 1970s, however, created new enabling conditions. Specifically, when the Portuguese Armed Forces Movement (AFM) overthrew the Caetano régime in 1974 and put an end to Portuguese fascism, the door to unhindered development of anti-colonial nationalism was opened, though the phenomenon found expression primarily among a small group of Portuguese-educated East Timorese élites in the capital city of Dili (Jolliffe 1978; Dunn 1983; Taylor 1995).

The most economically well placed of these, including owners of large coffee estates such as the Carrascalao family, played a central role in the creation of the first prominent East Timorese political party, the Timorese Democratic Union (UDT). UDT originally favored continued, indefinite alliance with Portugal, but as popular support for independence grew in 1974–5 it shifted its position accordingly. Somewhat lesser Timorese élites, including various members of the bureaucracy and military, formed the second major political party, originally called the Association of Social Democratic Timorese (ASDT), but renamed the Revolutionary Front for an Independent East Timor, or Fretilin. Fretilin quickly overtook UDT in popularity, championing a broad vision of national liberation, modeled in part on the experiences of national liberation struggles in the former Portuguese colonies of Mozambique, Angola and Guinea-Bissau, and gaining support among the majority of rural Timorese through its programs in local agriculture, the extension of health services and literacy campaigns—the last of these being modeled on the ideas of Brazilian educator Paolo Freire. Among a large number of other, smaller parties formed in the wake of the AFM *coup*, the only important one was the Timorese Popular Democratic Association, or Apodeti, which had a small base of support, primarily among certain traditional local rulers (*liurai*). Apodeti's significance was owed not to its popular base, however, but to the fact that it was established with the help of Indonesian intelligence operatives and was immediately recognized by the Indonesian government as an East Timorese voice calling for incorporation into Indonesia (Taylor 1999, 28; Jolliffe 1978; Dunn 1983; Ramos-Horta 1996, 29–39).

At the same time as Indonesia attempted to set the stage for forcible incorporation of East Timor through the promotion of Apodeti, it also attempted to recruit support for integration among UDT members—successfully in the case of conservative UDT leaders such as Francisco Lopes da Cruz, who split from much of the rest of the UDT by 1975 and worked from then on as a spokesperson for the Suharto government. During 1975, Indonesian intelligence worked to convince UDT that Fretilin was communist and was plotting a *coup* to seize power ahead of elections scheduled for 1976, when the Portuguese had agreed to leave East Timor. The ploy worked, and UDT attempted a preemptive *coup* on 11 August 1975. UDT had very limited support compared to Fretilin, however, particularly within the ranks of the crucial Portuguese-trained army, and by the end of September 1975 its members had successfully put down the UDT *coup and* decisively won a brief civil war, in which UDT had already received some support from Indonesia. When the Indonesian government found that the strategy of support for UDT and Apodeti failed to bring the desired results, it resorted to outright invasion, landing tens of thousands of troops in Dili on 7 December 1975, and installing Apodeti as the ruling party, thus beginning a long and bloody process of invasion and occupation that has been painfully and carefully documented elsewhere (Budiardjo and Liong 1984; Dunn 1995; Taylor 1999).

Two aspects of the process leading up to Indonesian invasion are important to highlight here. First, though Suharto himself acquiesced in the invasion plans, he seems to have hesitated somewhat over fear of the international repercussions. Ultimately, it was the strong desire for invasion on the part of Indonesian military commanders such as Generals Ali Murtopo and Benny Murdani that proved decisive, and it was these military figures who reaped the largest benefits from the invasion, including not only the benefits of opportunities for military "glory" and attendant promotions, but significantly for the military élites the opportunity to monopolize East Timor's economic exports—something accomplished through the establishment of a company called P.T. Denok Hernandes International (Taylor 1999, 52–3, 59, 125–7).

Second, the US government's "big wink" towards Jakarta—the popular term used to describe the approving rhetoric of US President Gerald Ford and Secretary of State Henry Kissinger as they visited Suharto on the day before the invasion—was clearly vital to Indonesia's plans. The US and other major Western powers not only tacitly approved of Indonesia's invasion—considering it both inevitable and perhaps desirable, given Fretilin's "leftist" politics—but supplied crucial military and economic aid to Indonesia in support of the invasion and occupation throughout its entire duration. Though this seems to have had roots in the interests of various Western (especially Australian) oil companies in the oil and gas lying below the Timor Gap, which it was felt would be best served by Indonesian control, it also had especially strong roots in US geo-political interests, including both its general antipathy to leftist governments and its insistence on access by its submarines to the important Ombai–Wetar Straits (Taylor 1999; Aditjondro 1999).

The results of this fateful coincidence of interest between Indonesian military capitalists, Western oil companies and US geo-political powerbrokers was the negation of Timorese national independence and the repression that made this negation possible. In this sense, the global geo-political environment and the structural forces at work within it—though they were enabling of the emergence of Timorese nationalism—were fundamentally unfavorable to Timorese independence. The unfavorable climate was only reinforced by the timing of the emergence of East Timor's independence struggle. By 1975, global economic stagnation and the retreat of the US military from Southeast Asia had made US strategists necessarily more willing to look to conservative regional élites in carrying out general policies of communist containment. Suharto's régime was crucial in this regard not only because of Indonesia's own substantial population and natural resource base, but because of its significance to the region as a whole in the context of Communist Party victories elsewhere in Southeast Asia. Moreover, by 1975 Indonesia's economy was increasingly being reoriented around nationalist policies in which state oil revenues played a central role (Robison 1986; Winters 1996). This made the lure of East Timor's potential resource wealth more important to Indonesian leaders, and not only for its own sake but because of the general dependence of the Indonesian economy on resources garnered in its outer islands—including forcibly incorporated territories like West Papua.

In this context, the class transformations underway within East Timorese society itself were of comparatively little moment. Strong support for Fretilin among Timorese peasants, and the social transformations for which Fretilin stood, were easily suppressed by Indonesian colonialism, even where this necessitated propping up archaic and unpopular leadership groups like certain of the *liurai* or comprador élites such as Lopes da Cruz. In spite of vast popular support for Timorese independence, the structural conditions were not ripe for successful struggle.

To say that imperial intervention and the global moment of structural power in which this was embedded during the 1970s prevented Timorese independence is not to say, however, that it destroyed Timorese resistance to colonization. Though Fretilin was largely dismantled by 1978–9, and its military arm, Falintil, reduced to rather desperate survival strategies, the brutality of the Indonesian invasion in fact catalyzed even deeper support for independence among most Timorese and provided an atmosphere conducive to Fretilin's continuation. In this context, Falintil expanded its already relatively inclusive strategy of liberation struggle to include all Timorese, regardless of class background or party affiliation, a move consecrated by specific changes introduced under the leadership of Xanana Gusmão in 1983. These changes helped broaden the already wide social basis for

Fretilin's national liberation project and pulled much of the UDT leadership back into alliance with the Fretilin leadership (Niner 2001, 20).

In addition, the incorporation of East Timor into Indonesia produced another unpredicted change. While East Timor today is seen as a Catholic country, the Portuguese had in fact been distinctly unsuccessful in converting most of the Timorese, and as of the time of Portugal's departure only some 200,000 out of 650,000 East Timorese were Catholic, the rest clinging to various religious views that are typically characterized as animist (Ramos-Horta 1996, 2). Paradoxically, it was the process of incorporation into predominantly Islamic Indonesia that turned most Timorese into Catholics. There were two reasons for this. First, the official Indonesian state ideology of Pancasila requires that all Indonesians officially belong to one of the five major world religions, Christianity being one of these. Thus, many Timorese animists officially adopted Christianity while essentially retaining much of their animist belief. Second, the Catholic Church became, under Indonesian occupation, the only place in East Timor where one could both seek refuge from persecution and have some possibility of contact with the outside world. Thus, many Timorese came to see the Catholic Church as a vital institutional location of political struggle. In this process, the Timorese who entered the Church managed to turn it from a conservative and often effete organization into an institution much more reminiscent of the Latin American Church under the influence of liberation theology (Ramos-Horta 1996, 205; Kohen 1999).

Indeed, the Catholic Church in East Timor might be seen as exhibiting a specific face of the internationalization process—the internationalization of a "cultural" yet "statist" institution (in both Gramsci's and Althusser's sense) that is the site of complex social struggles involving different elements of "civil society." From this perspective, Indonesian occupation had contradictory effects: on the one hand, it negated the internationalization of national liberation struggle on the model of the former Portuguese colonies in Africa, a form of internationalization stimulated by the rise of the AFM; on the other hand, the Indonesian occupation simultaneously generated an alternative form of internationalization in the Timorese struggle by driving Timorese resistance into the Catholic Church.

The longer-term political consequences of this transformation in the Catholic Church have been extremely significant. By the 1990s, the majority of East Timorese were officially regarded as Catholic, and leading Church figures such as Bishop Carlos Belo had become internationally recognized spokespersons for the struggle in East Timor (Carey 1995a, 10). This made the Timorese independence struggle more internationally visible, and enabled it to win increasingly substantial support from Catholic Church groups abroad. The significance of this process was highlighted by Belo's receipt of the Nobel Peace Prize in 1996 (along with Jose Ramos-Horta), an important event in calling international attention to the plight of the Timorese and a serious political condemnation of Indonesia's occupation.

The fruition of East Timorese independence struggle in the 1990s

The changes in the structure of the Timorese resistance that occurred during the 1980s and 1990s were never adequate by themselves to produce independence, however. It was only in the period of the second turning point, from the middle of the 1990s to the present, that transformations in structural power relations became enabling of independence. In standard accounts of this process, the starting point is the Santa Cruz massacre of 1991. This massacre occurred when Indonesian troops killed hundreds of Timorese at a funeral for a young boy who had been killed days earlier. The massacre was filmed and photographed by international media, and the event was thus another international black eye for the Indonesian state. Yet a linear narrative according to which the Santa Cruz massacre was the beginning of the end of Indonesian colonialism would be far too simple. As cases like those of Israeli massacres of Palestinians have shown, there is no level of international embarrassment that by itself necessitates a change of course. Indeed, the fact that it took another five years even for Belo and Ramos-Horta to be recognized by the Nobel committee is indicative of the extremely slow pace and contingency of change.

Indeed, as Ramos-Horta himself notes, changes in patterns of superpower behavior leading to a negotiated settlement are necessary for successful consummation of liberation struggles (Ramos-Horta 1996, 206). Such changes did, in fact, begin to occur in the 1990s, based on transformations of the global geopolitical economy (including within Indonesia). These transformations occurred independently of events in East Timor but eventually—and inadvertently—helped create the impetus for Indonesian withdrawal.

The most general of these transformations was the end of the Cold War and the ascendance of the United States to the position of sole global superpower. With this change, the US government began to look somewhat less tolerantly upon the deviations from neo-liberal economic policy practiced by various of its Asian Cold War allies and pushed for greater economic liberalization. This reflected the increasing intensity of international capitalist competition and the desire of the US government to open new opportunities for US capitalists in a region of the world regarded as dynamic and important to the global economy. The major targets of Washington's drive for liberalization in Asia were Japan and South Korea, but Southeast Asian states such as Indonesia were also encouraged to reduce the roles of their states within the domestic economy and to loosen regulations on capital flows (Tanter 2001, 198–9). Though the Indonesian state did not undo the many forms of "cronyism" that tied the state to Suharto's family interests, it did open new investment opportunities for foreigners and generally followed a US-backed liberalization strategy similar to those followed by other Southeast Asian states, a shift in orientation that was further necessitated by the decline in oil revenues the state had suffered since the 1980s (Bello 1998; Robison 2001).

These policies contributed to a more specific change in Indonesia's post-Cold War position, generated out of the economic crisis that broke in 1997. The vulnerability of Indonesia's economy to sudden withdrawal of foreign capital—a vulnerability caused by the very liberalization measures advocated in Washington—has been credited with responsibility for the crisis (Winters 2000). In the wake of the crisis, the US government and the International Monetary Fund (IMF) used Indonesia's difficulties as an opportunity to demand yet further concessions from the Indonesian state, insisting in particular that Suharto rescind certain state projects that had benefited his family members. When Suharto hesitated in this, the IMF threatened to withhold crucial loans, destabilizing his régime and giving incentive to the political opponents who eventually succeeded in producing his ouster during 1998 (MacIntyre 1999; Higgott 2000; Robison and Rosser 2000). Though Washington and the IMF had probably not set out to depose Suharto, they were no longer afraid to make demands that might lead to this, given the aging dictator's limited value in a post-Cold War era and the obstacles his nepotism placed in the way of expanded opportunities for foreign capital.

The fall of the Suharto régime was to prove a crucial moment in the struggle for independence in East Timor, and pro-independence activists in East Timor quickly seized the opportunity by intensifying their struggles in the immediate aftermath of Suharto's decision to step down (Taylor 1999, xvi; Kammen 2001, 170–1). Yet there was one more general political economic change that had occurred in the 1980s and 1990s which was to prove crucial to subsequent events, and without which even Suharto's ouster might not have proven decisive. The economic boom that occurred in Indonesia from the mid-1980s to the mid-1990s produced important changes in the class structure of Indonesia. Indonesia's economy has remained more dependent than those of other Southeast Asian countries on primary exports, but a huge influx of Japanese and South Korean investment has created a much more extensive manufacturing base and a larger class of Indonesians whose fortunes depend on manufacturing and related tertiary activities (Hill 2000; Robison 2001). This transformation affected Suharto's family itself, and made him less beholden to the specific interests of military capitalists in the 1990s than he had been in the 1970s. In short, Indonesia had begun to move away from being deeply enmeshed in processes of "primitive accumulation" to having substantial industrial capitalist interests.

The transformation was well-represented by the rising importance in the 1990s of Josef Habibie, a technocrat who favored state promotion of higher value added industry and was widely regarded as Suharto's "right hand man," yet had long-standing conflicts with military capitalists over control of economic resources within the state (Kingsbury 1998; Robison 2001). From his position as Vice President, Habibie was promoted to the Presidency by Suharto on the latter's way out in 1998. The assumption of most political observers was that Habibie would not do anything that Suharto wouldn't, but Habibie turned out to be even less committed to East Timor than Suharto, and proved willing to challenge the interests of the military over this (Taylor 1999, xvii; Kingsbury 2000, 69–70). While the general context of Habibie's willingness to allow a referendum is easy enough to comprehend, the specifics of his decision remain more opaque. According to Australian political scientist and Indonesia analyst Damien Kingsbury, Habibie may have believed that he could turn what was intended to be only an interim Presidency into a longer stay in office by combining his support among the Jakarta-based capitalist élites and bureaucrats with international recognition for moving to resolve the problem that had come to be called "the gravel in Jakarta's shoes" (Kingsbury 2000, 70; Anderson 1998, 131). Whatever the precise reasons, Habibie announced in January of 1999, following a suggestion to this effect by Australian Prime Minister John Howard, that he would allow a referendum in East Timor on whether or not to accept special sovereignty, with rejection understood to imply a vote for independence.

The Australian state's role in this process also deserves some attention. The acquiescence of Australia in Indonesia's occupation, and its eventual recognition of Indonesian incorporation (coupled with its negotiation of oil concessions), played a crucial role in the ability of the Indonesian state to carry out and maintain the occupation. Despite this, popular opinion in Australia has always run fairly strongly in favor of East Timorese independence—or at least in the direction of condemnation of Indonesia for its brutality. This has been abetted by both racism (the fear of the "yellow peril") and by the nostalgic attachment of many Australian World War II veterans to their "mates" in East Timor, who fought with them against the Japanese in World War II—at a cost of some 60,000 Timorese lives, when the Australians retreated (Taylor 1995, 32). Howard's régime was willing to capitalize politically on this sentiment in proposing the referendum. Essentially, the Indonesian crisis allowed Howard's Liberal government to move more strongly in promoting an end to the potentially regionally destabilizing East Timor occupation, while being less fearful of alienating an Indonesian leadership that was in turmoil and focused on many other issues (Huntley and Hayes 2001, 179–83).

When Habibie accepted Howard's suggestion and proposed the referendum, the response of the Indonesian military (TNI), and its commander General Wiranto, was one of outrage. It thus immediately mobilized the militia groups that it had used for years to intimidate East Timorese independence advocates, leading to the thousands of pre-referendum killings already mentioned. Throughout 1999, as the killings escalated, the US government and its international allies continued to supply and support Indonesia, while refusing to demand that it allow international peacekeeping forces into East Timor for the referendum (Chomsky 2001, 128, 136–7; Nairn 2001). This clearly indicates that though the US government was willing to challenge the Indonesian government on issues of interest to its own investors—and even to countenance the removal of Suharto—it in no way intended to promote Timorese independence or to challenge the Indonesian military on this issue.

In this context, the referendum went ahead without proper security, allowing the well-known, horrifying aftermath. Here too, it is important to avoid linear narratives that assume an inevitable outcome to the process. It appears that the goal of the TNI and its militias—part of a plan developed well before the referendum—was first to intimidate Timorese into voting for autonomy, failing that to disrupt the entire vote and failing that to cause complete havoc and destruction, perhaps being able to drive most Timorese into the mountains and to draw a new border between East and West Timor, claiming the richest coffee growing lands in the process (Kingsbury 2000, 73; Kammen 2001, 185–7). There was no inevitability to the curtailing of this savage plan, and if not for a huge outpouring of condemnation in Indonesia and elsewhere—including a massive strike by workers in Australia, who refused to handle Indonesian cargo and successfully encouraged international labor solidarity in this embargo—it is conceivable that the plan might have worked. Both international solidarity efforts and the sensitive position of the Clinton administration—which had worked hard to justify military interventions in Somalia and Kosovo on human rights grounds and thus could not argue effectively against such intervention in the case of East Timor—led to the US government's decision to force international peacekeepers upon an antagonistic Indonesian military. In short succession, the US government announced termination of military shipments, the World Bank President insisted upon an end to the slaughter and the IMF stopped delivery of the latest tranche of its structural adjustment loans (Taylor 1999, xxxii–xxxiii). The response of the Habibie régime was immediate, allowing the entry of international peacekeeping forces and terminating both the slaughter and Indonesia's quarter century of colonialism in East Timor. After this, the vote of the Indonesian parliament to recognize the result of the referendum was largely a formality.

It was thus in 1999 that transformations in structural power relations necessary for Timorese independence finally crystallized. From the account given here, it is obvious that nothing was inevitable in this, and not only the protracted struggle of the Timorese but a series of contingencies and perhaps outright accidents contributed to the final result. The multiplicity of factors that led to independence indicate the deeply overdetermined character of the process. Yet at the same time, it is evident that among the processes that were crucial to the eventuation of independence were structural transformations in the geo-political economy. It is these transformations that mark perhaps the major difference between 1975 and 1999. Though East Timorese nationalism was extremely young in 1975, popular sentiment for independence was overwhelming even before Indonesia invaded. Indonesian invasion and occupation did nothing to change this, nor did the survival strategies of the Timorese substantially change sentiment, though these strategies were crucial to the maintenance of resistance. Resistance could only achieve its final goal when the geo-political economic forces that had constrained it had been transformed—as they had been by 1999.

In this regard, it is not only the changes in US capitalists' interests, the changes in the Indonesian class structure, or the Indonesian economic crisis that matter. It is also important that changes in the global economic and geo-political situation between 1975 and 1999 had made any pretence of a threat to capitalism from a Fretilin government ludicrous. Moreover, the Fretilin government that formed with independence will inevitably be a far different one from the one that set out to transform East Timor in the mold of African national liberation struggles during 1974–5. The policies that Fretilin will implement are likely to be much more receptive to a kind of integration into the global capitalist economy that wouldn't have been considered desirable in 1975 (Mariano Saldanha 2001). Thus, the independence of East Timor has been procured not only at an enormous cost in human life, but also quite possibly at the cost of any opportunity to implement the kinds of development strategies and social policies that made Fretilin popular in the 1970s. Whether or not this means that Timorese independence will fail to deliver the poor into something other than an independent impoverishment is yet to be determined, but it would be both premature and callous to assume that nothing better will come to the people of East Timor as a result of independence. What matters for the analysis here is simply that the most important immediate goal of the liberation struggle has been attained, something that could not have occurred without a combination of truly heroic tenacity on the part of the Timorese and important shifts in structural relations of power—generally working far beyond the scale of the Timorese struggle itself—over which they had little or no control.

CONCLUSION

Every case of national liberation struggle has its own particularities, and national liberation struggles as a whole are distinct from many other forms of resistance in the degree to which they are able to mobilize coalitions across class, gender, and sometimes even racial and ethnic lines. Yet precisely because they are a form of broad-based collective action, national liberation struggles like those of East Timor help illustrate important general issues regarding the conditions under which resistance can escalate into something more than opposition and attain major goals of struggle.

The theoretical perspective that I have presented here suggests that approaches such as the structurationist perspective developed in the 1980s, while perhaps useful for analyzing the details of how humans knowingly *reproduce* social structures, may not be particularly relevant to the issue of when and how resistance struggles are able to *transform* social structures. Structure and agency can be constructed as highly abstract categories whose relationship to one another poses theoretical issues, but the questions that have often been asked under this rubric about the conditions that enable human agents to change social structures are better seen as socio-spatial scale issues. If structure is merely the agency of large collectivities of humans, exercised across time and across complex spatial networks, then the question of when and how given subsets of such collectivities can act to change them is a question of when and how a large enough portion of the collectivity might act in ways that intentionally or unintentionally enable the changes pursued by the subset in question.

Moreover, in concrete cases like those of East Timor, the reasons why various members of the collectivity act in ways that either reproduce or change structure are not especially complicated to discern. Interests that motivate behavior—including but not limited to class interests—may be socially and historically constructed, but rarely are they difficult to identify for given social groups in specific contexts. Thus, for example, the interests that can be seen as driving the crucial actions of both foreign investors in Indonesia and major Indonesian industrialists and state officials are neither surprising nor, for the most part, opaque—and indeed most have been consciously identified and openly asserted by the actors in question. To be sure, both the actions of given individuals and the ways in which the actions of different groups in a larger collectivity interact to either offset each other or crystallize into substantial forces for change cannot be easily predicted. Nonetheless, once a particular interaction of forces has in fact crystallized in specific ways, the processes at work can often be readily explained without recourse to theories that assume a problematic or complicated relationship between structure and agency.

It is such relatively parsimonious explanations of action in the context of structure that are at the core of Marxist analyses of structural power. Marxists argue that groups of social actors in given contexts have interests connected to—though not defined exclusively by—their position in class structures. Through overdetermined processes, the interest-based activities of these groups sometimes interact to produce social changes that were not necessarily intended by all or even a majority of the actors but result from the ways the actions of different groups involved in class and class-relevant social struggles crystallize. Thus, the enabling conditions for Timorese liberation included actions by groups indirectly connected to—but not involved in or intending to affect—the Timorese struggle. These included foreign investors in Indonesia and IMF and US officials, seeking changes in Indonesia's governmental structures to enhance prospects for foreign investment and ownership; Indonesian capitalists, professionals, student activist groups and others, seeking changes in Indonesian government policies for their own various reasons, including the desire for more democratization; and the numerous social actors inside and outside of Indonesia whose actions in pursuit of profits, export growth, higher wages, and the like, unintentionally produced the Indonesian economic crisis. East Timorese liberation could not have occurred without the remarkable and tenacious resistance struggle of the Timorese, but nor could it have occurred without the transformations of structural power occasioned by the activities of these other actors, which were not focused on East Timor and which could not be controlled by the Timorese.

Insofar as this kind of account of the relationship between structure and given acts of resistance is generalizable, the implications are both important and chastening. Resistance can achieve its goals and result in substantial social changes; but the conditions under which it can do so may not be either predictable or subject to any meaningful form of control by groups involved in resistance struggle. It is for this reason that while some acts of resistance succeed in attaining longer-term goals, most fail. The list of existing national liberation struggles that to date have not achieved their aims and are unlikely to do so anytime soon—including the struggles of Palestinians, Kurds, West Papuans, Acehnese, Tibetans and Shans—is long indeed.

The point, however, is not to encourage skepticism about the prospects of national liberation struggle or other forms of resistance—nor, for that matter, is it to endorse any particular form of resistance struggle. The point here, rather, is to note that since the necessary conditions of successful struggle include structural transformations not under the control of resistance groups, awareness of structural constraints and potential openings is crucial to resistance strategy. It is for this reason that actors in class struggles and national liberation struggles—from US labor organizers in the 1930s to Vietnamese revolutionaries in the 1950s and 1960s—have paid careful attention to the opportunities created by economic crises and changing configurations of geo-political power. Resistance struggles cannot control such developments, but by being alert to their evolution they can construct strategies and time actions in ways that maximize impact. The East Timorese activists who intensified their struggles as the economic crisis in Indonesia grew understood this and made good use of the opening.

ACKNOWLEDGEMENTS

I would like to thank Adam Tickell and two anonymous referees for helpful comments on an earlier version of this paper. I would also like to thank the various East Timor independence activists who have shared their insights about the liberation struggle with me, including Aaron Goodman, Kevin Nugent, Drew Penland, David Webster, and Al Wong. Any errors are my responsibility and not theirs.

NOTES

1 Elsewhere (Glassman 2003), I have presented an analysis of the roots of Althusser's concept of "overdetermination," showing its roots in the non-reductionist, non-economistic strategic thinking of Lenin and Mao. In my view, an understanding of how Althusser grounds his analysis in this kind of theoretical tradition argues against readings of his work as structuralist in the same sense as Saussure's linguistics or Levi Strauss' anthropology.

2 This is precisely how the matter was recently put by an early exponent of structuration approaches, Alan Pred, in a session during the 2001 meetings of the Association of American Geographers. The view being proposed here is also compatible theoretically with the claim made by Roy Bhaskar, Margaret Archer, Andrew Sayer and other critical realists that reality is "stratified" and that social structures are "emergent" phenomena, not reducible to aggregated individuals (e.g. Sayer 1984, 2000; Bhaskar 1989 1993; Archer 1995). Though useful in its own way, I do not pursue this critical realist argument here since it is directed at theoretical issues regarding agency in general rather than at the more specific issues of the agency I am addressing, having to do with classes and other social collectivities involved in struggles for change.

3 It is worth noting here the gist of the famed passage in Marx's *Eighteenth Brumaire* in which he makes the statement, problematized by structurationists, that "Men make their own history, but they do not make it just as they please" (Marx 1977b, 300). The constraint upon human action that Marx immediately goes on to cite is not a fundamentally "economic" constraint but rather the "spirits of the past" that various social actors conjure up in (vainly) attempting to understand periods of revolutionary crisis. He then counsels—in what might within the stiff categories of liberal social science be construed as an act of "culturalist" exhortation—that "The social revolution of the nineteenth century cannot draw its poetry from the past, but only from the future" (Marx 1977b, 302).

4 This is more fully the case if one considers that other forms of power-laden social relations—such as gender relations—can themselves be seen as structured. Thus, there is structural gender power, and this interpenetrates structures of class power within various social institutions, including households, workplaces and states (Walby 1990).

5 Archer develops a "morphogenetic" approach to realist social theory that draws on Bhaskar's TMSA, among other sources. Like Bhaskar's critical realism, Archer's morphogenetic approach emphasizes that social structure is an "emergent property" of society, not reducible to the sum of individual agents' actions. While I am in general sympathy with this approach, discussion of emergent properties is beyond the scope of my argument here, and also leads in the direction of attempting to resolve debates about structure and agency through the abstract, theoretical approach characteristic especially of Bhaskar's work. As mentioned earlier, I prefer here to approach such debates through a tactical empirical engagement.

6 The evidence in this section is drawn entirely from secondary sources, though some of my understanding of the situation in East Timor is based on discussions with various East Timorese, including the former Acting Rector of the University of East Timor and now Minister of Education, Armindo Maia. My understanding has also been influenced by conversations over the years with a large number of East Timor solidarity activists in North America. Most of the secondary sources cited are by North American or Australian authors. The majority of these works are themselves first-hand accounts, based on interviews, observations and/or ethnographic fieldwork in East Timor. Some information is also drawn from Timorese sources that have been translated into English, including the accounts of the Timorese resistance given by Jose Ramos-Horta (1996) and Constancio Pinto (Pinto and Jardine 1997), and various other accounts are based on interviews and research in Indonesia, especially regarding Indonesian military strategy and policies.

REFERENCES

Aditjondro, G. J. 1999 *Is oil thicker than blood?: a study of oil companies' interests and western complicity in Indonesia's annexation of East Timor*, Nova Science, Commack NY.

Agnew, J. and Corbridge, S. 1995 *Mastering space: hegemony, territory and international political economy*, Routledge, London.

Althusser, L. 1977 *For Marx*, Verso, London.

Althusser, L. and Balibar, E. 1979 *Reading capital*, Verso, London.

Anderson, B. 1998 "Gravel in Jakarta's shoes," in Anderson, B. ed. *The spectre of comparisons: nationalism, Southeast Asia, and the world*, Verso, London, 131–8.

Archer, M. S. 1995 *Realist social theory: the morphogenetic approach*, Cambridge University Press, Cambridge.

Bello, W. 1998 "East Asia: on the eve of the Great Transformation?," *Review of International Political Economy* 5: 424–44.

Bhaskar, R. 1989 *Reclaiming reality*, Verso, London.

—— 1993 *Dialectic: the pulse of freedom*, Verso, London.

Block, F. 1987 *Revising state theory: essays in politics and postindustrialism*, Temple University Press, Philadelphia.

Budiardjo, C. and Liong, L. S. 1984 *The war against East Timor*, Zed Books, London.

Carey, P. 1995a "Introduction: The forging of a nation: East Timor," in Carey, P. E. R. and Bentley, G. C. eds, *East Timor at the crossroads: the forging of a nation*, University of Hawai'i Press, Honolulu, 1–18.

—— 1995b "Introduction," in Cox, S. and Carey, P. eds, *Generations of resistance: East Timor*, Cassell, London, 9–55.

Carey, P. E. R. and Bentley, G. C. 1995 *East Timor at the crossroads: the forging of a nation*, University of Hawai'i Press, Honolulu.

Chomsky, N. 2001 "East Timor, the United States, and international responsibility: 'green light' for war crimes," in Tanter, R., Selden, M., and Shalom, S. R. eds, *Bitter flowers, sweet flowers: East Timor, Indonesia, and the world community*, Rowman & Littlefield, Lanham MD, 127–47.

Dunn, J. 1983 *Timor, a people betrayed*, Jacaranda Press, Milton.

—— 1995 "The Timor affair in international perspective," in Carey P. E.. R. and Bentley, G. C. eds, *East Timor at the crossroads: the forging of a nation*, University of Hawai'i Press, Honolulu, 59–72.

Giddens, A. 1979 *Central problems in social theory: action, structure and contradiction in social analysis*, Macmillan Press, London.

—— 1984 *The constitution of society: outline of the theory of structuration*, University of California Press, Berkeley.

Glassman, J. 2003, "Rethinking overdetermination, structural power, and social change: a critique of Gibson-Graham, Resnick, and Wolff," *Antipode* 35(4): 678–98.

Gramsci, A. 1971 *Selections from the prison notebooks*, International Publishers, New York.

Gunn, G. C. 2001 "The five-hundred-year Timorese funu," in Tanter, R., Selden, M. and Shalom, S. R. eds, *Bitter flowers, sweet flowers: East Timor, Indonesia, and the world community*, Rowman & Littlefield, Lanham MD, 3–14.

Higgott, R. 2000 "The international relations of the Asian economic crisis: a study in the politics of resentment," in Robison, R., Beeson, M., Jayasuriya, K., and Kim, H.-R. eds, *Politics and markets in the wake of the Asian crisis*, Routledge, London, 261–82.

Hill, H. 2000 *The Indonesia economy*, Cambridge University Press, Cambridge.

Huntley, W. and Hayes, P. 2001 "East Timor and Asian security," in Tanter, R., Selden, M., and Shalom, S. R. eds, *Bitter flowers, sweet flowers: East Timor, Indonesia, and the world community*, Rowman & Littlefield, Lanham MD, 173–85.

Isaac, J. 1987 *Power and Marxist theory: a realist view*, Cornell University Press, Ithaca NY.

Jolliffe, J. 1978 *East Timor: nationalism and colonialism*, University of Queensland Press, St. Lucia.

Kammen, D. 2001 "The trouble with normal: the Indonesian military, paramilitaries, and the final solution in East Timor," in Anderson B. ed., *Violence and the state in Suharto's Indonesia*, Cornell University Southeast Asia Program Publications, Ithaca NY, 156–88.

Kingsbury, D. 1998 *The politics of Indonesia*, Oxford University Press, Melbourne.

—— 2000 "The TNI and the militias," in Kingsbury, D. ed., *Guns and ballot boxes: East Timor's vote for independence*, Monash Asia Institute, Clayton, 69–80.

Kohen, A. S. 1999 *From the place of the dead: the epic struggles of Bishop Belo of East Timor*, St. Martin's Press, New York.

MacIntyre, A. 1999 "Political institutions and the economic crisis in Thailand and Indonesia," in Pempel, T. J. ed., *The politics of the Asian economic crisis*, Cornell University Press, Ithaca NY, 143–62.

Mariano Saldanha, J. 2001 "The transition of a small war-torn economy into a new nation: economic reconstruction of East Timor," in Tanter, R., Selden, M. and Shalom, S. R. eds, *Bitter flowers, sweet flowers: East Timor, Indonesia, and the world community*, Rowman & Littlefield, Lanham MD, 229–42.

Marx, K. 1977a "Comments on Adolph Wagner," in McLellan, D. ed., *Karl Marx: selected writings*, Oxford University Press, Oxford, 581–2.

—— 1977b "The Eighteenth Brumaire of Louis Bonaparte," in McLellan, D. ed., *Karl Marx: selected writings*, Oxford University Press, Oxford, 300–25.

—— 1977c "Preface to *A critique of political economy*," in McLellan, D. ed., *Karl Marx; selected writings* Oxford University Press, Oxford, 388–92.

Nairn, A. 2001 "US support for the Indonesian military: congressional testimony," in Tanter, R., Selden, M. and Shalom, S. R. eds, *Bitter flowers, sweet flowers: East Timor, Indonesia, and the world community*, Rowman & Littlefield, Lanham MD, 163–72.

Niner, S. 2001 "A long journey of resistance: the origins and struggle of CNRT," in Tanter, R., Selden, M. and Shalom, S. R. eds, *Bitter flowers, sweet flowers: East Timor, Indonesia, and the world community*, Rowman & Littlefield, Lanham MD, 15–29.

Ohmae, K. 1995 *The end of the nation state: the rise of regional economies*, Free Press, New York.

Pinto, C. and Jardine, M. 1997 *East Timor's unfinished struggle: inside the Timorese resistance*, South End Press, Boston MA.

Poulantzas, N. 1975 *Classes in contemporary capitalism*, NLB, London.

—— 1978 *State, power, socialism*, Verso, London.

Ramos-Horta, J. 1996 *Funu: the unfinished saga of East Timor*, Red Sea Press, Asmara, Eritrea.

Robinson, G. 2001 "With UNAMET in East Timor—an historian's personal view," in Tanter, R., Selden, M., and Shalom, S. R. eds, *Bitter flowers, sweet flowers: East Timor, Indonesia, and the world community*, Rowman & Littlefield, Lanham MD, 55–72.

Robison, R. 1986 *Indonesia: the rise of capital*, Allen & Unwin, Winchester MA.

—— 2001 "Indonesia: crisis, oligarchy and reform," in Rodan, G., Hewison, K. and Robison, R. eds, *The political economy of South-East Asia: conflicts, crises, and change*, Oxford University Press, Melbourne, 104–37.

Robison, R. and Rosser, A. 2000 "Surviving the meltdown: liberal reform and political oligarchy in Indonesia," in Robison, R., Beeson, M., Jayasuriya, K., and Kim, H.-R. eds, *Politics and markets in the wake of the Asian crisis*, Routledge, London, 171–91.

Rupert, M. 2000 *Ideologies of globalization: contending visions of a new world order*, Routledge, London.

Sayer, R. A. 1984 *Method in social science: a realist approach*, Routledge, London.

—— 2000 *Realism and social science*, Sage, London.

Swyngedouw, E. 1997 "Neither global nor local: "glocalization" and the politics of scale," in Cox, K. R. ed., *Spaces of globalization: reasserting the power of the local*, Guilford Press, New York, 137–66.

Tanter, R. 2001 "East Timor and the crisis of the Indonesian intelligence state," in Tanter, R., Selden, M. and Shalom, S. R. eds, *Bitter flowers, sweet flowers: East Timor, Indonesia, and the world community*, Rowman & Littlefield, Lanham MD, 189–208.

Tanter, R., Selden, M., and Shalom, S. R. eds, 2001 *Bitter flowers, sweet flowers: East Timor, Indonesia, and the world community*, Rowman & Littlefield, Lanham MD.

Taylor, J. G. 1995 "The emergence of a nationalist movement in East Timor," in Carey, P. E. R. and Bentley, G. C. eds, *East Timor at the crossroads: the forging of a nation*, University of Hawai'i Press, Honolulu, 21–41.

—— 1999 *East Timor: the price of freedom*, Zed Books, London.

Taylor, P. J. and Flint, C. 2000 *Political geography: world-economy, nation-state, and locality*, Prentice Hall, Harlow.

Thompson, E. P. 1978 *The poverty of theory and other essays*, Merlin Press, London.

Thrift, N. 1983 "On the determination of social action in space and time," *Environment and Planning D: Society and Space* 1: 23–57.

Walby, S. 1990 *Theorizing patriarchy*, Basil Blackwell, Oxford.

Winters, J. 1996 *Power in motion: capital mobility and the Indonesian state*, Cornell University Press, Ithaca NY.

—— 2000 "The financial crisis in Southeast Asia," in Robison, R., Beeson, M., Jayasuriya, K., and Kim, H.-R. eds, *Politics and markets in the wake of the Asian crisis*, Routledge, London, 34–52.

PART 3

Ethics-based goals

Introduction

Featherstone's "maps of grievance" discussed in Box 2 on pp. 271–72 lay the groundwork for what we call an ethics-based worldview. This worldview is concerned with the production (as such) of new political imaginaries of becoming. As in the discussions and readings on justice, we are indicating here what seems to be an aspect of social-political struggles and not only their ultimate goals: *thus, to be ethical in struggle is to be disposed or oriented toward it in an open, intensely self-reflexive way*. Openness might mean having to alter the trajectory that a particular struggle for rights or justice has taken; it might produce an analysis of how certain notions of justice or rights have become a means of domination and oppression; it might mean a knife-edge attempt to preempt notions of rights and justice altogether. (We might recall, for example, Marx's argument that in bourgeois society the right to equal participation in exchange was defined in such a way as to be nonexistent inside the factory gate where surplus value is produced. One might say, then, of Marx's critique of rights that he is concerned with the *ethics* of rights.)

Putting it bluntly, ethics, for us, and as divined in the readings collected for this final part, is about subjecting received and accepted notions of the good, the just, the right to some sort of radical or immanent critique so that the limits and boundaries of accepted notions become clear and their warrant is tested against the "outsides" they might produce or against certain desiderata that have yet to see the light of day. In our view, ethics is about the development of new concepts that potentially open up new political practices, and new or revised alliances and identities for those politics. And, as more than one contributor notes, ethical positions are themselves sites of contestation and struggle.

The role of critical research in furthering the struggle for ethical engagements is extraordinarily wide. If there is a common theme here it is just that ethics have to be cultivated, and to be cultivated particular kinds of resources also need to be developed. Perhaps it is most useful to think of the pieces here as defining what some of those resources might be. What these works offer are a critical reflection on the very nature of the social and the political; an argument for a flexible, more ethical form of oppositional consciousness within social struggles and movements; and an account of a "community" project that was structured in such a way as to encourage new understandings of local–global connections and possibilities among participants.

DEFINING THE SOCIAL ONTOLOGY, CLARIFYING THE POLITICAL: A CRITICAL REFLECTION

A number of these points are made clear in Chantal Mouffe's "Post-Marxism: democracy and identity" (Chapter 19). (Recall that Mouffe's work played a central role in a couple of the other essays collected in this anthology.) Her point of entry is the argument that "no center of subjectivity precedes the subject's identifications" (p. 333). In other words, human beings do not make connections with the world and with others on the basis of a *pre*-made self. We are not *first* coherent, rational agents, who *then* form social (or ecological) relationships or who *then* enter into social contracts, political agreements, and so on. Rather, from the get-go, selves are constituted inside of and never apart from relationships. (There are affinities with the contributions of Nancy Fraser (Chapter 4), Iris Marion Young (Chapter 5), and Sarah Whatmore (Chapter 7).) Yet, Mouffe, a political theorist at University of Westminster, London, and her sometime collaborator Ernesto Laclau also argue that social relationships (the "social objectivity" in her terminology) must involve exclusions in order to constitute themselves. This does not mean Mouffe desires exclusions but it does mean "any social objectivity is ultimately political and that it has to show the traces of the exclusion which governs its constitution" (see p. 333). A social entity (e.g. a nation-state or ethnic group) cannot be defined on the basis of a pure separation from some other entity; its "internal" characteristics are contingent upon that which it is in relationship to. And being intrinsically in relationship to some other entity means that it *could therefore be different from what it is at any one time.* There is a traffic in values and meanings back and forth between entities—this is their very condition of possibility. In very simple terms: no black without white; no hetero without homo; no Europe without not-Europe; and so on. The question that then follows is: What kind of relationships between entities are desirable?

Mouffe, a scholar activist who was involved in the social movements of the 1960s, prefaces her answer to this question with a statement about the unavoidability of power relations: "power should not be conceived of as an external relation taking place between two preconstituted identities, but rather as constituting the identities themselves" (see p. 333). Any existing social arrangement (as between different social movements, different social identities of race or class, the governing and the governed, and so on) has no source of legitimacy other than the power relations on which it is grounded. Even the identities for which certain rights are sought (gender-based rights, citizenship rights, race-based rights, etc.) have bound within them the traces of a power relation within which they have been formed (e.g. the power imbalance between individuals and states that guarantee rights). Indeed, this is part of the reason for the skepticism toward rights that many activists have voiced. That power is unavoidable is decisive for Mouffe: "if we accept that relations of power are constitutive of the social, then the main question of democratic politics is not how to eliminate power but how to constitute forms of power that are compatible with democratic values" (see p. 334). A democracy founded on such values is what Mouffe calls a radical or pluralistic democracy. This is a democracy that refuses to accept any social order that suppresses conflict, falsely imposes consensus, or conceives of society as a single organic body. Instead, "a democratic society makes room for the expression of conflicting interests and values"—it actually necessitates both dissent and social-political institutions through which dissent can be expressed. For this reason Mouffe argues that the survival of democracy "depends on collective identities forming around clearly differentiated positions, as well as on the possibility of choosing between real alternatives" (see p. 335). The urge for consensus, she implies, is overrated: "When the agonistic dynamic of the pluralist system is hindered because of a lack of democratic identities with which one could identify, there is a risk that this will multiply confrontations over essentialist identities and nonnegotiable moral values" (see p. 336). In other words, as real alternatives are brought into being there is an ethical responsibility to fight against their slippage into an essentialist ontology through which any one alternative poses as the one real alternative. No one alternative gets to posit itself as the one and only, say American identity, or British identity, or class identity, or gay identity, etc. Indeed, what ethics calls for is a mode of political engagement that pushes against exclusion. And there is a further implication. Real choices must be made available but any one choice must always be recognized as possibly becoming different. Therefore, what is ethically desirably is social relations that sustain a field of alternatives but also forestall the hardening of any one of those alternatives.

BOX 3: SPACES OF POWER IN GEO-POLITICAL ENCOUNTERS: FROM JUSTICE TO ETHICS AND BACK AGAIN

In reflecting upon Mouffe's arguments the tendency might be to apply them toward the political arrangements within nation-states or perhaps polities within nation-states. Such a tendency would have to be avoided (not that Mouffe entirely falls prey to it), since democratic arrangements are not desirable within states only. This is a strong theme in the work of David Slater, a geographer at Loughborough University, Loughborough, England. His task in, for example, "Spatialities of Power and Postmodern Ethics—Rethinking Geopolitical Encounters" is to rethink the boundedness of states and the meaning of their sovereignty (Slater, 1997). His work can be read alongside that of James M. Blaut, encountered in Chapter 2. (Readers can also refer to Box 2 on p. 125, highlighting the work of Chandra Mohanty.) Slater aims: to contextualize issues of justice and power internationally, including a recognition of the West's geopolitical power over the non-West; to challenge the idea of the West as a self-contained entity and any political ethic that rests on such an idea; and to offer some analytical tools with which to disrupt the West's "ethnocentric universalism," i.e. especially Western modernity as the universal desideratum. At the same time, he cautions that these tasks need to be undertaken without romanticizing the non-West. As he puts it, "[c]learly, the locally or regionally particular can be as violently oppressive as the centrally or globally universal . . . The covert celebration of the local or the particular can be analytically and politically disarming" (Slater, 1997: 57).

Slater invites us to resist thinking of the West as a self-sufficient center out of which emanates a modern culture of individual freedoms, diversity, and equality. Thinking of the West in these terms "goes together with a silence on the Western diffusion of structures of inequality" (Slater, 1997: 60). Structures of inequality are in ample supply both within the West and outside it, as evinced by anti-colonial and anti-imperial struggles. And despite a legacy of Western self-criticism—it is certainly the case that the West has produced a class of intellectuals who have exposed its hypocrisies—it is yet another form of silencing to enclose criticism of the West within the West itself. In the place of a self-sufficient West, he calls for a refurbished conception of the geo-political relationships between West and non-West, global North and global South, First World and Third World.

Slater's primary example is of the U.S. and its geopolitical history, from U.S. national origins through to the Cold War. (He might have chosen other "national" histories as well, but recall his point that the dominance of the West must be recognized and reckoned with: The playing field is not level and the U.S. has a history of geopolitical entanglements that has helped to make the playing field uneven in the first place.) The essay devotes particular attention to the idea of manifest destiny and the history of U.S. subversion of sovereignties in the Caribbean and the Pacific. (He pays particular attention to Cuba.) Why dwell on the past? Geopolitical origins matter because around them are formed important ideas concerning what we think we are doing in the world and how we ought to respond to others. Too often U.S. self-conceptions begin with a post-World War II U.S. in mind. In such a conception the U.S. becomes the model state, the exclusive force for good in the world. Through such a conception it is easy for the U.S.'s past entanglements and subversions of sovereignty to be written out of its self-conception. Instead, Slater argues, U.S. geopolitical origins deny Americans the fantasy that they are (collectively speaking) an autonomous subject; reckoning with the past calls U.S. subjects into a position of responsibility. Slater states:

> I am arguing here that questions of justice and equality are difficult to separate from a consideration of the ethics of responsibility, and that responsibility to the self and to the other is a subject of increasing interest and dispute. In other words, in contrast to that position in which one implicitly moves from a series of already constituted entities and practices to the call for a more just, more responsible, world, I am suggesting that the postmodern sensibility [à la Mouffe and others] can help

us reemphasize the need to destabilize and deconstruct given starting points and rethink what we might mean by responsibility to otherness.

(Slater, 1997: 68)

While Slater brings the U.S. to task, his argument is also a resolutely non-utopian, non-romanticizing one that refuses to reify the non-West as always and already better than the West. Much like Mouffe, Slater endorses a "radical democracy" that would refuse any group an a priori, pure moral superiority. With a refurbished, more open sense of its own past, however, the U.S. might be better equipped to enter into the sphere of more ethical encounters.

DEFINING A MORE FLEXIBLE, ETHICAL MOVEMENT CONSCIOUSNESS

In a sense, Slater's arguments, described above in Box. 3, call for an ethics that requires the cultivation of a global sense of place, in which places and peoples are simply never understood apart from the connections they have with other places and peoples (see Massey, 1991). While he is primarily motivated by the First World's history of intrusion into the Third World, Chapter 20 by Chela Sandoval takes up the reverse theme, the Third World's (struggle for) presence in the First World. Sandoval, a cultural theorist and Chicana Studies scholar at University of California, Santa Barbara, takes up the issue through an analysis of U.S. feminism and its fracturing along color lines. These are lines whose very existence is due to U.S. global entanglements with Africa, Latin America, and Asia. Although it is not her purpose to review the history of these entanglements, they are of great moment within U.S. feminism. Sandoval's purpose is to depict feminism as a social movement whose diversity has been vigorously struggled over. Her essay helps us to think about what an open feminist identity and feminist politics might be, including feminism's destabilizing relationship to other oppositional movements. (She is thereby also usefully read alongside Mouffe.)

Toward these ends Sandoval usefully summarizes the history of feminist movements in the U.S., giving special attention to the different forms of oppositional consciousness they have seized upon, sometimes too narrowly (e.g. an equal rights-based consciousness). In introducing the generic phrase "oppositional consciousness" Sandoval means to draw her reader's attention to the different tactics available to many liberation movements in capitalist society. She makes three arguments that are of interest here. The first is simply that the history of U.S. feminist political struggles reveals that these struggles are structured by the opportunities for opposition within liberal, capitalist "democracy," opportunities that other movements have at times also availed themselves of. Examples include (feminist) quests for equal rights, separatism, or even socialism. That is to say, "Any social order which is hierarchically organized into relations of domination and subordination creates particular subject positions within which the subordinated can legitimately function." At the same time, "These subject positions, once self-consciously recognized by their inhabitants, can become transformed into more effective sites of resistance to the current ordering of power relations" (see p. 345). Second, Sandoval argues that each of the tactics used by feminism has had in its own way a troubling relationship to women of color. Thus the history of feminism, while a history of changing political tactics and movement formation, is also a history of ethnic and racial closure and exclusion, even as this closure has been a condition of possibility for U.S. Third World feminism itself. Third, the history of U.S. Third World feminism, and its practice of oppositional consciousness, is therefore not easily captured by the dominant narrative of the feminist social movement, a history of linear change form one form of oppositional consciousness to the next. The salient fact of Third World feminism, however, is its flexibility regarding movement tactics: a specific form of oppositional consciousness is therefore in evidence. Sandoval terms this form of oppositional consciousness "differential consciousness." Differential consciousness refers to the ability to alter political tactics, to shift from one tactical register to another as the situation demands. The effect, and the reality, is to undermine "the appearance of the mutual exclusivity of oppositional strategies of consciousness" (see p. 346).

U.S. Third World feminists rarely adopted "the kind of fervid belief systems and identity politics that tend to accompany their construction under hegemonic understanding" (see p. 346). And "What U.S. Third World feminism demands is a new subjectivity, a political revision that denies any one ideology [e.g. quest for rights or a separate sphere of difference] as the final answer, while instead positing a *tactical subjectivity* with the capacity to recenter depending upon the kinds of oppression to be confronted" (see p. 347). Sandoval does not argue that U.S. (white) feminism would have been more inclusive had it adopted a differential consciousness. This probably would not have been possible, since exclusion from the dominant forms of feminism was a condition of possibility for U.S. Third World feminism in the first place. Rather, Sandoval argues that differential consciousness has value as a resource for the present, for any political situation that calls for openness. (Of course it has had value for U.S. Third World feminist too.) If Chantal Mouffe asks for a politics that will distinguish among choices rather than suppress the differences among them, and if she asks that any one political choice, path, or identification (including social identities) remain open to variation and alteration, to the multiplicity within it, then Sandoval's differential consciousness explores the implication of such an argument for the case of U.S. Third World feminism.

ENGAGING "COMMUNITIES" IN NEW UNDERSTANDINGS OF LOCAL AND GLOBAL CAPACITIES

A persistent theme in critical geographic research and writing is that it can be useful to understand that the world might be different from what we think it is. This is an important aspect of Slater's rethinking of the West's and the U.S.'s geopolitical past, in Box 3 on pp. 327–28. This is to say, in effect, these geographical entities might be different than we think they are and, if so, let's make it matter to what we do now. A differently comprehended past, Slater argues, opens the door to more ethical political encounters. J. K. Gibson-Graham ventures into similar territory. (Gibson-Graham is the merged authorial identity of geographers Julie Graham and Katherine Gibson, respectively of the University of Massachusetts, Amherst, and Australia National University, Sydney.) "Imagine," they implore, "that something is happening in the world that's not about an actual measurable phenomenon called 'globalization'" (see p. 355). For Gibson-Graham globalization is simply an impossibility. They do not deny there are powerful forces afoot that link economically, socially, and culturally disparate locations around the globe. But they insist that these forces cannot in fact make every place over in the image of some common denominator. As they put it, "If we can accept that it is impossible to subsume every individual being, place, and practice to a universal law, whether it be the law of the father [a phrase taken from certain psychoanalytic theories], or the market, or a geopolitical formation, then it will follow that the local cannot be fully interior to the global, nor can its inventive potential be captured by any singular meaning" (see p. 355). If globalization is impossible, and inventive potential has multiple forms and expressions other than the economic, narrowly defined, then there are grounds to incite ethical struggle. She outlines three registers of struggle.

First is the necessity to "*recognize particularity and contingency*" (see p. 356). This carries a double meaning. On the one hand, what is commonly called globalization (or "development" or "neoliberalism") is only a particular understanding of the economy that calls upon people and places to behave in ways conducive to the spread of market forces and ideologies. On the other hand, and precisely because we are alert to the particularity of globalization, things could be otherwise: there are other ways of organizing production and consumption. Second, Gibson-Graham calls for a struggle to "*respect difference and otherness*, between localities but also within them" (see p. 357). She understands locality as a site, *a priori*, for encounters between strangers: localities *are* settings for such encounters. (The body/person itself can be a locality too, in a similar sort of way. Bodies/persons can experience utter strangeness and difference, as when they are thrown out of work under the terms of "globalization." This will become clearer in a moment when I describe Gibson-Graham's specific research projects.) Of course, we have seen that exercising responsibility toward others who are unassimilable to "us" is a practice that can generate as many problems as it solves–recall what we termed the situation of justice in the previous discussion on justice-based struggles (see pp. 267–73). Just the same, Gibson-Graham's injunction perhaps helps us to see how struggles for justice can entail an ethical struggle; justice can't be pursued without ethical practice. Finally, Gibson-Graham argues for the necessity of cultivating "capacity." Economic "globalization" channels people toward a very narrow identity, constructing subjects "as 'citizens' of capitalism: they are entrepreneurs, or employees, or would-be employees; they are investors in capitalist firms; they are consumers of (capitalist) commodities." In contrast,

> the ethical practice of subject formation requires cultivating our capacities to imagine, desire, and practice noncapitalist ways to be. An ethics of the local would undermine ideas of individual self-sufficiency, fostering the affective acknowledgment of interdependence as a basis for some sort of "communism." It would produce citizens of the diverse economy.
>
> (see p. 357)

Respecting difference and otherness, then, and cultivating capacity, are not so much about a cultural outlook (though they can be that). They are about political economy. In fact, though controversially, Gibson-Graham argue that the "diverse economy" is in certain respects already here. As they briefly discusses in the essay (also see their books *The End of Capitalism (As We Knew It)* [1996] and *A Postcapitalist Politics* [2006]), there are a great variety of acts and labors through which useable goods and services are produced, circulated, and consumed that are not strictly speaking accountable to or identifiable as capitalist. These acts and the social wealth they generate need to be both recognized (for an understanding of the economy as different from what it "is") and broadened/deepened (for purposes of producing new, communal localities and relations between localities). But, they note, this will not likely happen unless the capacity is cultivated to be different from what capital "requires" of people. (We are not in full agreement with Gibson-Graham's binary framework of capitalist/non-capitalist formations [see Henderson 2004], but her notion of the diverse economy is no less interesting and ethically useful for that.)

The cultivation of capacity is the special objective of Gibson-Graham's research projects in Australia, Asia, the Pacific, and the U.S., although they pursue this objective in conjunction with the other two registers of struggle. The paper anthologized here discusses their projects in the Latrobe Valley of southeastern Australia and the Pioneer Valley of Massachusetts (U.S.). Both prove to be very tough cases, since "in both these regions, globalization sets the economic agenda—we are all being asked to become better subjects of capitalist development (though the path to such a becoming does not readily present itself) and to subsume ourselves more thoroughly to the global economy" (see p. 358). Gibson-Graham describe confronting a "patent lack of desire" for anything different from capitalist globalization, even as it is not producing the effects many people in these regions desire, especially in the Latrobe Valley, which has experienced pronounced deindustrialization and job layoffs.

It is this lack of desire that Gibson-Graham wanted to investigate and to reverse. The article describes a series of focus groups and field trips through which they began to see some progress toward these goals. The focus groups began as a set of discussions eliciting people's perceptions of the economic identity of the region, its successes and failures, its strengths and weaknesses, existing and untapped resources and skills, and so on. The discussions (conducted with community members and community researchers) revealed the very deep investments people had in the dominant economy and the sense of victimization at its failure. Over time, however, focus group members began to break away from their identity as economic "citizens" and began to see themselves, each other, and their regions in more diverse, hopeful ways. Gibson-Graham argue that this reorientation happens in a very visceral sort of way; it is as much about feeling, emotion, and affect as it is about rational argument and thought. This is itself a direct reflection of one of the ways that the dominant economy of capitalism works in the first place, by truly producing subjects who learn to desire it, who come to see their own interests as thoroughly stemming from it, and who develop emotional, affective attachments to it (see Althusser 1971; Lukács 1971). Gibson-Graham also describe communal events based around cooking, eating, and field trips (to a community garden and to a conference on worker cooperatives). These events—more fundamental even than Featherstone's convergence spaces or maps of grievance—also produced new orientations and sensibilities concerning what might be possible:

> we've tried to make our conversations and gatherings entirely pleasurable . . . and also loose and light—not goal-oriented or tied to definitions and prescriptions of what "a left alternative" should be. Over the course of

> the projects, without prompting, the community researchers and their interviewees began to express practical curiosity (as opposed to moral certainty) about alternatives to capitalism.
>
> (see p. 366)

This is emphatically an activism without a blueprint, a struggle to create an ethic, an openness to a different social and personal becoming. This is an activism that consists of creating the conditions through which people might "encounter themselves differently—not as waiting for capitalism to give them their places in the economy but as actively constructing their economic lives" (see p. 365). For Gibson-Graham this is not just about developing a new set of concepts concerning capitalism, globalization, region, and economy. It is about structuring new experiences through which new practical sensibilities and desires might come to the surface, become part and parcel of making one's livelihood, but then never settle as *identities*. The difficulty of this is one of Gibson-Graham's points: conventional notions of the economy and conventional thinking about economic identities have proved to be extraordinarily resilient. We have in Gibson-Graham, then, an implicit commentary on the essay by Mouffe that began this last part of the book (Chapter 19). There is nothing automatic about the extension of the chain of equivalence. If different entities never encounter each other except through the power relations that bring them into being in the first place, *new entities*, as links in this chain, have to be actively forged. It is often said that in order to change our situations, to change society (which always also means to change our geographies), we must start from where we are, look into our present situations, and grasp their latent possibilities. The injunction to grasp the present in a new way surely means—as Gibson-Graham and indeed many authors in this volume forcefully argue—changing ourselves too.

19

Post-Marxism

Democracy and identity[1]

Chantal Mouffe

from *Environment and Planning D: Society and Space*, 1995, 13: 259–265

In recent decades, the willingness to rely on categories like human nature, "universal reason," and "rational autonomous subject" has increasingly been put into question. From diverse standpoints, very different thinkers have criticized the idea of a universal human nature, of a universal canon of rationality through which that nature could be known, as well as the possibility of a universal truth. Such a critique of rationalism and universalism, which is sometimes referred to as "postmodern," is seen by authors like Jürgen Habermas as a threat to the modern democratic ideal. They affirm that there is a necessary link between the democratic project of the Enlightenment and its epistemological approach and that, as a consequence, to find fault with rationalism and universalism means undermining the very basis of democracy. This explains the hostility of Habermas and his followers towards the different forms of post-Marxism, poststructuralism, and postmodernism.

I am going to take issue with such a thesis and argue that it is only by drawing all the implications of the critique of essentialism—which constitutes the point of convergence of all the so-called "posties"—that it is possible to grasp the nature of the political and to reformulate and radicalize the democratic project of the Enlightenment. I believe that it is urgent to realize that the universalist and rationalist framework in which that project was formulated has today become an obstacle to an adequate understanding of the present stage of democratic politics. Such a framework should be discarded and this can be done without having to abandon the political aspect of the Enlightenment which is represented by the democratic revolution.

We should, on this subject, follow the lead of Hans Blumenberg who, in his book *The Legitimacy of the Modern Age* (1983), distinguishes two different logics in the Enlightenment, one of "self-assertion" (political) and one of "self-grounding" (epistemological). According to him those two logics have been articulated historically but there is no necessary relation between them and they can perfectly be separated. It is therefore possible to discriminate between what is truly modern—the idea of "self-assertion"—and what is merely a "reoccupation" of a medieval position, that is, an attempt to give a modern answer to a premodern question. In Blumenberg's view, rationalism is not something essential to the idea of self-assertion but a residue from the absolutist medieval problematic. This illusion of providing itself with its own foundations which accompanied the labor of liberation from theology should now be abandoned and modern reason should acknowledge its limits. Indeed, it is only when it comes to terms with pluralism and accepts the impossibility of total control and final harmony that modern reason frees itself from its premodern heritage.

This approach reveals the inadequacy of the term "postmodernity" when it is used to refer to a completely different historical period that would signify a break with modernity. When we realize that rationalism and abstract universalism, far from being constitutive of modern reason, were in fact reoccupations of premodern positions, it becomes clear that to put them into question does not imply a rejection of modernity but a coming to terms with the potentialities that were inscribed in it since the beginning. It also helps us to understand why the critique of the epistemological aspect of the Enlightenment does not put its political aspect of self-assertion into question but, on the contrary, can help to strengthen the democratic project.

THE CRITIQUE OF ESSENTIALISM

One of the fundamental advances of what I have called the critique of essentialism has been the break with the category of the subject as a rational transparent entity which could convey a homogeneous meaning on the total field of her conduct by being the source of her actions. Psychoanalysis has shown that, far from being *organized* around the transparency of an ego, personality is structured on a number of levels which lie outside the consciousness and rationality of the agents. It has therefore discredited the idea of the necessarily unified character of the subject. Freud's central claim is that the human mind is necessarily subject to a division between two systems, one of which is not and cannot be conscious. The self-mastery of the subject, a central theme of modern philosophy, is precisely what he argues can never be reached. Following Freud and expanding his insight, Lacan has shown the plurality of registers—the Symbolic, the Real, and the Imaginary—that penetrate any identity, and the place of the subject as the place of the lack which, though represented within the structure, is the empty place that at the same time subverts and is the condition of the constitution of any identity. The history of the subject is the history of her identifications and there is no concealed identity to be rescued beyond the latter. There is thus a double movement. On the one hand, a movement of decentering which prevents the fixation of a set of positions around a preconstituted point. On the other hand, and as a result of this essential nonfixity, an opposite movement; the institution of nodal points, partial fixations which limit the flux of the signified under the signifier. But the dialectics of nonfixity–fixation is possible only because fixity is not pregiven, because no center of subjectivity precedes the subject's identifications.

I think that it is important to stress that such a critique of essential identities is not limited to a certain current in French theory, but is found in the most important philosophies of the 20th century. For instance, in the philosophy of language of the later Wittgenstein we also find a critique of the rationalist conception of the subject that indicates that the latter cannot be the source of linguistic meanings since it is through participation in different language games that the world is disclosed to us. We encounter the same idea in Gadamer's philosophical hermeneutics in the thesis that there exists a fundamental unity between thought, language, and the world, and that it is within language that the horizon of our present is constituted. A similar critique of the centrality of the subject in modern metaphysics and of its unitary character can be found in different forms in several other authors and this allows us to affirm that, far from being limited to poststructuralism or postmodernism, the critique of essentialism constitutes the point of convergence of the most important contemporary philosophical currents.

ANTI-ESSENTIALISM AND POLITICS

In *Hegemony and Socialist Strategy* (Laclau and Mouffe, 1985) we have attempted to draw the consequences of this critique of essentialism for a radical conception of democracy by articulating some of its insights with the Gramscian conception of hegemony. This led us to put the question of power and antagonism and their ineradicable character at the center of our approach. One of the main theses of the book is that social objectivity is constituted through acts or power. This means that any social objectivity is ultimately political and that it has to show the traces of exclusion which governs its constitution; what, following Derrida, we have called its "constitutive outside." But, if an object has inscribed in its very being something other than itself; if, as a result, everything is constructed as difference, its being cannot be conceived as pure "presence" or "objectivity." This indicates that the logics of the constitution of the social is incompatible with the objectivism and essentialism dominant in social sciences and liberal thought.

The point of convergence—or rather mutual collapse—between objectivity and power is what we have called "hegemony." This way of posing the problem indicates that power should not be conceived as an external relation taking place between two preconstituted identities, but rather as constituting the identities themselves. This is really decisive. Because, if the "constitutive outside" is present within the inside as its "always real possibility," then the inside itself becomes a purely contingent and reversible arrangement. (In other words, the hegemonic arrangement cannot claim any source of validity other than the power basis on which it is grounded.) The structure of mere possibility of any objective order, which is revealed by its mere hegemonic nature, is shown in the forms assumed by the subversion of the sign, that is, of the relation signifier–signified. For instance, the signifier "democracy" is very different when fixed to a certain signified in a discourse that articulates it to anti-communism and when it is fixed to another signified in a discourse that makes it part of the total meaning of antifascism. As there is no common ground between those conflicting articulations, there is no way of subsuming them under a deeper objectivity which would reveal its true and deeper essence. This explains the constitutive and irreducible character of antagonism.

The consequences of these theses for politics are far-reaching. For instance, according to such a perspective, political practice in a democratic society does not consist in defending the rights of preconstituted identities, but rather in constituting those identities themselves in a precarious and always vulnerable terrain. Such an approach also involves a displacement of the traditional relations between "democracy" and "power." For a traditional socialist conception, the more democratic a society is, the less power would be constitutive of social relations. But, if we accept that relations of power are constitutive of the social, then the main question of democratic politics is not how to eliminate power but how to constitute forms of power that are compatible with democratic values. To acknowledge the existence of relations of power and the need to transform them while renouncing the illusion that we could free ourselves completely from power is what is specific to the project of "radical and plural democracy" that we are advocating.

Another distinct characteristic of our approach concerns the question of the de-universalization of political subjects. We try to break with all forms of essentialism. Not only the essentialism which penetrates to a large extent the basic categories of modern sociology and liberal thought and according to which social identity is perfectly defined in the historical process of the unfolding of being; but also with its diametrical opposite: a certain type of extreme post-modern fragmentation of the social which refuses to give the fragments any kind of relational identity. Such a view leaves us with a multiplicity of identities without any common denominator and makes it impossible to distinguish between differences that exist but should not exist and differences that do not exist but should exist. In other words, by putting an exclusive emphasis on heterogeneity and incommensurability, it impedes us to recognize how certain differences are constructed as relations of subordination and should therefore be challenged by a radical democratic politics.

DEMOCRACY AND IDENTITY

After having given a brief outline of the main tenets of our anti-essentialist approach and of its general implications for politics, I now would like to address some specific problems concerning the construction of democratic identities. I am going to examine how such a question can be formulated within the framework which breaks with the traditional rationalist liberal problematic and that incorporates some crucial insights of the critique of essentialism. One of the main problems with the liberal framework is that it reduces politics to the calculus of interests. Individuals are presented as rational actors moved by the search for the maximization of their self-interest. That is, they are seen as acting in the field of politics in a basically instrumentalist way. Politics is conceived through a model elaborated to study economics as a market concerned with the allocation of resources where compromises are reached among interests defined independently of their political articulation. Other liberals, those who rebel against this model and who want to create a link between politics and morality, believe that it is possible to create a rational and universal consensus by means of free discussion. They believe that by relegating disruptive issues to the private sphere, a rational agreement on principles should be enough to administer the pluralism of modern societies. For both types of liberals, everything that has to do with passions, with antagonisms, everything that can lead to violence is seen as archaic and irrational; as residues of a bygone age where the "sweet commerce" had not yet established the preeminence of interest over passions.

But this attempt to annihilate the political is doomed to failure because it cannot be domesticated in this way. As was pointed out by Carl Schmitt, the political can derive its energy from the most diverse sources and emerge out of many different social relations: religious, moral, economic, ethnic, or other. The political has to do with the dimension of antagonism which is present in social relations, with the ever-present possibility of an "us–them" relation to be constructed in terms of "friend–enemy." To deny this dimension of antagonism does not make it disappear; it only leads to impotence in recognizing its different manifestations and in dealing with them. This is why a democratic approach needs to come to terms with the ineradicable character of antagonism. One of its main tasks is to envisage how it is possible to defuse the tendencies to exclusion which are present in all constructions of collective identities.

To clarify the perspective that I am putting forward, I propose to distinguish between "the political" and "polities." By "the political," I refer to the dimension of antagonism that is inherent in all human society, antagonism that, as I said, can take many different forms and can emerge in diverse social relations. "Polities" refers to the ensemble of practices, discourses, and institutions which seek to establish a certain order and to organize human coexistence in conditions which are always potentially conflictual because they are affected by the dimension of "the political." This view, which attempts to keep together the two meanings of "polemos" and "polis," from where the idea of politics comes, is, I believe, crucial if we want to be able to protect and consolidate democracy.

In examining this question the concept of the "constitutive outside" to which I have referred earlier is particularly helpful. As elaborated by Derrida, its aim is to highlight the fact that the creation of an identity implies the establishment of a difference, difference which is often constructed on the basis of a hierarchy; as between form and matter, black and white, man and woman, etc. Once we have understood that every identity is relational and that the affirmation of a difference is a precondition for the existence of any identity, that is, the perception of something "other" that will constitute its "exterior," then we can begin to understand why such a relation may always become the breeding ground for antagonism. Indeed, when it comes to the creation of a collective identity, basically the creation of an "us" by the demarcation of a "them," there is always the possibility of that "them and us" relationship becoming one of "friend and enemy," that is, becoming antagonistic. This happens when the "other," who up until now had been considered simply as different, starts to be perceived as someone who puts into question our identity and threatens our existence. From that moment on, any form of "us and them" relationship, whether it be religious, ethnic, economic, or other, becomes political.

It is only when we acknowledge this dimension of "the political" and understand that "politics" consists in domesticating hostility and in trying to defuse the potential antagonism that exists in human relations, that we can pose the fundamental question for democratic politics. This question, pace the rationalists, is not how to arrive at a rational consensus reached without exclusion, or in other words how to establish an "us" which would not have a corresponding "them." This is impossible because there cannot exist an "us" without a "them." What is at stake is how to establish this "us–them" discrimination in a way that is compatible with pluralist democracy.

In the realm of politics, this presupposes that the "other" is no longer seen as an enemy to be destroyed, but as an "adversary," somebody with whose ideas we are going to struggle, but whose right to defend those ideas we will not put into question. We could say that the aim of democratic politics is to transform an "antagonism" into an "agonism." The prime task of democratic politics is not to eliminate passions, not to relegate them to the private sphere in order to render rational consensus possible, but to mobilize these passions in a way that promotes democratic designs. Far from jeopardizing democracy, agonistic confrontation is in fact its very condition of existence.

Modern democracy's specificity lies in the recognition and legitimation of conflict and the refusal to suppress it by imposing an authoritarian order. Breaking with the symbolic representation of society as an organic body—which is characteristic of the holist mode of social organization—a democratic society makes room for the expression of conflicting interests and values. For that reason pluralist democracy demands not only consensus on a set of common political principles but also the presence of dissent and institutions through which such divisions can be manifested. This is why its survival depends on collective identities forming around clearly differentiated positions, as well as on the possibility of choosing between real alternatives. The blurring of political frontiers between right and left, for instance, impedes the creation of democratic political identities and fuels disenchantment with political participation. This prepares the ground for various forms of populist politics articulated around nationalist, religious, or ethnic issues. When the agonistic dynamic of the pluralist system is hindered because of a lack of democratic identities with which one could identify, there is a risk that this will multiply confrontations over essentialist identities and nonnegotiable moral values.

Once it is acknowledged that any identity is relational and defined in terms of difference, how can we defuse the possibility of exclusion that it entails? Here again the notion of the "constitutive outside" can help us. By stressing the fact that the outside is constitutive, it reveals the impossibility of drawing an absolute distinction between interior and exterior. The existence of the other becomes a condition of the possibility of my identity since, without the other, I could not have an identity. Therefore, every identity is irremediably destabilized by its exterior and the interior appears as something always contingent. This questions every essentialist conception of identity and forecloses every attempt to conclusively define identity or objectivity. Inasmuch as objectivity always depends on an absent otherness, it is always necessarily echoed and contaminated by this otherness. Identity cannot, therefore, belong to one person alone, and no one belongs to a single identity. We may go further and argue that, not only are there no "natural" and "original" identities, since every identity is the result of a constituting process, but this process itself must be seen as one of permanent hybridization and nomadization. Identity is, in effect, the result of a multitude of interactions which take place inside a space, the outlines of which are not clearly defined. Numerous feminist studies and research inspired by the "postcolonial" perspective have shown that this process is always one of "over-determination," which establishes highly intricate links between the many forms of identity and a complex network of differences. For an appropriate definition of identity, we need to take account of both the multiplicity of discourses and the power structure which affects it, as well as the complex dynamic of complicity and resistance which underlines the practices in which this identity is implicated. Instead of seeing the different forms of identity as allegiances to a place or as a property, we ought to realize that they are what is at stake in any power struggle.

What we commonly call "cultural identity" is both the scene and the object of political struggles. The social existence of a group needs such conflict. It is one of the principal areas in which hegemony is exercised, because the definition of the cultural identity of a group, by reference to a specific system of contingent and particular social relations, plays a major role in the creation of "hegemonic nodal points." These partially define the meaning of a "signifying chain" allowing us to control the stream of signifiers, and temporarily to control the discursive field.

Concerning the question of "national" identities—so crucial again today—the perspective based on hegemony and articulation allows us to come to grips with the idea of the national, to grasp the importance of that type of identity instead of rejecting it in the name of anti-essentialism or as part of a defense of abstract universalism. It is very dangerous to ignore the strong libidinal investment that can be mobilized by the signifier "nation" and it is futile to hope that all national identities could be replaced by so-called "post conventional" identities. The struggle against the exclusive type of ethnic nationalism can only be carried out by articulating another type of nationalism, a "civic" nationalism expressing allegiance to the values specific to the democratic tradition and the forms of life that are constitutive of it.

Contrary to what is sometimes asserted, I do not believe that—to take the case of Europe, for instance—the solution is the creation of a "European" identity, conceived as a homogeneous identity which could replace all other identifications and allegiances. But if we envisage it in terms of "aporia," of a "double genitive" as suggested by Derrida in *The Other Heading* (1992), then the notion of a European identity could be the catalyst for a promising process, not unlike what Merleau-Ponty called "lateral universalism," which implies that the universal lies at the very heart of specificities and differences, and that it is inscribed in respect for diversity. Indeed, if "we conceive this European identity as a "difference to oneself," then we are envisaging an identity which can accommodate otherness, which demonstrates the porosity of its frontiers and opens up towards that exterior which makes it possible. By accepting that only hybridity creates us as separate entities, it affirms and upholds the nomadic character of every identity.

I submit that, by resisting the ever-present temptation to construct identity in terms of exclusion, and by recognizing that identities comprise a multiplicity of elements, and that they are dependent and interdependent, a democratic politics informed by an anti-essentialist approach can defuse the potential for violence that exists in every construction of collective identities and create the conditions for a truly "agonistic" pluralism. Such a pluralism is anchored in the recognition of the multiplicity within oneself and of the contradictory positions that this multiplicity entails. Its acceptance of the other does not merely consist of tolerating differences, but in positively celebrating them because it acknowledges that, without alterity and otherness, no identity could ever assert itself. It is also a pluralism that valorizes diversity and dissensus, recognizing in them the very condition of possibility, of a striving democratic life.

NOTE

1 Presented at a session on "Post-Marxism, Democracy, and Identity," organized by the Socialist, Urban, and Political Specialty Groups at the Annual Meeting of the Association of American Geographers, San Francisco, CA, 1 April 1994.

REFERENCES

Blumenberg, H. 1983 *The Legitimacy of the Modern Age* (MIT Press, Cambridge, MA).

Derrida, J. 1992 *The Other Heading: Reflections on Today's Europe* (Indiana University Press, Bloomington, IN).

Laclau, E. and Mouffe, C. 1985 *Hegemony and Socialist Strategy: Towards a Radical Democratic Politics* (Verso, London).

20

U.S. third world feminism

The theory and method of oppositional consciousness in the postmodern world[1]

Chela Sandoval

from *Genders*, 1991, 10: 1–24

The enigma that is U.S. third world feminism has yet to be fully confronted by theorists of social change. To these late twentieth-century analysts it has remained inconceivable that U.S. third world feminism might represent a form of historical consciousness whose very structure lies outside the conditions of possibility which regulate the oppositional expressions of dominant feminism. In enacting this new form of historical consciousness, U.S. third world feminism provides access to a different way of conceptualizing not only U.S. feminist consciousness but oppositional activity in general; it comprises a formulation capable of aligning such movements for social justice with what have been identified as world-wide movements of decolonization.

Both in spite of and yet because they represent varying internally colonized communities, U.S. third world feminists have generated a common speech, a theoretical structure which, however, remained just outside the purview of the dominant feminist theory emerging in the 1970s, functioning within it—but only as the unimaginable. Even though this unimaginable presence arose to reinvigorate and refocus the politics and priorities of dominant feminist theory during the 1980s, what remains is an uneasy alliance between what appear on the surface to be two different understandings of domination, subordination, and the nature of effective resistance—a shot-gun arrangement at best between what literary critic Gayatri Spivak characterizes as a "hegemonic feminist theory"[2] on the one side and what I have been naming "U.S. third world feminism" on the other.[3] I do not mean to suggest here, however, that the perplexing situation that exists between U.S. third world and hegemonic feminisms should be understood merely in binary terms. On the contrary, what this investigation reveals is the way in which the new theory of oppositional consciousness considered here and enacted by U.S. third world feminism is at least partially contained, though made deeply invisible by the manner of its appropriation, in the terms of what has become a hegemonic feminist theory.

U.S. third world feminism arose out of the matrix of the very discourses denying, permitting, and producing difference. Out of the imperatives born of necessity arose a mobility of identity that generated the activities of a new citizen-subject, and which reveals yet another model for the self-conscious production of political opposition. In this essay I will lay out U.S. third world feminism as the design for oppositional political activity and consciousness in the United States. In mapping this new design, a model is revealed by which social actors can chart the points through which differing oppositional ideologies can meet, in spite of their varying trajectories. This knowledge becomes important when one begins to wonder, along with late twentieth-century cultural critics such as Fredric Jameson, how organized oppositional activity and consciousness can be made possible under the co-opting nature of the so-called "postmodern" cultural condition.[4]

The ideas put forth in this essay are my rearticulation of the theories embedded in the great oppositional practices of the latter half of this century especially in the United States—the Civil Rights movement, the women's movement, and ethnic, race, and gender liberation movements. During this period of great social activity, it became clear to many of us that oppositional social movements which were weakening from internal divisions over strategies, tactics, and aims would benefit by examining philosopher Louis Althusser's theory of "ideology and the ideological state apparatuses."[5] In this now fundamental essay, Althusser lays out the principles by which humans are called into being as citizen-subjects who act—even when in resistance—in order to sustain and reinforce the dominant social order. In this sense, for Althusser, all citizens endure ideological subjection.[6] Althusser's postulations begin to suggest, however, that "means and occasions"[7] do become generated whereby individuals and groups in opposition are able to effectively challenge and transform the current hierarchical nature of the social order, but he does not specify how or on what terms such challenges are mounted.

In supplementing Althusser's propositions, I want to apply his general theory of ideology to the particular cultural concerns raised within North American liberation movements and develop a new theory of ideology which considers consciousness not only in its subordinated and resistant yet appropriated versions—the subject of Althusser's theory of ideology—but in its more effective and persistent oppositional manifestations. In practical terms, this theory focuses on identifying forms of consciousness in opposition, which can be generated and coordinated by those classes self-consciously seeking affective oppositional stances in relation to the dominant social order. The idea here, that the subject-citizen can learn to identify, develop, and control the means of ideology, that is, marshal the knowledge necessary to "break with ideology" while also speaking in and from within ideology, is an idea which lays the philosophical foundations enabling us to make the vital connections between the seemingly disparate social and political aims which drive yet ultimately divide liberation movements from within. From Althusser's point of view, then, the theory I am proposing would be considered a "science of oppositional ideology."

This study identifies five principal categories by which "oppositional consciousness" is organized, and which are politically effective means for changing the dominant order of power. I characterize them as "equal rights," "revolutionary," "supremacist," "separatist," and "differential" ideological forms. All these forms of consciousness are kaleidoscoped into view when the fifth form is utilized as a theoretical model which retroactively clarifies and gives new meaning to the others. Differential consciousness represents the strategy of another form of oppositional ideology that functions on an altogether different register. Its power can be thought of as mobile—not nomadic but rather cinematographic: a kinetic motion that maneuvers, poetically transfigures, and orchestrates while demanding alienation, perversion, and reformation in both spectators and practitioners. Differential consciousness is the expression of the new subject position called for by Althusser—it permits functioning within yet beyond the demands of dominant ideology. This differential form of oppositional consciousness has been enacted in the practice of U.S. third world feminism since the 1960s.

This essay also investigates the forms of oppositional consciousness that were generated within one of the great oppositional movements of the late twentieth century, the second wave of the women's movement. What emerges in this discussion is an outline of the oppositional ideological forms which worked against one another to divide the movement from within. I trace these ideological forms as they are manifested in the critical writings of some of the prominent hegemonic feminist theorists of the 1980s. In their attempts to identify a feminist history of consciousness, many of these thinkers believe they detect four fundamentally distinct phases through which feminists have passed in their quest to end the subordination of women. But viewed in terms of another paradigm, "differential consciousness," here made available for study through the activity of U.S. third world feminism, these four historical phases are revealed as sublimated versions of the very forms of consciousness in opposition which were also conceived within post-1950s U.S. liberation movements.

These earlier movements were involved in seeking effective forms of resistance outside of those determined by the social order itself. My contention is that hegemonic feminist forms of resistance represent only other versions of the forms of oppositional consciousness expressed within all liberation movements active in the United States during the later half of the twentieth century. What I want to do here is systematize in theoretical form a theory of oppositional consciousness as it comes embedded but hidden within U.S. hegemonic feminist theoretical tracts. At the end of this essay, I present the outline of a corresponding theory which engages with these hegemonic feminist theoretical forms while at the same time going beyond them to produce a more general theory and method of oppositional consciousness.

The often discussed race and class conflict between white and third world feminists in the United States allows us a clear view of these forms of consciousness in action. The history of the relationship between first and third world feminists has been tense and rife with antagonisms. My thesis is that at the root of these conflicts is the refusal of U.S. third world feminism to buckle under, to submit to sublimation or assimilation within hegemonic feminist praxis. This refusal is based, in large part, upon loyalty to the differential mode of consciousness and activity outlined in this essay but which has remained largely unaccounted for within the structure of the hegemonic feminist theories of the 1980s.

Differential consciousness is not yet fully theorized by most contemporary analysts of culture, but its understanding is crucial for the shaping of effective and ongoing oppositional struggle in the United States. Moreover, the recognition of differential consciousness is vital to the generation of a next "third wave" women's movement and provides grounds for alliance with other decolonizing movements for emancipation. My answer to the perennial question asked by hegemonic feminist theorists throughout the 1980s is that yes, there *is* a specific U.S. third world feminism: it is that which provides the theoretical and methodological approach, the "standpoint" if you will, from which this evocation of a theory of oppositional consciousness is summoned.

A BRIEF HISTORY

From the beginning of what has been known as the second wave of the women's movement, U.S. third world feminists have claimed a feminism at odds with that being developed by U.S. white women. Already in 1970 with the publication of *Sisterhood Is Powerful*, black feminist Francis Beal was naming the second wave of U.S. feminism as a "white women's movement" because it insisted on organizing along the binary gender division male/female alone.[8] U.S. third world feminists, however, have long understood that one's race, culture, or class often denies comfortable or easy access to either category, that the interactions between social categories produce other genders within the social hierarchy. As far back as the middle of the last century, Sojourner Truth found it necessary to remind a convention of white suffragettes of her female gender with the rhetorical question "ar'n't I a woman?"[9] American Indian Paula Gunn Allen has written of Native women that "the place we live now is an idea, because whiteman took all the rest."[10] In 1971, Toni Morrison went so far as to write of U.S. third world women that "there is something inside us that makes us different from other people. It is not like men and it is not like white women."[11] That same year Chicana Velia Hancock continued: "Unfortunately, many white women focus on the maleness of our present social system as though, by implication, a female dominated white America would have taken a more reasonable course" for people of color of either sex.[12]

These signs of a lived experience of difference from white female experience in the United States repeatedly appear throughout U.S. third world feminist writings. Such expressions imply the existence of at least one other category of gender which is reflected in the very titles of books written by U.S. feminists of color such as *All the Blacks Are Men, All the Women Are White, But Some of Us Are Brave*[13] or *This Bridge Called My Back*,[14] titles which imply that women of color somehow exist in the interstices between the legitimated categories of the social order. Moreover, in the title of bell hooks' 1981 book, the question "Ain't I a Woman" is transformed into a defiant statement,[15] while Amy Ling's feminist analysis of Asian American writings, *Between Worlds*,[16] or the title of the journal for U.S. third world feminist writings, *The Third Woman*,[17] also calls for the recognition of a new category for social identity. This in-between space, this third gender category, is also explored in the writings of such well-known authors as Maxine Hong Kingston, Gloria Anzaldua, Audre Lorde, Alice Walker, and Cherrie Moraga, all of whom argue that U.S. third world feminists represent a different kind of human—new "mestizas,"[18] "Woman Warriors" who live and are gendered "between and among" the lines,[19] "Sister Outsiders"[20] who inhabit a new psychic terrain which Anzaldua calls "the Borderlands," "la nueva Frontera." In 1980, Audre Lorde summarized the U.S. white women's movement by saying that "today, there is a pretense to a homogeneity of experience covered by the word SISTERHOOD in the white women's movement. When white feminists call for 'unity,' they are mis-naming a deeper and real need for homogeneity." We began the 1980s, she says, with "white women" agreeing "to focus upon their oppression as women" while continuing "to ignore difference." Chicana sociologist Maxine Baca Zinn rearticulated this position in a 1986 essay in *Signs*, saying that "there now exists in women's studies an increased awareness of the variability of womanhood" yet for U.S. feminists of color "such work is often tacked on, its significance for feminist knowledge still unrecognized and unregarded."[21]

How has the hegemonic feminism of the 1980s responded to this other kind of feminist theoretical activity? The publication of *This Bridge Called My Back* in 1981 made the presence of U.S. third world feminism impossible to ignore on the same terms as it had been throughout the 1970s. But soon the writings and theoretical challenges of U.S. third world feminists were marginalized into the category of what Allison Jaggar characterized in 1983 as mere "description,"[22] and their essays deferred to what Hester Eisenstein in 1985 called "the special force of poetry,"[23] while the shift in paradigm I earlier referred to as "differential consciousness," and which is represented in the praxis of U.S. third world feminism, has been bypassed and ignored. If, during the 1980s, U.S. third world feminism had become a theoretical problem, an inescapable mystery to be solved for hegemonic feminism, then perhaps a theory of difference—but imported from Europe—could subsume if not solve it. I would like to provide an example of how this systematic repression of the theoretical implications of U.S. third world feminism occurs.

THE GREAT HEGEMONIC MODEL

During the 1980s, hegemonic feminist scholars produced the histories of feminist consciousness which they believed to typify the modes of exchange operating within the oppositional spaces of the women's movement. These feminist histories of consciousness are often presented as typologies, systematic classifications of all possible forms of feminist praxis. These constructed typologies have fast become the official stories by which the white women's movement understands itself and its interventions in history. In what follows I decode these stories and their relations to one another from the perspective of U.S. third world feminism, where they are revealed as sets of imaginary spaces, socially constructed to severely delimit what is possible within the boundaries of their separate narratives. Together, they legitimize certain modes of culture and consciousness only to systematically curtail the forms of experiential and theoretical articulations permitted U.S. third world feminism. I want to demonstrate how the constructed relationships adhering between the various types of hegemonic feminist theory and consciousness are unified at a deeper level into a great metastructure which sets up and reveals the logic of an exclusionary U.S. hegemonic feminism.

The logic of hegemonic feminism is dependent upon a common code that shapes the work of such a diverse group of thinkers as Julia Kristeva, Toril Moi, Gerda Lerna, Cora Kaplan, Lydia Sargent, Alice Jardine, or Judith Kegan Gardiner. Here I follow its traces through the 1985 writings of the well-known literary critic Elaine Showalter;[24] the now classic set of essays published in 1985 and edited by Hester Eisenstein and Alice Jardine on *The Future of Difference* in the "women's movement"; Gale Greene and Coppelia Kahn's 1985 introductory essay in the collection *Making a Difference: Feminist Literary Criticism*;[25] and the great self-conscious prototype of hegemonic feminist thought encoded in Allison Jaggar's massive dictionary of feminist consciousness, *Feminist Politics and Human Nature*, published in 1983.

Showalter's well-known essay "Towards a Feminist Poetics" develops what she believes to be a three-phase "taxonomy, if not a poetics, of feminist criticism."[26] For Showalter, these three stages represent succeedingly higher levels of women's historical, moral, political, and aesthetic development.

For example, according to Showalter, critics can identify a first phase "feminine" consciousness when they detect, she says, women writing "in an effort to equal the cultural achievement of the male culture." In another place, feminist theorist Hester Eisenstein concurs when she writes that the movement's early stages were characterized by feminist activists organizing to prove "that differences between women and men were exaggerated, and that they could be reduced" to a common denominator of sameness.[27] So, too, do historians Gayle Greene and Coppelia Kahn also claim the discovery of a similar first-phase feminism in their essay on "Feminist Scholarship and the Social Construction of Woman."[28] In its first stage, they write, feminist theory organized itself "according to the standards of the male public world and, appending women to history" as it has already been defined, left "unchallenged the existing paradigm." Matters are similar in political scientist Allison Jaggar's book *Feminist Politics and Human Nature*. Within her construction of four "genera" of feminist consciousness which are "fundamentally incompatible with each other" though related by a metatheoretical schema, the first phase of "liberal feminism" is fundamentally concerned with "demonstrating that women are as fully human as men."[29]

In the second phase of this typology, shared across the text of hegemonic feminist theory, Showalter claims that female writers turn away from the logics of the "feminine" first phase. Under the influence of a second "feminist" phase, she states, writers work to "reject" the accommodation of "male culture," and instead use literature to "dramatize wronged womanhood."[30] Elsewhere, Eisenstein also insists that first-phase feminism reached a conclusion. No longer were women the same as men, but, rather, "women's lives WERE different from men, and . . . it was precisely this difference that required illumination."[31] In Greene and Kahn's view, feminist scholars turned away from the "traditional paradigm" of first-phase feminism, and "soon extended their enquiries to the majority of women unaccounted for by traditional historiography, 'in search of the actual *experiences* of women in the past,'" asking questions about "the quality of their daily lives, the conditions in which they lived and worked, the ages at which they married and bore children; about their work, their role in the family, their class and relations to other women; their perception of their place in the world; their relation to wars and revolutions."[32] If women were not like men, but fundamentally different, then the values of a patriarchal society had to be transformed in order to accommodate those differences. Jaggar argued that it was during this second phase that feminists undermined "first-phase liberal feminism" by turning toward Marxism as a way of restructuring a new society incapable of subordinating women.[33]

In Showalter's third and, for her, final "female" phase of what I see as a feminist history of consciousness, Showalter argues that "the movement rejected both earlier stages as forms of dependency" on men, or on their culture and instead turned "toward female experience as a source of a new, autonomous art."[34] It is in this third phase, Eisenstein asserts, that "female differences originally seen as a source of oppression appear as a source of enrichment."[35] Under the influence of this third-phase feminism, women seek to uncover the unique expression of the essence of "woman" which lies underneath the multiplicity of her experiences. Eisenstein reminds us that this feminism is "woman-centered," a transformation within which "maleness"—not female-ness—becomes "the difference" that matters: now, she says, "men were the Other."[36] Greene and Kahn also perceive this same third-phase feminism within which "some historians of women posit the existence of a separate woman's culture, even going so far as to suggest that women and men within the same society may have different experiences of the universe."[37] Jaggar's typology characterizes her third-phase feminism as an "unmistakably twentieth century phenomenon" which is the first approach to conceptualizing human nature, social reality, and politics "to take the subordination of women as its central concern." Her third-phase feminism contends that "women naturally know much of which men are ignorant," and takes as "one of its main tasks . . . to explain why this is so." Jaggar understands this third phase as generating either "Radical" or "Cultural" feminisms.[38]

Now, throughout what can clearly be viewed as a three-phase feminist history of consciousness, as white feminist Lydia Sargent comments in her 1981 collection of essays on *Women and Revolution*, "racism, while part of the discussion, was never successfully integrated into feminist theory and practice." This resulted, she writes, in powerful protests by women of color at each of these three phases of hegemonic feminist praxis "against the racism (and classism) implicit in a white feminist movement, theory and practice."[39] The recognition that hegemonic feminist theory was not incorporating the content of U.S. third world feminist "protests" throughout the 1970s suggests a structural deficiency within hegemonic feminism which prompted certain hegemonic theorists to construct a fourth and for them a final and "antiracist" phase of feminism.

The fourth category of this taxonomy always represents the unachieved category of possibility where the differences represented by race and class can be (simply) accounted for, and it is most often characterized as "socialist feminism." Eisenstein approaches her version of fourth-phase feminism this way: "as the women's movement grew more diverse, it became *forced* [presumably by U.S. feminists of color] to confront and to debate issues of difference—most notably those of race and class."[40] Jaggar laments that first-phase liberal feminism "has tended to ignore or minimize all these differences" while second-phase Marxist feminism "has tended to recognize only differences of class," and the third-phase "political theory of radical feminism has tended to recognize only differences of age and sex, to understand these in universal terms, and often to view them as determined biologically." By contrast, she asserts, a fourth-phase "socialist feminism" should recognize differences among women "as constituent parts of contemporary human nature." This means that the "central project of socialist feminism" will be "the development of a political theory and practice that will synthesize the best insights" of the second- and third-phase feminisms, those of the "radical and Marxist traditions," while hopefully escaping "the problems associated with each." Within Jaggar's metatheoretical schema socialist feminism represents the fourth, ultimate, and "most appropriate interpretation of what it is for a theory to be impartial, objective, comprehensive, verifiable and useful."[41]

Socialist feminist theorist Cora Kaplan agrees with Jaggar and indicts the previous three forms of hegemonic feminism—liberal, Marxist, and radical—for failing to incorporate an analysis of power relations, beyond gender relations, in their rationality. Most dominant feminist comprehensions of gender, she believes, insofar as they seek a unified female subject, construct a "fictional landscape." Whether this landscape is then examined from liberal, psychoanalytic, or semiotic feminist perspectives, she argues, "the other structuring relations of society fade and disappear, leaving us with the naked drama of sexual difference as the only scenario that matters." For Kaplan, differences among women will only be accounted for by a new socialist feminist criticism which understands the necessity of transforming society by coming "to grips with the relationship between female subjectivity and class identity."[42] Unfortunately, however, socialist feminism has yet to develop and utilize a theory and method capable of achieving this goal, or of coming to terms with race or culture, and of thus coming "to grips" with the differences existing between female subjects. Though continuing to claim socialist feminism as "the most comprehensive" of feminist theories, Jaggar allows that socialist feminism has made only "limited progress" toward such goals. Rather, she regretfully confesses, socialist feminism remains a "commitment to the development" of "an analysis and political practice" that will account for differences among and between women, rather than a commitment to a theory and practice "which already exists."[43] Finally, Jaggar grudgingly admits that insofar as socialist feminism stubbornly "fails to theorize the experiences of women of color, it cannot be accepted as complete."[44]

We have just charted our way through what I hope to have demonstrated is a commonly cited four-phase feminist history of consciousness consisting of "liberal," "Marxist," "radical/cultural," and "socialist" feminisms, and which I schematize as "women are the same as men," "women are different from men," "women are superior," and the fourth catchall category, "women are a racially divided class." I contend that this comprehension of feminist consciousness is hegemonically unified, framed, and buttressed with the result that the expression of a unique form of U.S. third world feminism, active over the last thirty years, has become invisible outside of its all-knowing logic. Jaggar states this position quite clearly in her dictionary of hegemonic feminist consciousness when she writes that the contributions of feminists of color (such as Paula Gunn Allen, Audre Lorde, Nellie Wong, Gloria Anzaldua, Cherrie Moraga, Toni Morrison, Mitsuye Yamada, bell hooks, the third world contributors to *Sisterhood Is Powerful*, or the contributors to *This Bridge*, for example) operate "mainly at the level of description," while those that are theoretical have yet to contribute to any "unique or distinctive and comprehensive theory of women's liberation."[45] For these reasons, she writes, U.S. third world feminism has not been "omitted from this book" but rather assimilated into one of the "four genera" of hegemonic feminism I have outlined earlier.

U.S. third world feminism, however, functions just outside the rationality of the four-phase hegemonic structure we have just identified. Its recognition will require of hegemonic feminism a paradigm shift which is capable of rescuing its theoretical and practical expressions from their exclusionary and racist forms. I am going to introduce this shift in paradigm by proposing a new kind of taxonomy which I believe prepares the ground for a new theory and method of oppositional consciousness. The recognition of this new taxonomy should also bring into view a new set of alterities and another way of understanding "otherness" in general, for it demands that oppositional actors claim new grounds for generating identity, ethics, and political activity.

Meanwhile, U.S. third world feminism has been sublimated, both denied yet spoken about incessantly, or, as black literary critic Sheila Radford Hill put it in 1986, U.S. third world feminism is "used" within hegemonic feminism only as a "rhetorical platform" which allows white feminist scholars to "launch arguments for or against" the same four basic configurations of hegemonic feminism.[46] It is not surprising, therefore, that the writings of feminist third world theorists are laced through with bitterness. For, according to bell hooks in 1982, the sublimation of U.S. third world feminist writing is linked to racist "exclusionary practices" which have made it "practically impossible" for any new feminist paradigms to emerge. Two years before Jaggar's *Feminist Politics and Human Nature*, hooks wrote that although "feminist theory is the guiding set of beliefs and principles that become the basis for action," the development of feminist theory is a task permitted only within the "hegemonic dominance" and approval "of white academic women."[47] Four years later Gayatri Spivak stated that "the emergent perspective" of hegemonic "feminist criticism" tenaciously reproduces "the axioms of imperialism." Clearly, the theoretical structure of hegemonic feminism has produced enlightening and new feminist intellectual spaces, but these coalesce in what Spivak characterizes as a "high feminist norm" which culminates in reinforcing the "basically isolationist" and narcissistic "admiration" of hegemonic feminist thinkers "for the literature of the female subject in Europe and Anglo America."[48]

We have just charted our way through a four-phase hegemonic typology which I have argued is commonly utilized and cited—self-consciously or not—by feminist theorists as the way to understand oppositional feminist praxis. I believe that this four-phase typology comprises the mental map of the given time, place, and cultural condition we call the U.S. white women's movement. From the perspective of U.S. third world feminism this four-category structure of consciousness as presently enacted interlocks into a symbolic container which sets limits on how the history of feminist activity can be conceptualized, while obstructing what can be perceived or even imagined by agents thinking within its constraints. Each category of this typology along with the overriding rationality that relates the categories one to the other is socially constructed, the structure and the network of possibilities it generates are seen by feminists of color as, above all, *imaginary* spaces which, when understood and enacted as if self-contained, rigidly circumscribe what is possible for feminists and their relations across their differences. Hegemonic feminist theoreticians and activists are trapped within the rationality of this structure, which sublimates or disperses the theoretical specificity of U.S. third world feminism.

Despite the fundamental shift in political objectives and critical methods which is represented by hegemonic feminism, there remains in its articulations a limited and traditional reliance on what are previous, *modernist* modes of understanding oppositional forms of activity and consciousness. The recognition of a specific U.S. third world feminism demands that feminist scholars extend their critical and political objectives even further. During the 1970s, U.S. feminists of color identified common grounds upon which they made coalitions across profound cultural, racial, class, and gender differences. The insights perceived during this period reinforced the common culture across difference comprised of the skills, values, and ethics generated by subordinated citizenry compelled to live within similar realms of marginality. During the 1970s, this common culture was reidentified and claimed by U.S. feminists of color, who then came to recognize one another as countrywomen—and men—of the same psychic terrain. It is the methodology and theory of U.S. third world feminism that permit the following rearticulation of hegemonic feminism, on its own terms, and beyond them.

TOWARD A THEORY OF OPPOSITIONAL CONSCIOUSNESS

Let me suggest, then, another kind of typology, this one generated from the insights born of oppositional activity beyond the inclusive scope of the hegemonic women's movement. It is important to remember that the form of U.S. third world feminism it represents and enacts has been influenced not only by struggles against gender domination, but by the struggles against race, class, and cultural hierarchies which mark the twentieth century in the United States. It is a mapping of consciousness in opposition to the dominant social order which charts the white and hegemonic feminist histories of consciousness we have just surveyed, while also making visible the different ground from which a specific U.S. third world feminism rises. It is important to understand that this typology is not necessarily "feminist" in nature, but is rather a history of oppositional consciousness. Let me explain what I mean by this.

I propose that the hegemonic feminist structure of oppositional consciousness be recognized for what it is, reconceptualized, and replaced by the structure which follows. This new structure is best thought of not as a typology, but as a "*topography*" of consciousness in opposition, from the Greek word "topos" or place, insofar as it represents the charting of realities that occupy a specific kind of cultural region. The following topography delineates the set of critical points around which individuals and groups seeking to transform oppressive powers constitute themselves as resistant and oppositional subjects. These points are orientations deployed by those subordinated classes which have sought subjective forms of resistance other than those forms determined by the social order itself. They provide repositories within which subjugated citizens can either occupy or throw off subjectivities in a process that at once both enacts and yet decolonizes their various relations to their real conditions of existence. This kind of kinetic and self-conscious mobility of consciousness is utilized by U.S. third world feminists as they identify oppositional subject positions and enact them *differentially*.

What hegemonic feminist theory has identified are only other versions of what I contend are the various modes of consciousness which have been most effective in opposition under modes of capitalist production before the postmodern period, but in their "feminist" incarnations. Hegemonic feminism appears incapable of making the connections between its own expressions of resistance and opposition and the expressions of consciousness in opposition enacted amongst other racial, ethnic, cultural, or gender liberation movements. Thus, I argue that the following topography of consciousness is not necessarily "feminist" in nature, but represents a history of oppositional consciousness.

Any social order which is hierarchically organized into relations of domination and subordination creates particular subject positions within which the subordinated can legitimately function.[49] These subject positions, once self-consciously recognized by their inhabitants, can become transformed into more effective sites of resistance to the current ordering of power relations. From the perspective of a differential U.S. third world feminism, the histories of consciousness produced by U.S. white feminists are, above all, only other examples of subordinated consciousness in opposition. In order to make U.S. third world feminism visible within U.S. feminist theory, I suggest a topography of consciousness which identifies nothing more and nothing less than the modes the subordinated of the United States (of any gender, race, or class) claim as politicized and oppositional stances in resistance to domination. The topography that follows, unlike its hegemonic feminist version, is not historically organized, no enactment is privileged over any other, and the recognition that each site is as potentially effective in opposition as any other makes possible another mode of consciousness which is particularly effective under late capitalist and postmodern cultural conditions in the United States. I call this mode of consciousness "differential"—it is the ideological mode enacted by U.S. third world feminists over the last thirty years.

The first four enactments of consciousness that I describe next reveal hegemonic feminist political strategies as the forms of oppositional consciousness most often utilized in resistance under earlier (modern, if you will) modes of capitalist production. The following topography, however, does not simply replace previous lists of feminist consciousness with a new set of categories, because the fifth and differential method of oppositional consciousness has a mobile, retroactive, and transformative effect on the previous four forms (the "equal rights," "revolutionary," "supremacist," and "separatist" forms) setting them into new processual relationships. Moreover, this topography compasses the perimeters for a new theory of consciousness in opposition as it gathers up the modes of ideology-praxis represented within previous liberation movements into the fifth, differential, and postmodern paradigm.[50] This paradigm can, among other things, make clear the vital connections that exist between feminist theory in general and other theoretical modes concerned with issues of social hierarchy, race marginality, and resistance. U.S. third world feminism, considered as an enabling theory and method of differential consciousness, brings the following oppositional ideological forms into view:

1 Under an "equal rights" mode of consciousness in opposition, the subordinated group argue that their differences—for which they have been assigned inferior status—are only in appearance, not reality. Behind their exterior physical difference, they argue,

is an essence the same as the essence of the human already in power. On the basis that all individuals are created equal, subscribers to this particular ideological tactic will demand that their own humanity be legitimated, recognized as the same under the law, and assimilated into the most favored form of the human in power. The expression of this mode of political behavior and identity politics can be traced throughout the writings generated from within U.S. liberation movements of the post-World War II era, Hegemonic feminist theorists have claimed this oppositional expression of resistance to social inequality as "liberal feminism."

2 Under the second ideological tactic generated in response to social hierarchy, which I call "revolutionary," the subordinated group claim their *differences* from those in power and call for a social transformation that will accommodate and legitimate those differences, by force if necessary. Unlike the previous tactic, which insists on the similarity between social, racial, and gender classes across their differences, there is no desire for assimilation within the present traditions and values of the social order. Rather, this tactic of revolutionary ideology seeks to affirm subordinated differences through a radical societal reformation. The hope is to produce a new culture beyond the domination/subordination power axis. This second revolutionary mode of consciousness was enacted within the white women's movement under the rubric of either "socialist" or "Marxist" feminisms.

3 In "supremacism," the third ideological tactic, not only do the oppressed claim their differences, but they also assert that those very differences have provided them access to a superior evolutionary level than those currently in power. Whether their differences are of biological or social origin is of little practical concern; of more importance is the result. The belief is that this group has evolved to a higher stage of social and psychological existence than those currently holding power; moreover, their differences now comprise the essence of what is good in human existence. Their mission is to provide the social order with a higher ethical and moral vision and consequently a more effective leadership. Within the hegemonic feminist schema "radical" and "cultural" feminisms are organized under these precepts.

4 "Separatism" is the final of the most commonly utilized tactics of opposition organized under previous modes of capitalist development. As in the previous three forms, practitioners of this form of resistance also recognize that their differences have been branded as inferior with respect to the category of the most human. Under this mode of thought and activity, however, the subordinated do not desire an "equal rights" type of integration with the dominant order, nor do they seek its leadership or revolutionary transformation. Instead, this form of political resistance is organized to protect and nurture the differences that define it through complete separation from the dominant social order. A Utopian landscape beckons these practitioners . . . their hope has inspired the multiple visions of the other forms of consciousness as well.

In the post-World War II period in the United States, we have witnessed how the maturation of a resistance movement means not only that four such ideological positions emerge in response to dominating powers, but that these positions become more and more clearly articulated. Unfortunately, however, as we were able to witness in the late 1970s white women's movement, such ideological positions eventually divide the movement of resistance from within, for each of these sites tend to generate sets of tactics, strategies, and identities which historically have appeared to be mutually exclusive under modernist oppositional practices. What remains all the more profound, however, is that the differential practice of U.S. third world feminism undermines the appearance of the mutual exclusivity of oppositional strategies of consciousness; moreover, it is U.S. third world feminism which allows their reconceptualization on the new terms just proposed. U.S. feminists of color, insofar as they involved themselves with the 1970s white women's liberation movement, were also enacting one or more of the ideological positionings just outlined, but rarely for long, and rarely adopting the kind of fervid belief systems and identity politics that tend to accompany their construction under hegemonic understanding. This unusual affiliation with the movement was variously interpreted as disloyalty, betrayal, absence, or lack: "When they were there, they were rarely there for long" went the usual complaint, or "they seemed to shift from one type of women's group to another." They were the mobile (yet ever present in their "absence") members of this particular liberation movement. It is precisely the significance of this mobility which most inventories of oppositional ideology cannot register.

It is in the activity of weaving "between and among" oppositional ideologies as conceived in this new topological space where another and fifth mode of oppositional consciousness and activity can be found.[51] I have named this activity of consciousness "differential" insofar as it enables movement "between and among" the other equal rights, revolutionary, supremacist, and separatist modes of oppositional consciousness considered as variables, in order to disclose the distinctions among them. In this sense the differential mode of consciousness operates like the clutch of an automobile: the mechanism that permits the driver to select, engage, and disengage gears in a system for the transmission of power.[52] Differential consciousness represents the variant, emerging out of correlations, intensities, junctures, crises. What is differential functions through hierarchy, location, and value—enacting the recovery, revenge, or reparation; its processes produce justice. For analytic purposes I place this mode of differential consciousness in the fifth position, even though it functions as the medium through which the "equal rights," "revolutionary," "supremacist," and "separatist" modes of oppositional consciousness became effectively transformed out of their hegemonic versions. Each is now ideological and tactical weaponry for confronting the shifting currents of power.

The differences between this five-location and processual topography of consciousness in opposition, and the previous typology of hegemonic feminism, have been made available for analysis through the praxis of U.S. third world feminism understood as a differential method for understanding oppositional political consciousness and activity. U.S. third world feminism represents a central locus of possibility, an insurgent movement which shatters the construction of any one of the collective ideologies as the single most correct site where truth can be represented. Without making this move beyond each of the four modes of oppositional ideology outlined above, any liberation movement is destined to repeat the oppressive authoritarianism from which it is attempting to free itself and become trapped inside a drive for truth which can only end in producing its own brand of dominations. What U.S. third world feminism demands is a new subjectivity, a political revision that denies any one ideology as the final answer, while instead positing a *tactical subjectivity* with the capacity to recenter depending upon the kinds of oppression to be confronted. This is what the shift from hegemonic oppositional theory and practice to a U.S. third world theory and method of oppositional consciousness requires.

Chicana theorist Aida Hurtado explains the importance of differential consciousness to effective oppositional praxis this way: "by the time women of color reach adulthood, we have developed informal political skills to deal with State intervention. The political skills required by women of color are neither the political skills of the White power structure that White liberal feminists have adopted nor the free spirited experimentation followed by the radical feminists." Rather, "women of color are more like urban guerrillas trained through everyday battle with the state apparatus." As such, "women of color's fighting capabilities are often neither understood by white middle-class feminists" nor leftist activists in general, and up until now, these fighting capabilities have "not been codified anywhere for them to learn."[53] Cherrie Moraga defines U.S. third world feminist "guerrilla warfare" as a way of life: "Our strategy is how we cope" on an everyday basis, she says, "how we measure and weigh what is to be said and when, what is to be done and how, and to whom . . . daily deciding/risking who it is we can call an ally, call a friend (whatever that person's skin, sex, or sexuality)." Feminists of color are "women without a line. We are women who contradict each other."[54]

In 1981, Anzaldua identified the growing coalition between U.S. feminists of color as one of women who do not have the same culture, language, race, or "ideology, nor do we derive similar solutions" to the problems of oppression. For U.S. third world feminism enacted as a differential mode of oppositional consciousness, however, these differences do not become "opposed to each other."[55] Instead, writes Lorde in 1979, ideological differences must be seen as "a fund of necessary polarities between which our creativities spark like a dialectic. Only within that interdependence," each ideological position "acknowledged and equal, can the power to seek new ways of being in the world generate, as well as the courage and sustenance to act where there are no charters."[56] This movement between ideologies along with the concurrent desire for ideological commitment are necessary for enacting differential consciousness. Differential consciousness makes the second topography of consciousness in opposition visible as a new theory and method for comprehending oppositional subjectivities and social movements in the United States.

The differential mode of oppositional consciousness depends upon the ability to read the current situation of power and to self-consciously choose and adopt the ideological form best suited to push against its configurations, a survival skill well known to oppressed peoples.[57] Differential consciousness requires grace, flexibility, and strength: enough strength to confidently commit to a well-defined structure of identity for one hour, day, week, month, year; enough flexibility to self-consciously transform that identity according to the requisites of another oppositional ideological tactic if readings of power's formation require it; enough grace to recognize alliance with others committed to egalitarian social relations and race, gender, and class justice, when their readings of power call for alternative oppositional stands. Within the realm of differential consciousness, oppositional ideological positions, unlike their incarnations under hegemonic feminist comprehension, are tactics—not strategies. Self-conscious agents of differential consciousness recognize one another as allies, country-women and men of the same psychic terrain. As the clutch of a car provides the driver the ability to shift gears, differential consciousness permits the practitioner to choose tactical positions, that is, to self-consciously break and reform ties to ideology, activities which are imperative for the psychological and political practices that permit the achievement of coalition across differences. Differential consciousness occurs within the only possible space where, in the words of third world feminist philosopher Maria Lugones, "cross-cultural and cross-racial loving" can take place, through the ability of the self to shift its identities in an activity she calls "world traveling."[58]

Perhaps we can now better understand the overarching Utopian content contained in definitions of U.S. third world feminism, as in this statement made by black literary critic Barbara Christian in 1985, who, writing to other U.S. feminists of color, said: "The struggle is not won. Our vision is still seen, even by many progressives, as secondary, our words trivialized as minority issues," our oppositional stances "characterized by others as divisive. But there is a deep philosophical reordering that is occurring" among us "that is already having its effects on so many of us whose lives and expressions are an increasing revelation of the INTIMATE face of universal struggle."[59] This "philosophical reordering," referred to by Christian, the "different strategy, a different foundation" called for by hooks are, in the words of Audre Lorde, part of "a whole other structure of opposition that touches every aspect of our existence at the same time that we are resisting." I contend that this structure is the recognition of a five-mode theory and method of oppositional consciousness, made visible through one mode in particular, differential consciousness, or U.S. third world feminism, what Gloria Anzaldua has recently named "la conciencia de la mestiza" and what Alice Walker calls "womanism."[60] For Barbara Smith, the recognition of this fundamentally different paradigm can "alter life as we know it" for oppositional actors[61] In 1981, Merle Woo insisted that U.S. third world feminism represents a "new framework which will not support repression, hatred, exploitation and isolation, but will be a human and beautiful framework, created in a community, bonded not by color, sex or class, but by love and the common goal for the liberation of mind, heart, and spirit."[62] It has been the praxis of a differential form of oppositional consciousness which has stubbornly called up Utopian visions such as these.

In this essay I have identified the hegemonic structure within which U.S. feminist theory and practice are trapped. This structure of consciousness stands out in relief against the praxis of U.S. third world feminism, which has evolved to center the differences of U.S. third world feminists across their varying languages, cultures, ethnicities, races, classes, and genders. I have suggested that the "philosophical reordering" referred to by Christian is imaginable only through a new theory and method of oppositional consciousness, a theory only visible when U.S. third world feminist praxis is recognized. U.S. third world feminism represents a new condition of possibility, another kind of gender, race and class consciousness which has allowed us to recognize and define differential consciousness. Differential consciousness was utilized by feminists of color within the white women's movement; yet it is also a form of consciousness in resistance well utilized among subordinated subjects under various conditions of domination and subordination. The acknowledgment of this consciousness and praxis, this thought and action, carves out the space wherein hegemonic feminism may become aligned with different spheres of theoretical and practical activity which are also concerned with issues of marginality. Moreover, differential consciousness makes more clearly visible the equal rights, revolutionary, supremacist, and separatist forms of oppositional consciousness, which when kaleidoscoped together comprise a new paradigm for understanding oppositional activity in general.

The praxis of U.S. third world feminism represented by the differential form of oppositional consciousness is threaded throughout the experience of social marginality. As such it is also being woven into the fabric of experiences belonging to more and more citizens who are caught in the crisis of late capitalist conditions and expressed in the cultural angst most often referred to as the postmodern dilemma. The juncture I am proposing, therefore, is extreme. It is a location wherein the praxis of U.S. third world feminism links with the aims of white feminism, studies of race, ethnicity, and marginality, and with postmodern theories of culture as they crosscut and join together in new relationships through a shared comprehension of an emerging theory and method of oppositional consciousness.

NOTES

1 This is an early version of a chapter from my book in progress on "Oppositional Consciousness in the Postmodern World." A debt of gratitude is owed the friends, teachers, and politically committed scholars who made the publication of this essay possible, especially Hayden White, Donna Haraway, James Clifford, Ronaldo Balderrama, Ruth Frankenberg, Lata Mani (who coerced me into publishing this now), Rosa Maria Villafahe-Sisolak, A. Pearl Sandoval, Mary John, Vivian Sobchak, Helene Moglan, T. de Lauretis, Audre Lorde, Traci Chapman and the Student of Color Coalition. Haraway's own commitments to social, gender, race, and class justice are embodied in the fact that she discusses and cites an earlier version of this essay in her own work. See especially her 1985 essay where she defines an oppositional postmodern consciousness grounded in multiple identities in her "A Manifesto for Cyborgs: Science, Technology, and Socialist Feminism in the 1980s," *Socialist Review*, no. 80 (March 1985). At a time when theoretical work by women of color is so frequently dismissed, Haraway's recognition and discussion of my work on oppositional consciousness have allowed it to receive wide critical visibility, as reflected in references to the manuscript that appear in the works of authors such as Sandra Harding, Nancy Hartsock, Biddy Martin, and Katherine Hayles. I am happy that my work has also received attention from Caren Kaplan, Katie King, Gloria Anzaldua, Teresa de Lauretis, Chandra Mohanty, and Yvonne Yarboro-Bejarano. Thanks also are due Fredric Jameson, who in 1979 recognized a theory of "oppositional consciousness" in my work. It was he who encouraged its further development.

This manuscript was first presented publicly at the 1981 National Women's Studies Association conference. In the ten years following, five other versions have been circulated. I could not resist the temptation to collapse sections from these earlier manuscripts here in the footnotes; any resulting awkwardness is not due to the vigilance of my editors. This essay is published now to honor the political, intellectual, and personal aspirations of Rosa Maria Villafane-Sisolak, "West Indian Princess," who died April 20, 1990. Ro's compassion, her sharp intellectual prowess and honesty, and her unwavering commitment to social justice continue to inspire, guide, and support many of us. To her, to those named here, and to all new generations of U.S. third world feminists, this work is dedicated.

2 Gayatri Spivak, "The Rani of Sirmur," in *Europe and Its Others*, ed. F. Barker, vol. 1 (Essex: University of Essex, 1985), 147.

3 Here, U.S. third world feminism represents the political alliance made during the 1960s and 1970s between a generation of U.S. feminists of color who were separated by culture, race, class, or gender identifications but united through similar responses to the experience of race oppression.

The theory and method of oppositional consciousness outlined in this essay are visible in the activities of the recent political unity variously named "U.S. third world feminist," "feminist women of color," and "womanist." This unity has coalesced across differences in race, class, language, ideology, culture, and color. These differences are painfully manifest: materially marked physiologically or in language, socially value laden, and shot through with power. They confront each feminist of color in any gathering where they serve as constant reminders of their undeniability. These constantly speaking differences stand at the crux of another, mutant unity, for this unity does not occur in the name of all "women," nor in the name of race, class, culture, or "humanity" in general. Instead, as many U.S. third world feminists have pointed out, it is unity mobilized in a location heretofore unrecognized. As Cherrie Moraga argues, this unity mobilizes "between the seemingly irreconcilable lines—class lines, politically correct lines, the daily lines we run to each other to keep difference and desire at a distance," it is between these lines "that the truth of our connection lies." This connection is a mobile unity, constantly weaving and reweaving an interaction of differences into coalition. In what follows I demonstrate how it is that inside this coalition, differences are viewed as varying survival tactics constructed in response to recognizable power dynamics. See Cherrie Moraga, "Between the Lines: On Culture, Class and Homophobia," in *This Bridge Called My Back, A Collection of Writings by Radical Women of Color*, ed. Cherrie Moraga and Gloria Anzaldua (Watertown, MA: Persephone Press, 1981), 106.

During the national conference of the Women's Studies Association in 1981, three hundred feminists of color met to agree that "it is white men who have access to the greatest amount of freedom from necessity in this culture, with women as their 'helpmates' and chattels, and people of color as their women's servants. People of color form a striated social formation which allow men of color to call upon the circuits of power which charge the category of 'male' with its privileges, leaving women of color as the final chattel, the ultimate servant in a racist and sexist class hierarchy. U.S. third world feminists seek to undo this hierarchy by reconceptualizing the notion of 'freedom' and who may inhabit its realm." See Sandoval, "The Struggle Within: A Report on the 1981 NWSA Conference," published by the Center for Third World Organizing, 1982, reprinted by Gloria Anzaldua in *Making Faces Making Soul, Haciendo Caras* (San Francisco: Spinsters/Aunt Lute, 1990), 55–71. See also "Comment on Krieger's *The Mirror Dance*," a U.S. third world feminist perspective, in *Signs* 9, no. 4 (Summer 1984): 725.

4 See Fredric Jameson's "Postmodernism, or the Cultural Logic of Late Capitalism," *New Left Review* 146 (July–August 1984). Also, note 50, this chapter.

5 Louis Althusser, "Ideology and Ideological State Apparatuses (Notes Towards an Investigation)," in *Lenin and Philosophy and Other Essays* (London: New Left Books, 1970), 123–73.

6 In another essay I have identified the forms of consciousness encouraged within subordinated classes which are resistant—but not self-consciously in political *opposition*—to the dominant order. In Althusser's terms, the repressive state apparatus and the ideological state apparatus all conspire to create subordinated forms of *resistant* consciousness that I characterize as "human," "pet," "game," and "wild." The value of each of these subject positions is measured by its proximity to the category of the most-human; each position delimits its own kinds of freedoms, privileges, and resistances. Whatever freedoms or resistances, however, their ultimate outcome can only be to support the social order as it already functions. This four-category schema stems from the work of the anthropologist Edmund Leach, who demonstrates through his examples of English and Tibeto-Burman language categories that human societies tend to organize individual identity according to perceived distance from a male self and then into relationships of exchange Leach characterizes as those of the "sister," "cousin," or "stranger." He suggests that these relationships of value and distance are replicated over and over again throughout many cultures and serve to support and further the beliefs, aims, and traditions of whatever social order is dominant. Edmund Leach, "Anthropological Aspects of Language: Animal Categories and Verbal Abuse," in *New Directions in the Study of Language*, ed. Eric Lenneberg (Cambridge: MIT, 1964), 62.

7 Althusser, "Ideology," 147.

8 Francis Beal, "Double Jeopardy: To Be Black and Female," in *Sisterhood Is Powerful: An Anthology of Writings from the Women's Liberation Movement*, ed. Robin Morgan (New York: Random House, 1970), 136.

9 Sojourner Truth, "Ain't I a Woman?" in The *Norton Anthology of Literature by Women* (New York: Norton, 1985), 252.

10 Paula Gunn Allen, "Some Like Indians Endure," in *Living the Spirit* (New York: St. Martin's Press, 1987), 9.

11 Toni Morrison, in Bettye J. Parker, "Complexity: Toni Morrison's Women—an Interview Essay," in *Sturdy Black Bridges: Visions of Black Women in Literature*, ed. Roseanne Bell, Bettye Parker, and Beverly Guy-Sheftall (New York: Anchor/Doubleday, 1979).

12 Velia Hancock, "La Chicana, Chicano Movement and Women's Liberation," *Chicano Studies Newsletter* (February–March 1971).

The sense that people of color occupy an "in-between/outsider" status is a frequent theme among third world liberationists who write both in and outside of the United States. Rev. Desmond Mpilo Tutu, on receiving the Nobel Prize, said he faces a "rough passage" as intermediary between ideological factions, for he has long considered himself "detribalized." Rosa Maria Villafane-Sisolak, a West Indian from the Island of St. Croix, has written: "I am from an island whose history is steeped in the abuses of Western imperialism, whose people still suffer the deformities caused by Euro-American colonialism, old and new. Unlike many third world liberationists, however, I cannot claim to be descendent of any particular strain, noble or ignoble. I am, however, 'purely bred,'—descendent of all the parties involved in that cataclysmic epoch. I . . . despair, for the various parts of me cry out for retribution at having been brutally uprooted and transplanted to fulfill the profit-cy of 'white' righteousness and dominance. My soul moans that part of me that was destroyed by that callous righteousness. My heart weeps for that part of me that was the instrument . . . the gun, the whip, the book. My mind echoes with the screams of disruption, desecration, destruction." Alice Walker, in a controversial letter to an African-American friend, told him she believes that "we are the African and the trader. We are the Indian and the Settler. We are oppressor and oppressed . . . we are the mestizos of North America. We are black, yes, but we are 'white,' too, and we are red. To attempt to function as only one, when you are really two or three, leads, I believe, to psychic illness: 'white' people have shown us the madness of that." And Gloria Anzaldua, "You say my name is Ambivalence: Not so. Only your labels split me." Desmond Tutu, as reported by Richard N. Osting, "Searching for New Worlds," *Time Magazine*, October 29, 1984. Rosa Maria Villafane-Sisolak, from a 1983 journal entry cited in *Haciendo Caras, Making Face Making Soul*, ed. Gloria Anzaldua; Alice Walker, "In the Closet of the Soul: A Letter to an African-American Friend," *Ms. Magazine* 15 (November 1986): 32–5; Gloria Anzaldua, "La Prieta," *This Bridge Called My Back: A Collection of Writings by Radical Women of Color* (Watertown, MA: Persephone Press, 1981), 198–209.

13 Gloria Hull, Patricia Bell Scott, and Barbara Smith, *All the Women Are White, All the Blacks Are Men, But Some of Us Are Brave: Black Women's Studies* (New York: Feminist Press, 1982).
14 Cherrie Moraga and Gloria Anzaldua, *This Bridge Called My Back: A Collection of Writings by Radical Women of Color* (Watertown, MA: Persephone Press, 1981).
15 bell hooks, *Ain't I a Woman: Black Women and Feminism* (Boston: South End Press, 1981).
16 Amy Ling, *Between Worlds* (New York: Pergamon Press, 1990).
17 Norma Alarcon, ed., *The Third Woman* (Bloomington, IN: Third Woman Press, 1981).
18 See Alice Walker, "Letter to an Afro-American Friend," *Ms. Magazine*, 1986. Also Gloria Anzaldua, *Borderlands, La Frontera: The New Mestiza* (San Francisco: Spinsters/Aunt Lute, 1987).
19 Maxine Hong Kingston, *The Woman Warrior* (New York: Vintage Books, 1977); Cherrie Moraga and Gloria Anzaldua, *The Bridge Called My Back: A Collection of Writings by Radical Women of Color.*
20 Audre Lorde, *Sister Outsider* (New York: The Crossing Press, 1984).
21 Maxine Baca Zinn, Lynn Weber Cannon, Elizabeth Higginbotham, and Bonnie Thornton Dill, "The Costs of Exclusionary Practices in Women's Studies," in *Signs: Journal of Women in Culture and Society* 11, no. 2 (Winter 1986): 296.
22 Alison Jaggar, "Feminist Politics and Human Nature," uncorrected proof (New York: Rowman and Allanheld, 1983), 11.
23 Hester Eisenstein, *The Future of Difference* (New Brunswick, NJ: Rutgers University Press, 1985), xxi.
24 Elaine Showalter, ed., *The New Feminist Criticism: Essays on Women, Literature and Theory* (New York: Pantheon Books, 1985). See especially the following essays: "Introduction: The Feminist Critical Revolution," "Toward a Feminist Poetics," and "Feminist Criticism in the Wilderness," 3–18, 125–43, and 243–70.
25 Gayle Greene and Copelia Kahn, eds., *Making a Difference: Feminist Literary Criticism* (New York: Methuen, 1985). See their chapter "Feminist Scholarship and the Social Construction of Woman," 1–36.
26 Showalter, *The New Feminist Criticism*, 128.
27 Eisenstein, *The Future of Difference*, xvi.
28 Gayle Greene and Copelia Kahn, eds., *Making a Difference: Feminist Literary Criticism* (New York: Methuen, 1985), 13.
29 Jaggar, *Feminist Politics*, 37.
30 Showalter, *The New Feminist Criticism*, 138.
31 Eisenstein, *The Future of Difference*, xviii.
32 Greene and Kahn, *Making a Difference*, 13.
33 Jaggar, *Feminist Politics*, 52.
34 Showalter, *The New Feminist Criticism*, 139.
35 Eisenstein, *The Future of Difference*, xviii.
36 Ibid., xix.
37 Greene and Kahn, *Making a Difference*, 14.
38 Jaggar, *Feminist Politics*, 88.

Like U.S. hegemonic feminism, European feminism replicates this same basic structure of feminist consciousness. For example, Toril Moi and Julia Kristeva argue that feminism has produced "three main strategies" for constructing identity and oppositional politics. They represent feminist consciousness as a hierarchically organized historical and political struggle which they schematically summarize like this:

1 Women demand equal access to the symbolic order. Liberal feminism. Equality.
2 Women reject the male symbolic order in the name of difference. Radical feminism. Femininity extolled.

3 (This is Kristeva's own position.) Women reject the dichotomy between masculine and feminine as metaphysical.

Toril Moi, *Sexual/Textual Politics: Feminist Literary Theory* (New York: Methuen, 1985), 12. Note that the second category here combines the second and third categories of U.S. feminism, and the third category dissolves "the dichotomy between masculine and feminine" altogether. Luce Irigaray is considered a "radical feminist" according to this schema.

39 Lydia Sargent, *Women and Revolution: A Discussion of the Unhappy Marriage of Marxism and Feminism* (Boston: South End Press, 1981), xx.
Indeed we can see how these "protests" pressed hegemonic feminist theory into recentering from its one "phase" to the next. This hegemonic typology of feminist consciousness, that women are the same as men, that women are different from men, and that women are superior, was challenged at its every level by U.S. third world feminists. If women were seen as the same as men—differing only in form, not in content—then feminists of color challenged white women for striving to represent themselves as versions of the male, and especially of the dominant version of the "successful" white male. When the class of women *recognized and claimed* their differences from men, then, as feminists of color pointed out, these differences were understood, valued, and ranked according to the codes and values of the dominant class, race, culture, and female gender. The response to this challenge is the third phase, which sees any feminist expression as as valid as any other as long as it is an expression of a higher moral and spiritual position, that of "woman." But U.S. feminists of color did not feel at ease with the essence of "woman" that was being formulated. If ethical and political leadership should arise only from that particular location, then for U.S. feminists of color, who did not see themselves easily identifying with any legitimized form of female subject, Sojourner Truth's lingering question "Ain't I a woman?" rang all the more loudly. This schema of forms does not provide the opportunity to recognize the existence of another kind of woman—to imagine another, aberrant form of feminism. We could go so far as to say that each hegemonic feminist expression generates equivalent forms of racist ideology.

40 Eisenstein, *The Future of Difference*, xix (emphasis mine).

41 Jaggar, *Feminist Politics*, 9.

42 Cora Kaplan, "Pandora's Box: Subjectivity, Class and Sexuality in Socialist Feminist Criticism," in *Making a Difference, Feminist Literary Criticism*, ed. Gayle Greene and Copelia Kahn (New York: Methuen, 1985), 148–51.

43 Jaggar, *Feminist Politics*, 123.

44 Ibid., 11.

45 Ibid.

46 Sheila Radford-Hall, "Considering Feminism as a Model for Social Change," in *Feminist Studies/Critical Studies*, ed. Teresa de Lauretis (Bloomington: Indiana University Press, 1986), 160.

47 bell hooks, *Feminist Theory from Margin to Center* (Boston: South End Press, 1984), 9.

48 Gayatri Chakravorty Spivak, "Three Women's Texts and a Critique of Imperialism," *Critical Inquiry* 12 (Autumn 1985): 243–61.

49 In another essay I characterize such legitimated idioms of subordination as "human," "pet," "game," and "wild."

50 The connection between feminist theory and decolonial discourse studies occurs within a contested space claimed but only superficially colonized by first world theorists of the term "postmodernism." Within this zone, it is generally agreed that Western culture has undergone a cultural mutation unique to what Frederic Jameson calls "the cultural logic of late capital." There is, however, profound *disagreement* over whether the new cultural dominant should be opposed or welcomed. Jameson's essay on postmodernism, for example, is a warning which points out how the new cultural dominant creates a citizen who is incapable of any real oppositional activity, for all novelty, including opposition, is welcomed by its order. Forms of oppositional consciousness, he argues, the "critical distance" available to the unitary subjectivities of a Van Gogh or a Picasso under previous modernist conditions, are no longer available to a postmodern subject. The critical distance by which a unitary subjectivity could separate itself from the culture it lived within, and which made parodic aesthetic expression possible, has become erased, replaced by an "exhiliratory" superficial affect, "schizophrenic" in function, which turns all aesthetic representations into only other examples of the plethora of difference available under advanced capital social formations. Given these conditions, Jameson can only see the first world citizen as a tragic subject whose only hope is to develop a new form of opposition capable of confronting the new cultural conditions of postmodernism. For Jameson, however, the catch is this: "There may be historical situations in which it is not possible at all to break through the net of ideological constructs" that make us subjects in culture and this is "our situation in the current crises." Jameson's own attempt to propose a new

form of "cognitive mapping" capable of negotiating postmodern cultural dynamics dissipates under the weight of his hopelessness, and, in my view, his essay coalesces into a eulogy to passing modes of Western consciousness.

What Jameson's essay does not take into account, however, is the legacy of decolonial discourse which is also permeating the cultural moment first world subjects now inherit. In the intersections between the critical study of decolonial discourse and feminist theory is a form of consciousness in opposition once only necessary to the socially marginalized citizen, but which postmodern cultural dynamics now make available to all first world citizens. The content of this form of oppositional consciousness is rather naively celebrated and welcomed by other (primarily white, male) first world theorists of postmodernism. But whether welcoming or rejecting the variously construed meanings of the new cultural dominant, both camps share the longing for a regenerated hope and new identity capable of negotiating the crumbling traditions, values, and cultural institutions of the West; in the first example by celebrating a passing modernist form of unitary subjectivity, in the second by celebrating an identity form whose contours are comparable to the fragmenting status of present Western cultural forms.

Interesting to certain third world scholars is the coalescing relationship between these theories of postmodernism (especially between those which celebrate the fragmentations of consciousness postmodernism demands) and the form of differential oppositional consciousness which has been most clearly articulated by the marginalized and which I am outlining here. The juncture I am analyzing in this essay is that which connects the disoriented first world subject, who longs for the postmodern cultural aesthetic as a key to a new sense of identity and redemption, and the form of differential oppositional consciousness developed by subordinated and marginalized Western or colonized subjects, who have been forced to experience the aesthetics of a "postmodernism" as a requisite for survival. It is this constituency who are most familiar with what citizenship in this realm requires and makes possible.

The juncture between all of these interests is comprised of the differential form of oppositional consciousness which postmodern cultural conditions are making available to all of its citizenry in an historically unique democratization of oppression which crosses class, race, and gender identifications. Its practice contains the possibility for the emergence of a new historical moment—a new citizen—and a new arena for unity between peoples. See Jameson, "Postmodernism," 53–92.

51 Gloria Anzaldua writes that she lives "between and among" cultures in "La Prieta," *This Bridge Called My Back*, 209.

52 Differential consciousness functioning like a "car clutch" is a metaphor suggested by Yves Labissiere in a personal conversation.

53 Aida Hurtado, "Reflections on White Feminism: A Perspective from a Woman of Color" (1985), 25, from an unpublished manuscript. Another version of this quotation appears in Hurtado's essay, "Relating to Privilege: Seduction and Rejection in the Subordination of White Women and Women of Color," in *Signs* (Summer 1989): 833–55.

54 Moraga and Anzaldua, xix. Also see the beautiful passage from Margaret Walker's *Jubilee* which enacts this mobile mode of consciousness from the viewpoint of the female protagonist. See the Bantam Books edition (New York, 1985), 404–7.

55 Gloria Anzaldua, "La Prieta," *This Bridge Called My Back*, 209.

56 Audre Lorde, "Comments at the Personal and the Political Panel," Second Sex Conference, New York, September 1979. Published in *This Bridge Called My Back*, 98. Also see "The Uses of the Erotic" in *Sister Outsider*, 58–63, which calls for challenging and undoing authority in order to enter a Utopian realm only accessible through a processual form of consciousness which she names the "erotic."

57 Anzaldua refers to this survival skill as "la facultad, the capacity to see in surface phenomena the meaning of deeper realities" in *Borderlands, La Frontera*, 38.

The consciousness which typifies la facultad is not naive to the moves of power: it is constantly surveying and negotiating its moves. Often dismissed as "intuition," this kind of "perceptiveness," "sensitivity," consciousness if you will, is not determined by race, sex, or any other genetic status, neither does its activity belong solely to the "proletariat," the "feminist," nor to the oppressed, if the oppressed is considered a unitary category, but it is a learned emotional and intellectual skill which is developed amidst hegemonic powers. It is the recognization of "la facultad" which moves Lorde to say that it is marginality, "whatever its nature . . . which is also the source of our greatest strength," for the cultivation of la facultad creates the opportunity for a particularly effective form of opposition to the dominant culture within which it is formed. The skills required by la facultad are capable of disrupting the dominations and subordinations that scar U.S. culture. But it is not enough to utilize them on an individual and situational basis. Through an ethical and political commitment, U.S. third world feminism requires the development of la facultad to a methodological level capable of generating a political strategy and identity politics from which a new citizenry arises.

Movements of resistance have always relied upon the ability to read below the surfaces—a way of mobilizing—to resee reality and call it by different names. This form of la facultad inspires new visions and strategies for action. But there is always the danger that even the most revolutionary of readings can become bankrupt as a form of resistance when it becomes reified, unchanging. The tendency of la facultad to end in frozen, privileged "readings" is the most divisive dynamic inside of any liberation movement. In order for this survival skill to provide the basis for a differential and unifying methodology, it must be remembered that la facultad is a process. Answers located may be only temporarily effective, so that wedded to the process of la facultad is a flexibility that continually woos change.

58 Maria Lugones, "Playfulness, World-Traveling, and Loving Perception," from *Hypatia: A Journal of Feminist Philosophy* 2, no. 2 (1987).

Differential consciousness is comprised of seeming contradictions and difference, which then serve as tactical interventions in the other mobility that is power. Entrance into the realm "between and amongst" the others demands a mode of consciousness once relegated to the province of intuition and psychic phenomena, but which now must be recognized as a specific practice. I define differential consciousness as a kind of anarchic activity (but with method), a form of ideological guerrilla warfare, and a new kind of ethical activity which is being privileged here as the way in which opposition to oppressive authorities is achieved in a highly technologized and disciplinized society. Inside this realm resides the only possible grounds of unity across differences. Entrance into this new order requires an emotional commitment within which one experiences the violent shattering of the unitary sense of self, as the skill which allows a mobile identity to form takes hold. As Bernice Reagon has written, "most of the time you feel threatened to the core and if you don't, you're not really doing no coalescing." Citizenship in this political realm is comprised of strategy and risk. Within the realm of differential consciousness there are no ultimate answers, no terminal Utopia (though the imagination of Utopias can motivate its tactics), no predictable final outcomes. Its practice is not biologically determined, restricted to any class or group, nor must it become static. The fact that it is a process capable of freezing into a repressive order—or of disintegrating into relativism—should not shadow its radical activity.

To name the theory and method made possible by the recognition of differential consciousness "oppositional" refers only to the ideological effects its activity can have under present cultural conditions. It is a naming which signifies a realm with constantly shifting boundaries which serve to delimit, for differential consciousness participates in its own dissolution even as it is in action. Differential consciousness under postmodern conditions is not possible without the creation of another ethics, a new morality, which will bring about a new subject of history. Movement into this realm is heralded by the claims of U.S. third world feminists, a movement which makes manifest the possibility of ideological warfare in the form of a theory and method, a praxis of oppositional consciousness. But to think of the activities of U.S. third world feminism thus is only a metaphorical avenue which allows one conceptual access to the threshold of this other realm, a realm accessible to all people.

59 Barbara Christian, "Creating a Universal Literature: Afro-American Women Writers," *KPFA Folio*, Special African History Month Edition, February 1983, front page. Reissued in *Black Feminist Criticism: Perspectives on Black Women Writers* (New York: Pergamon Press, 1985), 163.

60 Alice Walker coined the neologism "womanist" as one of many attempts by feminists of color to find a name which could signal their commitment to egalitarian social relations, a commitment which the name "feminism" had seemingly betrayed. See Walker, *In Search of Our Mother's Gardens: Womanist Prose* (New York: Harcourt Brace Jovanovich, 1983), xi–xiii. Anzaldua, *Borderlands, La Nueva Frontera*.

61 bell hooks, "Feminist Theory: From Margin to Center," 9; Audre Lorde, "An Interview: Audre Lorde and Adrienne Rich," held in August 1979, *Signs* 6, no. 4 (Summer 1981); and Barbara Smith, *Home Girls: A Black Feminist Anthology*, xxv.

62 Merle Woo, *This Bridge Called My Back*, 147.

21

An ethics of the local

J. K. Gibson-Graham

from *Rethinking Marxism*, 2003, 15.1 (January): 49–74

PREAMBLE

Imagine that something is happening in the world that's not about an actual, measurable phenomenon called "globalization." I am thinking here of the ascendancy of what might be called the "global imaginary" and its implications for how we feel, act, and identify. Perhaps a global régime is consolidating itself not so much through institutional initiatives but through subjects who experience themselves as increasingly subsumed to a global order—enter here the world economic system, known also as the market, or neoliberalism, or capitalism. Becoming part of the imagined global community involves our subjection to this order, our (re)constitution not primarily as national citizens but as economic subjects—productive or less so, competitive or not, winning or losing on the economic terrain.

It's not an emotionally neutral process. As the nation loses its simple and secular primacy, our familiar social container erodes—its walls become permeable, its stitching unravels. Inevitably, we are exposed. The government that once protected us from the world economy now hurls us up against it. Its rhetoric of competitiveness draws on the self-centeredness of community while abjuring its progressive and ethical force.

But there is more than one aspect to this experience. When we are laid open to global forces, we confront ourselves differently. As the nation loosens its hold on us, we encounter new possibilities of community. In this moment it is possible to ask what is possible —besides economic victimhood and social incivility. Can we find other ways to be? Can we be other than what globalization makes of us? These questions are challenging ones that ask for daily practices of learning to live differently. I hear them as a call for an "ethics of the local."

GLOBAL/LOCAL

Globalization discourse situates the local (and thus all of us) in a place of subordination, as "the other within" of the global order. At worst, it makes victims of localities and robs them of economic agency and self-determination. Yet in doing so globalization suggests its own antidote, particularly with respect to the economy: imagine what it would mean, and how unsettling it would be to all that is now in place, if the locality were to become the active subject of its economic experience.

In the discursive context of globalization, attempts to restore identity and capacity to localities assume moral force and political priority. But such attempts cannot succeed if the local is *necessarily* confined and constrained by the global. A less obvious, less predictable, less binary relation must obtain if we are to know the local as a space of freedom and capacity. *The impossibility of a global order* must be affirmed as a truth and reaffirmed as a truism. If we can accept that it is impossible to subsume every individual being, place, and practice to a universal law, whether it be the law of the father, or the market, or a geopolitical formation, then it will follow that the local cannot be fully interior to the global, nor can its inventive potential be captured by any singular imagining.

Impossible though a global order may be, there are afoot in the world today concerted efforts to produce global integration: the World Trade Organization, the Multilateral Agreement on Investment, IMF structural adjustment plans tethering individual societies to a global capitalist economy (and constructing the latter in the process). Critics have pointed to the violence inherent in such projects and the manifold erasures and suppressions that are enacted in their pursuit. In Seattle, demonstrators against the World Trade Organization became advocates for the peoples and practices that are violated when a global (economic) regime is imposed. They might also be seen as practitioners of an ethics of the local. Such an ethics is grounded in the necessary failure of a global order, which is the negative condition of an affirmation of locality.

A local ethic proffers respect, not just for difference and autonomy but for self understood as capability. Yet this is only a part of the story. In volume 2 of the *History of Sexuality*, Foucault distinguishes the two elements of every morality. The first element is the code, or the principles. But the second and often more important element is the cultivation of the ethical person. According to Foucault, the "relationship with the self . . . is not simply 'self-awareness' but self-formation as an 'ethical subject'" (1985, 28) and there is "no forming of the ethical subject without 'modes of subjectivation' and . . . 'practices of the self that support them" (1985, 28). In the story that follows I adopt Foucault's conception of morality as a template and a guide. I begin with simple principles, familiar to all, and then trace a complex, idiosyncratic, and highly social process of (re)subjectivation—involving practices of forming the ethical local subject that I have used in my research. The first moment yields clean and under-specified abstractions, without which we could not begin to orient ourselves, while the second embroils us in the dirt and danger of location, interpersonal engagement, and the labors of becoming.

FIRST, PRINCIPLES

The task of convening principles for a local ethics is to some extent a negative one. It involves countering not globalization itself (involving interchange between spatially separated processes or entities) but the meanings of globalization that come to bear on social possibility.[1] For the global is not merely a geographical scale that subsumes and subordinates the local; it has become a sign as well for universality and sameness/unity.[2] In this light, the preoccupations of recent social theory, where any number of thinkers enjoin us to *recognize particularity and contingency, honor difference and otherness*, and *cultivate local capacity*, can be read as appropriate guidelines for an ethics of locality. These three familiar injunctions (constituting almost a postmodern social mantra) gain force from what they are posed against. Each affirms a subordinate term, each values what globalization discourse (in some of its forms) threatens to endanger, each redresses an imbalance of emphasis in triumphalist accounts of globalization. As principled abstractions, they have not only the deconstructive energy to unsettle global certainties but the instrumental potential to transform local subjects through inculcatory practices. And despite their familiarity, they have not usually been treated as codified norms for a practical ethics.[3] In this sense, they have not yet reached their potential for performative efficacy.

(1) *Recognize particularity and contingency*. This principle establishes parity between global and local, existence and possibility. It bids us acknowledge that the global universal is a projection, on a world scale, of a local particularity. Thus, "development" is the historical experience of capitalist industrialization in a few regions that has become a description of a universal trajectory and a prescription for economic and social intervention in all the world's nations. Similarly, "neoliberalism" is an approach to economic regulation that emerges from a single economic tradition, presenting a *particular* understanding of the economy, presuming a *particular* economic subject, and focusing on enhancing *particular* types of economic practices—capitalist market practices, to be precise. As a hegemonic particularity, it has set the global regulatory agenda for the past decade or more, obscuring and often destroying local economic practices devalued as traditional or parochial, or invisible as nonmarket and noncapitalist. "Human rights"—again, emerging from a locality, that home of a small portion of humankind called the West. Now threatening to install itself as a universal discourse of liberation, obliterating other notions of justice and violating other visions of society and humanity.[4] The list goes on, or could.

But recognizing particularity (in all the "universals" that have migrated imperially from local to global scale) entails another cognizant move—the recognition of contingency. The universal/global is not only particular/local in its origins but is subject to the movements of history. It has been installed (perhaps by force) and can therefore be removed. "Things could be otherwise" is the positive implication of contingency and the sign of political possibility. What might a politics of the "otherwise" be? How might a "local" politics participate in constructing different universals and new communities?[5] Here we could examine the contingent economy for unexpected political subjects and opportunities.

Under the mantle of contingency, the economy loses its status as logical essence and foundational instance of globalization (Madra and Amariglio 2000). Stripped of inevitability, it becomes a domain of potentiality and a space for the unfolding of creative engagements. We find an enlarged political field where economic necessity once reigned and a range of options where narrow economic dictates once held sway.

(2) *Respect difference and otherness*, between localities but also within them. This principle affirms that locality need not be a parochial enclave but can be instead a place where we exercise our responsibility to the Other, understood as unassimilable, as absolute alterity: "To maintain an ethical bond with the Other . . . is to see the self in relation to something 'it cannot absorb' . . . the Other must remain a stranger 'who disturbs the being at home with oneself'" (Shapiro 1999, 63–5, quoting Levinas). Locality is the place where engagement with the stranger is enacted. In the words of Jean-Luc Nancy (1991), it is the place of exposure, of one to another singularity. It is also the crossroads where those who have nothing in common (all of us) meet to construct community (Lingis 1994).

Resonating with the principle of respect is Esteva and Prakash's call for a radical pluralism, in which the discourse of "human rights" is brought down from its pedestal and placed "amidst other significant cultural concepts that define 'the good life' in a pluriverse" (1998, 119). Human rights advocates are being asked not to withdraw from discussions of local justice, but instead to participate dialogically and generously, with "the openness to be hospitable to the otherness of the other" (1998, 128).

The discourse of globalization, with its overt or implicit celebration of capitalist dominance, prompts the question of what *respect for difference and otherness* might mean for the economy. What if we were to call for recognition of economic diversity? What if we were to offer full and free acknowledgment to economic subjects and practices that are not, or cannot be, subsumed to capital? What would a language of economic difference be, and what kind of practices would it usher into visibility?[6]

In *The End of Capitalism*, J. K. Gibson-Graham (1996) observes that, although difference has become an important and even central value in many dimensions of social existence, in the economic dimension we are still prisoners of the "same." Capitalism is the name of the economy of sameness; if noncapitalist forms of economy are seen to exist at all, they are understood as subordinated to or contained by capitalism.

But another story could be told, concocted from the writings of feminist economic theorists, or from economic anthropology, or from theories and chronicles of the informal economy. This rich narrative of a highly differentiated economy could undermine the capitalocentric imaginary; and it could also function as part of the imaginative infrastructure for cultivating alternative economic subjects and practices (Gibson-Graham *et al.* 2000, 2001). This brings us to the final principle, which reminds us that as local subjects we need to:

(3) *Cultivate capacity*. Here I am thinking very generally of the capacity to modify ourselves, to become different, and more specifically of the capacity to enact a new relation to the economy. In the discourse of globalization, the economy is something that does things to us and dictates our contours of possibility. It is not the product of our performance and creativity. Globalization discourse represents localities as economically dependent, not so much actors as acted upon, receiving the effects of economic forces as though they were inevitable. In the face of this representation, the urgent ethical and political project involves radically repositioning the local subject with respect to the economy.

Globalization discourse constructs its subjects as "citizens" of capitalism: they are entrepreneurs, or employees, or would-be employees; they are investors in capitalist firms; they are consumers of (capitalist) commodities. Given the impoverished field of economic possibility, the ethical practice of subject formation requires cultivating our capacities to imagine, desire, and practice noncapitalist ways to be. An ethics of the local would undermine ideas of individual self-sufficiency, fostering the affective acknowledgment of interdependence as a basis for some sort of "communism." It would produce citizens of the diverse economy.

CULTIVATING THE ETHICAL SUBJECT: THE POLITICS OF RESEARCH

I want to turn now to thinking about how we as local subjects might cultivate ourselves in accordance with the principles of a local ethics, and to describe as a vehicle for that cultivation process a multicontinental program of research that is attempting to create social and discursive spaces in which ethical practices of self-formation can occur. In introducing that research program, I invoke the term "politics"—because I see these practices of resubjectivation or making ourselves anew as ultimately (if not simply) political (Connolly 1999).[7]

The research projects I will describe are focused on transforming ourselves as local economic *subjects*, who are acted upon and subsumed by the global economy, into subjects with economic *capacities*, who enact and create a diverse economy through daily practices both habitual (and thus unconscious) and consciously intentional. But these practices of self-transformation rely on an initial and somewhat difficult move. If we are to cultivate a new range of *capacities* in the domain of economy, we need first to be able to see noncapitalist activities and subjects (including ones we admire) as visible and viable in the economic terrain. This involves supplanting representations of economic sameness and replication with images of economic difference and diversification.

Feminist economic theorists have bolstered our confidence that such a re-presentation is both possible and productive. Based on a variety of empirical undertakings, they argue that the noncommodity sector (in which unpaid labor produces goods and services for nonmarket circulation) accounts for 30–50 percent of total output in both rich and poor countries (Ironmonger 1996). According to the familiar definition of capitalism as a type of commodity production, this means that a large portion of social wealth is noncapitalist in origin. And even the commodity sector is not necessarily capitalist—commodities are just goods and services produced for a market. Slaves in the antebellum U.S. South produced cotton and other commodities, and in the contemporary United States, worker-owned collectives, self-employed people, and slaves in the prison industry all produce goods and services for the market, but not under capitalist relations of production.[8] Arguably, then, less than half the total product of the U.S. economy is produced under capitalism. From this perspective, referring to the U.S. or any economy as capitalist is a violent act of naming that erases from view the heterogeneous complexity of the economy.

Working against this process of erasure, our research is trying to produce a discourse of economic difference as a contribution to the ethical and political practice of cultivating a diverse economy. In projects under way in Australia, Asia, the Pacific, and the United States, we are attempting to generate and circulate an alternative language of economy, one in which capitalism is not the master signifier, the dominant or only identity in economic space. This eclectic language, emerging from conversations both academic and popular, provides the conceptual infrastructure for representing economic subjects and multiplying economic identities (Gibson-Graham 2002).

Two of our projects have moved beyond the planning and early implementation phase and are beginning to reveal their specificity as ethical practices and political experiments.[9] One is based in the Latrobe Valley in southeastern Australia (Cameron and Gibson 2001). The other is under way in the Pioneer Valley of Massachusetts, the region that stretches north–south along the Connecticut River in the northeastern United States (Community Economies Collective 2001). While the Latrobe Valley is a single-industry region (based on mining and power generation) with a recent history of downsizing and privatization, the Pioneer Valley mixes agriculture, higher education, and recognized economic alternatives, supplementing this unusual mixture with a small manufacturing sector that is suffering the lingering effects of deindustrialization. In both these regions, globalization sets the economic agenda—we are all being asked to become better subjects of capitalist development (though the path to such a becoming does not readily present itself) and to subsume ourselves more thoroughly to the global economy.

The two research projects provide a social context for Foucault's second moment of morality—cultivating the ethical subject—which involves working on our local/regional selves to become something other than what the global economy wants us to be. But what actual processes or techniques of self (and other) invention do we have at our disposal? Foucault is not forthcoming here, at the microlevel of actual practices. And when we embarked on these projects we did not imagine how difficult the process of resubjectivation would be. In both the United States and Australia, for example, we have come up against the patent lack of desire for economic difference in the regions where we are working. We have encountered instead the fixation of desires upon capitalism: individuals want employment as wageworkers; policymakers want conventional economic development. It was only after months of resistance, setbacks, and surprising successes that we could see the deeply etched contours of existing subjectivities and the complexity of the task of "re-subjecting" we were attempting to engage in. Invaluable in helping us to conceptualize and negotiate this complexity was the work of William Connolly. Whereas we had stumbled through the process of cultivating alternative economic subjects, Connolly's work on self-artistry and micropolitics allowed us retrospectively to see steps and stages, techniques and strategies.

Connolly is concerned with the subject as a being that is already shaped and as one that is always (and sometimes deliberately) becoming. In his view, active self-transformation—working on oneself in the way that Foucault has described—functions as a micropolitical process that makes macropolitical settlements possible. If we are to succeed in promoting a diverse economy and producing new subjects and practices of economic development, there must be selves who are receptive to such an economy and to transforming themselves within it. How do we nurture the micropolitical receptivity of subjects to new becomings, both of themselves and of their economies?

Micropolitics can be understood as an "assemblage of techniques and disciplines that impinge on the lower registers of sensibility and judgment without necessarily or immediately engaging the conscious intellect" (Connolly 2001, 33). One object of such a politics is what Connolly calls the "visceral" domain where "thought-imbued intensities below the reach of feeling" (1999, 148) dispose the individual in particular ways, with a seldom acknowledged impact on macropolitical interactions. In a discussion of the public sphere, where he argues that the visceral register cannot be excluded from public discourse and the process of coming to public consensus, Connolly puts forward a set of norms for discourse across differences (2001, 35–6). Instead of attempting to tame or exclude the body, reducing public discourse to rational argument, he advocates developing an appreciation of "*positive possibilities in the visceral register of thinking and discourse*" as a way of beginning to creatively produce and respond to the emergence of new identities. This appreciation of positive possibilities in the body, he suggests, might be supplemented by an "*ethic of cultivation*" that works against the bodily feelings of panic experienced when naturalized identities are called into question. And rather than expecting people to transcend their differences in order to be or behave like a community, he suggests the possibility of a "generous *ethos of engagement*" between constituencies in which differences are honored and bonds are forged around and upon them. All these attitudes and practices could make possible *ethically sensitive, negotiated settlements* between potentially antagonistic groups and individuals in the construction of communities.

We are drawn to Connolly's italicized arsenal of stances and strategies because they take into account the stubborn, unspoken bodily resistances that stand in the way of individual becoming and social possibility; and at the same time they acknowledge the visceral register of discourse as a positive resource for social creativity. For us, retrospectively, they offer a "cultivator's manual" for the ethical practice of cultivating different local economic subjects—subjects of capacity rather than debility, subjects whose range of economic identifications exceeds the capitalist order. Though Connolly did not intend them this way, for us they have become a way of organizing our narrative of local resubjectivation in the Latrobe and Pioneer Valleys.

FINDING POSITIVE POSSIBILITIES IN THE VISCERAL REGISTER: OPENINGS TO THE DIVERSE ECONOMY

The Economy haunts and constrains us as social beings—we find our life pathways and visions of social possibility hemmed and hampered by its singular capitalist identity. Intellectually, and in our bodily dispositions, we encounter daily a higher economic power, now burgeoning laterally as the "global economy." For local subjects, and for all of us as subjects of economic discourse with its relentless realism and drumbeat repetitiveness, it is not easy to access the possibilities that lie outside dominant narratives and images of Economy.

In beginning to construct the diverse economy as a set of possibilities for economic subjects, in both the Latrobe and Pioneer Valleys we started with the familiar capitalist economy that was seen to hold hostage the economic and social fate of each region.

Early on, we held focus groups that attempted to access the local countenance of Economy and begin to shift it from center stage or at the very least to create an opening for such a shift. The first focus groups were held in 1997 with business and community leaders in the Latrobe Valley (Gibson *et al.* 1999). When we asked them to talk about the social and economic changes that had occurred in the Valley over the last decade, the participants produced relatively uniform and well-rehearsed stories centered upon dynamics in the formal economy, especially the privatization of the State Electricity Commission (SEC) in the face of state debt and the pressures of globalization. Words like "victimization," "disappointment," "pawns," and "powerless" anchored these narratives in a sea of negativity and the moods of the speakers ranged from energetic anger to depressed resignation.

But when they were later asked to consider the strengths of the region and the capacities of the community to cope with change, an unmatched set of stories emerged, conveyed in that halting manner of speech that accompanies cognitive activity. Participants spoke of artistic ingenuity and enterprise, of contributions made by migrants from non-English-speaking backgrounds and intellectually challenged residents, of the potential to revalue unemployed people as a regional asset.[10] The knowledgeable and authoritative "voice" associated with discussions of downsizing and restructuring gave way to a more speculative and tentative tone. Moods began to lighten, and expressions of surprise and curiosity displaced dour agreement.

Though the stories were initially slow in coming, one example sparked another and soon they were tumbling out over and around each other. For us, these stories began to map the contours of a relatively invisible diverse economy. No longer simply abandoned by capital, the region became populated by numerous examples of community-based economic alternatives that held the potential for a very different vision of regional development.

At the end of the session one participant noted the shift that had occurred in his own understanding and sense of possibility—a shift that had resulted from being placed in a different relation to the formal economic "identity" and familiar downbeat narrative of the Valley:

> The interesting thing and rather ironic is that a bureaucratic organization like the Council or like the State Government or a welfare organization might organize a panel to sit around and discuss the sorts of things that we have discussed, and . . . they probably wouldn't have achieved as much as we have achieved today. Because the information that I've gained just from hearing everybody talk . . . it's been absolutely precious. And it hasn't come about as a consequence of some bureaucracy wanting to solve problems but rather as we are pawns in another exercise [i.e. our research project]. I'm actually going away from here with more than I came with.
>
> (Local government official)

Over the course of a two-hour conversation, the participants had moved from an emotionally draining but unsurprising narrative of regional destruction at the hands of the SEC, to outbreaks of raw emotion occasioned by retelling this painful story in the sympathetic and energizing presence of witnesses/listeners, to open, even exuberant responses to our questions about counterstories and alternative activities. What we perceived as a "positive possibility in the visceral register" was the intersubjectively energized disposition to be moved, the willingness not to be attached to a single and centered narrative or set of emotions.[11]

In a similar focus group in the Pioneer Valley in 1999, planners and business and community leaders were initially asked about the strengths and weaknesses, problems and successes of the regional economy. Again, familiar stories emerged, couched within the anxiety-ridden discourse of development in which every region is found wanting (and thus in need of economic intervention). The prescription was familiar: attracting "good" jobs by recruiting major capitalist employers—via subsidies and other inducements—to locate in the region.

But the discussion took an unsettling turn, as the participants reiterated several times that a requisite of economic development was a suitably educated and acculturated labor force. (That this was something entirely outside the control of these economic development specialists may partially explain why it repeatedly bubbled up out of the ambient sea of low-level anxiety.) Several people lamented the fact that the two-earner family, whether wealthy or impoverished, left no one at home to raise the children. Where was the appropriate labor force to come from if no one was fully engaged in producing it? At one moment the labor leader in the group recounted with muted horror the story of Conyers, Georgia, an affluent suburb of Atlanta where the largest outbreak of syphilis in the United States in decades had occurred among junior high school students, some of whom had as many as fifty partners. In Conyers, it seemed, economic development had betrayed its promise of social well-being, and indeed was undermining itself as a process (Healy 2000).

The tone of confidence that prevailed in discussions of "what the regional economy needs" had faded, and the productive anxiety of competent practitioners had given way to a confused and even despairing fearfulness. This was an instant in which we glimpsed the "role that the visceral register of intersubjectivity plays in moral and political life" (Connolly 1999, 27). While the participants, drawing on a longstanding intellectual tradition and buttressed by their social roles, authoritatively asserted the sufficiency of capitalist growth to the goal of producing economic and social development, on the visceral level they experienced untamed fears of society out of control, and a tacit shared recognition of the insufficiency of the capitalist economy (no matter how developed) to the task of sustaining a community—raising its children, reproducing its sociality. Perhaps this was not so much a "positive possibilit[y] in the visceral register of thinking and discourse" as an eruption, through the smooth surface of rational interchange, of vulnerability and the hope of solace. Each of these feelings involved a disposition to openness—in the place of the explicit closures and certainties of development, we encountered an unspoken, prerepresentational acknowledgment that capitalist economic development is a dependent rather than a self-sufficient process, and that social well-being has multiple wellsprings and determinants.

The Pioneer Valley discussion highlighted (though not explicitly) the interdependency that exists between formal capitalist economic practices and the workings of a neighborhood and household-based economy (Russo 2000). The Latrobe Valley discussion pointed to the various contributions that people seen as "marginal" to the mainstream economy and alternative community enterprises usually seen as "noneconomic" make to the functioning and well-being of a region. And despite the very mainstream notion of economy prevailing in both groups, the expressions of emotional openness to different understandings gave us confidence that the participants would be able to award saliency (at least in the visceral register) to a resignification of their region in economic terms. Their openness gave fuel to our desire to flesh out, through a community inventory, a diverse economy in which capitalist enterprises, formal paid wage labor, and market transactions occupy only the visible tip of the economic iceberg. By giving a place in the diverse economy to activities that are often ignored (collective enterprises, household and voluntary labor, transactions involving barter, sharing and gift giving, etc.), we hoped to refigure the identity and capacities of the regional economy. And by recognizing the particularity of people's economic involvements, including their multiple economic identities (in addition to being unemployed with respect to capitalist employment, for example, a person can be employed in household, neighborhood, and other noncapitalist activities), we were attempting to reframe the capacities of individuals. All these strategies of re-presentation would draw upon "positive possibilities in the visceral register" and potentially also give rise to affirmative affective and political stances, if the "negative possibilities" that also reside in the viscera could be diffused or transformed along the way.

EXPLORING AN ETHIC OF CULTIVATION: OPENING TO OTHER AND ALTERNATIVE SUBJECTIVITIES

Connolly finds in the body and, more specifically, the brain some of the factors that dispose us negatively (if unconsciously) to new situations and possibilities. He talks about "thought-imbued intensities that do not in themselves take the form of either conscious feelings or representations" (1999, 28) and finds one of their bodily locations in the amygdala, a small brain at the base of the cortex. Triggered by "signs that resemble a past trauma, panic, or disturbance," the amygdala transmits fear along the pathways of the brain with considerable energy and intensity. Recognizing the barrier that such a bodily function poses to new becomings, Connolly proposes cultivating or educating the amygdala, resistant though it may be to cognitive manipulation. Since amygdalic panic arises not just out of corporeal predispositions but out of experience (of pain or disturbance), he suggests that counter-experiences issuing from experimental self-artistry and intersubjective arts might play a part in attenuating that panic. They might even create a space for the creativity that the amygdala unleashes (in Connolly's speculative imagination) through its frictional interaction with the relatively staid and reflective brains, the cortex and hippocampus (1999, 29).

When we began our work in the Latrobe Valley with those who had been marginalized by economic restructuring, we often encountered hostility and anger, anchored in a deep sense of powerlessness. Introducing our project of economic resignification seemed to reactivate the trauma of retrenchment (especially for men in their forties and fifties who have found it impossible to secure alternative employment) and to reinforce the bleak future envisioned by young unemployed people for whom the expectation of jobs in the power industry and related service sectors has been dashed. Eve, one of the community researchers hired by the project, became quite skilled at dealing with the aversive reactions that emerged when the project was introduced, and with people's initial resistance when they were asked to portray themselves in terms of "assets and capacities" rather than needs and deficiencies.[12]

> One particular [older] gentleman in a literacy class was quite obviously very frustrated and pessimistic. He was quite vocal and kept presenting me with stumbling blocks. "Look what they have done. What are they going to do about it? What's the use? No one is going to be bothered. People will want to be paid." I tried to address his issues without being confrontational. I tried to be sympathetic and understanding. We talked a bit about the problems in our community. I agreed with what he had to say. . . . It was evident that we had to almost exhaust that line of thinking before moving on.

Eve found that she had to allow anger to be spoken before any movement could take place. This was a painful process, since much of the animosity was directed toward the researchers themselves as individuals associated with powerful institutions like the university or, even worse, the municipal government that cofunded the project. As Lenni, another researcher, remarked:

> In the end Eve would say, "Don't present yourself that you come from Monash [University]." She would present herself as a single parent, and I would present myself as an unemployed person, and automatically you would have that rapport with someone, cause you're on the level that they're on. It would be until you'd say that the project is sponsored by Monash Uni and the Latrobe Shire—that's when you'd get the political stuff.

Eve's strategy was to avoid engaging with the angry energy that arose from the narrative of "our" victimization at the hands of "them."[13] She suspected that further exploration would lead back to the evils of the SEC, the conservative state government, and ultimately the Economy. And despite the promise of transformative enunciation that is said to arise from denunciation,[14] we all sensed that this exercise in debunking would be unlikely to inspire creative thoughts and desires for alternative economies. Our tactics, then, involved moving away from the narratives that triggered the amygdalic reaction and trapped local community members in fear and fury, making it nearly impossible to think about things differently.

By speaking from their own experiences as individuals in difficulty, the community researchers were able to establish a rapport with other community members built upon shared identities and, at the same time, they were able to shift the discussion away from these limiting identifications. In conversation, they attempted to elicit the multiple identities of each person. Their questions about personal gifts and capacities introduced a new fullness into the agenda. No longer was a subject defined by deficiency or restricted to the subject position of economic victim. To return to Eve's difficult conversation with the older gentleman:

> [I found out that] he is very good with his hands and knows a bit about cars. I asked, hypothetically, if there were a group of single parents interested in learning about car maintenance, and if I could arrange a venue and possible tools, would he be interested in sharing his skills and knowledge? "Yeah. I'd do that, no worries," he said. I asked him would he expect to be paid for his time. "No. I wouldn't do it for money," he replied. I asked, "So do you think you'd get anything out of it yourself?" "Yeah. I suppose I'd get some satisfaction out of it 'cause I like to help people like yourself." So I really tried to turn it around and have him answer or resolve his own questions and issues.

Where the man had felt pain and anger associated with past experiences of Economy, under Eve's patient cultivation he has moved toward pleasure and happiness associated with a different economic way to be. Through her questioning his attention has shifted away from a powerful narrative of impotence and victimization into a hopeful scenario that positions him as skillful and giving, and endowed with an economic identity within a community economy. The threat represented by our project, by Eve herself, and by the formal Economy—all recalling the historic trauma of retrenchment that prompts the angry closing off to emotional and social novelty—has been neutralized through Eve's affective tutelage. And in the place of anger at remembered pain, there is a hint of joy in abilities seen in a new light, and a generosity of spirit that surprises with its unfamiliarity. These feelings that attend creative moments of becoming challenge and ultimately displace the more securely narrativized emotions of reproach, defensiveness, blame, and resentment associated with established economic identities.

Such a movement might appear to represent a very minor shift in the macropolitical scheme of things. But it is a requisite part of the larger political process of enacting a diverse regional economy, where individuals from very different backgrounds and life circumstances must move beyond fear and hatred to interact in inventive and productive ways. Here the story of Kara, one of the community researchers in the Pioneer Valley project, is exemplary. Kara initially was highly resistant to the project's goal of bringing mainstream and marginalized economic actors into conversation and collaboration, which she viewed as just a way of subsuming the latter (and the project) to the agenda of the powerful. She saw the mainstream people, whom she encountered through a video of our focus groups, as emissaries of the State, Capitalism, and Power, from which unholy Trinity she was hoping, and indeed planning, to entirely remove herself. Throughout the weekend training for community researchers she nursed and vociferously communicated her antipathy. But at the last moment, during the evaluation of the training, she had a moment of self-evaluation and opened to the possibility of productive engagement with those she saw as (threateningly) different from her: "I don't want to be so us–them," she said, "or to live in a world that is set up that way, emotionally and politically." Suddenly the mainstream types were rehumanized, and the possibility of working with them became a micropolitical opportunity. It was as though in that very moment Kara was working on herself "to attenuate the amygdalic panic that often arises when you encounter . . . identities" (Connolly 1999, 36) that call the naturalness or sufficiency of one's own identity into question. By engaging in a "selective desanctification of elements in [her] identity" (1999, 146)—in this case, a highly charged oppositional stance with respect to power understood as domination—Kara was able to open herself generously: to the humanity of others, to the possibility of being other than she was, to participating with those most different from herself (in her own antagonistic world-view) in constructing a diverse economy.[15]

FOSTERING AN ETHOS OF ENGAGEMENT: MULTIPLE OPPORTUNITIES

As a way of building upon the inventory of skills, capacities, and dreams that emerged from the initial conversations in the Latrobe Valley, we organized a range of events that drew people together in an action situation where crazy ideas and schemes could be freely thrown around without the pressure of a formal meeting régime and the expectation of concrete outcomes. Food-based events at which people made pizza or baked scones were particularly successful in getting people to meet, overcome the stultifications of shyness, begin to listen to one another, and build and transmit excitement. The focus upon food production as an end in itself produced its own outcomes: a meal that was consumed by its producers and unwitting involvement in the practice of collectivity. Without any expectation that a group with a common goal should form, these events provided a space for a range of people from many different backgrounds to experience what Jean-Luc Nancy (1991) might call "being-in-common." Or what Alphonso Lingis (1994) describes as the "community of those who have nothing in common" except that they die (and eat).

One of the community researchers, Lenni, described a moment of understanding prompted by these gatherings:

> I guess the crunch came when Eve was doing her food events and things like that. And just the mingling and talking to people, to me that was like the breakthrough of . . . this is what it's about, it's working with other people and listening to other people and getting that opportunity to listen to their dreams and things like that.

She went on to reflect upon how she had changed over the course of the project:

> If I give myself the time, I can listen to anyone. I had only ever dealt with people over the counter [before involvement in the project]—with commercial transactions. I'm not as critical as I was. Working with people from various backgrounds and abilities—I'm more tolerant. I've learnt to see the good in people. I had always been taught to be cautious and careful of people. My dad always used to say, "the only friend you've got is yourself." But the project is a place where you can relax and take people as they come. It offers the security to trust people.

Lenni is speaking here about the time afforded by the project for the transformative practice of listening. Without rushing, by affording space and time, a generous spirit was coaxed out of researchers and community members alike.

In addition to food-based and cooking events, field trips to alternative enterprises allowed people to spend time together, fostering mutual respect and engagement (Cameron 2000a). Two interested groups went on bus trips from the Valley into inner-city Melbourne to visit CERES, the community garden at Brunswick. While the garden itself made a strong impression upon the Valley visitors, it was almost as if the experience of the bus trip—of being cooped up together in a moving steel canister for hours at a time—produced an atmosphere of enchantment that became the life force of the garden project.[16] Said Jean:

> I sat up the back of the bus, knitting very quietly, trying to mind my own business, but Mario kept yacking in my ear all day. [Laughter] . . . They're just a mixed group that if they're trying to do so much work, trying to do something, you've got to find where you fit, what they're trying to do, if it's such a good cause. To me it's like a giant big social club.

For Jean, the bus trip and the CERES visit allowed her to see herself as part of the community garden project in the Latrobe Valley, although gardening is not her thing:

> Forget the gardening! . . . I'm taking a concrete square [where trailers used to be parked]. They all look at me as if I'm mad. "What am I going to do with a concrete square?" . . . I would like to bring my fairy garden[17] from out of my back yard that I won't open to the public, and give it to the public, or leave it behind when I go.

There is a feeling of hopeful surprise among people involved in the management committee and wider membership (retrenched workers, retirees, housewives, people of non-English-speaking backgrounds and unemployed youth) of the now incorporated, not-for-profit Latrobe Valley Community and Environmental Garden (LTVCEG). They are surprised to find themselves in an organization, and astonished that they have begun an enterprise with each other. A space has opened up for relations with others who are largely "other" to them—people with whom they have nothing in common—and a community economy is in the process of creation (Cameron 2000a). Listening to Jean again:

> It wouldn't matter if you were ten in here, or a hundred and ten, everybody's equal. They're sharing their morning teas, their coffee, have a laugh, have fun, get ideas, the youngies can come up with things too, you all learn from each other, you're coping with all types of people, from your hot-tempered stand over bully, to your type that if you say something to them they scream straight back at you. You've got to learn to deal with every type. It's good learning, it's . . . I don't know, there's just something right about this whole thing.

Framing this process in Connolly's terms, we can see the project as fostering what he calls a "generous ethos of engagement between a plurality of constituencies inhabiting the same territory" (1999, 36) and, we should add, not in the habit of speaking to one another. Rather than asking people to mute their differences or rise above them, or to leave substantial parts of themselves at home when entering the public arena, the generous ethos is accepting of a range of beings and behaviors, including the socially unacceptable. And the ethical practice of cultivation involves giving people multiple opportunities to encounter each other in pleasurable ways—creating spaces of engagement, offering activities and events that promote receptivity. As Jaime, a community gardener in the Pioneer Valley, would say, the garden is the community.[18]

An ethos of engagement is an aspect of a politics of becoming, where subjects are made anew through engaging with others. This transformative process involves cultivating generosity in the place of hostility and suspicion. But such affective predispositions are not displaced easily, which means that the process involves waiting as well as cultivating. One of the Pioneer Valley researchers reflected on the patience that must accompany actively fostering a different economy, and she came back to the relation of language and affect that we began with earlier. Not only does one need a language of the diverse economy, but one also needs trust among the potential subjects who may inhabit that economy and take on the task of building it together. And trust can only be engendered through multiple opportunities for engagement (the terms she uses are "conversations" and "relationships"):

> I think it comes back to the point that Sr. Annette [of the Pioneer Valley Project, a coalition of labor and churches] made, which is the knitting together is not just a language. It's creating contexts for that language to circulate . . . and so it's relationships and being patient enough to have conversations and talk to people . . . and even if only five people come out, you value their time and make something out of it . . . and that's where the knitting happens. Y'know, how difficult it is to create a context of trust where things can actually be built . . . and you've just got to be patient . . . and it's just a lot of talk . . . and the people that are doers, that are too impatient, you just hold a place at the table for them.

BACK TO THE BEGINNING: PRINCIPLES AS PRACTICES

Cultivating local capacity, respecting difference and otherness, recognizing particularity and contingency.[19] These three principles are tangled together now, after all we have been through, and difficult to distinguish. We have affirmed them in relation to the discourse of globalization, with its emphatic insistence that the world we share is a (capitalist market) economy. This unrelenting emphasis presses upon us, and the counterpressure we are impelled to exert traces the principled contours of a local ethics: working to undermine universals in the guise of economic commonalities; refusing unity brought about by economic inevitability; refiguring victimized subjects whose economic futures are bound into and bounded by capitalist development.

Starting with a practice of *respecting difference and otherness*, our two projects storied and inventoried the diversity of the local noncapitalist economy. Coming to a new language and vision of economy turned out to be an affirmation not only of difference but of economic capacity. The people engaged in our research conversations had a chance to encounter themselves differently—not as waiting for capitalism to give them their places in the economy but as actively constructing their economic lives, on a daily basis, in a range of noncapitalist practices and institutions. In this way they glimpsed themselves as subjects rather than objects of economic development, and development became transformed as a goal by giving it a different starting place, in an already viable diverse economy.

But there was more to the ethics of difference and otherness than enlivening economic diversity. Converting this principle into a practice of the self has involved us in nurturing local capacities for community. We are not speaking here of the community of commonality that "presumes subjects can understand one another as they understand themselves" (Young 1990, 302). Rather than convening people on the basis of presumed or constructed similarities, our projects seemed to foster communities of "compearance"[20] in which being together, or being-in-common, was both the ground and fullness of community. The awakening of a communal subjectivity did not emerge from common histories or qualities but from practices and feelings—of appreciation, generosity, desire to *do* and *be* with others, connecting with strangers (no matter who), encountering and transforming oneself through that experience:

> To be completely sincere . . . the greatest pride that I have working as a community leader is my being able to share and develop myself *within* the community. To meet the person I don't know. And for the people who never met me, didn't have the chance to meet me, that they meet me.
>
> (Jaime, Pioneer Valley)[21]

Linda Singer suggests that we understand community "as the call of something other than presence" (1991, 125), the call to becoming, one might say. And the capacity for becoming is the talent we have perhaps been most actively fostering—through individuals opening to one another, and to the inescapable fact of their "own existence as possibility or potentiality" (Agamben 1993, 43). Indeed, this is how we might summarize our practices of *cultivating local capacity*. Almost every meeting and engagement associated with the project stimulated desires for alternative ways to be, and each of these desires operated as a contagion or revealed itself as a multiplicity.

What emerged, for example, from the awakening of a communal subjectivity was a faint but discernible yearning for a communal (noncapitalist) economy. This was not an easy yearning to stimulate or cultivate. The ability to desire what we do not know, to desire a different relation to economy, requires the willingness to endanger what now exists and what we know ourselves to be.[22] Because they require a death of sorts, an offering up of the self to the unfamiliar, desires for existence outside the capitalist "order" are difficult to engender. When restructuring devastates a regional economy, unemployed workers may have little interest in economic alternatives. Instead they desire to be employed, to continue their social existence as workers. (As do we.) In the face of this fixation upon capitalism, we came to see that one of our tasks as researchers was to help set desire in motion again (not unlike the task of the Lacanian psychoanalyst). If we could release into fluidity desire that was stuck, perhaps some of it would manifest in perverse (noncapitalist) dreams and fantasies.

From the outset we saw our projects as "bringing desire into language," in part by constituting a new language of economy. But as we came late to understand, with the help of Foucault and Connolly, the subject is not constituted through language alone. It is formed through real practices that act upon the body (Foucault 1997, 277) or through "tactics or disciplines not entirely reducible" to the play of symbols (Connolly 1999, 193). These disciplines "fix dispositional patterns of desire" (Connolly 1995, 57) that become part of what we experience as subjection—to capitalism or commun(al)ism, or whatever the alternatives might be.

Perhaps it was our growing sense that language is not enough that inclined us toward bodily practices and sensations and away from the delights of wordiness. In any case, we've tried to make our conversations and gatherings entirely pleasurable (food has been one of their main ingredients) and also loose and light—not goal-oriented or tied to definitions and prescriptions of what "a left alternative" should be. Over the course of the projects, without prompting, the community researchers and their interviewees began to express practical curiosity (as opposed to moral certainty) about alternatives to capitalism. The Pioneer Valley researchers took a week-long trip to Cape Breton attend a conference on worker cooperatives, and spent three days listening to stories of workers who appropriated the surplus they produced and distributed it to sustain a community economy. Amid the hilarity in the dormitory and the van, during the sunlit walks, in restaurants and cafés, on the eleven-hour ferry ride, we explored and debated (desultorily) the virtues of worker cooperatives. Fears were spoken and then let go. By the end of the trip, we had produced several fantasies of communal enterprises and the social life they might enable, as a way of performing and acknowledging our temporary, satisfying collectivity. How are we to understand this unexpectedly pleasurable trip but as an experience of ethical self-formation, of working on the self, as Foucault would say (though without being aware of working)? Through the conversations in the van, the discourse of economic interdependence and community we had ingested for three days became transmuted into bodily desires and flows of energy directed toward a communal economy.

This brings us to the ethical practice of *recognizing particularity and contingency*. Our projects were attempts to build on the distinctiveness of a local economy rather than replacing a unique constellation of activities with a generic model of development. The infusion of particularity into development discourse was deeply destabilizing to economic certainty. It became possible to think the economy as a contingent space of recognition and negotiation rather than as an asocial body in lawful motion. But beyond thinking differently about the economy, what is the ethical practice of economic contingency? Ernesto Laclau helps us here, describing the political space opened up by current antinecessitarian thinking: "increasing the freeing of human beings through a more assertive image of their capacities, increasing social responsibility through the consciousness of the historicity of Being—is the most important possibility, a radically political possibility, that contemporary thought is opening up to us" (1991, 97–8). If the economy is a domain of historicity and contingency, other economies can be produced, and producing them is a project of politics. This suggests that we could move beyond capitalism and the economic politics of opposition "within" it. The ethics of contingency, Laclau implies, involves the cultivation of ourselves as subjects of freedom—self-believers in our economic capacities, responsible to our political abilities, conscious (we would add) of our potential to become something other than what we have heretofore chosen to be.

If recognizing contingency offers an enlarged domain of choice and responsibility, then the ethical practice of contingency is the cultivation of ourselves as choosers, especially in areas where choice has been understood as precluded to us. Implicated as we are in our identifications (because they are to some degree optional), we choose to be subjects of a capitalist economy, or we choose to work on ourselves—ethically, micropolitically, viscerally, intellectually—to forge some other way to be (Madra 2001).

Unavoidably we have had to think about the politics and ethics of our academic "locality." And here choice looms as a daily challenge: choice of the theorist, not to try to "get it right" but to pursue inventiveness;[23] not to think critically in a debunking mode (describing what something is and should not be) but instead ebulliently (Massumi 2001).

Finally, there is the process of writing. In Foucault's lexicon, writing is an ethical practice, a way that the self relates to itself. It is an intellectual discipline that allows us to consider "the possibility of no longer being, doing or thinking what we are, do, or think . . . seeking to give new impetus, as far and as wide as possible, to the undefined work of freedom" (1997, xxxv). For us writing is a practice of forming the hopeful subject—a left subject on the horizon of social possibility.

ACKNOWLEDGMENTS

We recognize Bill Connolly as our unwitting and unheralded coauthor and hope that he can recognize himself (and accept his positioning) in that description. Without his work this paper, and much of our current thinking, would not exist. We are grateful to Jack Amariglio and Yahya Madra for their invitation to participate in Rethinking Marxism's series on globalization and their intellectual and personal encouragement during the writing of the paper. We would also like to acknowledge the recent work of George DeMartino and David Ruccio on globalization, ethics, and politics, which provided both provocation and inspiration.

The research reported on here has been enabled on the Australian side by a grant from the Australian Research Council/Strategic Partnerships with Industry and Training as well as the indispensable coparticipation of Jenny Cameron; and on the U.S. side, by a grant from the National Science Foundation (Grant No. BCS-9819138) and the other members of the Community Economies Collective—coresearchers Brian Bannon, Carole Biewener, Jeff Boulet, Ken Byrne, Gabriela Delgadillo, Rebecca Forest, Stephen Healy, Greg Horvath, Beth Rennekamp, AnnaMarie Russo, Sarah Stookey, and Anasuya Weil. All worked long and hard and lovingly with us to produce the body of collaborative work that we have drawn upon in this paper. Finally, Julie gratefully acknowledges the Department of Human Geography, Research School of Pacific and Asian Studies at the Australian National University for providing space and other resources during the writing of the paper.

NOTES

1 See Dirlik (2000a) for a similar distinction between globalization as an historical process, which has been ongoing "since the origins of humanity," and globalization as a discourse or paradigm, "a self-consciously new way of viewing the world" (2000a, 3).

2 This despite the protestations of many theorists that globalization is both productive of, and accommodating to, heterogeneity and difference.

3 Here it is interesting to consider the argument of Hardt and Negri (2000) that these principles were once potent counters to modernity (and, in particular, modern sovereignty) but, with the passing of modern forms of rule, they have been robbed of moral relevance and political effectiveness. Indeed, for Hardt and Negri, theorists like Homi Bhabha who are still critiquing modernity and affirming these principles as the basis of a new postmodern politics of community are not only beating a dead horse but are unwittingly complicit in constructing the order of postmodern sovereignty, designated simply and terrifyingly as "Empire."

> What if these theorists are so intent on combating the remains of a past form of domination that they fail to recognize the new form that is looming over them in the present? . . . Power has evacuated the bastion they are attacking and has circled around to their rear to join them in the assault in the name of difference . . . Long live difference! Down with essentialist binaries!
> (Hardt and Negri 2000, 137–8)

Despite the best intentions, then, the postmodernist politics of difference not only is ineffective against but

> can even coincide with and support the functions and practices of imperial rule.
>
> (Hardt and Negri 2000 142–3)

Rather than being a threat to existing forms of power, Bhabha and others are "symptoms of the epochal shift we are undergoing, that is, the passage to Empire" (2000, 145). Their outmoded antimodernist critiques of binary hierarchies have been incorporated and subsumed by the postmodern imperial formation, which has devised new forms of hierarchy and domination.

While the sweeping scope of their pronouncements and the energetic affirmation of totality make me feel somewhat weary, I am also invigorated by Hardt and Negri. I can recognize my self-positioning and recommit to my various projects in the light of their very different one. As for the principles so easily dismissed by them, I am both less optimistic than they are (not believing that respect for difference and otherness have been embraced or enforced globally) and less pessimistic (not believing that they have done their work and are now disarmed and irrelevant). On the contrary, it seems to me that these principles have seldom been put into practice, and that the ethical process of cultivating subjects for whom these principles resonate has barely begun.

Stephen Healy takes a similar position, on the grounds that Empire—as Hardt and Negri define it—has not fully coalesced: "Insofar as this new discursive order has not yet solidified it becomes crucial for those of us who want to see a different world to be able to imagine other ways of representing difference" (Healy 2001, 103).

4 A municipal leader in Oaxaca, Mexico: "I can no longer do what is fair. Every time I try to bring justice to our community, applying our traditional practices to amend wrongdoings, a human rights activist comes to stop me" (Esteva and Prakash 1998, 110). Esteva and Prakash do not object to human rights *per se*, but to the ways in which they are currently being globalized.

5 Ernesto Laclau contends that the sustained critique of essentialist universals has created the space for the emergence of contingent universals—the latter do not conceal the political moment of their universalization. In Laclau's formulation, the universal is the politically mediated hegemony of a particularity (2000, 51). In a democratic context, universal values must come to the fore, but they are "not the values of a 'universal' group [such as the working class—JKGG], as was the case with the universalism of the past but, rather, of a universality that is the very result of particularism" (Laclau 1994, 5).

6 See Gibson-Graham *et al.* (2000, 2001) for two edited collections that explore economic difference in the dimension of class.

7 This research program has strong affinities with the work of Arturo Escobar (2001) and Arif Dirlik (2000b) on the politics of place.

8 There is a tendency to conflate all market-oriented (i.e. commodity) production with capitalism. We need to resist that tendency if we are to theorize economic difference in the market sphere, and to acknowledge the many types of economic organization that are compatible with commodity production.

9 Here it has become necessary to shift to the first person plural since the projects we are discussing are collective efforts involving large numbers of people (see acknowledgments on p. 367).

10 One participant gave the example of Whyalla, South Australia, where many people had been retrenched by the steel industry. Local planners came to see the unemployed as their major regional asset (rather than seeing them as a drain on the community) since unemployment benefits tended to be spent locally rather than on holiday travel out of state or on trips to the hairdresser in Adelaide.

11 Perhaps this willingness was an acknowledgment of what can never be entirely erased from consciousness—"the simple fact of one's own existence as possibility or potentiality" (Butler 1997, 130, quoting Agamben 1993, 43).

12 In the Latrobe Valley project, we were involved in adapting Kretzmann and McKnight's (1993) techniques of asset-based community development to a de-industrialized region, rather than the type of inner-city neighborhood for which the techniques were originally devised (see Cameron and Gibson [2001] for a resource kit that documents this process).

13 Antagonism and ressentiment are the common emotions of modernist politics—the fuel of revolutionary consciousness and action. But the political effectiveness of such emotions is questioned by many today. Wendy Brown, for example, in *States of Injury*, notes that the subject of this kind of affect becomes "deeply invested in its own impotence" and is more likely to "punish and reproach" than "find venues of self-affirming action" (1995, 71).

14 In, for example, Paulo Freire's "pedagogy of the oppressed" (1972).

15 Our examples here appear to confirm Foucault's observation that contemporary ethical practices of self-formation are addressed primarily to feelings. Feelings

are the "substance" of modern ethics; they are what we endeavor to form and transform. For the Greeks, by contrast, acts linked to pleasure were the substance of ethics; for the Christians the substance was desire (Foucault 1997, 263–9).

16 For an exploration of enchantment that has enchanted and inspired us, see Bennett (2001).

17 Jean's fairy garden is inhabited by garden sculptures (gnomes, flamingos, etc.) set up in scenes and distributed about the lawn.

18 Jaime reflects eloquently on his practices of cultivation: "It is necessary to strengthen new leaders, for them to do what I'm doing, so they continue forward, making a call to this community, so that . . . this, instead of being a community garden, that the whole city be a garden, and that the flowers be the people" (Community Economies Collective 2001, 26).

19 Those readers who are following our intersecting lists of principles and practices will notice that we have omitted the final element in Connolly's list: "*ethically sensitive, negotiated settlements* between chastened partisans who proceed from contending and overlapping presumptions while *jointly* coming to appreciate the unlikelihood of reaching rational agreement on several basic issues" (1999, 35). This is not because such settlements are excluded from the purview of our projects but because the projects are not far enough along for such settlements to have taken place.

20 This is Jean-Luc Nancy's word for a mode of being together that recognizes "no common being, no substance, no essence, no common identity" (1991, 1). It suggests that we are already in community, if we can only orient ourselves, affectively and cognitively, to the recognition and enjoyment of that experience.

21 As we consider the nascent communities in the two valleys, we are drawn to Linda Singer's essay on "a community at loose ends" and her suggestion that "community is not a referential sign but a call or appeal. What is called for is not some objective reference. The call of community initiates conversation, prompts exchanges . . . disseminates, desires the proliferation of discourse" (1991, 125).

22 Judith Butler asks what such a dangerous undertaking might involve: "What would it mean for the subject to desire something other than its continued 'social existence'? If such an existence cannot be undone without falling into some kind of death, can existence nevertheless be risked, death courted or pursued, in order to expose and open to transformation the hold of social power on the conditions of life's persistence?" (1997, 28–9).

23 See Gibson-Graham (1996, chap. 9) and Cameron (2000b) for a similar and more extended argument.

REFERENCES

Agamben, G. 1993. *The coming community*. Trans. M. Hardt. Minneapolis: University of Minnesota Press.

Bennett, J. 2001. *The enchantment of modem life: Crossing, energetics, and ethics*. Princeton, N.J.: Princeton University Press.

Brown, W. 1995. *States of injury: Power and freedom in late modernity*. Princeton, N.J.: Princeton University Press.

Butler, J. 1997. *The psychic life of power: Theories in subjection*. Stanford, Calif.: Stanford University Press.

Cameron, J. 2000a. "Asset-based community and economic development: Implications for the social capital debate." Paper presented as the Alison Burton Memorial Lecture, Royal Australian Planning Institute, Canberra.

——. 2000b. "Domesticating class." In *Class and its others*, ed. J. K. Gibson-Graham, S. Resnick, and R. Wolff. Minneapolis: University of Minnesota Press.

Cameron, J., and K. Gibson. 2001. *Shifting focus: Pathways to community and economic development—A resource kit*. Latrobe City Council and Monash University, Victoria.

Community Economies Collective. 2001. "Imagining and enacting noncapitalist futures." *Socialist Review* 28 (3&4): 93–135.

Connolly, W. 1995. *The ethos of pluralization*. Minneapolis: University of Minnesota Press.

——. 1999. *Why I am not a secularist*. Minneapolis: University of Minnesota Press.

——. 2001. "Brains, techniques and time: The ethics of nonlinear politics." Unpublished manuscript, Department of Politics, the Johns Hopkins University, Baltimore, Md.

DeMartino, G. 2000. *Global economy, global justice*. New York: Routledge.

Dirlik, A. 2000a. "Reconfiguring modernity from modernization to globalization." Paper presented at the conference on "Entangled Modernities," House of World Cultures and the Swedish Collegium for Advanced Studies in the Social Sciences, Berlin, December.

——. 2000b. "Place-based imagination: Globalism and the politics of place." In *Places and politics in an age of globalization*, ed A. Dirlik and R. Prazniak, 15–52. Boulder, Col.: Rowman and Littlefield.

Escobar, A. 2001. "Culture sits in places: Reflections on globalism and subaltern strategies of localization." *Political Geography* 20: 139–74.

Esteva, G., and M. S. Prakash. 1998. *Grassroots post-modernism.* London: Zed Books.

Foucault, M. 1985. *The history of sexuality.* Vol. 2, *The uses of pleasure.* New York: Pantheon.

——. 1997. *Ethics: Subjectivity and truth*, ed. P. Rabinow, trans. R. Hurley and others. New York: The New Press.

Freire, P. 1972. *Pedagogy of the oppressed.* New York: Penguin.

Gibson, K., J. Cameron, and A. Veno. 1999. "Negotiating restructuring: A study of regional communities experiencing rapid social and economic change." Australian Housing and Urban Research Institute Working Paper 11, AHURI, Brisbane, Queensland. Available at http://www.ahun.edu.au/pubs/work_paps/work pap11.html.

Gibson-Graham, J. K. 1996. *The end of capitalism (as we knew it): A feminist critique of political economy.* Oxford: Blackwell.

——. 2002. "Beyond global vs. local: Economic politics outside the binary frame." In *Placing scale*, ed. A. Herod and M. Wright, 25–60. Oxford: Blackwell.

Gibson-Graham, J. K., S. Resnick, and R. Wolff, eds 2000. Class and its others. Minnesota: University of Minnesota Press.

——. 2001. *Re/Presenting class.* Durham, N.C.: Duke University Press.

Hardt, M., and A. Negri. 2000. *Empire.* Cambridge, Mass.: Harvard University Press.

Healy, S. 2000. "Amplifying alternatives: Economic discourse and the politics of resignification in Western Massachusetts." Paper presented at the inaugural conference of the International Social Theory Consortium, Lexington, Kentucky, May.

——. 2001. "Colonialism and dimensionality: Memory, hybridity and postcoloniality." M.S. thesis, University of Massachusetts Amherst.

Ironmonger, D. 1996. "Counting outputs, capital inputs and caring labor: Estimating gross household output." *Feminist Economics* 2 (3): 37–64.

Kretzmann, J., and J. McKnight. 1993. *Building communities from the inside out: A path toward finding and mobilizing a community's assets.* The Asset-Based Community Development Institute, Institute for Policy Research, Northwestern University, Evanston, Illinois.

Laclau, E. 1991. "Community and its paradoxes: Richard Rorty's 'liberal Utopia.'" In *Community at loose ends*, ed. The Miami Theory Collective, 83–98. Minneapolis: University of Minnesota Press.

——. 1994. "Introduction" to *The making of political identities*, ed. E. Laclau. New York: Verso.

——. 2000. "Identity and hegemony: The role of universality in the constitution of political logics." In J. Butler, E. Laclau, and S. Zizek, *Contingency, hegemony, universality: Contemporary dialogues on the Left*, 45–89. New York: Verso.

Lingis, A. 1994. *The community of those who have nothing in common.* Bloomington: Indiana University Press.

Madra, Y. 2001. "Class, the unconscious, and hegemony." Unpublished paper, University of Massachusetts Amherst.

Madra, Y., and J. Amariglio. 2000. "Globalization under interrogation: An introduction." *Rethinking Marxism* 12 (4): 1–3.

Massumi, B. 2001. "Introduction: Concrete is as concrete doesn't." In *Parables for the virtual.* Cambridge, Mass.: Harvard University Press.

Nancy, J.-L. 1991. *The inoperative community.* Minneapolis: University of Minnesota Press.

Ruccio, D. 2000. "Rethinking globalization." Paper presented at "Marxism 2000," a conference at the University of Massachusetts at Amherst, September.

Russo, A. M. 2000. "'But I don't really feel that poor': Reimagining economic development in low-income urban communities." Paper presented at Marxism 2000; a conference at the University of Massachusetts Amherst, September.

Shapiro, M. J. 1999. "The ethics of encounter: Unreading, unmapping the imperium." In *Moral spaces: Rethinking ethics and world politics*, ed. D. Campbell and M. J. Shapiro, 57–91. Minneapolis: University of Minnesota Press.

Singer, L. 1991. "Recalling a community at loose ends." In *Community at loose ends*, ed. The Miami Theory Collective, 121–30. Minneapolis: University of Minnesota Press.

Young, I. M. 1990. "The ideal of community and the politics of difference." In *Feminism /postmodernism*, ed. L. Nicholson, 300–23. New York: Routledge.

Bibliography

Agnew, J., D. Livingstone, and A. Rogers, eds. *Human Geography: An Essential Anthology*. Malden, MA: Blackwell, 1996.

Althusser, Louis. "Ideology and Ideological State Apparatuses: Notes Toward an Investigation," in *Lenin and Philosophy*, trans. Ben Brewster. London: New Left Books, 1971, pp. 127–186.

Barndt, Deborah. *Naming the Moment: Political Analysis for Action.* Toronto: Jesuit Centre, 1989.

——. "Free Trade Offers 'Free Space' for Connecting Across Borders," in Roger Keil, Gerda R. Wekerle, and David V. J. Bell, eds., *Local Places in the Age of the Global City*. Montreal: Black Rose Books, 1996.

Barnes, Trevor and Derek Gregory, eds *Reading Human Geography: The Poetics and Politics of Inquiry.* Oxford: Oxford University Press, 1998.

Barrows, Harlan. "Geography as Human Ecology." *Annals of the Association of American Geographers* 13 (1911): 1–14.

Bebbington, Anthony. "Movements, Modernizations, Markets and Municipalities: Indigenous Organizations and Agrarian Strategies in Ecuador, Then and Now," in Richard Peet and Michael Watts, eds, *Liberation Ecologies: Environment, Development, Social Movements*. London: Routledge, 2004.

Benson, K. and Richa Nagar. "Collaboration as Resistance? Reconsidering Processes, Products, and Possibilities of Feminist Oral History and Ethnography." *Gender, Place and Culture* 13 (2006): 581–592.

Blaut, James M. "Jingo Geography: Part 1." *Antipode* 1 (1969): 10–13.

——. "Imperialism: The Marxist Theory and Its Evolution." *Antipode* 7 (1975): 1–19.

——. "Where Was Capitalism Born." *Antipode* 8 (1976): 1–11.

Blunt, Alison and Jane Wills. *Dissident Geographies: An Introduction to Radical Ideas and Practice.* Englewood Cliffs, NJ: Prentice Hall, 2000.

Brown, Andrew. "Developing Realistic Philosophy: From Critical Realism to Materialist Dialectics," in Andrew Brown, Steve Fleetwood, and John Michael Roberts, eds, *Critical Realism and Marxism.* London: Routledge, 2002, pp. 168–186.

Bunge, William. *Theoretical Geography.* Lund Studies in Geography Series C, no. 1. Lund: C.W.K. Gleerlup, 1966, 2nd ed.

Burton, Ian. "The Quantitative Revolution and Theoretical Geography." *Canadian Geographer* 7 (1963): 151–162.

Bystydzienski, Jill and Steven Schacht, eds. *Forging Radical Alliances across Difference: Coalition Politics for the New Millenium.* Lanham, MD: Rowman and Littlefield, 2001.

Cloke, P., C. Philo, and D. Sadler *Approaching Human Geography: An Introduction to Contemporary Theoretical Debates.* London: Guilford, 1991.

Esteva, Gustavo and Madhu Suri Prakash. *Grassroots Post-Modernism: Remaking the Soil of Cultures.* New York: Zed Books, 1998.

Featherstone, David. "Spatialities of Trans-national Resistance to Globalization: The Maps of Grievance of the Inter-continental Caravan." *Transaction of the Institute of British Geographers* 28(4) (2003): 404–421.

Fraser, Nancy. *Unruly Practices: Power, Discourse and Gender in Contemporary Social Theory.* Minneapolis: University of Minnesota Press, 1989.

Fraser, Nancy. "From Redistribution to Recognition? Dilemmas of Justice in a 'Postsocialist' Age." *New Left Review* 212 (1995): 68–93.

——. "A Rejoinder to Iris Young." *New Left Review* 223 (1997): 126–129.

——. "Rethinking Recognition." *New Left Review* 3 (2000): 107–120.

Gathorne-Hardy, Flora. "Accommodating Difference: Social Justice, Disability, and the Design of Affordable Housing," in Ruth Butler and Hester, eds, *Mind and Body Spaces: Geographies of Illness, Impairment, and Disability*. London: Routledge, 1999, pp. 240–255.

Geddes, P. "The Influence of Geographical Conditions on Social Development." *Geographical Journal* 12 (1898): 580–587.

Gibson-Graham, J. K. *The End of Capitalism (As We Knew It): A Feminist Critique of Political Economy.* Cambridge, MA: Blackwell, 1996.

——. *A Postcapitalist Politics.* Minneapolis: University of Minnesota Press, 2006.

Gilligan, Carol. *In a Different Voice*. Cambridge, MA: Harvard University Press, 1982.

Godlewska, A. and Neil Smith, eds. *Geography and Empire: Critical Studies in the Geography of Empire.* Oxford: Basil Blackwell, 1994.

Gould, Peter. "Geography 1957–1977: The Augean Period." *Annals of the Association of American Geographers* 69 (1979): 139–151.

Gramsci, Antonio. *Selections from the Prison Notebooks.* New York: International Publishers, 1971.

Gregory, Derek. *Geographical Imaginations.* Oxford: Blackwell, 1994.

Guthman, Julie. *Agrarian Dreams: The Paradox of Organic Farming in California.* Berkeley: University of California Press, 2004.

Haraway, Donna "Situated Knowledges: The Science Question in Feminism and the Privilege of Partial Perspective," in D. Haraway, *Simians, Cyborgs and Women.* London Free Association Press, 1991.

Hartshorne, Richard. *The Nature of Geography: A Critical Survey of Current Thought in the Light of the Past.* Lancaster, PA: Association of American Geographers, 1939.

Hartsock, Nancy. "Postmodernism and Political Change: Issues for Feminist Theory." *Cultural Critique* 14 (1989/90): 15–33.

Harvey, David. *Explanation in Geograpy.* London: Arnold, 1969.

——. *Social Justice and the City.* Baltimore: Johns Hopkins University Press, 1973.

——. *The New Imperialism.* Oxford: Oxford University Press, 2003.

Hayford, A. 1974. "The Geography of Women: An Historical Introduction." *Antipode* 6 (1974): 1–19.

Heidegger, Martin. *Being and Time.* New York: Harper and Row, 1962.

Henderson, George. "'Free Food, the Local Production of Worth, and the Circuit of Decommodification: A Value Theory of the Surplus." *Environment and Planning D: Society and Space* 22 (2004): 485–512.

Holt-Jensen, A. *Geography: History and Concepts*. Thousand Oaks, CA: Sage, 1999, 3rd edn.

Horvath, R. J. "The 'Detroit Geographical Expedition' Experience." *Antipode* 3 (1971): 73–85.

Hubbard, P., R. Kitchin, B. Bartley, and D. Fuller. *Thinking Geographically: Space, Theory and Contemporary Human Geography.* London: Continuum, 2002.

Huntington, Ellsworth. *The Character of Races, as Influenced by Physical Environment, Natural Selection and Historical Development.* New York: Scribner's, 1924.

Ingram, Mrill. "Biology and Beyond: The Science of 'Back to Nature' Farming in the United States." *Annals of the Association of American Geographers* 97(2) (2007): 298–312.

Jakobsen, Janet. *Working Alliances and the Politics of Difference: Diversity and Feminist Ethics.* Bloomington: Indiana University Press, 1998.

Johnston, R. J. *Geography and Geographers: Anglo-American Human Geography since 1945.* London: Arnold, 1997, 5th edn.

Kropotkin, P. "What Geography Ought to Be." *Nineteenth Century* 18 (1885): 940–956.

——. *Fields, Factories and Workshops.* Boston: Houghton Mifflin, 1899.

——. *Mutual Aid: A Factor of Evolution.* Harmondsworth: Penguin, 1902.

Kuhn, Thomas. *The Structure of Scientific Revolutions.* Chicago: University of Chicago Press, 1962.

Laclau, Ernesto and Chantal Mouffe. *Hegemony and Socialist Strategy*. London: Verso, 1985.
Lee, R. and David M. Smith, eds. *Geographies and Moralities: International Perspectives on Development, Justice, and Place*. Malden, MA: Blackwell, 2004.
Livingstone, David. N. *The Geographical Tradition.* Oxford: Blackwell, 1992.
Lukács, György. "Reification and the Consciousness of the Proletariat," in *History and Class Consciousness.* Cambridge, MA: MIT Press, 1971, pp. 83–222.
Marsh, George P. *Man and Nature; or Physical Geography as Modified by Human Action.* London: S. Low, Son and Marston, 1864.
Marx, Karl. *The Eighteenth Brumaire of Louis Bonaparte.* London: Allen and Unwin. 1926.
——. "Theses on Feuerbach," in *Early Writings.* Harmondsworth: Penguin, 1975.
——. *Capital, Volume 1.* New York: International Publishers, 1967.
Massey, Doreen. "A Global Sense of Place." *Marxism Today* (June, 1991): 24–29.
Mohanty, Chandra. "Under Western Eyes: Feminist Scholarship and Colonial Discourses." *Feminist Review* 30 (1988): 61–102.
Monk, Jan. "Classics in Human Geography Revisited. R. J. Johnston Geography and Geographers: Anglo American Geography since 1945. Commentary 2." *Progress in Human Geography* 31 (2007): 45–47.
Monk, Janice and Susan Hanson. "On Not Excluding Half of the Human in Human Geography." *Professional Geographer* 34(1) (1982): 11–23.
NLR. "Reinventing Geography: An Interview with David Harvey." *New Left Review* 4 (2000): 75–97.
O'Connor, James. *Natural Causes: Essays in Ecological Marxism.* New York and London: The Guilford Press, 1998.
Olwig, Kenneth R. "Recovering the Substantive Nature of Landscape." *Annals of the Association of American Geographers* 86 (1996): 630–653.
Peet, Richard. "The Social Origins of Environmental Determinism." *Annals of the Association of American Geographers* 75 (1985): 309–333.
——. *Modern Geographic Thought.* Oxford: Blackwell, 1998.
Pulido, Laura. "Development of the 'People of Color' Identity in the Environmental Justice Movement of the Southwestern United States." *Socialist Review* 26(4) (1996):151–180.
Ratzel, F. "The Territorial Growth of States." *Scottish Geographical Magazine* 12 (1896): 351–361.
Reclus, E. *Nouvelle Géographie universelle: La Terre et les hommes.* Paris: Hachette et Cie., 1876–94 (multiple volumes).
Rose, Gillian. *Feminism and Geography: The Limits of Geographical Knowledge.* Cambridge: Polity Press, 1993.
Sauer, Carl. "The Morphology of Landscape." *University of California Publications in Geography* 2 (1925): 19–54.
Schaefer, F. K. "Exceptionalism in Geography: A Methodological Examination." *Annals of the Association of American Geographers* 43 (1953): 226–249.
Semple, Ellen C. *Influences of the Geographic Environment on the Basis of Ratzel's System of Anthropogeography.* New York: Henry Holt, 1911.
Slater, David. "Spatialities of Power and Postmodern Ethics: Rethinking Geopolitical Encounters." *Environment and Planning D: Society and Space* 15 (1997): 55–72.
Smith, David M. *The Geography of Social Well-Being in the United States: An Introduction to Territorial Social Indicators.* New York: McGraw-Hill, 1973.
——. *Human Geography: A Welfare Approach.* London: Edward Arnold, 1977.
——. *Geography and Social Justice.* Oxford: Blackwell, 1994.
——. *Moral Geographies: Ethics in a World of Difference.* Edinburgh: Edinburgh University Press, 2000.
Stoddart, David. *On Geography and Its History.* Oxford: Blackwell, 1986.
Tivers, J. "How the Other Half Lives: The Geographic Study of Women." *Area* 10 (1978): 302–306.
Unwin, Tim. *The Place of Geography.* Harlow: Longman, 1992.
Whatmore, Sarah. *Hybrid Geographies: Natures Cultures Spaces.* Thousand Oaks, CA: Sage, 2004.

Vidal de la Blanche, P. *Principles of Human Geography*, trans M. Bingham. London: Constable, 1926.
Young, Iris Marion. 1990 *Justice and the Politics of Difference.* Princeton: Princeton University Press, 1990.
——. "Unruly Categories: A Critique of Nancy Fraser's Dual Systems Theory." *New Left Review* 222 (1997): 147–160.

Index

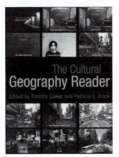